Arbuscular Mycorrhizal Fungi in Fruit Crop Production

The Editors

Dr. Sukhada Mohandas is an Emeritus Scientist of the Agriculture Research Service (ICAR) at Indian Institute of Horticultural Research,Bengaluru. She obtained her B.Sc (Hons), M.Sc. and Ph.D. from the Banaglore University, being a recipient of Gold Medal for obtaining Ist Rank in M.Sc. and Jawahar Lal Nehru Award (ICAR) for her doctoral thesis work. She has contributed extensively to the field of arbuscular mycorrhizal fungal (AMF) research. In a research carrier spanning more than 40 years, she devoted more than 25 years to mycorrhizal studies on fruit crops, publishing several widely cited papers in international and national journals. Dr. Sukhada is the pioneer in initiating arbuscular mycorrhizal work in fruit crops in the country. She studied the diversity of arbuscular mycorrhiza in several fruit crop species, multiplied dominant ones on finger millet roots and developed AMF inoculum production technique for large scale inoculation of fruit crops in the nursery and in the field. She has been recognised for her research with INSA Young Scientist Medal (1981), C.N. Patel Industrial Award (VASVIK foundation) (1993) and Panjabrao Deshmuk Outstanding Woman Agricultural Scientist Award (1998) besides obtaining Fellowships of several academies and Travel Awards by Government bodies.

Dr. P. Panneerselvam has done his graduation, post graduation and doctoral degrees from Tamil Nadu Agricultural University, Coimbatore. He was a Gold medalist in Ph.D, ICAR-JRF fellow in Post Graduation and received Prof. Dr. S. Kannaiyan and Dr. Surendar award for being the best student in Ph.D. His first appointment was in Central Coffee Research Institute, Chikmagalur where he served up to 2007. He joined Indian Institute of Horticultural Research, Bengaluru in 2007 and now continues his service as Scientist (Sr. Scale). In horticultural crops, his technology particularly Arka Microbial Consortium has made a great impact in vegetables, flower and fruits crops production. To produce disease, nematode free and healthy mycorrhized horticultural seedlings, he recently has developed "Soilless Arbuscular Mycorrhizal Fungal Inoculum Production technique" which is being patented. He has developed another bio-formulation i.e. "Actinobacterial Consortium" for nutrient and plant health management in horticultural crops production. He has published thirty five research papers in both international/ national peer reviewed journal and has authored fifteen proceeding papers, one books and ten book chapters. He was conferred "Fellow of CHAI 2015" by Confederation of Horticulture Association of India, New Delhi.

Arbuscular Mycorrhizal Fungi in Fruit Crop Production

Editors

Sukhada Mohandas

P. Panneerselvam

2016

Daya Publishing House®

A Division of

Astral International Pvt. Ltd.

New Delhi - 110 002

Cataloging in Publication Data–DK
Courtesy: D.K. Agencies (P) Ltd. <docinfo@dkagencies.com>
      Arbuscular mycorrhizal fungi in fruit crop production / editors,
Sukhada Mohandas, P. Panneerselvam.
     pages cm
     Includes bibliographical references.
     ISBN 9789351306863 (International Edition)
     1. Vesicular-arbuscular mycorrhizas. 2. Mycorrhizal fungi. 3.
Mycorrhizas in agriculture. 4. Fungi in agriculture. 5. Wood-decaying fungi.
6. Crop improvement.   I. Mohandas, Sukhada, editor. II. Panneerselvam, P.,
editor.
QK604.2.M92A73 2015   DDC 631.46   23

Published by	:	**Daya Publishing House**®
		A Division of
		Astral International Pvt. Ltd.
		– ISO 9001:2008 Certified Company –
		4760-61/23, Ansari Road, Darya Ganj
		New Delhi-110 002
		Ph. 011-43549197, 23278134
		E-mail: info@astralint.com
		Website: www.astralint.com
Laser Typesetting	:	**SSMG Computer Graphics, Burari-84**
Printed at	:	**Thomson Press India Limited**

Foreword

The arbuscular mycorrhizal fungi (AMF), which form a classical symbiotic association with root system of plants, play an important role in nutrient mobilization and protection against biotic and abiotic stress under the natural ecosystem. The information on occurrence of AMF, its symbiosis in large number of plant species, in crop growth and sustainability of crop yields and soil fertility is well documented through research papers and books during the last 60 years. However, in last two decades, interest in utilization of AMF as inoculation technology has been practical due to commercialization. Appropriate management of AMF-plant symbiosis would permit considerable reduction of chemical fertilizers, and pesticide inputs, for sustainability of plants particularly those of perennial horticultural plants. Since most of the fruit crops exhibit a considerable mycotrophic habit, maximum benefit could be obtained from careful selection of compatible host-fungus-substrate combinations. Further, mycorrhization of fruit crops in the nursery before transplanting would yield vigorous and healthy seedlings which would establish a viable fruit orchard. The performance of micropropagated plants, cuttings or artificial seeds could be greatly improved by ensuring a suitable mycorrhizal establishment at planting. Interaction between AM fungi and rhizobacteria have the potential for benefitting growth and health management in fruit crops through an integrated approach.

This book entitled "**Arbuscular Mycorrhizal Fungi in Fruit Crop Production**" edited by Dr Sukhada Mohandas and P. Panneerselvam, contains seven major areas is a unique blend of information on the role of AM fungi for sustainable fruit crops production and soil health management and includes a total package of information on inoculum production to seedling inoculation to field planting. Despite availability of abundant literature, an in depth comprehensive reference book on fruit crops production using AM fungi is not available. Therefore, I heartily congratulate the authors and the editors for envisioning the need for bringing out a book on this important subject and ensuring relevant and latest comprehensive information for use of researchers, students, entrepreneurs and farmers. This book is a rich source of inspiration for research and exploitation of the potential of AMfungi.

(Dr. J.H. KULKARNI)

Former Vice Chancellor,
University of Agricultural Sciences,
Dharwad

Preface

Fruits constitute a natural source of nutrients and vitamins in human diet. Fruit crop cultivation fetches high economic returns to the farmer compared to other crops. Presently, area under fruit crops in India is about 6.7 million hectare with a production of 76.4 million tons, which contributes 30% of share in total production and it is expected to reach 115 million tons by 2017. Practices followed in fruit crop cultivation play a major role in retaining or improving the natural characters of the fruits. Increasing use of chemical fertilizers and pesticides in cultivation system effects fruit quality. To meet the ever increasing demand for quality fruits there is a need to improve the technologies involved in fruit crop production. Beneficial plant-microbe interactions are considered the primary determinants of plant health, soil fertility and a sustainable production systems. Arbuscular mycorrhizal fungi (AMF) are one of the most promising amongst them. The word "Mycorrhiza" is given to a mutualistic association between a fungus (Myco) and the roots (rhiza) of the plants. This association is symbiotic because the relationship is advantageous for both organisms. The plant (the macrosymbiont) gains increased exploration of the rhizosphere with the intricate net of hyphae that increases the uptake of water and nutrients from the soil interphase. The fungus (the microsymbiont) uses the carbon provided by the plant for its physiological functions, growth and development. AMF symbiosis help the plants in improving (i) rooting and plant establishment; (ii) uptake of low mobile ions; (iii) nutrient cycling; (iv) enhanced plant tolerance to (biotic and abiotic) stress; (v) quality of soil structure; (vi) enhanced plant community.

The benefits of AMF inoculation in fruit crop cultivation have been evaluated by mostly pot culture trials and by a few field trials which have been published world wide. It is now evident that AMF are highly beneficial to fruit cultivation and AMF inoculation must be considered an essential factor for promoting health and productivity of fruit crops. But a comprehensive effort to bring the literature together and provide a guideline on how to make AM application more rewarding in fruit cultivation is the need of the hour. Hence an effort was made to collate information on different fruit crops benefitted by its use and aspects of integration of AMF in fruit crop cultivation system.

It is important to assess the diversity of AMF if we want to utilize the benefits of the association. Even though AMF is not specific in its colonisation they exhibit certain host preferences. Any AMF can colonize any suitable plant species, a single root system can support different AMF species, and roots of various plant species can be linked by the mycelium of a single species. The species richness is the only useful and real indicator for AMF species diversity. The species richness and diversity of AMF have been shown to be influenced by several soil factors and environmental adaptations indicating that the physiology and genetics of AM along with their responses to the host and edaphic conditions regulates their diversity. Knowing the diversity of AMF is absolutely necessary to develop any inoculation package for a particular fruit crop. Diversity of AMF associated with fruit crops in tropical, subtropical and temperate cropping system is discussed in part I of the book. The recent nomenclature for AMF followed by Redecker *et al.* (2013) is used only in part I. In other parts old nomenclature is used.

Response of different fruit crops to the AMF inocula have been tested in many parts of the world independently. These results strengthen the prospects for large scale utilization in crop cultivation. AMF have a striking effect on crop growth because of its hyphae which are present at 100 meters per cubic centimeter of soil which connects with the plants and aquire water and nutrients from soil which is not accessible to plant root. The rate of inflow of phosphorus into mycorrhizae can be up to six times that of the root hairs. AMF also have an indirect influence on plant growth because of their effects on soil structure stabilization, i.e. soil aggregate formation and humic substances accumulation. Glomalin related soil protein (GRSP) component is produced by AMF hyphae as a stable glue. GRSP binds to soil, producing a uniform aggregated structure composed of minerals and humus. Improvement of growth and nutrient uptake in fruit crops of the tropics, sub tropics and temperate region has been discussed in part II.

Enrichment of soil health depends on the availability of nutrients to the plants, which will help in effective cycling of the nutrients thereby creating a balance between influx and efflux of nutrients. Mycorrhizal hyphae are more efficient than plant roots in taking up phosphorus , nitrogen and micronutrients like Cu, Zn, Al, Mn and Mg and transporting to the plants. Enhancements in the acquisition of K, Ca and Mg are often observed in AMF colonized plants grown on acidic soils than neutral or alkaline soils. AMF influence microorganisms which mineralize organic matter and nitrification in the soil increasing thereby the availability of organic source of P and N which are efficiently absorbed by the plant through the hyphae of mycorrhizal fungi. Thus, mycorrhizal symbiosis offers a wholesome package of nutritional benefits and assists in maintenance of soil health. Nutrient cycling and soil dynamics are discussed in part III.

AM symbiosis plays significant role in protecting the plants against osmotic stresses by altering water relations of the plant with consequent effects on tissue hydration and plant physiology. AMF increase the ability of the root system to scavenge water in dry soil, resulting in less strain on foliage, and hence higher stomatal conductance and shoot water content at a particular soil water potential. AMF can help to overcome the problem of salinity stress. Plants growing in saline soils are subjected to physiological stresses. So far, studies on salt stress tolerance in mycorrhizal plants have suggested that AMF plants grow better due to improved mineral nutrition and physiological processes like photosynthesis or water use efficiency, the production of osmo regulators, higher K^+/Na^+ ratios and compartmentalization of sodium within some plant tissues.

Fruit crops usually experience crop loss to an extent of 20-30% due to infestation with pathogens. AMF associations have been suggested as biocontrol agents as they can reduce or even suppress, damage caused by soil-borne plant pathogens. Pathogen suppression by AMF involves changes in mycorrhizosphere microbial populations and their dominance. The colonization by AMF results in morphological, physiological and systemic changes to the root which makes the root stronger to resist the pathogens.

AMF can adapt to metal stress, and are able to assist plants in metal polluted soils. They are widely recognized as plant growth promoters that assists by increasing acquisition of several nutrients and also encouraging plant growth in soils contaminated by heavy metals (oids) through a series of mechanisms. Alleviation of abiotic and biotic stresses by AMF and remediation of soil of heavy metals is discussed in part IV.

AMF spores are a reservoir of beneficial bacteria, which help them in root colonization, spore germination and extraradical hyphal growth by producing stimulatory compounds or by influencing nutrients acquisition. These bacteria are generally called as mycorrhiza associated bacteria and are mostly fungi-specific but can also be host specific. The AMF spore associated bacteria are known to secrete metabolites, change soil pH, produce cell wall degrading enzymes and growth promoting substances for stimulation of mycorrhizal spore germination and colonization of host plants. In addition to supporting mycorrhizal colonization, these bacteria act as plant growth promoters, nutrient solubilizers and effective biocontrol agents in some of the fruit crop seedlings. Many findings strongly suggest the combined application of AMF with their associated bacteria for better performance of plant growth rather than single application. Beneficial attributes of some bacteria and fungi in synergistically boosting crop growth with AMF has been highly recognized in recent times and applied in crop production practices. Combined inoculations of these organisms have helped to improve the fruit quality and also to maintain a sustainable production system. Interaction of AMF with associated microbes in plant growth promotion is discussed in part V.

Production of AMF inocula is the most challenging task as they are obligate symbionts and depend on a living host for their multiplication. Methods of AMF inocula production have evolved considerably during the recent decade. They are produced in raised bed nurseries, earthen pots and other containers and on-farm. On-farm inoculum production is very beneficial for fruit crop cultivation. Many host species have been tested and found to be good for inoulum multiplication, Different substrates have been used in soil and substrate-based production technique. Substrate-free culture techniques (hydroponics and aeroponics) and *in-vitro* cultivation methods have all been attempted for the large-scale production of AMF.

Most successful use of AMF is in nurseries where micropropagated plantlets are hardened. Micropropagated plants usually face transplantation shock as they are not equipped enough to face the challenges of pathogens in the soil. AMF plays significant role in their pre and post-transplanting performance. They help in the development of a superior root system which help in better absorption and nutrients and hence survival of plants. They help in protecting plants against pathogen attack.

Fruit crops are raised directly through seeds, seedlings, cuttings, through grafting of scions of poplular varieties on rootstocks, using suckers, air layering and using micropropagated plantlets. As these crops involve a plant transplant stage, mycorrhization in the nursery before transplanting would yield vigorous and healthy seedlings which would establish a viable fruit orchard in different agro climatic conditions. The main criterion for inoculums application is the exact species of inoculums to be used, the number of healthy propagules in an inoculum and its quantity required for application. Details of the application of AMF in micropropagated fruit crops and an insight into the production of quality inoculum, information about its availability, methods of utilization in fruit crop nursery and its use in orchard rejuvenation, solving replant problem and restoration of soil structure are discussed in part VI.

Several studies at molecular level have resulted in correct identification of AMF spores and the taxon to which it belongs. Different molecular techniques have been used to study expression level of the genes responsible for agronomically important traits. The principles of AMF symbiosis at cellular level has also been worked out and is covered in part VII. .

Overall the book provides an insight into the usage of mycorrhizal technology in different fruit crop production systems in the world, its benefits in nutrient cycling/management and in biotic and abiotic stress management and the prospects of its application in nurseries and field crop. How a careful selection of right host/fungus/substrate combination is critical for the success of the symbiosis and an early establishment of the AMF status after sowing or transplanting which improves plant performance in horticultural practices. How AMF technology is most feasible for fruit crops which involve a transplant stage, where plants are produced in nursery beds, containers or by tissue culture. Fruit crops raised through seedlings, root stocks, suckers, micropropagated plants can be subjected to AM application at an early stage and get colonized with the fungus to reap the benefit of application. Finally it deals with inoculum production techniques for supply of quality inoculum in commercial horticultural plant production systems.

It is hoped that this book will be useful to students, researchers, fruit crop growers, industrialists who wish to adopt mycorrhizal technology as inoculation with AMF will surely become an integral part of most horticultural production systems in the future.

Sukhada Mohandas

P. Panneerselvam

Contents

Part III

Improvement in Soil and Plant Nutrition by Arbuscular Mycorrhizal Fungi

Part IV

Stress management by Arbuscular Mycorrhizal Fungi

Part V

Interaction of Arbuscular Mycorrhiza with other beneficial microbes

Part VI

AMF inoculum Production and Mycorrhization of Crop Plants

Part VII
Molecular Approaches

Part I

Arbuscular Mycorrhizal Fungal Diversity

Chapter 1

Diversity of Arbuscular Mycorrhizal Fungi in Fruit Cropping System

K. M. Rodrigues and B. F. Rodrigues*

Department of Botany, Goa University, Goa 403 206, India.

* Address for the correspondence: felinov@gmail.com

ABSTRACT

Arbuscular mycorrhizal fungi (AMF) form symbiotic relationship with more than 80% of terrestrial plants. AMF are reported to occur in almost all habitats. The diversity of AMF has significant ecological consequences because individual species or isolates vary in their potential to promote plant growth and adaptation to biotic and abiotic factors. AMF population in a given agro-ecosystem plays a crucial role in the structure of plant community and maintenance of ecosystem stability and development of sustainable agriculture. It is well known that many fruit crops are mycorrhizal. The species richness and diversity of AMF have been shown to be influenced by several soil factors and environmental adaptations indicating that the physiology and genetics of AMF along with their responses to the host and edaphic conditions regulates their diversity. This review summarizes the rich diversity of AMF associated with different fruit crops and indicates the need to identify indigenous AMF populations for their distinct roles in association with fruit crops so as to achieve maximum benefits of the mutualistic association.

Keywords: AMF inoculum, root colonization, endomycorrhizal biotechnology, phosphorus, horticulture.

1.0 Introduction

AMF are a group of microbes, associating with the roots of the majority of terrestrial plants including horticultural tree species (Guissou, 2009) and vegetables (Ortas, 2010), providing the host with nutrients and alleviating ecological stresses. To understand the benefits provided by AMF and to utilize those benefits, it is important to assess the diversity of AMF. Rarely one AMF species, but mostly three to nine AMF species of different genera occupy the host plant root system at the same time (Morton *et al.*, 1995). The AMF diversity does not follow but may regulate patterns of plant diversity i.e. if one of the indigenous AMF species becomes extinct in a habitat, there may be shifts in how plants acquire resources in that habitat (Allen *et al.*, 1995). The interspecific genetic diversity of AMF probably overrides the intraspecific diversity, which is why conservation and research should also deal with the adaptations of isolated indigenous AMF species (Friberg, 2001). Agricultural management and practices also affect the diversity, development and functioning of AMF along with soil parameters. Therefore, it is necessary to evaluate the possible agricultural benefits provided by different AMF species at different sites or habitats and also to take cautionary efforts when applying modern agricultural methods.

2.0 Classification of AMF

The most recent classification of *Glomeromycota* (**Table 1.1**) is based on a consensus of regions spanning rRNA genes: 18S (SSU), ITS1-5.8S-ITS2 (ITS), and/or 28S (LSU). The phylogenetic reconstruction underlying this classification is discussed in Redecker *et al.* (2013). The diversity of mycorrhiza reported in different fruit crops in this review is as per this nomenclature.

3.0 Mycorrhiza in fruit crops

Many fruit crops are dependent on AMF colonization for survival and growth (Covey *et al.*, 1981; Powell and Santhanakrishnan, 1986; Schubert and Cammarata, 1986). First evidence of the positive influence of AMF symbiosis on fruit crop production was provided by Menge *et al.* (1977), who demonstrated that the AMF inoculation is a pre-requisite for the establishment of *Citrus* species in biocide-treated nursery beds. Since then, a number of experiments have been carried out as reviewed by Miller *et al.* (1986); Nemec (1986); Gianinazzi *et al.* (1990 a b); Barea *et al.* (1993); Chang (1994); Lovato *et al.* (1995) and Varma and Schuepp (1995). The importance of AMF symbiosis in horticultural crop production has been evaluated in many fruit crops like *Citrus* (Menge *et al.*, 1978; Onkarayya and Sukhada, 1993), *Malus* (Plenchette *et al.*, 1981, 1983, Reich, 1988) and strawberry (*Fragaria* x *ananassa* Duchesne) (Hughes *et al.*, 1978; Williams *et al.*, 1992; Chávez and Ferrara-Cerrato, 1990; Vestberg *et al.*, 2000), banana,(1992) papaya (Sukhada,1994), mango

(Sukhada, 2012), and many more fruit crops. Moreover, mycorrhizal fruit trees are also known to be possessing enhanced tolerance to biotic and abiotic stresses (Menge *et al.*, 1978; Guillemin *et al.*, 1994a, 1994b). AMF association is an important component in sustainable agriculture. Beneficial role of AMF symbiosis in growth and development of some of the fruit crop is summarized in **Table 1.2.**

Table 1.1: Consensus classification of the Glomeromycota by Redecker et al. (2013).

Class	Order	Family	Genus
		Diversisporaceae	*Tricispora**
			*Otospora**
			Diversispora
			*Corymbiglomus**
			Redeckera
		Acaulosporaceae	*Acaulospora*
	Diversisporales	Sacculosporaceae*	*Sacculospora**
		Pacisporaceae	*Pacispora*
			Scutellospora
		Gigasporaceae	*Intraomatospora**
			*Paradentiscutata**
			Dentiscutata
			Cetraspora
			Racocetra
			Glomerales
Glomeromycetes		Claroideoglomeraceae	*Claroideoglomus*
			Glomeraceae
			Glomus
			Funneliformis
			Septoglomus
			Rhizophagus
			Sclerocystis
		Ambisporaceae	*Ambispora*
	Archaeosporales	Geosiphonaceae	*Geosiphon*
		Archaeosporaceae	*Archaeospora*
	Paraglomerales	Paraglomeraceae	*Paraglomus*

Table 1.2: Efficient AMF for some fruit crops.

S. No.	Plant species	Efficient AMF	AMF tested	Parameters evaluated	Workers
1.	Ananas comosus	Rhizophagus fasciculatus	R. fasciculatus, Glomus sp., Acaulospora sp., Scutellospora sp.	Plant growth, nutrition	Jaizme-Vega and Azcón (1995)
2.	Carica papaya	Rhizophagus fasciculatus	R.fasciculatus, Glomus sp., Acaulospora sp., Scutellospora sp.	Plant growth, nutrition	Jaizme-Vega and Azcón (1995)
3.	Malus domestica	Funneliformis mosseae	F. mosseae, G. macrocarpum	Root colonization, plant height, plant growth	Miller et al. (1989)
4.	Citrus jambhiri	Rhizophagus fasciculatus	R. fasciculatus, F. mosseae, Claroideoglomus etunicatum	Plant height, top weight, root weight, chlamydospores in soil	Nemec (1979)
5.	Persea americana	Rhizophagus fasciculatus	R. fasciculatus, Glomus sp., Acaulospora sp., Scutellospora sp.	Plant growth, nutrition	Jaizme-Vega and Azcón (1995)

AMF form obligate symbiotic association with host plants belonging to most of the families. Although AMF are not host specific, they exhibit certain host preferences. Gracias Flor (2005) assessed the AMF diversity in fruit trees from Goa, India and reported a rich diversity of AMF species in fruit trees. This study reported the occurrence of 15 AMF species belonging to three genera *viz., Acaulospora, Glomus* and *Scutellospora* indicating that these fungi play a vital role in the growth and survival of plant species and suggested the need for screening suitable efficient strains of AMF for each of the fruit trees studied for better sustenance, increased nutrient uptake and enhanced yield and productivity (**Table 1.3**).

Table 1.3: Studies on AMF diversity in fruit trees (Gracias Flor, 2005)

S.No.	Plant species	Family	AMF species	Percent colo-niz-ation	Spore density g-1 soil
1.	Psidium guajava L.	Myrtaceae	A. scrobiculata, F. geosporum and S. pellucid	43	52
2.	Artocarpus incisa L.	Moraceae	A. nicolsonii, R.fasciculatus, G. heterosporum and Glomus sp.	30	45
3.	Achras sapota L.	Sapotaceae	R. fasciculatus	59	30
4.	Annona squamosa L.	Anonaceae	A. scrobiculata, G. heterosporum, Glomus sp. and S.heterogama	25	53
5.	Artocarpus heterophyllus Lamk.	Moraceae	S. heterogama and Glomus sp.	56	46
6.	Mangifera indica L.	Anacardiaceae	F. constrictum, Glomus sp., S. heterogama, Racocetra gregaria and S. biornata	35	25
7.	Punica granatum L.	Punicacaea	R. fasciculatus and F. geosporum	90	3
8.	Ananas comosus (Linn.) Merr.	Bromeliaceae	A. scrobiculata and Racocetra gregaria	45	10
9.	Carica papaya L.	Caricaceae	G. heterosporum, S.heterogama and S. pellucid	20	20
10.	Musa paradisiaca L.	Musaceae	F. geosporum and Racocetra gregaria	80	14
11.	Citrus sinensis Osbeck.	Rutaceae	F. geosporum and G. multicaule	-	150
12.	Citrus limon Burm.f.	Rutaceae	F. geosporum and A. scrobiculata	-	15

The species richness and diversity of AMF have been shown to be influenced by several soil factors and environmental adaptation. The physiology and genetics of AMF and their responses to the host and edaphic conditions regulates their diversity. The host plant, crop rotation; soil pH, moisture, soil temperature, nutrient levels and carbon from their host plants, and interaction with other soil biota influence the diversity (**Table 1.4**). AMF are perceived as critical components of various ecosystems and play a crucial role in the maintenance of agro-ecosystem stability and sustainable agricultural development (Wang *et al.*, 2013). The potential of endomycorrhizal biotechnology has captivated the interest of horticulturists. Several workers have reported the diversity of AMF in different fruit cropsunder diverse habitats.   A review of the studies carried out in different fruit crops of the tropics, sub-tropics and temperate conditions is given here.

Table 1.4: AMF species diversity in fruit crops.

S. No.	Plant species	Family	AMF species	Reperence
1.	*Carica papaya*	Caricaeae	*Claroideoglomus etunicatum, C.claroideum, Paraglomus occultum, A. scrobiculata, Acaulospora* sp., *Glomus* sp., *Gigaspora* sp., and *Scutellospora* sp.	Trindade *et al.*, 2006; Khade and Rodrigues, 2008 a, b (**Plate 1.1**)
2.	*Citrus* species	Rutaceae	*Glomus macrocarpum, G. aggregatum, Funneliformis constrictum, F. mosseae, Rhizophagus fasciculatus, R. irregularis, Sclerocystis sinuosa, Glomus* sp., *Gigaspora* sp., *Sclerocystis* sp., *Acaulospora* sp., *Scutellospora* sp., and *Entrophospora* sp.	Menge *et al.*, 1977; Inserra *et al.*, 1980; Nemec, 1981; Camprubi and Calvet, 1996; Ferguson and Schenck, 1997; Wang *et al.*, 2013
3.	*Vitis vinifera*	Vitaceae	*R. fasciculatus, G. macrocarpum, G. microcarpum, Glomus* sp., *F. mosseae, Paraglomus/Archaeospora* sp., *Scutellospora* sp., *Gigaspora* sp. and *Sclerocystis* sp.	Balestrini *et al.*, 2010; Lumini *et al.*, 2010; Schreiner and Mihara, 2009; Nappi *et al.*, 1985
4.	*Persea americana*	Lauraceae	*Sclerocystis sinuosa, G. macrocarpum* and *F. constrictum*	Hass and Menge (1990)
5.	*Malus* species	Rosaceae	*Glomus* species, *G. aggregatum, G. macrocarpum, C. claroideum, R. irregularis, F. caledonium, F. constrictum, F. mosseae* and *Gi. margarita*	Miller *et al.*, 1985; Sumorok *et al.*, 2011
6.	*Musa* species	Musaceae	*Glomus* sp., *Acaulospora* sp. and *Gigaspora* sp.	Arias *et al.*, 1998; Khade, 1999; Jefwa *et al.*, 2010; Fogain and Njifenjou, 2002

Contd...

Table 1.4: Contd...

S. No.	Plant species	Family	AMF species	Reperence
7.	Passiflora edulis	Passifloraceae	R. clarus, G. spurcum, G. macrocarpum, G.invermaium, S. fulgida, S. pellucida, S. heterogama, A. colombiana and A. appendiculata	Soares et al. (2005)
8.	Litchi chinensis	Sapindaceae	Glomus, Gigaspora, Rhizophagus and Acaulospora.	Singh and Prasad (2006)
9.	Psidium guajava	Myrtaceae	Sclerocystis coremioides, S. calospora, Acaulospora sp. and F. mosseae	Janos, 1980; Kumuran and Azizah, 1995
10.	Actinidia deliciosa	Actinidiaceae	Paraglomus occultum, Gigaspora sp., Glomus sp. and R. clarus	Calvet et al. (1989)
11.	Phoenix dactylifera	Arecaceae	F. mosseae, F. constrictum, R. fasciculatus, G. aggregatum, G. macrocarpum, Acaulospora species and Scutellospora species	Bouamri et al. (2006)
12.	Olea europaea	Oleaceae	Glomus sp., Gl. versiforme, G. viscosum (BEG 126), Entrophospora sp., Gigaspora sp., Acaulospora sp., R. irregularis, R. irregularis (BEG 123), R. clarus, R. clarus (BEG 125), F. mosseae (BEG 124), A. colossica, Scutellospora sp., and S. heterogama	Kachkouch et al., 2012; Sghir et al., 2013; Calvente et al., 2004
13.	Annona squamosa	Anonaceae	Glomus sp. and Acaulospora sp.	Sarwade et al., 2011; Pindi, 2011
14.	Tamarindus indica	Fabaceae	Glomus sp., Acaulospora sp., and Scutellospora sp.	Bourou et al. (2010)

3.1 Apple (*Malus domestica* Borkh.)

Apple is the most widely cultivated temperate fruit crops over a wider geographical range reported to be endomycorrhizal (Bouwens, 1937). AMF most often colonize the roots of wild species viz., *M. communis* L. and *M. sylvestris* Mill. (Harley and Harley, 1987). Miller *et al.* (1985) carried out an extensive survey of AMF associated with apple from 18 sites in USA. Their survey reported that AMF belonging to *Glomus* occurred more commonly than *Acaulospora*, *Gigaspora* and *Sclerocystis*. Apple plants show significant growth enhancements when inoculated with AMF (Morin *et al.*, 1994). Schubert and Lubraco (2000) reported increased P uptake and enhanced plant growth in *M. pumila* L. upon association with *F. mosseae*. Sumorok *et al.* (2011) identified eight AMF species *viz.*, *G. aggregatum,F. caledonium*, *C. claroideum*, *F. constrictum*, *R. irregularis*, *G. macrocarpum*, *F. mosseae* and *G. margarita* in rhizosphere soil of apple variety 'Gold Milenium' from Poland. Miller *et al.* (1989) observed that *F. mosseae* exceeded *G. macrocarpum* in the rate of initial colonization, amount of maximum colonization and persistence of arbuscules and external hyphae in *M. domestica.*

3.2 Avocado (*Persea americana* Mill.)

Avocado has hairless mangrove-like roots that are highly dependent on mycorrhizal symbiosis (Alarcón *et al.*, 2001). Presence and identification of AMF has also been detectedin different isolated avocado orchards in Michoacan, Mexico (Bárcenas *et al.*, 2006). Hass and Menge (1990) carried out studies on AMF associated with avocado orchard soils from Israel and California (USA). They reported the difference in population and species composition of AMF. *S. sinuosa* and *G. macrocarpum* were commonly associated with avocado in Israel while *F. constrictum* was reported from California. The positive effect of mycorrhiza on avocado plants has been observed in seedlings and micropropagated plants.

3.3 Banana (*Musa* spp.)

AMF occur naturally in the rhizosphere, and the banana plants are highly mycorrhizal (Msiska, 2001). Arias *et al.* (1998) reported the dominance of *Glomus* followed by *Acaulospora* species in banana agro-ecosystem from Caribbean area of Costa Rica. Khade (1999) reported 17 AMF species from 7 varieties of *Musa* species sampled from three sites in North Goa, India and reported *Glomus* (13 species) as the dominant genus followed by *Acaulospora* (3 species) and *Gigaspora* (1 species). AMF species belonging to the genus *Glomus* and family Acaulosporaceae have been found to be most common from banana cultivations in Kenya, Uganda, Rwanda and Burundi (Jefwa *et al.*, 2010), similar with the findings of Adriano-Anaya *et al.* (2006) for 'Grande Naine' (AAA genome) banana plantations in Mexico. In Cameroon, an undescribed *Glomus* species isolated and cultured from banana plantation readily colonized the plants (Fogain and Njifenjou, 2002).

3.4 Citrus (*Citrus* spp.)

Menge *et al.* (1977) and Inserra *et al.* (1980) reported the widespread occurrence of AMF belonging to *Glomus, Gigaspora* and *Sclerocystis* in *Citrus* species. Several reports on colonization of *Citrus* roots by AMF have been published (Reed and Fremont, 1935; Neill, 1944; Sabet, 1945; Kruckelmann, 1975). Most *Citrus* species, such as *Sour orange, Trifoliate orange, Cleopatra mandarin, Swingle citrumelo* and *Carrizo citrange* have rare and short root hairs, and are fairly dependent on AMF that are mostly *Glomus* species (Davies and Albrigo, 1994). Nemec (1981) isolated the AMF in California and Florida and reported the occurrence of *Funneliformis constrictum, R. fasciculatus, G. macrocarpum* and *Sclerocystis sinuosa*. Camprubi and Calvet (1996) reported the presence of 13 AMF species *viz.,* 9 *Glomus* species and 1 species each of *Acaulospora, Gigaspora, Sclerocystis* and *Scutellospora*. Ferguson and Schenck (1997) reported the presence of 5 genera (*Acaulospora, Glomus, Gigaspora, Entrophospora* and *Scutellospora*) and a total of 34 species with *Acaulospora* species and *Glomus* species being most commonly associated with *Citrus* species in central and southern Florida. Wang *et al.* (2013) identified 18 AMF speciesbelonging to 3 different orders, Archaeosporales (1 species), Diversisporales (7 species) and Glomerales (10 species) from rhizosphere soils of *C. reticulata* Blanco (red tangerine) rootstock in hillside *Citrus* orchards. However, they observed that in all of the surveyed orchards, *Glomus aggregatum, F. mosseae* and *Rhizophagus irregularis* were the dominant AMF species. Canonical correspondence analysis revealed that the AMF community structure was significantly influenced by environmental factors, especially altitude, pH, soil moisture, and available nitrogen (N). Their data indicate that environmental factors are important in determining AMF root colonization, propagule numbers, and species diversity in *Citrus* orchards.

3.5 Cherimoya (*Annona cherimola* Mill.)

Azcón-Aguilar *et al.* (1994ab) reported a strong dependence of cherimoya on AMF symbiosis for optimal growth and development. The role of AMF establishment benefiting most of the plants produced *in-vitro* has been demonstrated (Lovato *et al.*, 1996). AMF inoculation appears as a critical factor for the development of the micro-propagated cherimoya plants, at the weaning stages (Azcón-Aguilar and Barea, 1996).

3.6 Date palm (*Phoenix dactylifera* L.)

First report on the presence of AMF in *P. dactylifera* was in the Crescent desert near Baghdad (Iraq) where it seemed to contribute to the plant mineral nutrition and supply water when it takes the place of root hairs as absorbing structures (Khudairi, 1969). Similar observations were recorded for date palm growing in oasis of Qassim, Saudi Arabia indicating the presence of mycorrhiza (Khaliel and Abou-Hailah, 1985). Bouamri *et al.* (2006) conducted a survey of AMF diversity and root colonization in *P. dactylifera* in arid areas of Southwestern Morocco. They recorded 10 AMF species *viz.*, *F. mosseae, R. fasciculatus, F. constrictum, G. aggregatum* and *G. macrocarpum*. Three *Acaulospora* species and two *Scutellospora* species remained unidentified. *F. constrictum* and *Acaulospora* sp. 2 were found to be most abundant AMF species. The benefits of inoculating*P. dactylifera* plants with AMF have been documented in earlier studies (Al-Whaibi and Khalil, 1994; Bouhired *et al.*, 1992; Jaiti *et al.*, 2007; Al-Karaki, 2000; Al-Karaki *et al.*, 2004). Bouhired *et al.* (1992) reported positive effect of *R. irregularis* on the growth of *P. dactylifera*.

3.7 Grapevine (*Vitis vinifera* L.)

Grapevine is another fruit crop that attracted the attention of mycorrhizologists (Hayman *et al.*, 1976). In field conditions, grapevine roots normally are colonized by AMF (Balestrini *et al.*, 2010; Schreiner and Mihara, 2009). Many workers studied the symbiotic association in grapevine from the morphological and physiological point of view (Bartschi and Garree, 1980; Oehl *et al.*, 2005; Lumini *et al.*, 2010). Grapevines appear to be reliant on AMF colonization for normal growth and development due to relatively coarse fine roots (Biricolti *et al.*, 1997; Linderman and Davis, 2001). In some cases, grapevines are even completely dependent on AMF

(Menge *et al.*, 1983). Little is known, however, of the species composition of the AMF communities that colonize grapevines in production vineyards, as most of the published studies have been carried out on small-scale experiments. AMF isolates previously encountered in vineyard soils and in grapevine roots are dominated by *Glomus* species (Balestrini *et al.*, 2010; Lumini *et al.*, 2010; Schreiner and Mihara, 2009) although *Paraglomus/Archaeospora* and *Scutellospora* species have also been reported as colonizers of grapevine roots (Schreiner and Mihara, 2009). Nappi *et al.* (1985) reported the widespread occurrence of AMF species (*R. fasciculatus*, *G. macrocarpum*, *G. microcarpum* and *F. mosseae*) in the vineyards of Piedmont (Italy). They also reported a low occurrence of *Gigaspora* and *Sclerocystis* species. AMF have an increasingly important role in vineyard production systems, as many vineyards receive little water and are planted on less fertile soils (Schreiner, 2005). AMF symbiosis of vines roots can result in increased growth (Linderman and Davis, 2001), enhanced nutrient uptake (Schreiner, 2007) and improved drought tolerance (Schreiner, 2007). As such, AMF constitute an integral and important component of the vineyard ecosystem and might have significant applications for sustainable agricultural ecosystems (Schreiner and Bethlenfalvay, 1995).

3.8 Guava *(Psidium guajava* L.) and **Sapota** *(Achras sapota* L.)

It is observed that field-grown guava plants are colonized by different AMF species *viz.*, *S. coremioides*, *S. calospora*, *Acaulospora* species and *F. mosseae* (Janos, 1980; Kumuran and Azizah, 1995). In guava cropping systems, the occurrence of AMF spore load was observed to be higher in top rhizosphere soil .i.e 15-20 cm (3.4 -9.6 number spores per gram soil in guava and 1.2 - 8.1 number spore per g soil in sapota respectively) as compared to 30-45 cm depth. Interestingly, the occurrence of AMF was observed to be higher in multiple cropping systems in both sapota and guava as compared to mono crop. Spores of four different genera belonging to *Glomus*, *Gigaspora*, *Acaulospora* sp. and *Sclerocystis* were commonly recorded.However, *Glomus* was found to be dominant irrespective of the age group of guava and sapota orchards (Saritha et *al.*, 2014b) **Plate 1.2.**

3.9 Kiwifruit *(Actinidia deliciosa* (A Chev) Liang et Ferguson)

Kiwifruit is normally colonized by AMF in the field, and growth enhancements can be obtained by inoculating AMF on kiwifruit seedling (Schubert *et al.*, 1987). Sampling of a commercial kiwifruit orchard in Spain showed high colonization of the root systems by naturally occurring AMF species *viz.*, *Paraglomus occultum*, *Gigaspora* sp., *Glomus* sp. and *clarus* (Calvet *et al.*, 1989). Schubert *et al.* (1990) screened AMF species and found *F. constrictum* to be less effective in enhancement of growth of *A. deliciosa* than *F. caledonium*, *Paraglomus occultum* and *G. versiforme*.

3.10 Litchi *(Litchi chinensis* Sonn.)

Coville (1921) speculated that AMF might be indispensable to *Litchi*, but available data concerning growth effects are equivocal (Menzel and Simpson, 1987). Pandey and Misra (1975) examined the effect of AMF inoculation on *Litchi* in pots of sterilized soil and recorded that all control plants (without AMF inoculation) did not survive. Although this suggested strong dependence by *Litchi* upon AMF for survival, it precluded determination of effects on growth. Singh and Prasad (2006) observed maximum colonization and spore population in *Litchi* orchards from Uttar Pradesh and reported colonization by AMF species belonging to the genera *viz., Glomus, Gigaspora, Rhizophagus* and *Acaulospora*.

3.11 Mango (*Mangifera indica* L.)

Few studies have been conducted on the response of *Mangifera indica* to AMF inoculation. Xiutang (1990) tested different species of *Glomus* on *M. indica* in pot culture and found that mineral uptake especially of P was improved due to the fungal colonization. Reddy and Bagyaraj (1994) studied different AMF for their association with 'Nekkare' mango in sterilised soil and found improved plant growth and shoot P content. Sukhada (1998) reported improved plant growth and nutrient content in shoots of rootstock totapuri inoculated with *F. mosseae* and *R. fasciculatus* in pot culture. Sukhada (2012) studied the diversity of AMF in seven root stocks of mango and found *Glomus* and *Acaulospora* to be the major genera in the rhizosphere with *R. fasciculatus* and *F. mosseae* as the predominant AMF species.

3.12 Olive (*Olea europaea* L.)

Olive plants are known to be associated with AMF (Roldán-Fajardo and Barea, 1985). Kachkouch *et al.* (2012) evaluated mycorrhizal status in *O. europaea* and also conducted the survey of AMF species in the olive grove soils of Morocco. They recorded spores belonging to *Glomus, Entrophospora, Gigaspora, Acaulospora* and *Scutellospora*, with *Glomus* being the dominant genus. Sghir *et al.* (2013) also surveyed *O. europaea* tree groves of Morocco and reported five AMF species *viz., R. irregularis, R. clarus, G. versiforme, A. colossica* and*heterogama*, with the genus *Glomus* being dominant. Calvente *et al.* (2004) reported experiments that support the potentiality of AMF inoculation for two olive cultivars *viz.*, Arbequina and Leccino. They also analysed the natural diversity of AMF in the root- associated soil from long-term established olive tree plantations and identified four distinguishable native AMF species *viz., R. irregularis* (BEG 123), *F. mosseae* (BEG 124), *R. clarus* (BEG 125) and *G. viscosum* (BEG 126). *R. irregularis* and *G. viscosum* isolated from the target agro-system were reported as the most effective fungi to improve the development of both the olive varieties. Thus exploitation of the natural diversity of AMF is necessary so as to formulate inoculum to be applied during the commercial nursery production of olive varieties.

3.13 Papaya (*Carica papaya* L.)

Papaya is shown to be highly responsive to *Glomus* under greenhouse as well as field conditions (Jaizme-Vega and Azcon, 1995). Trindade *et al.* (2006) studied 47 commercial plantations in Brazil for AMF associations in both field and nursery and observed that papaya roots showed considerable variation in AMF

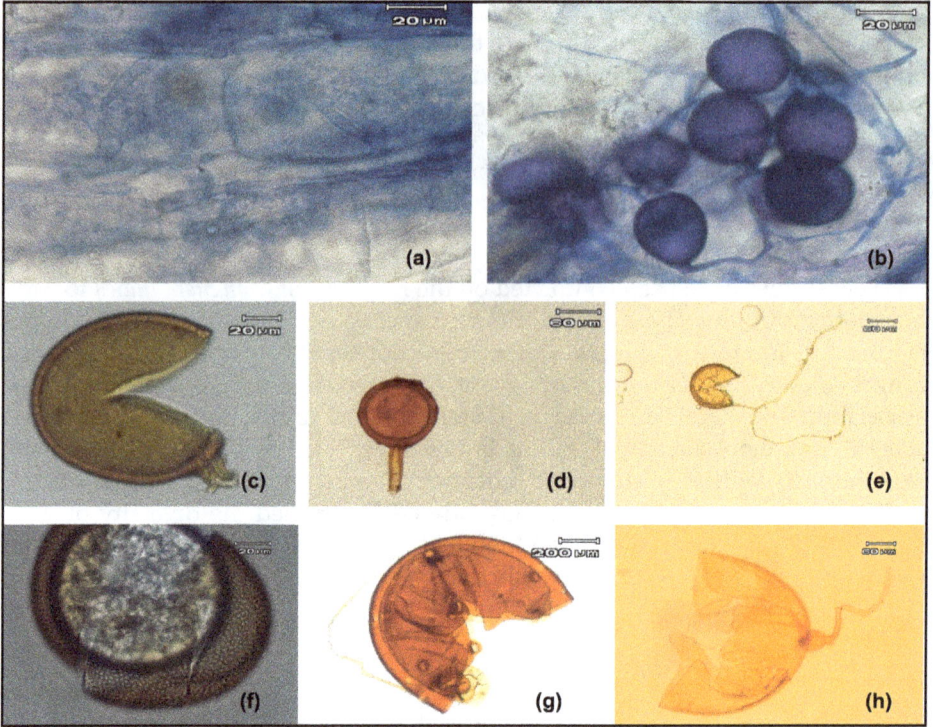

Plate 1.1: (a)Arbuscular colonization. (b) Vasicular colonization. (c) *Rhizophagus fasciculatus*. (d) *Funneliformis geococetra gragaria*. (h) *Scutellospora pellucida* isolated from the rhizospere of fruit crops.

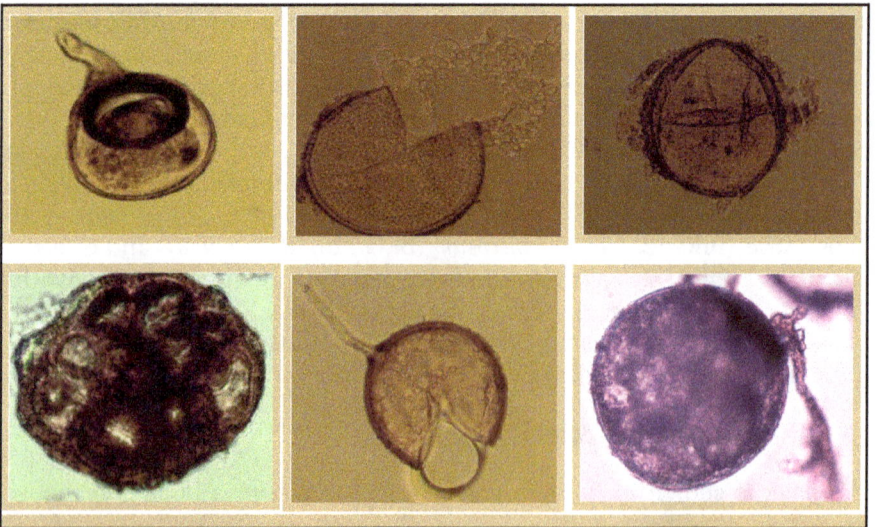

Plate.1.2: AMF spores isolated from the rhizosphere of Sapota orchard in Karnataka. Source: Saritha *et al.*, 2014.

colonization, ranging from 6% to 83%. They reported very low AMF colonization in nursery seedlings, while the field spore numbers varied from 34 to 444/30g of soil. A total of 24 AMF species were recorded with *Claroideoglomus etunicatum, Paraglomus occultum, Acaulospora scrobiculata* and *Gigaspora* sp. being the most common species. Khade and Rodrigues (2008 a, b) recorded 18 AMF species belonging to four genera *viz., Acaulospora, Glomus, Gigaspora* and *Scutellospora* in mono-culture plantation of *C. papaya. Claroideoglomus claroideum* was the most frequently occurring species and was recovered throughout the study period suggesting AMF are well established in *C. papaya* and they exhibit variations depending on edaphic factors and seasonal patterns in the weather.

3.14 Passion fruit *(Passiflora edulis* Sims. f. *flavicarpa* Deg.)

Soares *et al*. (2005) identified 9 native AMF species *viz., R. clarus, G. spurcum, S. fulgida, G. macrocarpum, G. invermaium, A. colombiana, S. pellucida, A. appendiculata* and *S.heterogama* from a passion fruit plantation in Brazil with *R. clarus* and *G. spurcum* being the most predominant species. Cavalcante *et al*. (2001) suggested that inoculation with AMF may reduce or eliminate the need for P fertilization in passion fruit. They investigated the response of passion fruit seedlings to inoculation with four AMF species *viz., Gi. albida, Gi.margarita, A. longula* and *S. heterogama* either as single species or mixed culture. Their study revealed that the relative mycorrhizal dependency (RMD) of the passion fruit cultivar used was influenced by the AMF species inoculated, soil sterilization and soil P level and that this crop obtained significant benefit from AMF inoculation, thus recommending the use of AMF for plant growth improvement even without P fertilization.

3.15 Sugar Apple *(Annona squamosa* L.)

Sarwade *et al*. (2011) reported the AMF association in *A. squamosa* from Maharashtra, India. They reported the association of *Glomus* and *Acaulospora* with *A. squamosa*. Pindi (2011) reported *Glomus* as the dominant genus.

3.16 Tamarind *(Tamarindus indica* L.)

Tamarind is normally associated with AMF (Tomlinson *et al.*, 1995). Positive effects of AMF on the growth of *T. indica* have been reported (Reena and Bagyaraj, 1990). The MD of*indica* was not more than 36% on inoculation with five species of AMF *viz., A. spinosa, F. mosseae, R. irregularis, G. aggregatum* and *R. manihotis* as reported in the study conducted by Guissou *et al*. (1998). In this study they observed the root colonization by *A. spinosa, G. aggregatum* and *R. manihotis* was greater with significant increase in tree growth. While, *R. irregularis* colonized well, but provided little growth benefit, *F. mosseae* colonized poorly and did not stimulate plant growth thereby suggesting that the effect of different AMF species on *T. indica* differed. Bourou *et al*. (2010) studied root colonization, the mycorrhizal inoculum potential and spore diversity of tamarind parklands in Senegal. Their results did not reveal a specific AMF strain associated with tamarind plants and found that trees from the Sahelian zone are colonized more by mycorrhizas (11.17%) than those in the Sudano-Sahel and Sudan zones (5.72 and 3.85 respectively). They also observed that the numbers of *Glomus* and *Acaulospora* spores were significantly higher than *Scutellospora* spores in all tamarind parklands studied.

4.0 Conclusion and future strategies

Distribution, diversity, abundance and functioning of AMF are primarily based upon root colonization and spore density or spore counts which further depend upon many environmental factors such as soil pH, moisture content of soil, soil temperature, nutrient levels, presence of other soil microorganisms, host specificity and agricultural practices. Some AMF species have been reported to occur frequently but display variations depending on the edaphic and seasonal patterns in weather. Such species can be used as a source of inoculum in selected crops to determine their relative efficiency and potential. AMF play a vital role in ecosystem restoration. In low nutrient ecosystems, AMF are a resource to rely on and should be cautiously maintained with more efforts to retain the indigenous AMF species. To maintain and increase the local AMF species diversity and its inoculum potential it is necessary to carry out studies on growth yield response of AMF in intercropping systems and under different fertilizer regimes. The indigenous AMF species having adaptations to different environmental conditions or stresses should also be studied and maintained. AMF are the keystone of the development of sustainable agriculture and as such there is a necessity to accelerate their incorporation as biofertilizers in agricultural production systems.

References

1. Al-Whaibi, M. H. and Khaliel, A. S., 1994. The effect of Mg on Ca, K and P content of date palm seedlings under mycorrhizal and nonmycorrhizal conditions. *Mycosci.* 35, 213 – 217.

2. Allen, E.B., Allen, M.F., Elm, D.J., Trappe, J.M., Molina, R. and Rincon, E., 1995. Patterns of regulation of mycorrhizal plant and fungal diversity. *Plant Soil* .170, 47- 62.

3. Al-Karaki, G.N., 2000. Growth of mycorrhizal tomato and mineral acquisition under salt stress. *Mycorrhiza.* 10, 51 – 54.

4. Alarcón, A., Almaraz, J.J., Ferrera-Cerrato, R., González-Chávez, M.C.A., Lara H., M.E., Manjarrez M., M.J., Quintero, L.R. and Santamaria, R.S., 2001. *Manual:* Technologia de hongos micorrizicos en la producción de especies forestales en vivero. Colegio de postgraduados-Pronare-Conafor. México, pp. 98.

5. Al-Karaki, G.N., McMichael, B. and Zak, J., 2004. Field response of wheat to arbuscular mycorrhizal fungi and drought stress. *Mycorrhiza.* 14, 263–269.

6. Adriano-Anaya, M.L., Solis-Dominguez, F., Gavito-Pardo, M.E. and Salvador- Figureueroa, M., 2006. Agronomical and environmental factors influence root colonization, sporulation and diversity of arbuscular

mycorrhizal fungi at a specific phenological stage of banana trees. *J. Agron.* 5, 11–15.

7. Arias, F., Balanco, F. A., Vargas, R. and Ferrer, R., 1998. Anatomical and morphological identification of predominant species of arbuscular mycorrhizae banana agro-ecosystem in the Caribbean area of Costa Rica. *Corbana.* 22(48), 61-75.

8. Azcón-Aguilar, C., Encina, C.L., Azcón, R. and Barea, J.M., 1994a. Effect of arbuscular mycorrhiza on the growth and development of micropropagated *Annona cherimola* plants. *Agric. Sci. Finn.* 3, 281 – 288.

9. Azcón-Aguilar,C.,Encina,C.L.,Azcón,R.andBarea,J.M.,1994b.Mycotrophy of *Annona cherimola* and the morphology of its mycorrhizae. *Mycorrhiza.* 4, 161 – 168.

10. Azcón-Aguilar, C. and Barea, J.M., 1996. Arbuscular mycorrhizas and biocontrol of soil-borne plant pathogens, an overview of the biological mechanisms involved. *Mycorrhiza.* 6, 457–464.

11. Bartschi, H. and Garree, J.P., 1980. Etude comparative de la répartition cytologique de quelques élements minéraux dans 1 'écoree de racines saines et d'endomycorhizes de *Vitis vinifera* L. C.R. *Acad.Se. Paris, t.290, Série D*, 919 – 922.

12. Barea, J.M., Azcon, R. and Azcon-Aguilar, C., 1993. Mycorrhiza and crops. In: Tommerup, I. (Ed.), *Advances in Plant Pathology, vol. 9, Mycorrhiza: A Synthesis*. Academic Press, London, pp. 167–189.

13. Bárcenas, O., A.E., Varela, F.L., Carreón, A., Y., Lara Ch., M.B.N., González, C., J.C. and Aguirre, P., S. 2006. Estudios sobre hongos micorrizógenos arbusculares en huertos de aguacate *Persea americana* Mill. (RANALES: LAURACEAS). In: Memoria del *XXIX Congreso Nacional de Control Biológico*. SMCB. Manzanillo, Col. pp 1 – 5.

14. Balestrini, R., Magurno, F., Walker, C., Lumini, E. and Bianciotto, V., 2010. Cohorts of arbuscular mycorrhizal fungi (AMF) in *Vitis vinifera*, a typical Mediterranean fruit crop. *Environ. Microbiol. Rep.* 2, 594–604.

15. Biricolti, S., Ferrini, F., Rinaldelli, E., Tamantini, I. and Vignozzi, N. 1997. VAM fungi and soil lime content influence rootstock growth and nutrient content. *Am. J. Enol. Vitic.* 48, 93–99.

16. Bouwens, H., 1937. Investigations about the mycorrhiza of fruit trees, especially of quince (*Cydonia vulgaris*) and of strawberry plants (*Fragaria vesca*). *Zentralbl. Bakt., Abt. II.* 97, 34 – 49.

17. Bouhired, L., Gianinazzi, S. and Gianinazzi-Pearson, V., 1992. Influence of endomycorrhizal inoculation on the growth of *Phoenix dactylifera*. In: Micropropagation, Root Regeneration and Mycorrhizas. Joint Meeting between COST 87 and COST 8.10, Dijon, France, pp. 53.

18. Bourou, S., Diouf, M. and Van Damme, P., 2010. Tamarind (*Tamarindus indica* L.) parkland mycorrhizal diversity within three agro-ecological zones of Senegal. *Fruits.* 65 (6), 377 – 385.

19. Calvet, C., Pera, J., Estaun, V. and Camprub, A., 1989. Vesicular-arbuscular mycorrhizae of kiwifruit in an agricultural soil: inoculation of seedlings and hardwood cuttings with *Glomus mosseae*. *Agronomie*. 9, 181 – 185.

20. Calvente, R., Cano, C., Ferrol, N., Azcon-Aguilar, C. and Barea, J.M., 2004. Analysing natural diversity of arbuscular mycorrhizal fungi in olive tree (*Olea europaea* L.) plantations and assessment of the effectiveness of native fungal isolates as inoculants for commercial cultivars of olive plantlets. *Appl. Soil Ecol*. 26, 11–19.

21. Camprubi, A., and Calvet, C., 1996. Isolation and screening of mycorrhizal fungi from citrus nurseries and orchards and inoculation studies. *Hort. Sci.* 31, 366–369.

22. Cavalcante, U.M.T., Maia, L.C., Costa, C.M.C and Santos, V.F., 2001. Mycorrhizal dependency of passion fruit (*Passiflora edulis* f. flavicarpa). *Fruits*. 56, 317 – 324.

23. Chang, D.C.N., 1994. What is the potential for management of vesicular-arbuscular mycorrhizae in horticulture. In: A.D. Robson, L.K. Abbott and N. Malajczuk (Eds.), *Management of Mycorrhizas in Agriculture, Horticulture and Forestry*. Kluwer, Dordrecht, pp. 187 – 190.

24. Chávez, M.C.G. and Ferrara-Cerrato, R., 1990. Effect of vesicular arbuscular mycorrhizae on tissue culture-derived plantlets of strawberry. *Hort. Sci.* 25, 903–905.

25. Coville, F. V., 1921. The litchi (*Litchi chinensis*) a mycorrhizal plant. Appendix VI. In: G W Groff. (Ed.), The Litchi and Lungan. *Canton Christian College/Orange Judd Company*, New York, pp 151–152.

26. Covey, R.P., Koch, B.L. and Larsen, H.J., 1981. Influence of vesicular arbuscular mycorrhizae on the growth of apple and corn in low-phosphorous soil. *Phytopathology*. 71, 712–7115.

27. Davies, F.S. and Albrigo, L.G., 1994. Citrus. Wallingford: *CAB International*.

28. Ferguson, J. J. and Schenck, N. C., 1997. Arbuscular mycorrhizal fungi associated with *Citrus* in Florida. *Soil & Crop Science Society of Florida Proceedings*. 56(0), 24– 27.

29. Fogain, R. and Njifenjou, S., 2002. Effect of mycorrhizal *Glomus* sp. on growth of plantain and on the development of *Radopholus similis* under controlled conditions. *Afr. Plant Prot.* 8, 1–4.

30. Friberg, S., 2001. Distribution and diversity of arbuscular mycorrhizal fungi in traditional agriculture on the Niger inland delta, Mali, West Africa. CBM:s Skriftserie 3, 53-80.

31. Gianinazzi, S., Gianinazzi-Pearson, V. and Trouvelot, A., 1990a. Potentialities and procedures for the use of endomycorrhizas with special emphasis on high value crops. In: J.M. Whipps and B. Lumsden (Eds.), Biotechnology of Fungi for Improving Plant Growth. *Cambridge University Press, Cambridge*, pp. 41 – 54.

32. Gianinazzi, S., Trouvelot, A. and Gianinazzi-Pearson, V., 1990b. Role and use of mycorrhizas in horticultural crop production. In: XXIII *International Horticulture Congress*, Florence, pp. 25 – 30.

33. Gracias Flor, J.P., 2005. Studies on arbuscular mycorrhizal (AM) fungal diversity in fruit trees. *M.Sc. dissertation*. pp. 1 – 47.

34. Guillemin, J.P., Gianinazzi, S., Gianinazzi-Pearson, V. and Marchal, J., 1994a. Contribution of arbuscular mycorrhizas to biological protection of micropropagated pineapple (*Ananas comosum* (L) Merr.) against *Phytophthora cinnamomi* Rands. *Agric. Sci. Finl.* 3, 241 – 251.

35. Guillemin, J.P., Gianinazzi, S., Gianinazzi-Pearson, V. and Marchal, J., 1994b. Control by arbuscular endomycorhizae of *Pratylenchus brachyurus* in pineapple microplants. *Agric. Sci. Finl.* 3, 253 – 262.

36. Guissou, T., Ba, A.M., Ouadba, J.M., Guinko, S. and Duponnois, R., 1998. Responses of *Parkia biglobosa* (Jacq.) Benth, *Tamarindus indica* L. and *Zizyphus mauritiana* Lam. to arbuscular mycorrhizal fungi in a phosphorus-deficient sandy soil, *Biol. Fertil. Soils.* 28, 194 – 198.

37. Guissou, T., 2009. Contribution of arbuscular mycorrhizal fungi to growth and nutrient uptake by jujube and tamarind seedlings in a phosphate (P)-deficient soil. *Afr.J. Biotechnol.* 3, 297–304.

38. Hayman, D.S., Barea, J.M. and Azcón, R., 1976. Vesicular-arbuscular mycorrhiza in southern Spain: its distribution in crops growing in soil of different fertility. *Phytopathol. Med.* 15, 1–6.

39. Harley, J.L. and Harley, E.L., 1987. A check-list of mycorrhiza in the British flora.*New Phytol.* 105, 1–102.

40. Hass, H.J. and Menge, J.A., 1990. VA-micorrhizal fungi and soil characteristics in avocado (*Persea americana* Mill) orchard soils. *Plant Soil.* 127, 207 – 212.

41. Hughes, M., Martin, L.W. and Breen, P.J., 1978. Mycorrhizal influence on the nutrition of strawberries. *J.Amer Soc.Hort Sci.* 103 (3), 179-181.

42. Inserra, R.N., Nemec, S. and Giudice, V. Lo., 1980. A survey of endomycorrhizal fungi in Italian citrus nurseries. *Riv. ortoflorofruttic. Ital.* 64(1), 83 – 88.

43. Janos, D.P., 1980. Vesicular arbuscular mycorrhizae affect lowland tropical rain forest plant growth. *Ecology.* 61, 151–162.

44. Jaizme-Vega, M.C. and Azcón, R., 1995. Response of some tropical and subtropical cultures to endomycorrhizal fungi. *Mycorrhiza.* 5, 213 – 217.

45. Jaiti, F.A., Meddich, A. and Hadrami, I. El., 2007. Effectiveness of arbuscular mycorrhizal fungi in the protection of date palm (*Phoenix dactylifera* L.) against bayoud disease. *Physiol. Mol. Plant Pathol.* 71,166 – 173.

46. Jefwa, J., Vanlauwe, B., Coyne, D., van Asten, P., Gaidashova, S., Rurangwa, E., Mwashasha, M. and Elsen, A., 2010. Benefits and potential

use of arbuscular mycorrhizal fungi (AMF) in banana and plantain (*Musa* spp.) systems in Africa. In: Dubois, T., (Eds.), Proceedings IC on Banana & Plantain in Africa. *Acta Hort.* 879, ISHS 2010.

47. Kachkouch, W., Touhami, A.O., Filali-Maltouf, A., El Modafar, C., Moukhli, A., Oukabli, A., Benkirane, R. and Douira, A., 2012. Arbuscular mycorrhizal fungi species associated with rhizosphere of *Olea europaea* L. in Morocco *J. Anim. Plant Sci.* 15(3), 2275 – 2287.

48. Khudairi, A.K., 1969. Mycorrhiza in Desert Soils. *BioSci.* 19(7), 598 – 599.

49. Khaliel, A.S. and Abou-Hailah, A.N., 1985. Formation of vesicular-arbuscular mycorrhiza in *Phoenix dactylifera* L. cultivated in Qassim region, Saudi Arabia. *Pak. J. Bot.* 17, 267 – 270.

50. Khade, S.W., 1999. Mycorrhizal association in different varieties of banana (*Musa* sp.) in soils of Goa. *M.Sc. dissertation.* pp. 1 – 36.

51. Khade, S.W., and Rodrigues, B.F., 2008a. Spatial variations in arbuscular mycorrhizal (AM) fungi associated with *Carica papaya* L. in a tropical agro-based ecosystem. *Biol.Agric. Hortic.* 26, 149 – 174.

52. Khade, S.W., and Rodrigues, B.F., 2008b. Ecology of arbuscular mycorrhizal fungi associated with *Carica papaya* L. in agro-based ecosystem of Goa, India. Trop. Subtrop. *Agroecosyst.* 8, 265 – 278.

53. Kruckelmann, H.W., 1975. Effects of fertilisers, soils, soil tillage, and plant species on the frequency of Endogone chlamydospores and mycorrhizal infection in arable soils. In: Sanders, F.E., Mosse, B., Tinker, P.B. (Eds.), Endomycorrhizas. London: *Academic Press*, pp. 511 – 515.

54. Kumaran, S. and Azizah, H., 1995. Influence of biological soil conditioner on mycorrhizal versus non-mycorrhizal guava seedlings. *Trop. Agric (Trin.).*72, 39–43.

55. Linderman, R.G. and Davis, A.A., 2001. Comparative response of selected grapevine rootstocks and cultivars to inoculation with different mycorrhizal fungi. *Am. J. Enol. Vitic.* 52, 8–11.

56. Lovato, P.E., Scheüpp, H., Trouvelot, A. and Gianinazzi, S., 1995. Application of arbuscular mycorrhizal fungi (AMF) in orchard and ornamental plants. In: Varma A., Hock. B (Eds.), Mycorrhiza structure, function, molecular biology and biotechnology. Springer, *Heidelberg*, pp. 521 – 559.

57. Lovato, P.E., Gianinazzi-Pearson, V., Trouvelot, A. and Gianinazzi, S., 1996. The state of art of mycorrhizas and micropropagation. *Adv. Hortic. Sci.* 10, 46–52.

58. Lumini, E., Orgiazzi, A., Borriello, R., Bonfante, P. and Bianciotto, V. 2010. Disclosing arbuscular mycorrhizal fungal biodiversity in soil through a land-use gradient using a pyrosequencing approach. *Environ. Microbiol.* 12, 2165–2179.

59. Menge, J.A., Lembright, H. and Johnson, E.L.V., 1977. Utilization of mycorrhizal fungi in citrus nurseries. *Proc. Int. Sot. Citric,* 1, 129 – 132.

60. Menge, J.A., Steirle, D., Bagyaraj, D.J., Johnson, E.L.V. and Leonard, R.T., 1978. Phosphorus concentrations in plants responsible for inhibition of mycorrhizal infection. *New Phytol.* 80, 575-578.

61. Menge, J.A., RAski, D.J., Lider, L.A., Johnson, E.L.V., Jones, N.O.;, Kissler, J.J., and Hemstreet, C.L. 1983. Intractions between mycorrhizal fungi, soil fumigation and growth of grapes in California. *Am. J. Enol. Viticult.* 34, 117-121.

62. Menzel, C.M., and Simpson, D.R., 1987. Lychee nutrition: a review. *Sci. Hortic.* 31, 195–224.

63. Miller, D.D, Domoto, P.A. and Walker, C., 1985. Colonization and efficacy of different endomycorrhizal fungi with apple seedlings at two phosphorus levels. *New Phytol.* 100, 393 – 402.

64. Miller, J.C., Jr., Rajzpakse, S. and Garber, R.K., 1986. Vesicular arbuscular mycorrhizae in vegetable crops. *Hortic. Sci.* 21, 974 – 984.

65. Miller, D.D., Bodmer, M. and Schuepp, H., 1989. Spread of endomycorrhizal colonization and effects on growth of apple seedlings. *New Phytol.* 111, 51 – 59.

66. Morin, F., Fortin, J.A., Hamel, C., Granger, R.L. and Smith, D.L., 1994. Apple rootstock response to vesicular–arbuscular mycorrhizal fungi in a high phosphorous soil. *J. Am. Soc. Hort. Sci.* 119, 578–583.

67. Morton, J.B., Bentivenga, S.P. and Bever, J.D., 1995. Discovery, measurement, and interpretation of diversity in arbuscular mycorrhizal fungi (Glomales, Zygomycetes). *Can. J. Bot.* 73 (Suppl. 1), S25-S32.

68. Msiska, Z., 2001. Arbuscular Mycorrhizal Fungi of Uganda banana plantation soils. *M.Sc. Thesis.* University of Pretoria, Pretoria.

69. Nappi, P., Jodice, R., Luzzati, N. and Corino, L., 1985. Grape vine root system and VA mycorrhiza in some soils of Piedomont (Italy). *Plant Soil.* 85, 205 – 210.

70. Neill, J.C., 1944. *Rhizophagus* in *Citrus. N. Z. J. Sci. Tech.* 25, 191 – 201.

71. Nemec, S., 1979. Response of six citrus rootstocks to three species of *Glomus,* a mycorrhizal fungus. *Citrus Ind. Mag.* 5, 5–14.

72. Nemec, S., 1981. Histo-chemical characteristics *of Glomus etunicatus* infection of *Citrus limon* fibrous roots. *Can. J. Bot.* 59, 609 – 617.

73. Nemec, S., Menge, J.A., Platt, R.G. and Johnson, L.V., 1981a. Vesicular-arbucular mycorrhizal fungi associated with citrus in Florida and California notes on their distribution and ecology. *Mycologia.* 73, 112–127.

74. Nemec, S., 1986. VA mycorrhizae in horticultural systems. In:, Ecophysiology of VA Mycorrhizal Plants. G.R. Satir, ed., CRC Press Enc., *CRC, Boca Raton, FL,* pp. 193 – 211.

75. Oehl, F., Sieverding, E., Ineichen, K., Ris, E.A., Boller, T. and Wiemken, A., 2005. Community structure of arbuscular mycorrhizal fungi at different soil depths in extensively and intensively managed agroecosystems. *New Phytol.* 165, 273–283.

76. Onkarayya, H. and Sukhada, M., 1993. Studies on the dependency of *Citrus* rootstocks to VAM inoculation in Afisol. *Advances in Horticulture and Forestry* .3, 81 – 91.

77. Ortas, I., 2010. Effect of mycorrhiza application on plant growth and nutrient uptake in cucumber production under field conditions. *Span. J. Agric. Res.* 8(SI), 116 – 122.

78. Pandey, S. and Misra, A.P., 1975. Mycorrhiza in relation to growth and fruiting of *Litchi chinesis* Sonn. *J. Indian Bot. Soc.* 54, 280 – 293.

79. Pindi, P. K., 2011. Mycorrhizal association of some agroforestry tree species in two social forestry nurseries. *Afr. J. Biotechnol.* 10(51), 10425 – 10430.

80. Plenchette, C., Furlan, V. and Fortin, J.A., 1981. Growth stimulation of apple trees in unsterilized soils under field conditions with VA mycorrhiza inoculum. *Can. J. Bot.* 59, 2003–2008.

81. Plenchette, C., Fortin, J.A. and Furlan, V., 1983. Growth responses of several plant species to mycorrhizae in a soil of moderate P-fertility. I. Mycorrhizal dependency under field conditions. *Plant Soil* .70, 199 – 209.

82. Powell, C.L. and Santhanakrishnan, P., 1986. Effect of mycorrhizal inoculation and phosphorus fertilizer on the growth of hardwood cuttings of kiwifruit (*Actinidia deliciosa* cv. Hayward) in containers. *N.Z. J. Agric. Res.* 29, 263 – 268.

83. Reed, H.S. and Fremont, T., 1935. Factors that influence the formation and development of mycorrhizal association in citrus roots. *Phytopathology* 25, 645 – 647.

84. Reich, L., 1988. Rates of infection and effects of five vesicular arbuscular mycorrhizal fungi on apple. *Can. J. Plant Sci.* 68, 233–239.

85. Reddy, B. and Bagyaraj, D.J., 1994. Selection of efficient vesicular arbuscular mycorrhizal fungi for inoculating the mango rootstock cultivar 'Nekkare'. *Sci. Horti.* 59 (1), 69–73.

86. Redecker, D., Schüßler, A., Stockinger, H., Stürmer, S., Morton, J. and Walker, C., 2013. An evidence-based consensus for the classification of arbuscular mycorrhizal fungi (*Glomeromycota*). *Mycorrhiza.* 23, 515 – 531.

87. Roldán-Fajardo, B.E. and Barea, J.M., 1985. Mycorrhizal dependency in the olive tree (*Olea europaea* L.). In: Gianinazzi-Pearson, V., Gianinazzi, S. (Eds.), Physiological and Genetical Aspects of Mycorrhizal. *INRA*, Paris, pp. 323–326.

88. Sabet, Y.S., 1945. Reaction of *Citrus* mycorrhiza to manorial treatment. *Proc.Egyptian Acad. Sc.* 1, 21 – 28.

89. Saritha, B., Panneerselvam, P., Sukhada, M., Sulladmath, V.V., Ravindrababu, P.and 2014b. Studies on host preference of *Glomus* spp and their synergistic effect on sapota (*Manilkara achras* (Mill) Forsberg) seedlings growth. *Pl. Arch.*(In Press).

90. Sarwade, P.P., Chandanshive, S.S., Kanade, M.B. and Bhale, U.N., 2011. Diversity of Arbuscular mycorrhizal (AM) fungi in some common plants of marathwada region. I. *A. M. U. R. E. 1*(12), 11 – 12.

91. Schubert, A. and Cammarata, S., 1986. Effect of inoculation with different endophytes on growth and P nutrition of grapevine plants grown in pots. In: Gianinazzi-Pearson, V., Gianinazzi, S., (Eds.), *Mycorrhizae: Physiology and Genetics INRA*, Paris, pp. 327 – 331.

92. Schubert, A., Cravero, M.C. and Mazzitelli, M., 1987. Vesicular-arbuscular mycorrhizae in field- and pot-grown kiwifruit (*Actinidia deliciosa*). *Adv. Hortic. Sci.* 1, 80 – 82.

93. Schubert, A., Mazzitelli, M., Ariusso, O. and Eynard, I., 1990. Effects of vesicular- arbuscular mycorrhizal fungi on micropropagated grapevines. Influence of endophyte strain, P fertilization and growth medium. *Vitis* .29, 5 – 13.

94. Schreiner, R.P. and Bethlenfalvay, G.J., 1995. Mycorrhizal interactions in sustainable agriculture. *Crit. Rev. Biotechnol.* 15, 271–285.

95. Schubert, A. and Lubraco, G., 2000. Mycorrhizal inoculation enhances growth and nutrient uptake of micropropagated apple rootstocks during weaning in commercial substrates of high nutrient availability. *Appl. Soil Ecol.* 15, 113–118.

96. Schreiner, R.P., 2005. Mycorrhizas and mineral acquisition in grapevines. In: Christensen, L.P., Smart, D.R. (Eds.), Proceedings of the soil environment and vine mineral nutrition symposium. pp. 49 – 60.

97. Schreiner, R.P., 2007. Effects of native and non native arbuscular mycorrhizal fungi on growth and nutrient uptake of 'Pinot noir' (*Vitis vinifera* L.) in two soils with contrasting levels of phosphorus. *Appl. Soil. Ecol.* 36, 205–215.

98. Schreiner, P.R. and Mihara, K.L., 2009. The diversity of arbuscular mycorrhizal fungi amplified from grapevine roots (*Vitis vinifera* L.) in Oregon vineyards is seasonally stable and influenced by soil and vine age. *Mycologia.* 101, 599–611.

99. Sghir, F., Chliyeh, M., Kachkouch, W., Khouader, M., Ouazzani Touhami A., Benkirane, R. and Douira, A., 2013. Mycorrhizal status of *Olea europaea* spp. oleaster in Morocco. *J. Appl. Biosci.* 61, 4478 – 4489.

100. Sieverding, E., 1991. Vesicular-Arbuscular Mycorrhiza Management in Tropical Agroecosystems. Deutsche Gesellschaft for Technische Zusammenarbeit (GTZ) GmbH, Eschbom.

101. Singh, R.P. and Prasad, V., 2006. Occurrence and population dynamics of vesicular arbuscular mycorrhizae in the Indian orchards of litchi *Litchi chinensis* Sonn., aonla *Phyllanthus emblica* L. and banana *Musa paradisiaca* L.. *Asian Journal of Bio Science.* 1, 154 – 156.

102. Soares, A.C.F., Martins, M.A., Mathias, L. and Freitas, M.S.M., 2005. Arbuscular mycorrhizal fungi and the occurrence of flavonoids in roots of passion fruit seedlings. *Sci. Agric.* (Piracicaba, Braz.) .62(4), 331 – 336.

103. Sukhada, M., 1992. Effect of VAM inoculation on plant growth nutrient level and root phosphatase activity in papaya (*Carica papaya* cv. Coorg Honey Dew). *Fert. Res.* 31, 263–267.

104. Sukhada, M., 1994. Utilization of vesicular mycorrhizal fungi in banana cultivation. In: Adholeya, A., and Singh, S. (Eds.), *Mycorrhizae biofertilizers for future. TERI Pub.*, New Delhi.

105. Sukhada, M., 1998. Inoculation response if mango to VAM fungi in glass house and field trials. In: *National Symposium on Mango Production and Export,* Lucknow, June 25–27, .

106. Sumorok, B., Sas Paszt, L., Głuszek, S., Derkowska, E. and Żurawicz, E., 2011. The effect of mycorrhisation and mulching of apple trees 'Gold Millennium' and blackcurrant bushes 'Tiben' on the occurrence of arbuscular mycorrhizal fungi. *J. Fruit Ornam. Plant Res.* 19(1), 35 – 49.

107. Sukhada, M., 2012. Arbuscular mycorrhizal fungi benefit mango (*Mangifera indica* L.) plant growth in the field. *Sci. Horticult.* 143, 43–48.

108. Tomlinson, H., Teklehaimanot, Z., Traoré, A. and Olapade, E., 1995. Soil amelioration and root symbioses of *Parkia biglobosa* (Jacq.) Benth. in West Africa. *Agroforest. Syst.* **30,** 145 – 159.

109. Trindade, A.V., Siqueira, J.O.and Stürme, S.L., 2006. Arbuscular mycorrhizal fungi in papaya plantations of Espírito santo and Bahia, Brazil. *Braz. J. Microbiol.* 37, 283 – 289.

110. Varma, A. and Schuepp, H., 1995. Infectivity and effectiveness of *Glomus intraradices* on micropropagated plants. *Mycorrhiza.* 5, 29 – 37.

111. Vestberg, M., Kukkonen, S., Neuvonen, E.L. and Uosukainen, M., 2000. Mycorrhizal inoculation of micropropagated strawberry—case studies on mineral soil and a mined peat bog. *Acta Hort.* 530, 297–304.

112. Wang, P., Shu, B., Wang, Y., Zhang, D.J., Liu, J.F. and Xia, R.X., 2013. Diversity of arbuscular mycorrhizal fungi in red tangerine (*Citrus reticulata* Blanco) rootstock rhizospheric soils from hillside citrus orchards. *Pedobiologia.* 56, 161–167.

113. Williams, S.C.K., Vestberg, M., Uosukainen, M., Dodd, J.C. and Jeffries, P., 1992. Effects of fertilizers and arbuscular mycorrhizal fungi on the *post vitro* growth of micropropagated strawberry. *Agronomie.* 12, 851–857.

114. Xiutang, L., 1990. The effect of v-a mycorrhiza on mango seedling absorbing nutrient and growth. *J. Guangxi Agric. Biol. Sci.* 4.

Chapter 2

Glomeromycotean Species Diversity in *Citrus* spp., *Vitis vinifera* and *Physalis peruviana*

Raúl Hernando Posada[1], Ewald Sieverding[2]*.

[1]*Zenkinoko SAS, Diag 151b 136A-75 Cs 93*

Bogotá – Colombia.

[2]*University of Hohenheim, Institute of Plant Production and Agroecology in the Tropics and Subtropics, University of Hohenheim, Garbenstr. 13, Stuttgart Hohenheim, Germany.*

* Address for the Correspondence: sieverdinge@aol.com.

ABSTRACT

Citrus spp. L. are economically important fruit trees in the tropics, subtropics and Mediterranean climates. *Vitis vinifera* has worldwide importance for table grape and wine production and has its distribution in climates between 40-50 degree latitude South and North of the Americas, Europe, Asia, Africa and Australia. *Physalis peruviana* (cape gooseberry) is a scrub and only of regional importance in South America, Africa and China. All three crops have in common that they depend under most ecological conditions on Arbuscular Mycorrhizal Fungi (AMF) for development and growth. Little is known however about the distribution of AMF species in these crops, factors affecting the presence and the diversity of AMF species under natural conditions, and the benefits these crops derive from different consortia of AMF fungi. Some reports of this association are available from *Citrus*, less on grapes and / or *Physalis* only one from Colombia, South

America. This article summarizes the information available in literature, mainly from South America but also considering investigations from North America, Europe and China.

Keywords: *AMF, species diversity, ecology, crop management, edaphic factors, AMF consortia*

1.0 Introduction

AMF are microbiological soil components, which are recognized to play an important role in the growth and productivity of many tropical and temperate crops (de Almeida *et al.*, 2003; Siqueira *et al.*, 1998; Vaast *et al.*,1996). AMF formation in roots is known to be affected by soil chemical and physical conditions, such as pH (Heijne *et al.*, 1996), soil texture and humidity (Andrade *et al.*, 2009; van der Heijden *et al.*, 1998), fluctuations in nutrients availability (Hodge, 2001; Hodge *et al.*, 2001), levels of soil water holding capacity (Augé, 2001; Porcel *et al.*, 2005), xenobiotic and heavy metal contents (Leyval *et al.*, 2002) and the altitude and inclination of soil (Ramírez, 2014).

Different studies indicate that extreme high or low values of edaphic properties, such as pH and soil humidity, can lead to non-favorable conditions for AMF development like growth, root and soil colonization, and efficiency for improving plant nutrient uptake and or plant growth (Camargo-Ricalde and Esperón-Rodríguez, 2005; Miller, 2000). Very few studies relate the influence of environmental or edaphic factors on AMF species diversity (Posada, 2011; Ramírez, 2014).

Within the frame of applied mycorrhizal research, the effect of inoculation with different AMF on the productivity of agronomic plant species have been tested. Inocula with identified AMF species compositions have been used in tests (Miranda *et al.*, 2011; Martínez,*et al.*, 2011), as well as inocula with native unidentified AMF consortia (Pellegrino *et al.*, 2011; Schreiner, 2007). Experiments with single, dual or multi-AMF species composition gave sometimes complementary or divergent results, even with the same plant-species. The results give indications of certain degree of preferences in fungus-plant species interactions which in turn depended on environmental or edaphic factors (Ramírez, 2014; Vandenkoooornhuyse *et al.*, 2002). Commercial inocula often showed improvement of plant growth when the soils were sterilized but showed less positive effects under natural non-sterilized conditions (Guana *et al.*, 2011; Santos *et al.*, 2001; Taylor and Harrier, 2000, 2001).

Few investigations have been developed to identify the native AMF species associated to citrus, grapes and cape gooseberry crops under field conditions (Ramírez, 2014). In addition, the composition of native AMF consortia was very seldom related to potential benefits of such consortia for the crop. To describe the composition of AMF species communities and to make comparisons among the communities, fungal species richness, number of AMF spores per sample unit and the evenness of distribution of the AMF species are commonly used as ecological indicators (de Souza *et al.*, 2013; Liu *et al.*, 2010; Purin *et al.*, 2006; Oehl *et al.*, 2004). We believe that species richness is the only useful and real indicator for AMF species diversity. Species richness (S) is a term that refers to the number of species in a community (Whitcomb and Stutz, 2007). Some researchers indicate that AMF species richness is important for the resilience capacity of ecosystems from negatively affecting practices, so as to conserve the AMF genetic diversity and to maintain their stability (Martínez-García *et al.*, 2011; Urcelay *et al.*, 2009). To estimate species richness AMF spores are extracted from the rhizospheric soil around specific plants and identified morphologically (Oehl *et al.*, 2003), or the AMF species are identified in roots by molecular biological methods (Öpik *et al.*, 2013). Recent investigations have shown that morphological methods may result in superior detection of AMF species than molecular methods (Wetzel *et al.*, 2014). Species richness determination requires that AMF bait cultures must be established from which spores are isolated from time to time in addition to isolating spores directly from field samples (Oehl *et al.*, 2003). Only about 70% of AMF species may be detected in field samples (Posada, 2011) and it has often been shown that species may be found in bait cultures only (from the same field soils) but not in the field, and *vice versa* (Oehl *et al.*, 2005). Species richness on basis of field samples alone, and evenness of distribution, as calculated from the number of spores in the soil, is misleading and erratic as sporulation of AMF species may have been absent or may be only at specific times of the year. In addition, spore formation can depend on plant growth stages and environmental and edaphic factors (Kang *et al.*, 2011; Oehl *et al.*, 2010; Shujie *et al.*, 2009; Wang *et al.*, 2011). We defer from using spore numbers as a valid indicator of dominance of a specific AMF species in the soil. Molecular methods have also problems in identifying correctly the AMF species richness, as often the fungus may not colonize the root at time of sampling. The sample size for molecular probes is often extremely small considering that sometimes only 6 cm roots are taken to define AMF species diversity per one ha land (Wetzel *et al.*, 2014).

We describe in this chapter the diversity of Glomeromycotean species in three fruit crops of which two are widely distributed worldwide (citrus and grapes) while Cape gooseberry has more regional distribution in South America, South Africa and China. *Citrus* is the denomination for a group of fruit tree species and hybrids of the genus *Citrus*, including

Citrus sinensis (orange), *Citrus limon* (lemon), *Citrus volkameriana* (Volkameriana), *Citrus* x limonia (Rangpur), *Citrus nobilis* (tangerine), *Citrus* x limón (rough lemon), *Citrus reticulate* (Sunki), *Citrus* x tangelo (Orlando), *Citrus aurantium* (bitter orange or sour orange), as well as for the genus *Poncirus* (*P. trifoliate*, trifoliate orange) and the genus *Fortunella*. Most of these species and hybrids are grown on root stocks resistant to some diseases, pests and environmental stresses. In the same sense, the bushy grapes are usually denominated as *Vitis vinifera* but are grown on different rootstocks like *V. vinifera* x *V. rotundifolia*, *V. berlandieri* x *V. riparia*, *V. berlandieri* x *V. rupestris* and *V.riparia* x *V.rupestris*. Of the 82 bushy *Physalis* spp. known from South America, *P. peruviana* (Cape gooseberry) is economically the most important and cultivated in South America, while *P. philadelphica* is grown in Central America and many *Physalis* spp. are ornamentals. While citrus and grapes have huge economical importance world wide, *Physalis* is of less importance. All three groups of plants have been reported to depend on AMF for growth as will be discussed below.

We present the diversity of Glomeromycotean AMF for the above 3 groups of plants in four different ways: 1) we indicate the AMF diversity under field conditions, 2) we show which factors influence the AMF species diversity in these cultivated crops, including the effect of soil chemical properties and/or plantation management practices, 3) we relate the growth of these plant species to different AMF species consortia, and 4) we make some conclusions.

2.0 AMF species diversity in *Citrus*, *Vitis* and *Physalis*

As explained above, there are two methods to define the AMF species diversity: spores are isolated from soil and AMF species are taxonomically identified by the spore morphology, or molecular methods are used probing spores or other infection structures of the fungus. In this article, we apply the taxo nomical classification of the Glomeromycotean species as proposed by Oehl *et al.*(2011d); latest names for new genera (Oehl *et al.*, 2011a; 2011b, 2011c) are also included.

Different studies conducted on citrus, grapes and cape gooseberry describe the arbuscular mycorrhizal diversity in soils. **Table 2.1** summarizes the findings of some studies conducted in America and Europe. Carrenho *et al.*(1998) developed a study on naturally established citrus plants at two sites of Sao Paulo state (Brazil), in 1998. From the soil samples of the rhizosphere of 50 plants (*Citrus sinensis / C. limon*), they isolated 13 AMF species, where *Gigasporara misporophora* was the most frequent, while the other species were regularly less frequent with an occurrence below 13%. Later, Focchi *et al.* (2004) found 26 AMF species in 36 soil samples from different citrus orchards and nurseries at Rio Grande do Sul (Brasil). In 12 years old citrus orchads in the same state of Brasil, 10 AMF spp were found (Dutra de Souza *et al*, 2002). Predominance of *Glomus* spp. and *Acaulospora* spp. were reported from different Brazilian orchards under various agronomic management systems (Dias da Silveira and dos Santos, 2007). These two genera appear to be predominant anyway in acidic soils of South America (Castillo *et al.*, 2006, 2010).

In South western USA, Fidelibus *et al.* (2000) identified 11 AMF species under *C. volcalmer*, in the localities of Borrego Springs, Padre Canyon, Verde River, Mesa and Yuma. They highlighted that in Yuma orchard more than > 80% of the total number of AMF spores were from one single species, *Paraglomus occultum*. Finally, in a more recent research from Zigui County, Southern China 18 different AMF species were found at different altitudes in red tangerine (*C. reticulata*) orchards (Wang *et al.*, 2013). In all of the surveyed orchards of red tangerine, *Glomus aggregatum*, *Funneliformis mosseae* and *Glomus intraradices* were the most frequent species.

Table.2.1: AMF species isolated from *Citrus* spp, *Vitis* spp rootstocks and *Physalis peruviana* in diferent parts of the world.

AMF species	Citrus	Crop Vitis	Physalis
Glomerales			
Albahypha drummondii			PP
Albahypha walker			PP
Claroideoglomus claroideum	UD, UC, CR	VS, VV	PP
Claroideoglomus etunicatum	UD, CV, CR	VS, VM, VO	PP
Claroideoglomus lamellosum		VO	
Claroideoglomus luteum			PP
Entrophospora infrequens	UD	VS, VV, VO	PP
Funneliformis caledonius	CR	VO	
Funneliformis coronatus			PP
Funneliformis geosporus	CS, CL, UD	VO	PP
Funneliformis monosporus			PP
Funneliformis mosseae	UD, CV, CR	VV, VM, VO	PP
Glomus intraradices	CV, CR	VV, VM, VB	PP
Glomus aggregatum	CR		PP
Glomus aureum		VM, VO	
Glomus australe	UD		
Glomus badium		VO	
Glomus brohultii			PP
Glomus clavisporum	CR		
Glomus coremioides		VB	
Glomus diaphanum		VS, VO	
Glomus fasciculatum	UD	VB, VO	PP
Glomus glomerulatum	UD		
Glomus heterosporum		VO	
Glomus indicum		VB	
Glomus (Rhizophagus) intraradices	CV, CR	VV, VM, VB	PP
Glomus invermaium	UC	VS	
Glomus iranicum		VB	
Glomus irregulare			PP
Glomus macrocarpum	CS, CL, UD, UC, CV	VS, VO	PP

Contd...

Table.2.1: Contd...

AMF species	Citrus	Crop Vitis	Physalis
Glomus microaggregatum	UD, CV	VM	
Glomus microcarpum			PP
Glomus proliferum			PP
Glomus rubiforme		VV	
Glomus sinuosum		VM, VM, VB	PP
Glomus spinuliferum		VO	
Glomus sp1	UD, CR	VV, VM	PP
Glomus sp2	UD, CR	VV	PP
Glomus sp3		VV	
Glomus sp4		VV	
Glomus sp5	CV		
Septoglomus constrictum	UC	VO	PP
Simiglomus hoi			PP
Viscospora viscosa		VB	
Diversisporales			
Acaulospora bireticulata	UD, UC, CR		
Acaulospora capsicula	UD		
Acaulospora denticulata			PP
Acaulospora elegans		VV	
Acaulospora foveata	CS, CL, UD		
Acaulospora lacunosa		VV	
Acaulospora laevis	CS, CL	VV	
Acaulospora longula			PP
Acaulospora mellea	UD	VS	
Acaulospora morrowiae			PP
Acaulospora rehmii	UD, CR	VV	PP
Acaulospora scrobiculata	UD, UC	VS	PP
Acaulospora sp1	UD, CR	VV	PP
Acaulospora sp2		VV	PP
Acaulospora sp3			PP
Acaulospora spinosa	CR		PP
Corymbiglomus tortuosum	CR		
Diversispora celata			PP

Contd...

Table.2.1: Contd...

AMF species	Citrus	Crop Vitis	Physalis
Diversispora eburnea	CV		
Diversispora spurca	CV	VM	
Diversispora versiformis	UD, CV	VO	PP
Kuklospora colombiana	UD, UC		PP
Tricispora nevadensis			PP
Ambispora appendicula	CS, CL		PP
Ambispora sp1		VM	PP
Intraspora schenckii	CR		
Archaeospora trappei	CV	VV, VM, VO	PP
Cetraspora gilmorei	CS, CL		
Cetraspora nodosa	UD		PP
Cetraspora pellucida			PP
Scutellospora pellucida	CS, CL, UD		
Claroideoglomus claroideum	UD, UC, CR	VS, VV	PP
Claroideoglomus etunicatum	UD, CV, CR	VS, VM, VO	PP
Claroideoglomus lamellosum		VO	
Claroideoglomus luteum			PP
Diversispora celata			PP
Diversispora eburnea	CV		
Diversispora spurca	CV	VM	
Diversispora versiformis	UD, CV	VO	PP
Entrophospora infrequens	UD	VS, VV, VO	PP
Pacispora chimonobambusae	CR		
Pacispora sp.			PP
Tricispora nevadensis			PP
Gigasporales			
Dentiscutata biornata	CS, CL		
Dentiscutata cerradensis	UD		
Dentiscutata nigra	CS, CL		
Fuscutata heterogama	CS, CL, UD, UC		
Gigaspora albida	CR		
Gigaspora ramisporophora	CS, CL		
Gigaspora sp1	UD		

Contd...

Table.2.1: Contd...

AMF species	Citrus	Crop Vitis	Physalis
Quatunica erythropus		VM	
Racocetra pérsica	CS, CL, UC		
Racocetra tropicana			PP
Scutellospora calospora	CS, CL	VV, VM	
Scutellospora sp1	CR		
Archaeosporales			
Ambispora appendicula	CS, CL		PP
Ambispora sp1		VM	PP
Archaeospora trappei		VV, VM, VO	PP
Intraspora schenckii	CR		
Intraspora sp.			PP
Paraglomerales			
Paraglomus laccatum			PP
Paraglomus occultum	UC, CV	VM, VO	PP

Conventions indicate the plant species from which the AMF was isolated and the reference of the study. CS = *Citrus sinensis* (Carrenho *et al.*, 1998), CL = *Citrus limón* (Carrenho *et al.*, 1998), UD = Undefined (Focchi *et al.*, 2004), UC = Undefined *Citrus* sp. (Dutra de Souza *et al.*, 2002), CV = *Citrus volcalmer* (Fidelibus *et al.*, 2000), CR = *Citrus reticulata* (Wang *et al.*, 2013), VO= *Vitis vinífera* (Oehl *et al.*, 2005), VV = *Vitis vinífera* (Schreiner and Mihara, 2009), VM = *Vitis vinífera* (Balestrini *et al.*, 2010), VB = *Vitis berlandieri* x *Vitis rupestris* (Radić *et al.*, 2014), PP = *Physalis peruviana* (Ramírez, 2014).

Grapes are grown almost world wide and the biota of soils where grapes are cultivated are diverse. A few studies investigated the diversity of AMF species in grapes. From Brasil, 8 AMF species were identified in vineyards (Ávila, 2004) which were managed organically or conventionally. Schreiner & Mihara (2009), investigated the AMF species compositions in 10 vineyards in Oregon. They reported seven AMF species belonging to Acaulosporaceae, eight phylotypes dominated by *Glomus* spp., a single phylotype inside Gigasporaceae and one from Archaeosporaceae. There were little differences in the species compositions, but always the 4 phylotypes were most frequent. In a compilation of information from the Mediterranean area realized by Balestrini *et al.*(2010)15 AMF species were commonly found in vineyards, a number similar to that reported by Oehl *et al.* (2005) from German vineyards. Radić *et al.*(2014) isolated up to 19 AMF species in Croatian vineyards. The number was clearly higher in organically managed plantations (9-19 spp.) than in conventional vineyards (5-11 spp.) (Radić *et al.*, 2014). It is interesting to note that the number of AMF species decreased from 0 to 70 cm soil depth more in vineyards than under grassland (Oehl *et al.*, 2005) although it is said that grapes can have a deep rooting system. A dramatic change was found in the species composition with more depth, as Gigasporales were more dominant in undisturbed deeper soil horizonts.

Physalis peruviana has its origin in the Andean area, and it has a wide altitudinal distribution. There is only one study on the diversity of AMF species in *Physalis* available (Ramírez, 2014) from which 46 AMF species were reported, as detailed in **Table 2.1**.

3.0 Factors affecting diversity

There are various factors reported in literature to affect AMF species diversity and propagule numbers in different ecosystems, and agro ecosystems (Kang *et al.*, 2011; Oehl *et al.*, 2010; Shujie *et al.*, 2009; Wang *et al.*, 2011). In general, however, it is not yet clear which are the main factors for the fungal diversity are. While some researchers relate diversity to the plant species community and the host plant itself, others are concluding that soil physical and chemical characteristics, like organic matter content, soil type, soil structure, water content, soil aeration, nutrient content (in particular P), soil pH, and content of heavy metals, or salts are responsible for low or high diversity of AMF species. Climate, and altitude of cultivation were also investigated, as well as farm management practices and intensity of farming with high or low farm input (often generalized as conventional farming vs. ecological farming) were claimed to influence the fungal diversity (Franke-Snyder *et al.*, 2001; Jefwa *et al.*, 2012; Piotrowski and Rilling, 2008; Verbruggen *et al.*, 2010).

Although the number of studies on the diversity in the investigated crops is rather low (**Table 2.1**), the three plant groups can be associated with a large number of AMF species, which were between 51 with *Citrus* and 43 and 46 with grapes and cape gooseberry respectively. Some AMF species were found, however, only with one of the three crops, e.g. some *Acaulospora* spp. were found in one but not the other crop (**Table 2.1**) as with *Claroideoglomus* and some specific *Glomus* spp. and in particular with species of the Gigasporales (*Gigaspora* spp., *Dentiscutata* spp. and *Racocetra* spp.) in citrus. On the other hand, there are some AMF species that were observed in all crops and thus appear to be very frequent in the Americas or even worldwide: these are species like *A. rehmii, A. scrobiculata, C. claroideum, C. etunicatum, D. versiforme, E. infrequens, F. mosseae, G. intraradices, G. fasciculatum, G. macrocarpum, P. occultum* and *S. constrictum*. Some of these species are easily recognized by their morphology of the spores so that there is high chance that these species were taxonomically correctly identified by different researchers. Oehl *et al.* (2003) and Castillo *et al.* (2010) divided thus the AMF species in specialists, those that occur specifically in specific crops and generalists that can be found under many different ecological conditions with many different crops. Such division of AMF species may also apply in this study. as those frequently found in the herein reported studies were also identified frequently in other studies with other agronomic crops, or in grasslands.

When more than 50% of the spores of an AMF species consortium belong to one single species, some researchers talk about dominance of this species (e.g. Fidelibus *et al.* (2000), in *Citrus*). These are often small spore AMF species that have the ability of quick sporulation and of thus dominating the AMF species consortium in spore numbers. This was described already in 1991 by Sieverding who observed such phenomena in cassava, a crop with huge photosynthetic capacity and which needs one year to grow until harvest. *Archeospora* spp. (with spores <80 µm diameter) and other small spore species like *P. occultum* and *A. longula* were frequently found to dominate up to 90% of the spore population in mono-cultured and weed-free cassava, already 3 months after planting. Such dominance is sometimes interpreted to be a preferential association of an AMF species with a crop, but virtually it can be any small-spored AMF species that is present at a given location. Such small spore species are worldwide found and are believed to be distributed by wind and they are often pioneer species (Oehl *et al.*, 2011e). However, Carrenho *et al.* (1998) believed that the dominance of the big-spored *G. ramisporophora and C. gilmorii* (Gigasporales), and *G. macrocarpum* in *C.sinensis* and *C. limon* in two locations of Brazil were related to the better adaptation of these AMF species to the plant species which gave them a competitive advantage within the inter specific competition within the AMF community. Gigasporales need, in general, a longer time for sporulation (Oehl *et al.*, 2009). All three crops, *Citrus*, grapes, and *Physalis* are perennial, and thus, the AMF species have time enough for sporulation. At least in the tropics and subtropics, *Citrus* and *Physalis* are evergreen and thus can supply year-round assimilates to AMF for spore production. It is possible that Gigasporales (as in **Table 2.1**: *Gigaspora, Cetraspora, Dentiscutata, Fuscutata, Racocetra, Quatunica* and *Scutellospora*) are more frequent in *Citrus* than in grapes or *Physalis*. Cultivation practices and agronomic practices are different, at least between *Citrus* and grapes. While the soil in the surroundings of the *Citrus* trunk is maintained clean by means of herbicide applications (without soil disturbance), grapes are cultivated between the rows by means of rotor tilling and ploughing which destroys the hyphal network of the AMF. Oehl *et al.* (2005) reported that *Gigasporales* and some other AMF species were not found in the upper soil horizon of vineyards but only in the deeper undisturbed soil layers. It is possible that the reason is the inability or lower ability to rebuild the destroyed hyphal network through hyphal anastomosis, as in particular reported for Gigasporales (de la Providencia *et al.*, 2005; Purin and Morton, 2011); this may give such AMF species a disadvantage when agronomic practices are applied which frequently destroy hyphal networks. Often, however, AMF species are isolated only from the higher soil horizon, and Oehl *et al.* (2005) discussed already that many fungi might find a refuge below the ploughed soil horizon. This would mean that not all AMF species are always detected in ecological studies due to erratic sampling techniques. Hence, if samples had been taken from deeper undisturbed horizon, the species diversity would likely have been higher.

Instead of preferential associations, we believe that specific AMF species in our three plantation crops may have been more favored by ecological factors like soils (Siqueira, 1986), climate (Koske, 1987), soil pH (Lambais and Cardoso, 1988; Siqueira *et al.*, 1986, Oehl *et al.*, 2010), and crop management practices (Oehl *et al.*, 2005, 2003). Carrenho *et al.*(1998) argued that in *Citrus* edaphic factors are important for AMF species indicating higher diversity with more organic matter and lower P-content. pH has an effect on the composition of the fungal species community.

This is in accordance with observations in other agronomic crops (Oehl *et al.*, 2010). Wang *et al.* (2013) found important effects of environmental factors on the AMF species diversity in tangerines. They observed a decrease in AMF species diversity with increasing altitude (>700 masl), and a significant effect of soil pH, organic matter, soil moisture and N-content of soils on the fungal community structure.

Edaphic factors, including the water content can be indicative of soil nutrient levels, but are not the unique factors for the diversity of AMF species and AMF species communities, as expressed by Ávila (2004) in his work with grapes in Brasil. *Glomus invermaium, G. macrocarpum, C. claroideum* and *C. etunicatum*were most frequent in environments with the lowest contents of P and K while the species *A. scrobiculata, E. infrequens* and *G. diaphanum* were present indistinct of soil characteristics. Balestrini *et al.*(2010) working with vineyards in the mediterraneum zone found data that reinforce the concept that the general AMF assemblage structure and composition in vineyards might be influenced more by soil type than by host plant features (age, vegetative stages) or management practices. Similarly with *Physalis* in Colombia, where the distribution and composition of species of the genera *Acaulospora, Scutellospora* and *Entrophospora* are dependant on the organic matter content (Ramírez, 2014). More *Acaulospora* spp. were found in lower pH soils confirming results of Álvarez-Sánchez *et al.* (2011); Castillo *et al.* (2006); Posada (2011) in other crops grown in acidic low P content soils.

In general, farm management practices appear to affect significantly the AMF species diversity (Franke-Snyder *et al.*, 2001; Posada, 2011; Purin *et al.*, 2006; Tian *et al.*, 2011): the higher the intensity of farming inputs, like soil preparation, mono-cropping, fertilization, pesticide use, the lower was the diversity of AMF species (Jansa *et al.*, 2003; Oehl *et al.*, 2003, 2010). In established older citrus orchards in Brazil, however, the composition of AMF species was not significantly affected by the agronomic management system, being conventional with some chemical inputs, or organic with lower inputs and in particular organic fertilizers (Focchi *et al.*, 2004). It must be mentioned that the input level in citrus plantations is not high anyway, and it appears that once citrus is established, any farm practice will not alter significantly the AMF diversity. Also, once a citrus plantation gets older, there is little understore growth of weeds or other plants due to shadiing effects, so that the density of plant species is low in conventional and ecologically managed plantations. Higher plant and root density is often the reason for a more diverse AMF species community (Oehl *et al.*, 2003). Similarily, Posada (2011) found little effect of any farming practices on the AMF species diversity in 8-12 years old coffee plantations in Colombia and Mexico. In vineyards, however, organic management practices had significant positive impact on the AMF species diversity, in particular due to more presence of bigger sized AMF species (Radić *et al.*, 2014). This may be related to the simple fact that in organically managed vineyards organic materials like farm manures are applied as fertilizers and that the soil is permanently covered by intercropped grasses to control soil erosion. Again, higher plant density will lead to higher AMF species diversity. *Physalis* is cultivated in Colombia almost always in a type of organic farm management as little if not no inputs are applied to this crop. This may have been the reason for the very high diversity of AMF species reported in this article (**Table 2.1**).

Some other factors influencing the AMF species diversities in *Citrus*, grapes and *Physalis* should be mentioned. Studies on the AMF species diversity in different citrus orchards in Brazil indicated that the age of plants favor the AMF species richness, mainly of species of *Glomus* and *Acaulospora* (Focchi *et al.*, 2004). The importance of sampling time was obvious in the *Physalis* study in Colombia where there are two pronounced rainy seasons and two dry seasons during the year. During the dry season spores of 46 species were found and only 31 species during the wet season. Spores of 15 AMF species were only isolated during dry season: *G. aggregatum, G. irregulare, G. sinuosum*, two unidentified *Glomus* spp., *F. coronatus, F. monosporus, S. hoi, S. constrictum, A. walkeri, T. nevadensis, Acaulospora* sp., *P. occultum* and *P. laccatum* (Ramírez, 2014). This may indicate something about the survival strategy of these species during dry seasons with lower availability of photosynthates from the host plants. Four other species were not affected by seasons: *G. macrocarpum, G. brohultii, G. intraradices* and *C. claroideum* which thus may indeed be generalist not only for all plant species but also able to suvive/sporulate under all envionmental conditions.

4.0 Benefit of AMF consortia for citrus, grapes and *cape gooseberry*

It is often said or assumed that a broad diversity of AMF species is more beneficial for a crop and its productivity than a less diverse population of AMF species (Van Der Heijden *et al.*, 1998). However, the benefits of single AMF species versus AMF species consortia, as found in fields, have been rarely investigated. The efficiency of AMF species usually is investigated in bioassays where different parameter of growth and productivity of mycorrhizal vs non-mycorrhizal plants are measured. Such bioassays are generally performed under more or less controlled conditions in glasshouses. Soils are often sterilized and then inoculated with a single AMF species or a consortium of different species. The "non-mycorrhizal" control receives a washing of the inoculum by passing through a sieve with openings smaller than AMF spores but bacteria and other microorganisms may pass. The consortia of AMF species are often unknown for their species composition, and in such cases researchers talk about native mycorrhiza. Sometimes the efficacy of the native AMF consortium was defined by comparing field soil non-sterilized with sterilized; such trials were carried out in pots and in glasshouses (Sieverding, 1991). The aim of all such inoculation trials was either to define the efficacy of the AMF species for several plant growth parameters, or to define the dependency of the plant on AMF for its growth and development. Plant dependency on mycorrhizal fungi is expressed as the ratio of plant growth parameter of a mycorrhizal (M-) plant to the data of the non-mycorrhizal (NM-) plant. Even though this comparison may be problematic because the physiology of a plant can completely change through mycorrhization (and thus the physiology of the NM-plant is different from that of the M-plant), the method can give a first indication about the dependency of the plant species on the AMF itself. From such trials it is also known since longer time that the response of a crop species may be slow or quick but in both cases the plant species may be highly mycotrophic (dependent on mycorrhiza). On the other hand facultative mycotrophic plants need mycorrhizal fungi only under low nutrient availability conditions, or under stress condition (Janos, 1980). It is well known from literature that *Citrus,Vitis* and Physalis are obligate mycotrophic. *Citrus sinensis* is a highly mycorrhiza dependent tree species for the uptake of nutrients (Melloni *et al.*, 2000), like *"Vitis* spp." growth is heavily dependent of AMF on nutrient depleted soils (Schreiner, 2007), and *Physalis peruviana* in saline soils (Miranda *et al.*, 2011). However, while *citrus* and *Physalis* were always responding positively to the inoculation with AMF, grapes, under some conditions did not respond or negatively respond to inoculation with AMF in experiments. Seldom, however, different corsortia of AMF species, as they occur in nature, compared with specific species for their efficacy and its ecological interest permits to know whether tree species need a broad consortia of different AMF species for optimum growth or whether a few or only one species may be sufficient for helping the plant.

Consortia of AMF can indeed consist in only one, a few or many (up to 30 or 40 different species (**Table 2.1).** Often, however, only consortia of a few species have been investigated. For example a consortium of *C. claroideum, G. diaphanum* and *G. albidum* benefited *Citrus volkameriana,* even in high P content soils (Alarcón and Ferrera-Cerrato, 2003), and in other experiments showed the same plant species (Alarcón *et al.*, 2003) and the same consortium along with *G. aggregatum* also increased photosynthetic activities in 25 days old plantlets. The consortia *G. albidum, G. clarum, G. manihotis* and *K. colombiana* were efficient with *Citrus sinensis* fertilized with organic matter (Dias da Silveira and dos Santos Freitas, 2007).

Benefits were also observed with two AMF strains like *G. leptotichum* and *G. gilmorei* with *Citrus* x limonia, *G. intraradices* and *G. clarum* with *C. sinensis* and tangerine in arrangement with conventional P fertilization (Dias da Silveira and dos Santos Freitas, 2007). Positive effects were all observed with different combinations of AMF species on different *Citrus* spp like: *Glomus clarum* and *G. intraradices* or *G. intraradices* and *G. etunicatum* with *Citrus sinensis, G. clarum* and *G. intraradices* with *Citrus volkameriana, G. clarum* and *G. intraradices* with tangerine, *G. clarum* and *G. intraradices* with Citrange Troyer, *G.clarum* and *G. intraradices* with *Citrus reticulatus* and *Citrus* x tangelo, *G. clarum* and *G. intraradices* with bitter orange, *G. clarum* and *G. intraradices* with trifoliate orange with organic P fertilization (Dias da Silveira and dos Santos Freitas, 2007).

Individual inoculations demostrated that specific single AMF species can be efficient on citrus, too. Rootstock of *Citrus sinensis* inoculated with *G. intraradices* expressed a major diameter, more dry weight and taller plants (Melloni *et al.*, 2000). *Citrus* x limonia showed benefits when inoculated with *G. etunicatum* or *G. intraradices,* while *C. sinensis* showed positive effects when inoculated with *G. margarita* under conventional P fertilization (Dias da Silveira and dos Santos Freitas, 2007). The same can be observed in *Citrus sinensis* with *A. morrowae,* in *Citrus* x limon with *G. clarum* or *A. morrowae,* in *Citrus volkameriana* with *G. etunicatum,* in Tangerine with *G. etunicatum,* and in Citrange Troyer with *G. intraradices,* cultivated in conditions of organic P fertilization (Dias da Silveira and dos Santos Freitas, 2007). Some experiments report the efficiency of native or indigenous AMF on different growth parameters, for example for *Citrus* x limonia and with *Citrus aurantium* under conventional P fertilization (Dias da Silveira and dos Santos Freitas, 2007, Ortas and Ustuner, 2014).

In grapevine, the AMF species consortia *F. mosseae, G. intraradices* and *S. calospora* increased plants in dry mass and nutrient uptake depending on the soil conditions (Schreiner, 2007). However, a single AMF species, *F. mosseae* was more effectively colonizing the roots of "pinoir noir" grape plants and was more efficient for the promotion of plant growth and nutrient uptake than a consortium of undefined native AMF species (Schreiner, 2007). Other evaluations showed that the effect of AMF species depended on the soil type: *Funneliformis mosseae, G. intraradices*, and *S. calospora* gave positive effects on growth of grapes in nutrient deficient soils, but were ineffective in nutrient rich soils (Schreiner, 2007). Single AMF species *C. etunicatum* and *S. heterogama* provided also better nutrition and greater vegetative growth of vine rootstock varieties when compared to the non-inoculated plants (Anzanello *et al.*, 2011). Micropropagated plantlets of grapes growing under greenhouse conditions, showed positive results to the inoculation with *G. fasciculatum* and *G. etunicatum* (Alarcón *et al.*, 2001). Plants inoculated with *G. fasciculatum* increased foliar dry weight and leaf area despite low root colonization, and increased 12 and 15 fold the P and N uptake. The effect was temperature dependent and inoculation was only positive above 20° C (Alarcón *et al.*, 2001). The rootstock *Vitis vinifera* x *Vitis rotundifolia* responded positively to *C. etunicatun*, while *Vitis berlandieri* x *Vitis riparia* and *Vitis berlandieri* x *Vitis rupestris* reacted positive to *C. etunicatum* and *S. heterogama* (Anzanello *et al.*, 2011). In some cases the inoculation with AMF was ineffective or decreased the growth parameter: rootstocks *Vitis berlandieri* x *Vitis riparia* and *Vitis riparia* x *Vitis rupestris* inoculated with *A. scrobiculata* and *G. clarum*, was inneffective for rooting and shoot growth, while the *S. heterogama* effect was negative for the growth of *Vitis berlandieri* x *Vitis riparia* rootstock (Büttenbender and Dutra de Souza, 2001).

Little investigation has been done in *Physalis* with respect to find the benefits of AMF species for the plant. The inoculation with *A. mellea* and *G. intraradices* replaced the effect of the synthetic fertilizer in terms of dry weight. The inoculated plants showed the best fruit production and quality after transplanting to the field (Guana *et al.*, 2011). An unidentified species of AMF, like a *Glomus* sp. had, however no effect on the growth of *Physalis*, indicating that under specific conditions some AMF species may be only ecological partners of *Physalis* but not mutualistic symbionts (Guana *et al.*, 2011).

5.0 Conclusions and future strategies

The diversity of Glomeromycotean species in citrus and grapes is generally lower than in *Physalis*, which may have several reasons: the sampling depth may have been insufficient as some AMF species may find a refuge below the ploughed/sampled soil horizon. Also, greater plant density and ground cover with other plants grown under the investigated trees can result in more diverse AMF species communities.

There is no indication that real "specialist" AMF species exist for *Citrus*, grapes and *Physalis*. The occurrence of a great number of species of Gigasporales in citrus was likely related to the undisturbed soil conditions below the tree and the all year long photosynthetic activity of the plant. "Generalist" AMF species, however, may exist species like *C. claroideum, C. etunicatum, F. mosseae, G. intraradices, G. macrocarpum,* were reported from most investigations and are known from many other crops. It is likely that they are commonly found because they tend to sporulate under all climatic conditions, during dry and wet seasons. This can mean that many more AMF species are "generalists" if the correct sampling methodology was applied, i.e. AMF species diversity is defined from field samples and from bait cultures inoculated with field samples.

It is likely that some AMF species are more frequent and dominate the AMF species consortia in number of spores because they have found ideal ecological niches to produce survival structures like spores. Such species can have small sized spores as well as big sized spores. There is however, no clear indication that this can be called a preferential association between e.g. such AMF species and *Citrus* spp., grapes or *Physalis*.

Edaphic factors and farm management practices (conventional or ecological farming) can play a role for lower or higher AMF species diversity in citrus and grapes. Higher species richness under organic farming can be related to higher plant species density through intercropping and thus often lower nutrient availability. However, in mature citrus plantations growth of weeds is little due to the shade below the dense leaf canopy. In dense plantation crops with little understory there may be no difference between conventional and organic farming in AMF species diversity.

It is by no means clear whether citrus, grapes or *Physalis* need a broad AMF species diversity to be productive in terms of biomass or fruit production. There is no methodology available so far, to define functions of different AMF species for one of the three crops. All glasshouse experiments are giving only some indications on whether a fungal species can promote the growth or not, at the beginning of the plant development. Some reports indicate that citrus may respond always positive to whatever AMF species is used for inoculation but in grapes and *Physalis* also negative responses to inoculation with some AMF species are known. All experiments carried out so far were done with a few fungal species, and some consortia with 2-4 species. Sometimes single AMF species were more beneficial for the plant growth than the native AMF fungal consortium. The so-called "generalists" of AMF species found in the natural community were often also the most effective in inoculation experiments, these were e.g. *F. mosseae, C. etunicatum, C. claroideum* and *G. intraradices*.

It is currently very unclear which function a specific AMF species carried out the at any of the different growth stages of the fruit trees investigated. It is thus questionable whether from an agro-ecological point of view we have to preserve a high AMF species diversity in citrus and grapes. More likely that we have to take care that some of those AMF species from which their positive function on plant growth are known, are maintained in the *Glomeromycotean* species community under field conditions. This may be possible by selecting and applying correct agricultural management practices as done by organic farming practices.

References

1. Alarcón, A., González-Chávez, María del C., Ferrera-Cerrato, R. and Villegas-Monter, A., 2001. Efectividad de *Glomus fasciculatum* y *Glomus etunicatum* en el crecimiento de plantulas de *Vitis vinifera* L. obtenidas por micropropagación. *Terra* .19, 29–35.

2. Alarcón, A., González-Chávez, María del C. and Ferrera-Cerrato, R., 2003. Crecimiento y fisiología de *Citrus volkameriana* Tan & Pasq en simbiosis con hongos micorrízicos arbusculares. *Terra Latinoam.* 21, 503–511.

3. Alarcón, F. and Ferrera-Cerrato, R., 2003. Aplicación de fósforo e inoculación de hongos micorrízicos arbusculares en el crecimiento y estado nutricional de *Citrus volkameriana* Tan and Pasq. *Terra Latinoam.* 21, 91–99.

4. Álvarez-Sánchez, J., Johnson, N.C., Antoninka, A., Chaudhary, V.B., Lau, M.K., Owen, S.M., Sánchez-Gallen, I., Guadarrama, P. and Castillo, S., 2011. Large-scale diversity patterns in spore communities of Arbuscular Mycorrhizal Fungi., In: Pagano, M.C. (Ed.), Mycorrhiza: Occurrence and Role in Natural and Restored Environments. *Nova Science Publishers,* Hauppauge, New York, NY, USA., pp. 33–50.

5. Andrade, S.A.L., Mazzafera, P., Schiavinato, M.A. and Silveira, A.P.D., 2009. Arbuscular mycorrhizal association in coffee. *J. Agric. Sci.* 147, 105–115.

6. Anzanello, R., Dutra de Souza, P.V. and Casamali, B., 2011. Fungos micorrízicos arbusculares (FMA) em porta-enxertos micropropagados de videira. *Bragantia Campinas.* 70, 409–415.

7. Augé, R., 2001. Water relations, drought and vesicular- arbuscular mycorrhizal symbiosis. *Mycorrhiza.* 11, 3–42.

8. Ávila, A.L., 2004. Ocorrência de fungos micorrízicos arbusculares em cultivos de videira (*Vitis* sp.) sob manejo orgânico e convencional. Universidade federal do rio grande do sul. pp. 68

9. Balestrini, R., Magurno, F., Walker, C., Lumini, E. and Bianciotto, V., 2010. Cohorts of arbuscular mycorrhizal fungi (AMF) in *Vitis vinifera*, a typical Mediterranean fruit crop. *Environ. Microbiol. Rep.* 2, 594–604.

10. Büttenbender, D. and Dutra de Souza, P.V., 2001. Efeito de fungos micorrízicos arbusculares sobre o desenvolvimento degetativo de porta-inxeetos de videira, In: VII Viticulture and *Epicology Latin-American Congress. Montevideo* - Uruguay, p. 11.

11. Carrenho, R., Trufem, S.F.B. and Bononi, V.L.R., 1998. Arbuscular mycorrhizal fungi in *Citrus sinensis* / *C. limon* treated with Fosetyl-Al and Metalaxyl. *Mycol. Res.* 102, 677–682.

12. Camargo-Ricalde, S.L. and Esperón-Rodríguez, M., 2005. Efecto de la heterogeneidad espacial y estacional del suelo sobre la abundancia de esporas de hongos micorrizógenos arbusculares en el valle semiárido de Ţehuacán-Cuicatlán, México. *Rev. Biol. Trop.* 53, 339–352.

13. Castillo, C.G., Borie, F.R., Godoy, R., Rubio, R. and Sieverding, E., 2006. Diversity of arbuscular mycorrhizal plant species and fungal species in evergreen forest, deciduous forest and grassland ecosystems of Southern Chile. *J. Appl. Bot. Food Qual.* 80, 40–47.

14. Castillo, C., Rubio, R., Borie, F. and Sieverding, E., 2010. Diversity of arbuscular mycorrhizal fungi in horticultural production systems of southern Chile. *J. soil Sci.* 10, 407–413.

15. De Almeida, V.C., Nogueira, M.I., Guimarães, R.J. and Mourão J.M., 2003. Carbono da biomassa microbiana e micorriza en solo sub mata nativa e agroecossistemas cafeeiros. *Acta Sci. Agron.* 25, 147–153.

16. De la Providencia, I.E., de Souza, F.A., Fernández, F., Delmas, N.S., and Declerck, S., 2005. Arbuscular mycorrhizal fungi reveal distinct patterns of anastomosis formation and hyphal healing mechanisms between different phylogenic groups. *New Phytol.* 165, 261–71.

17. De Souza, R.G., da Silva, D.K.A., de Mello, C.M.A., Goto, B.T., da Silva, F.S.B., Sampaio, E.V.S.B. and Maia, L.C., 2013. Arbuscular mycorrhizal fungi in revegetated mined dunes. L. *Degrad. Dev.* 24, 147–155.

18. Dias da Silveira, A.P.and dos Santos Freitas, S., 2007. Microbiota do Solo e Qualidade Ambiental. *Instituto Agronómico,* Campinas. pp. 317

19. Dutra de Souza, P.V., Kroeff Schmitz, J.A., Santos de Freitas, R., Carniel, E. and Carrenho, R., 2002. Identificação e quantificação de fungos micorrízicos arbusculares autóctones em municípios produtores de citros no Rio Grande do Sul. Pesqui. *Agropecuária Bras.* 37, 553–558.

20. Fidelibus, M.., Martin, C.., Wright, G. and Stutz, J.., 2000. Effect of arbuscular mycorrhizal (AM) fungal communities on growth of "Volkamer" lemon in continually moist or periodically dry soil. *Sci. Hortic.* 84, 127–140.

21. Focchi, S.S., Kessler dal Soglio, F., Carrenho, R., Dutra de Souza, P.V. and Lovato, P.E., 2004. Fungos micorrízicos arbusculares em cultivos de citros sob manejo convencional e orgânico. Pesqui. *Agropecuária Bras.* 39, 469–476.

22. Franke-Snyder, M., Douds, D.D., Galvez, L., Phillips, J.G., Wagoner, P., Drinkwater, L. and Morton, J.B., 2001. Diversity of communities of arbuscular mycorrhizal (AM) fungi present in conventional versus low-input agricultural sites in eastern Pennsylvania , USA. *Appl. Soil Ecol.* 16, 35–48.

23. Guana, O.A., Rodríguez, A., Ramírez, M. and Roveda, G., 2011. Evaluación del efecto de la inoculación de plantas de uchuva con hongos formadores de micorrizas arbusculares. *Suelos ecuatoriales.* 41, 122–127.

24. Heijne, B., van Dam, D., Heil, G.W. and Bobbink, R., 1996. Acidification effects on vesicular-arbuscular mycorrhizal (VAM) infection, growth and nutrient uptake of established heathland herb species. *Plant Soil* .179, 197–206.

25. Hodge, A., 2001. Arbuscular mycorrhizal fungi influence decomposition of, but not plant nutrient capture from, glycine patches in soil. *New Phytol.* 151, 725– 734.

26. Hodge, A., Campbell, C.D. and Fitter, A.H., 2001. An arbuscular mycorrhizal fungus accelerates decomposition and acquires nitrogen directly from organic material. *Nature* .413, 297–299.

27. Janos, D.P., 1980. Vesicular-Arbuscular mycorrhizae affect lowland tropical rain forest plant growth. *Ecology* .61, 151.

28. Jansa, J., Mozafar, A., Kuhn, G., Anken, T., Ruh, R., Sanders, I.R. and Frossard, E., 2003. Soil tillage affects the community structure of mycorrhizal fungi in maize roots. *Ecol. Appl.* .13, 1164–1176.

29. Jefwa, J.M., Okoth, S., Wachira, P., Karanja, N., Kahindi, J., Njuguini, S., Ichami, S., Mung'atu, J., Okoth, P. and Huising, J., 2012. Impact of land use types and farming practices on occurrence of arbuscular mycorrhizal fungi (AMF) Taita-Taveta district in Kenya. *Agric. Ecosyst. Environ.* 157, 32–39.

30. Kang, J., Amoozegar, A., Hesterberg, D. and Osmond, D.L., 2011. Phosphorus leaching in a sandy soil as affected by organic and inorganic fertilizer sources. *Geoderma.* 161, 194–201.

31. Koske, R.E., 1987. Distribution of VA mycorrhizal fungi along a latitudinal temperature gradient. *Mycologia* .79, 55–68.

32. Lambais, M.R. and Cardoso, E.J.B.N., 1988. Avaliaçao da germinaçao de esporos de fungos micorrízicos vesículo-arbusculares e da colonizaçao micorrízica de *Stylosanthes guianensis* em solo ácido e distrífico. *Rev. Bras. Cienc. do Solo*.12,249-255.

33. Leyval, C., Joner, E.J., Del Val, C.and Haselbandter, D.A., 2002. Potential of arbuscular mycorrhizal fungi for bioremediation, In: Gianinazzi, S., Schüepp, H., Barea, J.M., Haselwandter, K. (Eds.), Mycorrhiza Technology in Agriculture, from Genes to Bioproducts. Birkäuser Verlag, London, Basel, Switzerland, pp. 175–186.

34. Liu, Z., Fu, B., Zheng, X. and Liu, G., 2010. Plant biomass, soil water content and soil N:P ratio regulating soil microbial functional diversity in a temperate steppe: A regional scale study. *Soil Biol. Biochem.* 42, 445–450.

35. Martínez-García, L.B., Armas, C., Miranda, J.D.D., Padilla, F.M. and Pugnaire, F.I., 2011. Shrubs influence arbuscular mycorrhizal fungi communities in a semi-arid environment. *Soil Biol. Biochem.* 43, 682–689.

36. Melloni, R., Nogueira, M.A., Freire, V.F. and Cardoso, E.J.B.N., 2000. Fosforo adicionado e fungos micorrízicos arbusculares no crescimento e nutriçao mineral de limoeiro-cravo [*Citrus limonia* (L.) Osbeck]. *Rev. Bras. Cienc. do Solo.* 24, 767–775.

37. Miller, S.P., 2000. Arbuscular mycorrhizal colonization of semi-aquatic grasses along a wide hydrologic gradient. *New Phytol.* 145, 145–155.

38. Miranda, D., Fischer, G. and Ulrichs, C., 2011. The influence of arbuscular mycorrhizal colonization on the growth parameters of cape gooseberry (*Physalis peruviana* L.) plants growth in a saline soil. *J. Soil Sci. Plant Nutr.* 11, 18–30.

39. Oehl, F., Sieverding, E., Ineichen, K., Mäder, P., Boller, T. and Wiemken, A., 2003. Impact of land use intensity on the species diversity of arbuscular mycorrhizal fungi in agroecosystems of Central Europe. *Appl. Environ. Microbiol.* 69, 2816–2824.

40. Oehl, F., Sieverding, E., Mäder, P., Dubois, D., Ineichen, K., Boller, T. and Wiemken, A., 2004. Impact of long-term conventional and organic farming on the diversity of arbuscular mycorrhizal fungi. *Oecologia.* 138, 574–583.

41. Oehl, F., Sieverding, E., Ineichen, K., Ris, E.-A., Boller, T. and Wiemken, A., 2005. Community structure of arbuscular mycorrhizal fungi at different soil depths in extensively and intensively managed agroecosystems. *New Phytol.* 165, 273–283.

42. Oehl, F., Sieverding, E., Ineichen, K., Mäder, P., Wiemken, A. and Boller, T., 2009. Distinct sporulation dynamics of arbuscular mycorrhizal fungal communities from different agroecosystems in long-term microcosms. *Agric. Ecosyst. Environ.* 134, 257–268.

43. Oehl, F., Laczko, E., Bogenrieder, A., Stahr, K., Bösch, R., van Der Heijden, M.G.A. and Sieverding, E., 2010. Soil type and land use intensity determine the composition of arbuscular mycorrhizal fungal communities. *Soil Biol. Biochem.* 42, 724–738.

44. Oehl, F., Alves da Silva, G., Sánchez-Castro, I., Goto, B.T., Costa Maia, L., Evangelista Vieira, H.E., Barea, J.M., Sieverding, E. and Palenzuela, J., 2011a. Revision of Glomeromycetes with entrophosporoid and glomoid spore formation with three new genera. *Mycotaxon* .117, 297–316.

45. Oehl, F., Da Silva, G.A., Goto, B.T., Costa Maia, L. and Sieverding, E., 2011b. Glomeromycota: two new classes and a new order. *Mycotaxon.* 116, 365–379.

46. Oehl, F., da Silva, G.A., Goto, B.T. and Sieverding, E., 2011c. Glomeromycota: three new genera and glomoid species reorganized. *Mycotaxon.* 116, 75–120.

47. Oehl, F., Sieverding, E., Palenzuela, J., Ineichen, K. and Alves da Silva, G., 2011d. Advances in Glomeromycota taxonomy and classification. *IMA Fungus* .2, 191–199.

48. Oehl, F., Schneider, D., Sieverding, E. and Burga C. A., 2011e. Succession of arbuscular mycorrhizal communities in the foreland of there treating Morteratsch glacierin the Central Alps. *Pedobiologia*. 54, 321– 331.

49. Öpik, M., Davison, J., Moora, M. and Zobel, M., 2013. DNA-based detection and identification of Glomeromycota: the virtual taxonomy of environmental sequences 1. *Botany*. 147, 135–147.

50. Ortas, I. and Ustuner, O., 2014. The effects of single species, dual species and indigenous mycorrhiza inoculation on citrus growth and nutrient uptake. *Eur. J. Soil Biol.* 63, 64–69.

51. Pellegrino, E., Bedini, S., Avio, L., Bonari, E. and Giovannetti, M., 2011. Field inoculation effectiveness of native and exotic arbuscular mycorrhizal fungi in a Mediterranean agricultural soil. *Soil Biol. Biochem.* 43, 367–376.

52. Piotrowski, J.S. and Rilling, M.C., 2008. Succession of Arbuscular Mycorrhizal Fungi: Patterns, Causes, and Considerations for Organic Agriculture. Adv. Agron. *Advances in Agronomy* .97, 111–130.

53. Porcel, R., Gómez, M., Kaldenhoff, R. and Ruiz-Lozano, J., 2005. Impairment of NtAQP1 gene expression in tobacco plants does not affect root colonisation pattern by arbuscular mycorrhizal fungi but decreases their symbiotic efficiency under drought. *Mycorrhiza*. 15, 417–423.

54. Posada, R.H., 2011. Comunidades de hongos de micorriza arbuscular y hongos solubilizadores de fósforo en cultivos de café (*Coffea arabica* L.) bajo diferentes tipos de manejo. *Dissertation Instituto de Ecologia A.C., Xalapa*, Mexico. pp. 140.

55. Purin, S., Filho, O.K. and Stürmer, S.L., 2006. Mycorrhizae activity and diversity in conventional and organic apple orchards from Brazil. *Soil Biol. Biochem*. 38, 1831–1839.

56. Purin, S. and Morton, J.B., 2011. In situ analysis of anastomosis in representative genera of arbuscular mycorrhizal fungi. *Mycorrhiza*. 21, 505–514.

57. Ramos, L.M., 2006. Efecto de hongos endofíticos sobre promoción de crecimiento en vitro plantas de banano y piña. Universidad de Zamorano, Honduras. pp. 42

58. Radić, T., Likar, M., Hančević, K., Bogdanović, I. and Pasković, I., 2014. Occurrence of root endophytic fungi in organic versus conventional vineyards on the Croatian coast. *Agric. Ecosyst. Environ.* 192, 115–121.

59. Ramírez, M.M., 2014. Evaluación de la diversidad de Hongos Formadores de Micorrizas Arbusculares (HFMA) y su relación con el establecimiento de simbiosis con *Physalis peruviana* L. *Universidad Nacional de Colombia*. pp. 237

60. Santos, B.A., Maia, L.C., Cavalcante, U.M.T., Correia, M.T.S. and Coelho, L.C.B.B., 2001. Effect of arbuscular mycorrhizal fungi and soil phosphorus level on expression of protein and activity of peroxidase on passion fruit roots. *Braz. J. Biol.* 61, 693–700.

61. Schreiner, P.R., 2007. Effects of native and nonnative arbuscular mycorrhizal fungi on growth and nutrient uptake of "Pinot noir" (*Vitis vinifera* L.) in two soils with contrasting levels of phosphorus. *Appl. Soil Ecol*. 36, 205–215.

62. Schreiner, R.P. and Mihara, K.L., 2009. The diversity of arbuscular mycorrhizal fungi amplified from grapevine roots (*Vitis vinifera* L.) in Oregon vineyards is seasonally stable and influenced by soil and vine age. *Mycologia* .101, 599–611.

63. Shujie, M., Yunfa, Q. and Lianren, Z., 2009. Aggregation stability and microbial activity of China s black soils under different long-term fertilisation regimes. New Zeal. *J. Agric. Res*. 52, 57–67.

64. Siqueira, J.O., 1986. Micorrizas: forma e funçao, In: Programas E Resumos Da IV Reuniaho Brasileira Sobre Micorrizas. Secretarias de Agricultura e do Meio Ambiente e Universidade de Sao Paulo, Sao Paulo, Brazil, pp. 44–70.

65. Siqueira, J.O., Mahmud, A.W. and Hubbell, D.H., 1986. Comportamento diferenciado de fungos formadores de micorrizas vesicular-arbuscular em relaçao à acidez do solo. *Rev. Bras. Cienc. do Solo*. 10, 11–16.

66. Sieverding, E., 1991. Vesicular-arbuscular mycorrhiza management in tropical agrosystems, first. ed. Deutsche Gesellschaft fur Technische Zusammenarbeit, Eschborn.

67. Siqueira, J.O., Saggin-Júnior, O.J., Flores-Aylas, W.W. and Guimarães, P.T.G., 1998. Arbuscular mycorrhizal inoculation and superphosphate application influence plant development and yield of coffee in Brazil. *Mycorrhiza*. 7, 293–300.

68. Taylor, J. and Harrier, L.A., 2000. A comparison of nine species of arbuscular mycorrhizal fungi on the development and nutrition of micropropagated *Rubus idaeus* L . cv . Glen Prosen (Red Raspberry). *Plant Soil* .225, 53–61.

69. Taylor, J. and Harrier, L.A., 2001. A comparison of development and mineral nutrition of micropropagated Fragaria×ananassa cv. Elvira (strawberry) when colonised by nine species of arbuscular mycorrhizal fungi. *Appl. Soil Ecol*. 18, 205–215.

70. Tian, H., Drijber, R.A., Niu, X.S., Zhang, J.L. and Li, X.L., 2011. Spatio-temporal dynamics of an indigenous arbuscular mycorrhizal fungal community in an intensively managed maize agroecosystem in North China. *Appl. Soil Ecol*. 47, 141–152.

71. Urcelay, C., Díaz, S., Gurvich, D.E., Chapin III, F.S., Cuevas, E. and Domínguez, L.S., 2009. Mycorrhizal community resilience in response to experimental plant functional type removals in a woody ecosystem. *J. Ecol*. 97, 1291–1301.

72. Vaast, P., Zasoski, R.J. and Bledsoe, C.S., 1996. Effects of vesicular-arbuscular mycorrhizal inoculation at different soil P availabilities on

growth and nutrient uptake of in vitro propagated coffee (*Coffea arabica* L.) plants. *Mycorrhiza* .6, 493–497.

73. Van Der Heijden, M.G.A., Klironomos, J.N., Ursic, M., Moutoglis, P., Streitwolf-Engel, R., Boller, T., Wiemken, A. and Sanders, I.R., 1998. Mycorrhizal fungal diversity determines plant biodiversity, ecosystem variability and productivity. *Nature.* 396, 69–72.

74. Vandenkooornhuyse, P., Husband, R., Daniell, T.J., Watson, I.J., Duck, J.M., Fitter, A.H. and Young, P.W., 2002. Arbuscular mycorrhizal community composition associated with two plant species in a grassland ecosystem. *Mol. Ecol.* 11, 1555–1564.

75. Verbruggen, E., Roling, W.F.M., Gamper, H.A., Kowalchuk, G.A., Verhoef, H.A. and van Der Heijden, M.G.A., 2010. Positive effects of organic farming on below-ground mutualists: large-scale comparison of mycorrhizal fungal communities in agricultural soils. *New Phytol.* 186, 968–979.

76. Wang, F.Y., Hu, J.L., Lin, X.G., Qin, S.W. and Wang, J.H., 2011. Arbuscular mycorrhizal fungal community structure and diversity in response to long-term fertilization: a field case from China. *World J. Microbiol. Biotechnol.* 27, 67–74.

77. Wang, P., Shu, B., Wang, Y., Zhang, D.J., Liu, J.F. and Xia, R.X., 2013. Diversity of arbuscular mycorrhizal fungi in red tangerine (*Citrus reticulata* Blanco) rootstock rhizospheric soils from hillside citrus orchards. *Pedobiologia (Jena).* 56, 161–167.

78. Wetzel, K., Da Silva, G., Matczinski, U., Oehl, F. and Fester, T., 2014. Superior differentiation of arbuscular mycorrhizal fungal communities from till and no-till plots by morphological spore identification when compared to T-RFLP. *Soil Biol. Biochem* .72, 88–96.

79. Whitcomb. S., Stutz, J.C., 2007. Assessing diversity of arbuscular mycorrhizal fungi in a local community: role of sampling effort and spatial heterogeneity. *Mycorrhiza* .17:429–437.

Part II

Crop Improvement by Arbuscular Mycorrhizal Fungi

Chapter 3

Improvement in Plant Growth and Nutrition of Tropical and Sub-tropical Fruit Crops

Poovarasan S and Sukhada Mohandas*

[1] *ICAR -Indian Institute of Horticultural Research, Hessaraghatta,*
Bengaluru 560089, India

**Address for the correspondence: sukhada.mohandas@gmail.com*

ABSTRACT

India is the second largest producer of fruit crops in the world. The consumption of fruits has increased several folds due to awareness of its health benefits amongst consumers. Cultivation packages for fruit crops concentrate on reduction in pesticide application and use of ecofriendly methods without compromising on yield. Several bioinoculants are being used in the fruit crop nursery and later in the field to harness the benefits for sustainable growth. Among the bio-inoculants, arbuscular mycorrhizal fungi (AMF) have wide range of application. AMF colonization contributes significantly to the world phosphate and carbon recycling and influences the development of primary terrestrial ecosystem. Chemotaxis ability of AMF hyphae helps it to colonize different hosts. AMF take up the products of the plants photosynthesis to the arbuscules and intra and extraradical hyphae. Hyphal network of mycorrhizae helps the plants to sustain in a limited nutrient supplied soil. The application of AMF in fruit crop cultivation will help to improve the fruit yield and maintain the soil nutrient content. The present review gives an insight into the efforts made by various studies to harness the benefits of the application of AMF in tropical and subtropical fruit cultivation and discusses the prospects of using AMF as a routine package of practice in the eco-friendly management of tropical and subtropical fruit crops.

Keywords: *Symbiosis, Rhizosphere, Hyphae Network, Glomalin, Growth Improvement.*

1.0 Introduction

Interaction between AMF belonging to the monophyletic phylum called as Glomeromycota and plant roots is probably the most widespread terrestrial symbiosis formed by 70-90% of land species. Frank was the first person to coin the term "Mykorrhizen" in the year 1885 for this association which means "fungal-root" in Greek. Allen (1991) defined a mycorrhiza as "a mutualistic symbiosis between plant and fungus localized in a root or root-like structure in which energy moves primarily from plant to fungus and inorganic resources move from fungus to plant". Symbiotic development results in the formation of tree-shaped sub cellular structures within plant cells. These structures, which are known as arbuscules (from the Latin 'arbusculum', meaning bush or little tree) are the main scavengers for nutrient exchange between the fungal and plant symbiotic partners. AMF intimately connects the plants with the hyphae network of the fungi, which can be in excess of 100 meters of hyphae per cubic centimeter of soil. The nonspecific interaction of AMF with host plant is highly compatible at both the structural and physiological levels. The beneficial relationship is very effective in physiological and bio- chemical processes (Smith and Gianinazzi-Pearson,1988). AMF colonization contributes significantly to the world phosphate and carbon recycling and influences the development of primary terrestrial ecosystem. Mycorrhiza by virtue of its extensive hyphal network provides increased plant nutrient supply by acquiring the nutrient forms that would not normally be available to plants. Root colonization by AMF can provide protection from parasitic fungi and nematodes. The fungus colonizes the host plant's roots inside the cortical tissues and the association may be either intracellular like AMF, or extracellular as in ectomycorrhizal fungi. Mycorrhizae influence soil microbial populations and exudates in the mycorrhizosphere and hyphosphere. AMF are considered to contribute to soil structure and carbon transportation in the soil between plant root and other organisms. Besides, micronutrients like Cu, Zn, Al, Mn and Mg are also absorbed by the hyphae efficiently and transported to the plants. AMF are most wide spread and ecologically important plant root fungal symbionts essential for the survival of more than 90% tropical plants species (Sukhada, 1998, 1992; Schreiner *et al.*, 2003).

2.0 AMF symbiosis and Rhizosphere

The plant rhizosphere plays an important role in mycorrhizal symbiosis. The root exudates influence the microbial population. The composition of root exudates, fungal exudates determines the diversity and abundance of microbes in the rhizosphere (Marschner and Timonen, 2004). The influence of AMF on plant root and shoot growth may also have indirect effect on the rhizosphere through the growth and degeneration of the hyphal network. Favorable soil matrix, temperature, carbondioxide concentration, pH and phosphorus concentration are essential for AMF spore germination (Sbrana and Giovannetti, 2005) and further growth is controlled by host root exudates and the soil phosphorus concentration. Low phosphorus concentrations in the soil favours hyphal growth and branching. As the concentration of exudates increases, the fungi produce more tightly clustered branches. At the highest concentration, arbuscules, the structures of phosphorus exchange are formed. AMF have chemotaxis abilities, which enable hyphal growth toward the roots of a potential host plant. This chemotactic fungal response to the host plants exudates is thought to increase the efficacy of host root colonization in low phosphorus soils (Nagahashi *et al.*, 1996). Molecular techniques have been used to further understand the signaling pathways, which occur between AMF and the plant roots. Tuomi *et al.* (2001) reported that exudates from potential host plant roots allow the AMF to undergo physiological changes, which allow it to colonize its host. AMF genes required for the respiration of spore carbon compounds are triggered and turned on by host plant root exudates (Abiala *et al.*, 2013).

3.0 Factors effecting mycorrhizal colonization

Climatic and edaphic factors such as light, temperature, rainfall, atmospheric CO_2, soil pH, moisture content, fertility level and density of inocula have significant influence on AMF and root colonization (Singh, 2005; Miller and Jackson, 1998; Li and Zhao, 2005). Light availability is positively correlated with mycorrhizal colonization (Koide and Mosse, 2004) as photosynthetic carbon requirement is very much essential for development of symbiosis, higher light levels can enhance the efficiency of photosynthesis, which can contribute more carbon compounds to AMF growth. The influence of temperature on AMF plants appears related to the exact fungal-host species combination, the development stage of the plant, temperature controlling fungal germination, photosynthesis and carbon flow to roots (Entry *et al.*, 2002). Soil pH influences AMF species composition, colonization and effectiveness (Van Aarle *et al.*, 2002; Hayman and Tavares, 1985). The response of AMF to soil pH seems to be dependent primarily on the fungal species (Hayman and Tavares, 1985). In China, *Glomus sp.* appears to dominate in alkaline and neutral soils while *Acaulospora* sporulate more abundantly in acid soils (Zhang *et al.*, 1998b; Gai *et al.*, 2006). Indication for probable influence of soil organic matter (SOM) on the AMF species diversity and root colonization has been found. Covacevich (2007) reported the frequency of occurrence of *G. mosseae* decreased with increasing organic matter content while *G. sinousum* and *G. taiwanese* were found only when SOM was less than 1.5%. Inverse correlation between AMF colonization and soil organic matter have also been reported (Miller and Jackson ,1998). Soil fertility status most especially P availability has significant influences on AMF root colonization, spore production, hyphal growth and response of plants to AMF inoculation are reduced by abundance of P in soil (Abbot and Robson, 1991; Bethlenfalvay, 1992; Covacevich, 2007). However, the magnitude of the effect of P is strongly dependent on the host plant dependency on mycorrhiza (Plenchette, 1983) and other environmental factors (Smith and Read, 1997). Management of AMF is very important for organic and low agriculture systems where soil phosphorus is generally low, although all agroecosystems can benefit by promoting AMF establishment. Some crops that are poor at seeking out nutrients in the soil are very dependent on AMF for phosphorus uptake. Heavy usage of phosphorus fertilizer can inhibit mycorrhizal colonization and growth. As the soil's phosphorus levels available to the plants increase, the amount of phosphorus also increases in the plant's tissues and carbon drain on the plant by the AMF symbiosis become non-beneficial to the plant (Grant *et al.*, 2005).

The rate of infection of AMF in plants is strongly influenced by amount of spore propagules. Low numbers of propagules in field soils may result in low level of colonization (Smith and Read, 1997). Conventional agriculture practices, such as tillage, heavy fertilizers and fungicides, poor crop rotations and selection for plants which survive these conditions, hinder the ability of plants to form symbiosis with AMF. Most agricultural crops can perform better and are more productive when well colonized by AMF (Abiala *et al.*, 2013). .

4.0 Glomalin

One of the compounds produced by AMF is a recalcitrant glycoprotein, glomalin (Wright *et al.* 1996). Concentrations of glomalin range from 2 to 15 mg g^{-1} of soil in temperate climates (Wright et al. 1996; Wright & Upadhyaya 1998) and is much higher in tropical soils. Glomalin molecule is a clump of small glycoproteins with iron and other ions attached. Glomalin related soil protein (GRSP) component is produced by AMF hyphae as stable glue. GRSP binds to soil, producing a uniform aggregated structure composed of minerals and humus. The hypha has an important role in soil aggregate stabilization and with carbon sequestration in the soil by helping to physically protect organic matter within aggregates. Increasing organic matter increases cation exchange capacity of soils. Primarily, these aggregates permit the soil to retain water better and facilitate root penetration. In addition, the aggregates reduce soil erosion and compaction while facilitating root hair adhesion, enhancing nutrient and water uptake. AMF uses carbon from the plant to produces GRSP. It is extremely tough and resistant to microbial decay and does not dissolve easily in water.

5.0 Plant AMF Nutrient Uptake

AMF being obligate symbionts are dependent on the plant for their carbon nutrition (Harley and Smith, 1983). AMF take up the products of the plants photosynthesis as hexoses, fructose and sucrose. The transfer of carbon from the plant to the fungi may occur through the arbuscules or intraradical hyphae (Pfeffer *et al.*, 1999). Inside the mycelium, hexose is converted to trehalose and glycogen. Trehalose and glycogen are carbon storage forms which can be rapidly synthesized and degraded and may buffer the intracellular sugar concentrations (Pfeffer *et al.*, 1999). The intraradical hexose enters the oxidative pentose phosphate pathway, which produces pentose for nucleic acids. Lipid biosynthesis also occurs in the intraradical mycelium. Lipids are then stored or exported to extraradical hyphae where they may be stored or metabolized. The breakdown of lipids into hexoses, known as gluconeogenesis, occurs in the extraradical mycelium (Pfeffer *et al.*,1999). Approximately 25% of the carbon translocated from the plant to the fungi is stored in the extraradical hyphae (Hamel, 2005). Up to 20% of the host plant's photosynthate carbon is understood to be transferred to the AMF (Pfeffer *et al.*, 1999) which means a considerable carbon investment in mycorrhizal network by the host plant and contribution to the below ground organic carbon pool. An increase in the carbon supplied by the plant to the AMF increases the uptake of phosphorus and the transfer of phosphorus from fungi to plant (Bucking and Shachar-Hill, 2005).The benefit of mycorrhizae to plants is mainly attributed to increase uptake of nutrients, especially phosphorus. This increase in uptake may be due to increased surface area of soil contact, increased movement of nutrients into mycorrhizae, a modification of the root environment and increased storage (Bolan, 1991). Mycorrhizal hyphae can be much more efficient than plant roots in taking up phosphorus to the root or via diffusion and hyphae reduce the distance required for diffusion, thus increasing uptake. The rate of inflow of phosphorus into mycorrhizae can be up to six times that of the root hairs (Bolan, 1991). In some cases, the role of phosphorus uptake can be completely taken over by the mycorrhizal network and all the plant's phosphorus may be of hyphal origin (Smith *et al.*, 2003). AMF are bestowed with high affinity phosphate transporter genes in them which help in better scavenging of nutrients in low phosphorus soil (Gordon-Weeks *et al.*, 2003). Moreover the micronutrients like Cu, Zn, Al, Mn and Mg are also absorbed by the hyphae efficiently and transported to the plants. Several studies have demonstrated the transport of inorganic nitrogen (N) by AMF (Johansen *et al.*, 1992; Hawkins *et al.*, 2000; Blanke *et al.*, 2005). Enhancements in the acquisition of K, Ca and Mg are often observed in AMF colonized plants grown on acidic soils than neutral or alkaline soils (Harrier and Watson, 2003). Zinc and Copper have been taken up by mycorrhiza in a deficient condition to increase plant yield (Gildon and Tinker, 1983; Kucey and Jarzen, 1987). There is evidence that AMF can inhibit Zinc and Manganese (Mn) uptake at toxic concentration in soil thus reducing adverse effect on host (Dueck *et al.*, 1986). The production of growth hormones such as IAA, gibberellin, cytokinin, auxin and growth regulators like Vitamin B by mycorrhizal fungi have also been documented (Barea and Azcon-Anguilar , 1983; Selvaraj, 1998; Manoharachary *et al.*, 2009). The nutrient management through the AMF symbiosis is more beneficial for the plant as well as the environment also. It has many advantages than the other crop managements which fully depend on synthetic chemicals. AMF potential can be best exploited in agricultural/horticultural crops, which could be cultivated well on poor fertility and marginal soils. However, it is essential to select the suitable efficient AMF strains for establishment and management of the economically important fruit crops by environmental friendly approach. Fortuna *et al.* (1996) recommended the use of infective and efficient species of AMF which promote rapid increase in plant growth. The present review highlights the role of AMF in management of growth and nutrient uptake of important fruit crops of the tropical and subtropical region.

Fruit crops commonly develop AMF relationships and exhibit a high degree of dependence on this symbiosis for normal development (Nemec, 1986). There are many research reports stating the beneficial output of AMF application on mango, banana, pomegranate, papaya, sapota, watermelon, jackfruit, lemon, custard apple, passion fruit, litchi, avocado, apple, citrus etc. The symbiotic relationship of AMF with fruit crops will facilitate to maintain the quality of crop as well as the soil fertility. The advantages of AMF inoculation will be more in perennial crops. There are a number of review papers which bring about the need of AMF inoculation in fruit crops. Naher *et al.* (2013) and Abiala *et al.* (2013) reviewed beneficial effects of mycorrhizal associations in crop plants. Jaizme- vega and azcon (1995) reported response of some tropical and subtropical plants to AMF. Benefit of AMF inoculation on tropical and subtropical fruit crops on plant growth and nutrient uptake is discussed herewith.

5.1 Apple (*Malus domestica* **Brokh***)*

Apple is an economically important fruit crop, which is predominantly a temperate crop though grown in subtropical regions of the world. Apple trees are very good symbiotic hosts for the AMF colonization, the AMF fungal inoculation in apple trees improves the fruit quality and economical value. Morin *et al.* (1994) tested the beneficial effect of AMF inoculation on four varieties of apple seedlings planted in soil containing P (644 kg bray/ ha^{-1}). The mycorrhizal seedlings grew taller with increased plant dry mass and leaf P content. G. *versiforme* inoculated plants generally had the best nutrient balance, the greatest final height and shoot biomass, and produced an extensive hyphal network. Apple cv *Malus pumila* Mill.var.*domestica* Schneid. (cv McIntosh, American Summer, Pearmain, Jonathan, Goldern Deliicious, Starking Delicious, Fuji, Mutsu and Red gold) were inoculated with *G.etunicatum* and *Gigaspora margarita* to study the effect on growth enhancement. AMF colonization upto 50% was noticed in Starking Delicious and Jonathan. Increase in plant height and weight and P concentration was observed in colonized plants (Matsubara *et al.*, 1996). Resendes *et al.* (2008) reported the efficiency of colonization of mycorrhizal and non-mycorrhizal fungi in the apple tree roots. The newly emerged roots were colonized with AMF which induced longer root growth and enhanced hypal network in the root rhizosphere. Forge *et al.*, 2001 studied the effect of AMF (6 types) colonization in apple (Ottawa) root stocks for growth and to control the nematode lesions under green house condition. The experimental results were not sufficient to differentiate the treated and control plants but the root stocks preinoculated with *G.mosseae* in field condition showed significant growth difference and enhanced leaf nutrient content (N, P, K, Mn, Cu) than the control and other five isolates. Schubert and Lubraco (2000) demonstrated AMF inoculation on apple seedlings at the time of transplantation in soil and three different peat based substrates (enriched with P fertilizer). After 112 days of inoculation the plants inoculated with *G.mosseae* showed significant level of P increases in plant tissue than the control plants. The depletion of P in the substrate was due to the intense root colonization and nutrient uptake, enhancement in plant growth.

5.2 Avocado (*Persea americana* Mill)

By nature the avocado plants have hairless mangrove-like roots (Salazar-Garcia, 2002) and hence are highly dependent on the mycorrhizal symbiosis for the better growth (Alarcón *et al.*, 2001). Menge *et al.* (1978) stated that avacados inoculated with beneficial mycorrhizal fungi have up to a 250 percent greater growth rate than non-mycorrhizal avocados in sterilized soil. Mycorrhizal avocados resist transplant shock because of better water absorption. Menge *et al.* (1980) studied the effect of AMF (*G. fasiculatum*) on the avocado cv. Topa Topa planted with AMF spores in the sterilized loamy sand soil in three different treatments (i) supplemented with all the nutrients (N, P, K, S, Ca, Mg, Cu, Zn, Mn, Fe, Mo, B) (ii) without any nutrients, (iii) containing only 10 x of P. The study revealed that the plants from the potting mix containing no nutrient had a significant growth difference than the plant that had all the nutrient and there was no difference found in P content between mycorrhizal and non-mycorrhizal plants grown in 10xP potting mix. Silveria *et al*, 2002 studied the effect of 6 different species of AMF on the avocado rootstocks which were grafted with Carmen scions. After seven month of inoculation the grafted plants which received (*Scutellospora heterogama, A. scorbiculata, G.etunicatum, G.clarum*) inoculums showed the enhanced nutrient uptake and higher accumulation of reserve substances, which favored vegetative growth of plants. Micropropagated avocado plants had poor rooting which severely affected the field survival rate at the time of transplantation. Preinoculation of seedlings with AMF facilitated the establishment of rooting. Banuelos *et al.*, 2013 studied the effect of 6 AMF (*Rhizophagus fasciculatum, Gigaspora margarita, Claroideoglomus etunicatum, Pacispora scintillans, Rhizophagus intraradices, A.laevis*) on avocado seedlings planted in the sterile and non sterile soil. The plants in sterile soil inoculated with *Rhizophagus fasciculatum* showed significant growth changes over plants in the non-sterile soil. The outcome of the study demonstrated the compatibility between AMF and host, the planting mixture (sterile/non-sterile) played a key role in the growth of avocado seedlings.

5.3 Banana (*Musa* spp.)

Banana and plantain are the fourth most important food crops all over the world and they promise food security by fruiting throughout year and giving income for the rural people (Roux *et al.*, 2008). Declerck *et al.* (1995) showed the great variation in dependency of seven banana cultivars on mycorrhizal colonization. They inoculated seven banana cultivars (*Musa acuminate*,AAA group) with two species of AMF (*Glomus mosseae* and *Glomus macrocarpum)* in a greenhouse experiment. Inoculated plants had generally greater shoot dry weight and shoot phosphorus concentrations compared to the non-inoculated plants.Among the seven banana cultivars. cv. *Williams* showed the highest relative mycorrhizal dependency (RMD) and cv. *Poyo* the lowest. For all the cultivars studied, inoculation with

G. macrocarpum resulted in the highest RMD values. Both root dry weight and root hair length or densities of the non-inoculated plants were inverserly correlated with the RMD values of cultivars.

Sukhada (1995) studied the potentiality of the AMF inoculation at the time of planting banana suckers (corms) in the field. The plants inoculated with 250 g of AMF containing around 50 chlamydospores g soil^{-1} of either *G.fasiculatum or G.mosseae* showed significant colonization in the roots by 5 months with increased nutrients (Zn, Cu) uptake. The growth parameters like bunch weight and number of fingers per bunch also increased significantly (**Plate 3.1**). Both the fungi studied were equally effective in increasing the plant growth. A number of reports have revealed that AMF application enhanced plant growth and better yield in banana (Jaizme-Vega *et al.*, 2004). Jaizme-vega and Pinochet (1997) found early inoculation of AMF contributing more to increase in host tolerance by enhancing the plant nutrient content and reducing the nematode infections in roots. The effect of *G.mosseae* at various levels of P fertilizers on the hardened banana plants under glass house condition was evaluated by Shashikala *et al.* (1999). The plants inoculated with *G.mosseae* with 50% P fertilizers showed an increased plant height, stem girth, number of leaves, and total biomass than the plant inoculated with 100% of P fertilizers in the absence of AMF. An *in-vitro* study revealed a gradual decrease of aluminum toxicity in banana plants when inoculated with AMF (*Glomus intraradicus*) and found enhanced shoot dry mass with increased uptake of nutrient and water (Rufyikiri *et al.*, 2000). Eswarappa *et al.* (2002) demonstrated the effect of AMF on banana cv. Elakkibale (syn. Neypoovan) cultivation in the farmers field. AMF application increased the yield in every harvest and the physical growth parameters like bunch weight, number of fruits and flowering period were comparatively more than the un-treated plants. *Glomus* sp. and *G. proliferum* stimulated plant growth and increased shoot P content in the inoculated banana plants under the green house condition (Declecrk *et al.*, 2002). Increased primary, secondary and tertiary root growth in the banana cv. Grande Naine plants were observed after the plants were transferred to media supplemented with pre germinated AMF (*G. intraradicus*) Koffi *et al.* (2009). Gaidashova *et al.* (2012) reported the occurrence of AMF sp. among the 18 banana genotypes collected from six different places of east Africa. The existence of AMF in the root was not highly specific between the banana genotypes (host). Other than the major sp. of (*Glomus*) AMF, minor spp. are also (*A. scrobiculata*) dominant and effective in growth improvement in banana plantation (Jefwa *et al.*, 2012). Sukhada *et al.* (2004) demonstrated on-farm inoculums production of AMF on finger millet, a common food crop in farmers' field around Bangalore, India (Plate 14.4). The farmers were also trained in the method of inoculating banana crop in the field with the fungi.

5.4 Citrus (*Citrus* spp.)

Citrus is an agriculturally important fruit crop and it ranks highest in production and is cultivated in nearly 50 countries in the world. As it is an economically important crop, the sustainability of the crop throughout the year is also important. Nemac (1978) studied the response of 6 different citrus root stocks (Sour orange, Cleopatra mandarin, Sweet orange, Rough lemon, Rangpur lime, and Carrizo citrange) to colonization by three different AMF in astatula fine sand subsoil under the green house condition. All the inoculated root stocks grew well with 21 (*G. etunicatus*) 8 (*G. mosseae*) 6 (*G. fasciculatum*) fold increase in plant growth than control plants. Menge *et al.* (1978 a, b) recorded the mycorrhizal dependency of several citrus cultivars. Inoculation of AMF helped plant growth and nutrient uptake in citrus orchids. Krikun *et al.* (1980) retrieved AMF (*G. mosseae*) from citrus trees growing in the soil of Negev region of Israel. Inoculation of citrus seedlings with the mycorrhizal fungus greatly increased the plant growth in low phosphorus soil. Response of Rough lemon to the AMF colonization was more than Sour orange and the AMF colonization showed higher intake of P, Cu and lower intake of N, K and Ca in leaves of inoculated plants. Vinayak and Bagyaraj (1990) studied the effectiveness of 18 different AMF against the trifoliate orange (*Poncirus trifoliate* an economically important citrus root stock). Of eighteen AMF, *G. velum, G. caledonicum, G. merredum, G. macrocarpum* and *Acaulospora* sp. were found most effective in improving plant height, stem girth and total biomass and uptake of P, Zn and Cu. The root stocks with these AMF improved the stem girth and facilitated the plant for budding within 14-15 months, when compared with controls which took 22-24 months. Syvertsen and Graham, (1990) tested the role of AMF colonization on the growth development, nutrient uptake and net CO_2 exchange in different aged citrus leaves. AMF colonization, improved the plants growth, N level in leaves and there was no difference found between the AMF and non AMF plants. Singh *et al.* (1993) described the role of AMF inoculation in rough lemon seedlings in nutrient uptake and phosphorous modification on its growth. The seedlings inoculated with *G. fasiculatum* prior to planting showed increased spore load and colonization with plant growth parameters like dry weight and total N, P, K content than the control seedlings, the plant nutrient uptake was not influenced by the normal P level of the soil. Onkarayya and Sukhada (1993) studied mycorrhizal dependency of seven root stocks of citrus. Rough lemon, Trifoliate orange, Troyer citrange, Carrizo Citrange and Rangpur Lime inoculated with *G. fasiculatum*. Growth parameters were assessed after 160 days of planting. The AMF infection recorded 69% improvement in Trifoliae Orange with increased shoot and root dry weight and increased nutrient uptake (N, P, K). Based on the increased biomass of the plants Trifoliate Orange was adjudged the most dependent root stock for AMF colonization followed by Rough Lemon, Troyer Citrange, Rangpur Lime and Carrizo Citrange. Darocha, (1995) verified the effects of phosphorus doses on inoculation with a mixed population of *A. morrowae, G. clarum* and *G. etunicatum* on growth and mineral nutrition of 'Cleopatra' mandarin seedlings. The inoculation promoted better plant growth, and increased the boron contents in the shoots. The inoculated plants presented larger growth and reached average height of 12.30 cm at four months after sowing, when they were ready to be transplanted. Root and shoot growth of the sour orange rootstocks (*Citrus aurantium* L.) and 'Carrizo' citrange (*Citrus sinensis* L.) were increased by AMF inoculation (*G. intraradices*) along with indole- 3-butyric acid (IBA) at 2 gl^{-1}. Reddy *et al.* (1995) reported the interactional effect of AMF *G.fasiculatum* with neem oil cakes against citrus nematode *Tylenchulus semipenetrans*. The plants inoculated with these inoculants showed positive growth response with improved resistance against the nematode infection. The application of IBA to the non-mycorrhizal seedlings was ineffective in increasing growth, but the application with mycorrhizae had a positive AM-IBA effect on seedling growth (Dutra *et al.*, 1996). The effect of the reduced P fertilization on the development of AMF and the quality of fruit was studied in citrus orchards. Satsuma mandarin trees which were inoculated with AMF grew larger and had better fruit quality as compared with non AMF control trees under low concentrations of applied phosphorus (P) condition (Shrestha *et al.*, 1996). The number of AMF spores in the soil and the percentage of AMF infection in the roots increased with the reduction of P, sugar content in the juice, the ratio of sugars to acids, and the a/b value of peel color increased as compared with P the control. The reduction of P also affected the content of carotenoids in the peel. The reduction of P fertilizer affected the growth of AMF and increased the percentage of AMF infection in citrus roots. Authors suggest that insoluble fertilizer source like bone dust would be useful for the propagation and maintenance of AMF. Camprubi and Calvet (1996) isolated AMF from citrus nurseries and orchards in the major citrus-growing areas of eastern Spain. The most common AMF found in citrus soils belonged to *Glomus* species. *G. mosseae, G. intraradices* were the most frequently associated AMF with citrus roots. Graham *et al.* (1998) found that mycorrhizae can produce negative crop responses when phosphorus availability is sufficient in agricultural soils as the fungi start parasitising the plant when there is surplus P. Fidelibus *et al.* (2000) analyzed the role of mycorrhizae collected from the different citrus orchards (Mesa and Yuma, AZ, USA, and undisturbed North American Sonoran Desert and Chihuahuan Desert soils) for the growth enhancement of volkamer lemon seedlings planted in continuously moist and periodically dry soil. Plants in the continuously moist soil had greater growth than the plants from the dry soil and there was no significant difference in P content of both plants. Frequent water input did not affect the AMF colonization in the seedlings. Mixed AMF species (Mesa, AZ, USA, and undisturbed North American Sonoran Desert and Chihuahuan Desert soils) significantly improved root and shoot growth of the plant than the single inoculum. Bhosale and Navale, (2006) studied the response of Rangpur Lime to the AMF colonization in the P deficient soil. The plants inoculated with *G.mosseae* and mix of *G.epigaeum, G.mosseae,* and *G.calospora* recorded the maximum height, number of leaves, root development and the mycorrhizal dependency of the plants and better nutrient uptake. Various glasshouse and field experiments proved that inoculation with AMF increased the growth and ion uptake in citrus plants, enhanced the tolerance to drought, salt stress and also improved the quality of fruit (Wu and Zou, 2009; Wu *et al.*, 2010a, b).

5.5 Custard Apple (*Annona squamosa* L.)

Custard-apple is one of the best tropical fruit, and it is useful for making several food products such as jelly, jam, conserves, sherbet, syrup, tarts and fermented drinks etc (Ojha *et al.*, 2005). It has notable medicinal properties, its leaves and young fruits exhibit insecticidal activity. AMF provide manifold advantage to the host plant besides increasing nutrient and water uptake (Bohra *et al.*, 2007). Game *et al.* (2006) studied the effect of three species of AMF

(*G. epigaeum, G. mosseae* and *Gigaspora calospora*) on the custard apple seedlings under pot culture condition. The seedlings which received the mixed inoculum showed maximum nitrogen and phosphorous uptake over the control plants. Ojha *et al.* (2008) evaluated the role of *G. fasiculatum* in growth enhancement on custard apple seedlings. Significantly enhanced plant height, fresh and dry weight of the roots and shoots were observed in *G. fasciculatum* treated plants compared to the respective controls. They attributed the quantum of herbage in AMF treated plants to enhanced uptake of essential mineral nutrients. Game *et al.*, 2009 reported the effect of AMF (*G. epigaeum, G. mosseae* and *Gigaspora calospora*) in the form of single inoculation/mixed form for the growth improvement of custard apple seedlings. Mixed form of *G. epigaeum* and *G. mosseae* had showed enhanced shoot and root growth, total bio-mass after 180 days of inoculation. The dependency of plants to AMF was observed with *G. epigaeum* and *G. mosseae* inoculation.

5.6 Date Palm (*Phoenix dactylifera* L.)

Date palm is a crop cultivated in the hottest region of the world. It is mainly cultivated on the soil with less nutrient and more water evaporation, And hence is highly dependent on the more input of chemical fertilizers and continuous irrigation. Date palm cultivation with mycorrhizal application will help to reduce the cultivation cost. Abo- Rekab *et al.* (2010) have studied the role of AMF colonization on date palm seedlings planted on (after acclimatization period) soil fertilized with full strength fertilizer (NPK) at three different concentration for two weeks Enhanced growth was noticed in parameters like plant height (cm), number of leaves/ plantlet, length of root (cm), and number of roots per plantlet 2.5 gl^{-1} as compared to control treatment. Aqqua *et al.* (2010) studied effect of core inoculation of saprophytic fungi on AMF colonization in the date palm seedlings for the growth enhancement under green house condition. The plants which received double inoculation of *G.mosseae* and *Trametes versicolor* recorded increased nitrate reductase (NR) and glutamine synthetase (GS) activities than the control plants.

5.7 Grape (*Vitis vinifera* L.)

Grape is one of the important commercially grown fruit crops in the country. The response of grape vine to AMF colonization has been well documented by many researchers. Possingham and Obbink (1971) reported the natural occurrence of AMF in grape vine roots and analyzed the AMF role in the grape vine growth improvement. The seedlings with mycorrhizal colonization improved the plant total dry mass, length of shoot and root. The increased phosphorus and sulphur were observed in the live mycorrhizal inoculated plant leaves. Menge *et al.* (1983) discussed the application of AMF in fumigated soil for the growth improvement of grape vine and found that mycorrhizal inoculation induced the growth improvement in grape vine seedlings planted in soil fumigated with methyl bromide than the control plants without any mycorrhizal inoculation. Karagiannidis *et al.* (1995) studied the influence of three AMF *G. fasciculatus, G. mosseae* and *G. macrocarpus* on the shoot and root dry weight, shoot nutrient concentration of P, K, Ca, Mg, Zn, Mn and Fe as well as the total P uptake in three grapevine rootstocks (41 B, 110 R, 5 BB) and cv. Razaki table grape vines (syn. Dattier de Beyrouth). *G. macrocarpus* inoculated vines showed an increased shoot and root dry weight. Mycorrhizal colonization increased the shoot P concentration but did not affect the levels of K, Ca, Mg and Zn and low concentration of Mn and Fe (M/NM down to 0.2). Grapevine rootstocks (*Vitis berlandieri* x *Vitis riparia*, cv. SO_4) grown in pots containing sterilized soil with low P level was evaluated by inoculation with AMF (*G.mosseae*) by Petgen *et al.* (1998). After 6 weeks of growth, mycorrhizal colonization of roots was highest in the inoculated soil with increasing distance from the inoculum band. When the inoculum was placed in the top soil, the shoot dry weight and the leaf blade Zn and P concentrations significantly increased in mycorrhizal plants as compared to nonmycorrhizal plants. When the inocula was placed in 36-45 cm soil depth, leaf blade Zn and Cu concentrations increased in mycorrhizal plants, but shoot dry weight was not affected. They concluded that a locally restricted mycorrhizal colonization of the root system was sufficient to increase growth and nutrient uptake of grape rootstocks. Motosugi *et al.* (2002) compared effect of *Gigaspora margarita* on growth and leaf mineral content of tetraploid and diploid grape root stocks. Shoot and root growth in tetraploid AMF inoculated grape was significantly higher than inoculated diploid grape rootstocks. AMF inoculation in the grape vine nursery beds to study induction of the root development before transplant to the field was tested by Aguin *et al.* (2004).The grape vine plant cutting inoculated with *G.aggregatum* changed root morphology and increased the branching of first-order lateral roots. When rooted cuttings were transplanted to pots with soil sufficient in P, indigenous AMF induced a significant growth enhancement in the inoculated rootstocks. Hare Krishna *et al.* (2006) studied the effect of AMF inoculation in grape vine seedlings at the time of hardening, either in single or in mixed form. The plantlets inoculated with *A.laevis* spore alone and in mixed form showed increased shoot height, root length, shoot and root dry weights, leaf number, leaf area and physiological parameters such as relative water content and photosynthetic rate over control. Paul Schreiner (2007) evaluated the role of AMF for the growth improvement and nutrient uptake in grape vine seedlings (cv. Pinot noir) planted in two types of soil (Chehalis soil and Jory series soil) containing different levels of P. The plants inoculated with AMF mix (*G. mosseae, G. intraradices,* and *S. calospora*) showed high AMF dependency for growth and P uptake in jory soil and it was not dependent on AMF when it was grown in Chehalis soil. Cheng and Baumgartner (2004) studied cover crops that are planted in between vineyard rows to reduce soil erosion, increase soil fertility, and improve soil structure. Roots of both grapevines and cover crops form mutualistic symbioses with AMF, and interconnected by AM hyphae. Nutrient transfers from cover crops to grapevines through AMF links were studied. Evidence of AMF-mediated N-15 transfer from cover crops to grapevines 5 and 10 days after labeling was significantly higher from grass to the grapevine than from legume to the grapevine. Anguine *et al.* (2004) studied the effects of inoculation of three grapevine rootstocks on root morphology and growth. Results indicated that inoculation with the AMF *G. aggregatum* of grapevine cuttings changed root morphology and increased branching of lateral roots and caused significant growth enhancement in two of the inoculated rootstocks. The influence of three commercial AMF strains (*G. intraradious, G.mosseae, G. fasciculatum* and a mixture of them) on growth and biochemical status of four grapevine varieties (Shahroodi, Asgari, Keshmeshi and Khalili) under greenhouse conditions were studied by Efthekari *et al.* (2010). Most growth related parameters (vine length, shoot length and leaf area) were enhanced following mycorrhization but root length and number of leaves were not significantly affected by any fungal intervention. Treated plants typically showed changes in their biochemical status. The total chlorophyll content (especially "b") and total root and shoot phenols were raised in treated plants. The effect of four AMF (*G. etunicatum, G. caledonium, G.clarum, G.mosseae* and mixed inoculum) on the nutritional status of four grapevine rootstocks (420 A, 41 B, 1103 P, and 'Rupestris du Lot') was studied by Caglar and Bayram (2006). Ten months after inoculation, grapevine rootstocks were colonized by the AMF (*G. etunicatum* and *G. clarum*) up to 47.0 to 64.1%. The leaf areas of 41 B, 420 A, and 1103 P rootstocks increased significantly. The AMF increased leaf P, but not N and K concentrations. There was a two to four fold increase in leaf total sucrose concentrations of the grapevine rootstocks with certain inocula compared with the control. Almaliotis (2008) studied mycorrhizal colonization, soil fertility and P nutrition in 53 vineyards of cv. Victoria. The roots of all samples were colonized by AMF (*Glomus*). There was a negative correlation between soil P and the number of mycorrhiza spores/100 g of soil, root colonization arbuscules/cm of root and vesicles/cm of root, but positively correlated with leaf P. Also, the number of spores/100g of soil was negatively correlated with soil K, Mg and Cu concentration, soil electrical conductivity and organic matter content, but positively correlated with soil Ca concentration. Furthermore, root colonization (%) was negatively correlated with soil K and Cu concentration as well as with soil electrical conductivity. Ozdemir *et al.*, 2010 evaluated the effect of AMF (*G.mosseae* and *G.intraradices*) colonization on the grape vine cuttings (Cv. 5BB, 1613 C, 41 B and Early Cardinal) at the time of transplantation into the soil. The dormant cuttings inoculated with AMF at 1000 spores / plant at 50mm depth showed that *G.mosseae* inoculation improved shoot growth and the cuttings inoculated with *G.intraradices* showed enhancement in root growth, leaf P and Zn concentrations. Shedalcek *et al.* (2013) tested the effect of AMF colonization for the enhancement of micro and macro element content in the grape vine (cv. Pinot Noir and Blaufränkisch) leaves. After the two year experiment the plant inoculated with AMF showed the significant increase in Ca, N, K and Mn level in the mycorrhizal treated plants than the control. Taylor *et al.* (2013) used high throughput sequencing of the large subunit rDNA to analyze the diversity of AMF growing in a vineyard and found over 40 different taxa, but these communities differed based on host plant identity. These differences were apparent even after accounting for differences in soil chemical properties and differences in host plant diversity between vinerows and interrows, indicating that *Vitis* preferentially interacts with a subset of the viticultural fungal community

5.8 Guava (*Psidium guajava* L.)

Guava is economically important fruit crop which has the ability to grow in both tropical and subtropical climates and it is well known for their vitamin C richness and other anti-oxidant properties. Cultivation of guava with the application of AMF will enhance the plant physical properties like height, leaves per plant, fruit quality, and nutrient content of the leaf. Kumaran and Azizah (1995) tested the growth promotion activity of AMF on guava seedling with biological soil conditioner (BSC). The non-mycorrhizal plants which received more BSC showed increased growth of shoot and root and plant dry matter, but the mycorrhizal plant showed the more dry matter yield along with plant growth than the non-mycorrhizal plants. The more mycorrhizal dependency was observed at lowest level of BSC. Estrada-Luna *et al.* (2000) studied the role of AMF inoculation in the micropropagated guava seedlings to avoid the transplantation damage and extra nutrient absorption from soil.

The plants inoculated with AMF mix survived after transplantation and they showed significant shoot/root growth, increased leaf production (for the enhanced photosynthesis) and increased mineral levels of P, Mg, Cu, and Mo in the shoot and leaf than the non-mycorrhizal plants. Roots of inoculated guava plantlets were heavily colonized with arbuscules, vesicles and endospores. Guava plantlets were highly mycotrophic with a mycorrhizal dependency index of 103%. Shrivastava *et al.* (2001) studied the inoculation effect of AMF (*G.mosseae)* on guava cv. Sardar against *Fusarium* infection under field condition. Compared with control and other treatment, the grown up plants (10 year old) inoculated with 5.0 kg of inoculum had showed significant growth difference and resistance against the infection. The effect of AMF colonization on balady guava trees grown in newly reclaimed sandy calcareous soil was studied by Ibrahim *et al.* (2010). The trees inoculated with AMF (*G.mosseae, G.fasiculatum, G, aggregatum*) showed significant increase in plant height, number of leaves per branch, leaf nitrogen, phosphorus, potassium, calcium, and magnesium contents than the non-mycorrhizal plants.

The AMF inoculation improved root formation in micropropagated plants at the time of acclimatization on soil. Campos *et al.* (2013) tested the AMF inoculation for the growth enhancement and protection against root nematode (*Meloidogyne enterolobii*) infection in guava seedlings. The seedlings inoculated with AMF (*Gigaspora albida, G. etunicatum* or *A.longula*) showed increased plant growth, total dry mass and number of fruits. The interaction of nematode with plant roots gradually decreased due to AMF colonization. El-Shamma *et al.* (2013) studied Balady Guava trees grown in calcareous soil in Egypt. Application of slow release fertilizers along with AMF improved AMF infection percentage as well as all the growth parameters tested. Panneerselvam *et al.* (2013) studied the growth promotional effect of *Glomus mosseae* spores on guava seedlings. Compared with uninoculated control treatment the seedlings inoculated with *G.mosseae* showed significant increase in seedling height, stem girth, plant total dry mass and total leaf area.

5.9 Jackfruit (*Artocarpus heterophyllus* L.)

Jackfruit is an ancient fruit crop cultivated in south side of the Asia, and familiar for its texture and taste. The flesh of fruit is rich in antioxidant and vitamin C. The sustainability of this plant is naturally so strong and the chances for the AMF application towards the growth improvement in jackfruit are limited. But the micro propagated jack fruit seedlings do not have the ability to adapt in outer environment because of the poor root growth. Sivaprasad *et al.* (1995) studied the effect of five different AMF (G. *fasiculatum*, G. *etunicatum*, *G.constrictum*, *G.mosseae*, A. *morroweae*) inoculation in the micropropagated jackfruit plants at the time of transplantation into the soil. Out of five isolates the plants inoculated with *G.fasiculatum* showed 88% of mycorrhizal colonization after 150th day of inoculation over control plants which showed 31.3% of mycorrhizal colonization. Plants inoculated with *G.mosseae* showed growth enhancement with increased plant height (5.85cm), and fresh shoot weight (4.38g). The AMF colonization increased phenol content in root tissues over the control plants.

5.10 Litchi (*Litchi chinensis* Sonn.)

Janos *et al.* (2010) found that inoculation of litchi (*Litchi chinensis* Sonn.) with indigenous South Florida AMF improved leaflet expansion after 120 days of inoculation, and subsequently enhanced shoot height and leaf production but did not affect stem diameter, net CO_2 assimilation, or survival. At harvest mycorrhizal plants had 39% higher shoot dry weight than control plants. Leaflets of inoculated plants had higher concentrations of P, K, Cu, and Zn, and lower concentrations of Ca, Mg, and Mn than those of control plants. Litchi growth was enhanced by Mycorrhiza even though phosphorus was not limiting for growth. Rawat *et al.*, 2013 studied the effectiveness of efficient AMF (*Gigaspora albida*, G. *intraradices* and A. *scrobiculata*) for the growth promotion in transplanted air-layered litchi seedling to the pot. The effectiveness of the AMF was studied in terms of infective propagules (IP). Out of different concentration of IP (100-800) 800IP recorded the maximum plant growth parameters like number of lateral roots (21.00), length of lateral roots (14.00 cm),number of leaflets (28.12), thickness of the stem (3.58cm), plants shoot and root length and total nutrient uptake.

5.11 Mango (*Mangifera indica* L.)

The king of fruits, Mango is one of the important fruit crops of India. India is the highest producer of mango in the world. Under the total area of fruit crops farming, mango occupies 22% cultivation in 2.3 million hectares with 48% of the world's output. Ever since research work began on studying the AMF colonization of mango and its response on crop inoculation, the results are encouraging enough. A pot culture experiment carried out by Xiutang (1990) found enhanced mineral and phosphorous uptake in mango after the *Glomus* colonization. Reddy (1991) tested AMF inoculation along with its helper bacteria *Azotobacter Chrococcum* on mango plant growth found that it showed the positive effect on the crop and concluded that effective growth regulators like gibberllic acid secreted by the AMF have induced early seed germination, improved stem girth and leaf area in inoculated mango plants. Reddy and Bagyaraj, (1994) studied the AMF colonization of mango root in unsterilized soil and reported improved plant growth and accumulation of phosphorous in shoot tip of "Nekkare" mango. Combination of *Glomus mosseae* and *Glomus fasiculatum* improved the nutrient content in the root stocks of Totapuri followed by the enhanced plant growth. Kamble *et al.* (2009) subjected a local mango variety in pot culture condition to evaluate the combinational effect of three AMF, *G. epigaeum*, GM and *Gigaspora calospora* (GC) on plant growth and root colonization and found it highly beneficial in terms of growth improvement.

Sukhada (2012a) studied the AMF spore load in the rhizosphere of three year old mango rootstocks and root colonization at 15 cm, and 30 cm depths. Mycorrhizal spores were highest in Totapuri root stock followed by Bappakai, Olour and Peach and Vellakulamban at 15 cm depth. Spores belonged to the genera *Glomus* and *Acaulospora* and few other genera. All the rootstock seedlings responded to mycorrhizal inoculations, the response being higher in Vellakullamban and Totapuri rootstocks. Mycorrhizal rootstocks showed varied intensity of improvement in plant height, growth and nutrient content compared to non-mycorrhizal in pot culture (**Plate 3.2.a.b**). Sukhada (2012a) reported the results of an 8-year field study on mycorrhizal mango rootstock cv. Totapuri grafted with scions of hybrids Arka Aruna and Arka Puneeth. Under field conditions, rootstock cv Totapuri inoculated with 250 g of AMF *G.mosseae* or *G.fasciculatum* inoculum containing around 50 spores g-1 soil and scions of mango hybrids Arka Aruna and Arka Puneeth grafted on them produced shoots earlier compared to non-mycorrhizal plants. Within two years of yearly application of AMF clear differences in growth performance of mycorrhizal and of non-mycorrhizal plants were observed (**Plate 3.2.a**). Plant growth studied in terms of number of branches, available soil P, leaf P, Zn and Cu improved significantly in AMF colonized plants compared to un-inoculated plants at the end of second year (**Table 3.1**). This trend continued in the 8th year of sampling. The root acid and alkaline phosphatase activity was higher in six month old Arka Puneeth grafted on AMF colonized Totapuri rootstock (**Table 3.2**).

5.12 Papaya (*Carica papaya* L.)

Papaya is a tropical fruit, well noticed for its high nutrient content and other medicinal value. The nutrient management in papaya by applying AMF is well documented by many researchers. Papaya crop responds excellently to AMF inoculation. Papaya cv "Coorg Honey Dew" inoculated plants with AMF inoculums *G.mosseae* or *G.fasciculatum* showed remarkable increase in plant height, number of leaves per seedlings and enhanced plant dry matter than control uninoculated under pot culture conditions (Sukhada, 1989). These studies were repeated in field conditions under different p levels (50, 75 and 100% of the recommended dose) (Sukhada, 2012b). Around 50 g of AMF inoculation was mixed with 500 g of soil (sand:soil:fym) in the nursery raised in poly bags. When the seedlings were colonized upto 40% they were transferred to field in pits which were filled with a 1:2 (v/v) mix of farmyard manure:soil and 250 g of *G. mosseae* or *G. fasciculatum* inoculum. Two successive inoculations with the same AMF inocula were done at an interval of 6 months to ensure good colonisation of roots. AMF-inoculated plants performed better than uninoculated control plants at all levels of P applied. *G. mosseae* was more effective at improving plant growth, fruit yield, and P and Zn contents than *G. fasciculatum* at the 75% and 50% P-levels. Cu contents increased at all P-levels in *G. fasciculatum* colonized plants. Total soluble solid contents showed marginal improvements at the 75% P level with both fungi. Carotene contents increased significantly in *G. mosseae* colonized plants at the 50% and 75% P-levels, and in *G. fasciculatum*-colonized plants at the 75% P-level. (**Plate 3.3 and 3.4**).

Uptake of phosphorus by plants through the AMF was studied (Sukhada *et al.* 1995) by using 32P radio-labeled phosphorus. The increased uptake of soluble 32P labeled phosphorous was observed in mycorrhizae inoculated plants than the control plants. Reddy *et al.* (1996) inoculated thirteen different AMF to 50 days old papaya cv. Solo plants individually. After 90 days of inoculation there was improved growth parameters like plant height total leaf area, plant total dry biomass and P and Zn content in the shoot and root observed in *G. mosseae* inoculated plants along with 76 % of AMF colonization. Papaya plants inoculated with AMF in alkaline soil performed better, but were limited in the acidic soil cultivation (Clark, 1997). Martins, (2000) found that when papaya plants were colonized by AMF in infertile soil, the plant growth and soil improvement were observed in AMF colonized papaya plants. Effective AMF colonization in papaya could promote the plant growth by reducing input of P fertilizers till the fruiting period (Mamatha *et al.*, 2002). Nagarajappa *et al.* (2003) reported that two month- old papaya seedlings inoculated prior to transplant with three species of AMF (*G. fasciculatum, Sclerocystis. dussii, A. leavis*) responded excellently to drought by withdrawing water. Significantly higher chlorophyll 'b' and total chlorophyll content was noticed in plants inoculated with *G. fasciculatum* and *S. dussii* before and 10 days after drought imposition. Higher relative water content of leaf was observed in *G. fasciculatum* inoculated plants. Significantly higher soil moisture percent was also noticed in *G. fasciculatum* inoculated plants before and after drought imposition. Significantly lowest proline was also observed in *G. fasciculatum* and *A. leavis* inoculated plants after 10 and 20 days respectively. Trindade *et al.* (2006) studied population of AMF in papaya roots in 47 papaya plantations in Espirito Santo and Bahia state of Brazil and found AMF colonization, ranging from 6% to 83% and the AMF Colonization rates were most influenced by available soil P, pH and clay content. AMF colonization of nursery seedlings was very low in most samples. Field spore numbers varied from 34 to 444/30g of soil. All Glomerales families were represented and 24 fungal species were identified. *G. etunicatum, P. occultum, A. scrobiculata* and *Gigaspora* sp. were the most common species. Khade and Rodrigues (2009) studied the effect of selective AMF on the papaya cv. Surya seedlings. Out of four treatments, the plants inoculated with *G. mosseae* showed most significant root colonization and enriched plant growth than the control and other treatments. Vazquez -Hernández *et al.* (2011) studied the effect of inoculating AMF, *G. mosseae* and *Entrophospora colombiana* (EC) on papaya. The results showed that both mycorrhizae species increased the number of fruits and yield in papaya plants by 41.9 and 105.2% for

GM and 22.1 and 44.1% for EC, respectively, with respect to control plants. *G. mosseae* significantly increased plant height, sugar content, firmness, color, and ripening process of mycorrhized plant fruitsWeight loss of mycorrhized plant fruits was considerably less than that of the control. Inoculation of papaya with AMF is recommended, particularly with *G. mosseae* since it increases yield, and fruit weight (45.1%), furthermore, it reduced fruit weight loss during ripening.

5.13 Pomegranate *(Punica grantum* L.)

Pomegranate is an important fruit crop of subtropical and tropical regions of the world. It is well known for its nutritive value and is rich in vitamins such as folic acid, vitamin C and numerous antioxidants (Gil *et al.*, 2000). AMF application helps plants to sustain the adverse condition under field conditions (Lee and Wetzstein, 1988; Schubert *et al.*, 1990; Louro *et al.*, 1999). Rupnavar and Navale (2000) have studied the effect of AMF (*Glomus epigaeum ; Glomus versiformea; Glomus mosseae; Glomus calospora*) inoculation on growth of pomegranate in pot culture. *G.mosseae* and mixture of *G.mosseae+G.epigaeum+G.calospora* inoculation recorded the maximum plant height, root length, plant total dry mass and percentage of colonization on the roots than the control plants. Sukhada *et al.* (2000) studied the effect of mycorrhizal inoculation on pomegranate cultivar "Ganesh' under pot culture and field conditions and found the treatment beneficial to plant growth (**Plate 3.5**). When the inoculation was made along with *Azospirillum* and phosphate soluble microorganisms the benefits were augmented. Nursery and field experiments were carried out to assess the effectiveness of selected N_2-fixing bacteria and AMF alone or in combination, on the growth and biomass production of *P. granatum* under Thar desert conditions by Aseri *et al.* (2008).   In both experiments, the combined treatment of *A. chroococcum* and *G. mosseae* was found to be the most effective. Besides enhancing the rhizosphere microbial activity and concentration of various metabolites and nutrients, these bioinoculants helped in better establishment of pomegranate plants under field conditions. A significant improvement in the plant height, plant canopy, pruned material and fruit yield was evident in 5-year-old pomegranate plants in field conditions.

5.14 Passion Fruit *(Passiflora edulis* **Sims.** f. *flavicarpa* **Deg.**)

Pomegranate is an important fruit crop of subtropical and tropical regions of the world. It is well known for its nutritive value and is rich in vitamins such as folic acid, vitamin C and numerous antioxidants (Gil *et al.*, 2000). AMF application helps plants to sustain the adverse condition under field conditions (Lee and Wetzstein, 1988; Schubert *et al.*, 1990; Louro *et al.*, 1999). Rupnavar and Navale (2000) have studied the effect of AMF (*Glomus epigaeum ; Glomus versiformea; Glomus mosseae; Glomus calospora*) inoculation on growth of pomegranate in pot culture. *G.mosseae* and mixture of *G.mosseae G.epigaeum+G.calospora* inoculation recorded the maximum plant height, root length, plant total dry mass and percentage of colonization on the roots than the control plants. Sukhada *et al.* (2000) studied the effect of mycorrhizal inoculation on pomegranate cultivar "Ganesh' under pot culture and field conditions and found the treatment beneficial to plant growth (**Plate 3.5**). When the inoculation was made along with *Azospirillum* and phosphate microorganisms the benefits were augmented. Nursery and field experiments were carried out to assess the effectiveness of selected N_2 fixing bacteria and AMF alone or in combination, on the growth and biomass production of *P. granatum* under Thar desert conditions by Aseri *et al.* (2008). In both experiments, the combined treatment of *A. chroococcum* and *G.mosseae* was found to be the most effective. Besides enhancing the rhizosphere microbial activity and concentration of various metabolites and nutrients, these bioinoculants helped in better establishment of pomegranate plants under field conditions. A significant improvement in the plant height, plant canopy, pruned material and fruit yield was evident in 5-year-old pomegranate plants in field conditions.

5.15 Pineapple (*Ananas comosus* L.)

Pineapple is a tropical plant and the most economically important plant in the Bromeliaceous family. It contains high level of manganese, vitamin C and many proteolytic enzymes. The poor performance of micropropagated plants on field transplanting is a great concern. Lovato *et al.*, 1992 have studied the inoculation effect of AMF (*Glomus* sp., *G. intraradices*) on 3 varieties of micro-propagated pineapple seedlings (Queen Tahiti, Smooth Cayenne and Spanish) transplanted in acid and alkaline soil. The *Glomus* sp showed enhanced growth promotion by increasing total leaf area, shoot fresh and dry mass of the inoculated seedlings in acid soil. Guillemen *et al.*, 1992 compared the efficiency of AMF (*G. clarum* (LPA16), *S. pellucida* (LPA20), *Glomus* sp. (LPA21), *Glomus* sp. (LPA22) and *Glomus* sp. (LPA25)) infection in 3 varieties of pineapple seedlings which planted in acid soil. Out of all the isolates *Glomus* sp. (LPA21) showed significant growth enhancement on 2 varieties (Queen Tahiti, Smooth Cayenne) like total leaf area and total plant dry matter followed by *Glomus* sp. (LPA 25) which showed significant growth promotion activity on Spanish variety. The mycorrhizal dependency was highly influenced by the type of inoculants. El-Fiki *et al.*, 2007b have studied the effective AMF spore concentration for the growth enhancement of growth in pineapple seedlings. For pre-inoculation the shoots were dipped in different concentration (0.8-6.8 x 106/ml) of spore suspension. The lowest concentration 0.85 x 106/ml spores showed plant growth promotion over control plants. Improved spore colonization was observed at lowest concentration in host plant shoots than other concentration.

5.16 Persimmion (*Diospyros kaki* Thunb.)

Incesu *et al.* (2014) tested the effects of five AMF species (*Glomus mosseae, G. clarium, G. etunicatum, G. caledonium and G. intraradices*) on plant growth, chlorophyll concentration and chlorophyll fluorescence (*Fv'/Fm'*) in Persimmion, *D. kaki* and *D. virginiana* rootstocks under greenhouse conditions. Mycorrhizal inoculations increased shoot and root dry weights compared with the non-inoculated plants. Plants inoculated with *G. etunicatum* showed the highest total plant dry weight. Highest leaf chlorophyll concentration was measured in plant inoculated with *G. caledonium*. Results of chlorophyll fluorescence were similar with all the AMF inoculations, however significantly differed from non-inoculated plants. The results demonstrated the benefit potential of the mycorrhizal inoculations in persimmon production.

5.17 Sapota (*Manilkara zapota* L.)

Sapota fruits are well known for their delicious taste, vitamin content (A, B1, B2 and C) and minerals (phosphorus, calcium, potash, iron, magnesium and sodium. The cultivation of sapota seedlings has major problems mainly slow growth and poor root formation. AMF inoculation of sapota enhance the seedlings growth under nursery condition. Sreeramulu *et al.* (1998) studied the effect of 11 AMF on *M. hexandra* in pot culture assay. Out of 11 isolates, they found *G.intraradices* improved plant height and total dry weight after the 12 months of inoculation. Khanam (2007) reported the three year observation of occurrence of AMF spores in the rhizosphere of 19 fruit crops. Results of the study revealed the gradual increase of AMF spore load (697.3/100g of soil) in the sapota rhizosphere in every year of observation period. Saritha (2014,under publication) found a benefit of 15-20% in plant growth and nutrient uptake in sapota cv Cricket Ball inoculated with AMF *G.mosseae* **(Plate 3.6).**

5.18 Strawberry (*Fragaria×ananassa* Duchesne)

Strawberry is a widely grown hybrid species of the genus *Fragaria*. It is cultivated worldwide for its characteristic aroma, bright red color, juicy texture, and sweetness. AMF colonization in strawberry roots is economically viable to produce healthy plants with low fertilizers input or no fertilizers. Williams *et al.*, (1992) tested the AMF (*G.intraradices, G.geosporum* and undescribed *Glomus* sp.) colonization effect in strawberry plants (cv. Senga Sengana) planted in peat-sand-vermiculture substrate by supplying commercial fertilizers (Osmocote/rock phosphate) at recommended level. The plants inoculated with both fertilizers showed the significant growth promotion activity like increased shoot, root growth and total plant dry mass by means of mycorrhiza colonization. All 3 fungal species significantly increased the stolon number per plant when compared with the non-mycorrhizal controls. Sebastia *et al.* (1999) have tested the difference in relative water content between the mycorrhizal and non-mycorrhizal plants under the axenic well-watered conditions. The plants inoculated with *G. intraradices* showed increased relative water content with enhanced nutrient uptake in the micropropagated strawberry plants than the non-mycorrhizal propagated plants. Taylor and Harrier (2001) compared the effect of 9 different AMF colonization in the strawberry roots for the growth enhancement. Out of nine isolates the plants colonized by *G. clarum* and *G. rosea*, had significantly increased concentrations of 7 and 9 mineral nutrients respectively within the shoot tissue. Douds *et al.* (2008) tested the effect of AMF inoculation on the yield of strawberry (cv.Chandler) fruits on-farm. At the time of harvesting the plants colonized by the AMF showed 17% increase in fruit yield over un-inoculated control plants. Matsubara *et al.* (2009) examined the effect of AMF inoculation for the enhancement of plant growth and free amino acid level in the strawberry plant (cv Nohime) with and without addition of P. The results showed the plants inoculated with *G. mosseae* increased the plant shoot and root dry matter in the (-P) soil than the plants colonized by the *G.aggregatum*. The levels of essential amino acids (serine, glutamic acid, glycine, leucine, and alanine) also increased in the both mycorrhizal plants but the P supplement didn't play any major role in the amino acid enhancement by the colonization of mycorrhiza. Morales *et al.*, (2010) reported the effect of AMF colonization on strawberry fruit quality with different level of nitrogen application. The mycorrhizal plants showed significant increases of major phenolic compounds and mineral uptake than the non-mycorrhizal plants at concentration of 6 mmol L^{-1} N.

5.19 Watermelon (*Citrullus lanatus* (Thunb.)

Watermelon is an economically important crop which belongs to the cucurbitaceae family mainly known for its water content (91 %) and richness in vitamin C. The rooting of seedling under the field condition is an important task for the better yield. Inoculation of AMF during watermelon cultivation facilitated the enhanced root colonization for the improved nutrient uptake. Westphal *et al.* (2008) studied the effective of AMF colonization on the roots of transplanted watermelon seedlings for the field stability, early fruit yield and disease suppression under some environmental stress condition. The seedlings inoculated with AMF spores showed early fruiting with early establishment, but it didn't show any disease suppression. Sheng *et al.* (2012) have tested the importance of AMF colonization (*G.versifomea*) in maintaining the soil fertility and continue the watermelon monoculture. Chen *et al.* (2013) have tested the inoculation effect of AMF *G.versiformea* on grafted and non-grafted watermelon seedlings in a continuous cropping soil. The colonization of *G.versiformea* in non grafted and grafted seedlings could activate the defensive enzyme activities which enabled the seedling roots to produce rapid response to harsh conditions, and thus, improve the capability of watermelon seedling against continuous cropping difficulty.

6.0 Conclusion

Inoculation of fruit crops with AMF is a very promising strategy to improve plant growth, nutrient application and yield. During crop cultivation the inoculation needs to be done at the seedling stage and thereafter while transplanting. Careful selection of compatible host/fungus/substrate combination is critical for an early establishment of the AMF status, after sowing or at transplanting. This is a key factor to improve plant performance in horticultural practices. AMF application is more appropriate mainly for crops which are produced in nursery and later transplanted to field.

A healthy level of nutrients will be maintained by regular application of AMF inocula to the soil and the input of chemical fertilizers can be reduced substantially to make cultivation of fruit crops economical. As the soil health also improves by AMF application, the quality of the fruits will be maintained. AMF inocula can be produced on- farm without much hastle and used in fruit nurseries followed by field application. It is necessary to repeat the application once in 6 months or yearly depending upon the soil conditions. No-till management practices should be followed to allow AMF to grow during the cropping season. Tillage disrupts the hyphal network that produces glomalin. Disruption of the hyphal network also decreases the number of spores and hyphae. Use of cover crops is advisable to maintain living roots for the fungi to colonize. Adequate phosphorus levels should be maintained for crops, and not over-application of P , because high levels depress the activity of these fungi.

7.0 Future strategies

There is an urgent need to identify the dominant mycorrhizal species in different areas and multiply them for inoculation. Response of different cultivars to different AMF species needs to be tested and the best one selected for different cropping conditions. There is a need to improve and widely apply analytical methods to evaluate characteristics such as, relative field mycorrhizal dependency, soil mycorrhizal infectivity, and mycorrhizal reseptivity of soil. Little is known about the effects of environmental changes on AMF abundance, activity and the impact of these changes on the ecosystem services.

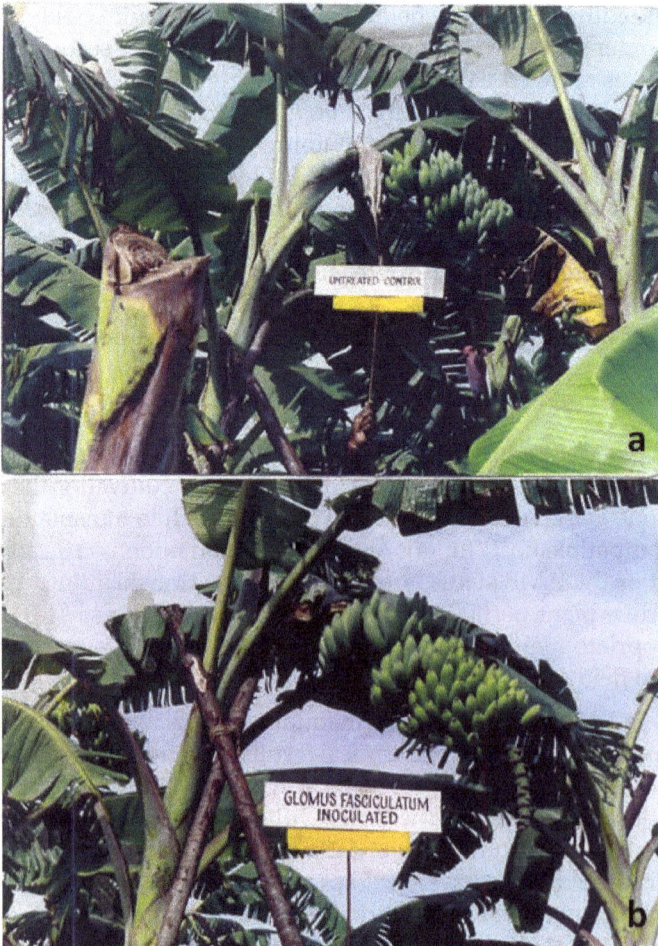

Plate: 3.1 Banana cv Elakkibale syn Neypoovan (AB) inoculated with *Glomus fasciculatum* under field conditions (b). Uninoculate plants (a) (Sukhada, 2004)

Table 3.1: Effect of AM inoculation on different plant attributes in mango cultivars grafted on mycorrhizal root stocks under field conditions two years after establishment. (Reproduced from Sukhada, 2012a)

Treatment/ Cultivar	Time taken for shoot appearance after grafting (days)	Colonisation of fungus %	Number of branches	Available soil P (ppm)	Leaf P (ppm)	Leaf Zn (ppm)	Cu (ppm)	Yield per plant Kg	TSS Brix
Arka Aruna									
G. mosseae +Half P	16	65	455	162	412	32	48.5	10	16
G. fasciculatum +Hal P	18	58	557	145	432	27	52	9	16
Control 1+Half P	28	12	560	111	403	31	35	10	15
Control 2+Full P	22	14	672	171	515	28	80.5	13	16
Arka Puneeth									
G. mosseae +Half P	14	52	387	106	333	31	66.5	16	14
G. fasciculatum Half P	15	48	561	104	540	26	31.3	18	18
Control 1+Half P	26	15	530	100	301	29	35.0	13	11
Control 2+Full P	20	18	821	142	443	31	46.8	16	14
LSD 5%	8.25	15.4	23.8	10.3	67.8	4.6	8.9	4.56	5.67

Control 1, plants that have received 50% of P fertilizer (112 g per plant) and full dose of N andK.

Control 2, plants that have received 100% of P fertilizer (224 g per plant) and full dose of N and K.

AMF inoculated plants have received 50% P (112 g per plant) and full dose of N and K.

Table 3.2: Leaf phosphorus, zinc, copper and yield in mango hybrids inoculated with AM fungi in the 8th year.(Reproduced form Sukhada, 2012a)

Treatment	Soil Phos- phorus (ppm)	Leaf P (ppm)	Copper (ppm)	Zinc (ppm)	Yield kg plant⁻¹
Arka Aruna					
G. *mosseae* + half P	425	305	46	30	23
G. *fasciculatum* + half P	486	352	30	29	24
Control 1(half P)	390	305	35	30	16
Control 2(full P)	650	450	52	31	22
Arka Puneeth					
G. *mosseae* + half P	600	350	52	30	14
G. *fasciculatum* + half P	634	390	27	42	16
Control 1(half P)	390	278	38	28	8
Control 2(full P)	680	425	45	29	15
	219.38	156.52	21.89	16.88	7.03

Control 1, plants that have received 50% of P fertilizer (112 g per plant) and full dose of N and K.

Control 2, plants that have received 100% of P fertilizer (224 g per plant) and full dose of N and K.

AM inoculated plants have received 50% P (112 g per plant) and full dose of N and K.

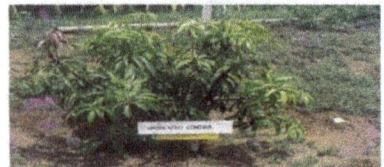

VAM inoculation in mango

Uninoculated mango plants

A   B

Plate: 3.2 Response of Totapuri rootstock grafted with Arka Aruna mango variety inoculated with AMF (*G.mosseae*) under field condition, a- Un-inoculated grafted rootstock, b- *G.mosseae* inoculated rootstock. Two (A) and four years (B) old in field. (Sukhada, 2012a).

Plate: 3.3 Papaya cv Coorg Honey Dew inoculated with AMF *Glomus mosseae* in nursery and there after in the field (b). Uninoculated control (a). (Sukhada, 2012b)

      A. Pot culture   **B.** Control uninoculated   **C.** *G.fasciculatum* inoculated

Plate: 3.4 Growth enhancement in the AMF (*G.fasiculatum and G.mosseae*) inoculated pomegranate cv. Ganesh plants in A. pot condition, a- control (un-inoculated plants), b- Plants inoculated with *G.fasiculatum* and c- Plants inoculated with *G.mosseae*. Field inoculated plants B.& C (.Sukhada *et al.* 2000)

          A   **B**

Plate: 3.5 Arbuscules of AMF *G.mosseae* colonizing roots of Papaya plants in field (A) and Chlamydospores of the fungus attached to the roots (B).

Control   *Glomus.mosseae* inoculated

Plate: 3.6. Sapota plants inoculated with AMF (Source: Saritha PhD work under publication)

References

1. Abbot,L.K. and A.D. Robson.,1991. Factors influencing the occurrence of vesicular-arbuscular mycorrhizas, *Agric. Ecosys. Environ.*, 35, 121-150.

2. Abo- Rekab, Z.A., Darwesh, R.S.S and Hassan, N.,2010. Effect Of Arbuscular Mycorrhizal Fungi, Npk Complete Fertilizers On Growth And Concentration Nutrients Of Acclimatized Date Palm Plantlets. *Mesopotomia J. of Agric.* 38 (1).

3. Abohatem, M., Chakrafi, F., Jaiti, F., Dihazi, A and Baaziz, M.,2011. Arbuscular Mycorrhizal Fungi Limit Incidence of *Fusarium oxysporum* f.sp. *albedinis* on Date Palm Seedlings by Increasing Nutrient Contents, Total Phenols and Peroxidase Activities. *The Open Horticulture Journal.* 4, 10-16.

4. Abiala, M.A., Popoola, O.O., Olawuyi, O.J., Oyelude, J.O., Akanmu, A.O., Killani, A.S., Osonubi, O and Odebode, A.C., 2013. Harnessing the Potentials of Vesicular Arbuscular Mycorrhizal (VAM) Fungi to Plant Growth – A Review. *Int. J. Pure Appl. Sci. Technol.* 14(2), 61-79.

5. Adriana, M., Orivaldo, J.Y.M., Júnior, S., Lima-Filho, J.M., Natoniel, F and Maia, M.L.C.,1999. Effect of arbuscular mycorrhizal fungi on the acclimatization of micropropagated banana plantlets. *Mycorrhiza.* 9,119–123.

6. Aguin, O., Mansilla, J.P., Vilarino, A and Sainz, M.J., 2004. Effects of mycorrhizal inoculation on root morphology and nursery production of three grapevine rootstocks. *Am. J. Enol. Vitic.* 55, 108–111.

7. Aizme-Vega, M.C and Pinochet, J.,1997. Growth response of banana to three mycorrhizal fungi in Pratylenchus goodeyi infested soil. *Nematropica.* 27, 69-76.

8. Ajay Pal and Sonali Pandey.,2014. Role of Glomalin in Improving Soil Fertility, Department of Botany, JECRC University, Jaipur.303905, India. *International Journal of Plant & Soil Science* 3(9): 1112-1129.

9. Alarcon, A., Almaraz, J.J., Ferrera-Cerrato, R., González-Chavez, M.C.A., Lara, H., Manjarrez, M., Quintero, L.R and Santamaria, R.S., 2001. Manual: Tecnologia de hongos micorrizicos en la produccion de especies forestales envivero. Colegio de Postgraduados Pronare-Conafor. México. pp. 98.

10. Almaliotis, D., Karagiannidis, N., Chatzissavvidis, C., Sot ropoulos, T and Bladenopoulou, S.,2008. Mycorrhizal colonization of table grapevines (cv. 'Victoria') and its relationship with certain soil parameters and plant nutrition. *Agrochimica*. 52, 129–136.

11. Anjos, E.C.T., Cavalcante, U.M.T., Santos, V.F.S and Maia, L.C., 2005. Produção de mudas de maracujazeiro-doce micorrizadas em solo desinfestado e adubado com fósforo. *Pesquisa Agropecuária Brasileira*.40, 345-351.

12. Anjos, E.C.T., Cavalcante, U.M.T., Gonçalves, D.M.C., Pedrosa, E.M.R., Santos, V.F and Maia, L.C., 2010. Interactions between an arbuscular mycorrhizal fungus *(Scutellospora heterogama)* and the root-knot nematode (Meloidogyne incognita) on sweet passion fruit *(Passiflora alata)*. *Braz. Arch Biol. Technol*. 53,801–809.

13. Aqqua, K., Ocampo, J.A., Garcıa Romera, I and Qaddoury, A.,2010. Effect of Saprotrophic Fungi on Arbuscular Mycorrhizal Root Colonization and Seedlings Growth in Date Palm under Greenhouse Conditions; *Acta Hort*. 882, 891-898.

14. Aseri, G.K., Jain, N., Panwar, J., Rao, A.V and Megwal, P.R.,2008. Biofertilizers improve plant growth, fruit yield, nutrition, andmetabolismand rhizosphere enzyme activities of Pomegranate *(Punica granatum* L.) in Indian Thar Desert. *Sci. Hortic*. 117, 130-135.

15. Azcon-Aguilar,C., Barcelo, A., Vidal, M.T and De-le Vina, G., 1992. Further studies on the influence of mycorrhizae on growth and development of micropropagated avocado plants. *Agronomie*. 12, 837-840.

16. Barea, J.M and Azcon-aguilar, C., 1983. Mycorrhizas and their significance in nodulating nitrogen-fixing plants. *Advances in Agronomy*. 36, 1-54.

17. Banuelos, J., Trejo, D., Lara, L., Gavito, M and Carreón, Y., 2013. Effects of seven different mycorrhizal inoculum in Persea Americana in sterile and non-sterile soil. *Tropical and Subtropical Agroecosystems*. 16, 423 – 429.

18. Bethlenfalvay, G.I., 1992. Vesicular-arbuscular mycorrhizal fungi in nitrogen-fixing legumes: Problems and prospects, *Methods in Microbiology*. 24, 375-389.

19. Bhosale, D.M and Navale, A.M., 2006. Response to rangpurlime sedlings to VA-mycorhizal inoculation. *International Journal of Plant Sciences*. 2, 202-204.

20. Bolan, N.S., 1991. A critical review on the role of mycorrhizal fungi in the uptake of phosphorus by plants. *Plant and Soil* .134, 89–207.

21. Bohra, A., Mathur, N., Bohra, S., Singh, J and Vyas, A., 2007. Influence of AM fungi on physiological changes in *Terminalia arjuna* L.: An endangered tree of Indian thar desert. *Indian Forester*. 133(11), 1558-1562.

22. Blanke,V., C. Renke, M. Wagner, K. Fuller, M. Held, A.J. Kuhn and F. Bruscot., 2005. Nitrogen supply affects arbuscular mycorrhizal colonization of *Artemisia vulgaris* in a phosphate polluted field sites, *New Phytol.*, 166, 981-992.

23. Bucking,H and Shacker-Hill,Y., 2005.Phosphate uptake, transport and transfer by arbuscular mycorrhizal fungus is increased by carbohydrate availability, *New Phytol.*, 165, 889- 912.

24. Camprubi, A and Calvet, C., 1996. Isolation and screening of mycorrhizal fungi from citrus nurseries and orchards and inoculation studies. *HortScience*. 31,366-369.

25. Cavalcante, V.M.T., Maia, L.C., Nogueira, R.J.M.C and Santos, V.F., 2001. Physiological responses of yellow passion fruit (*Passiflora edulis Sims. f. flavicarpa*) seedlings inoculated with arbuscular mycorrhizal fungi under water stress. *Acta Botanica Brasilica*, Feira de Santana. 15(4), 379-390.

26. Caglar, S and Bayram, A., 2006. Effects of Vesicular-Arbuscular Mycorrhizal (VAM) fungi on the leaf nutritional status of four grapevine rootstocks. *Eur. J. Hort. Sci.* 71,109–113.

27. Castellanos-Morales, V., Villegas, J., Wendelin, S., Vierheilig, H., Ederc, R and Cardenas-Navarro, R., 2010. Root colonisation by the Arbuscular mycorrhizal fungus *Glomus intraradices* alters the quality of strawberry fruits (*Fragaria × ananassa* Duch.) at different nitrogen levels, *J Sci Food Agric*, 90, 1774–1782.

28. Campos, M.A.S., Barbosa., da Silva, F.S., Yano-Melo, A.M., Franklin, de Melo, N., Pedrosa, E.M and Maia, L.C., 2013. Responses of Guava Plants to Inoculation with Arbuscular Mycorrhizal Fungi in Soil Infested with *Meloidogyne enterolobii*. *Plant Pathol. J.* 29(3), 242-248.

29. Cheng,X and Baumgartner, K., 2004. Survey of arbuscular mycorrhizal fungal communities in northern California vineyards and mycorrhizal colonization potential of grapevine nursery stock. *HortScience*. 39, 1702–1706.

30. Chen, K.E., Ji-qing, S., Run-jin, L and Min, L.I., 2013. Effects of arbuscular mycorrhizal fungus on the seedling growth of grafted watermelon and the defensive enzyme activities in the seedling roots. *Chinese Journal of Applied Ecology*. (1), 135-141.

31. Clark, R.B., 1997. Arbuscular mycorrhizal adaptation, spore germination, root colonization, and host plant growth and mineral acquisition at low pH. *Plant and Soil*. 19,15-22.

32. Covacevich,F., H.E.A.Echeverria and L.A.N.Aguirrezabal., 2007.Soil available phosphorus status determines indigeneous mycorrhizal colonization of field and glasshouse-grown spring wheat from Argentina, *Applied Soil Ecology.* 35, 1-9.

33. Darocha, M.R., Decarvalhocorrea,G and Deoliveira,E.,1995. Effect of vesicular-arbuscular mycorrhizal infection and phosphorus fertilization on 'Cleopatra' mandarin root-stock. *Pesquisa Agropecuaria Brasileira.* 30(10),1253-1258.

34. Declerck. S.,Plenchette.C. and Strullu D.G.,1995. Mycorrhizal Dependency Of Banana *(Musa acuminata, Aaa Group)* Cultivar. *Plant and Soil.* 176, 183-187.

35. Declerck, S., Risede, J.M and Delvaux, B., 2002. Greenhouse response of micropropagated bananas inoculated with *in vitro* monoxenically produced arbuscular mycorrhizal fungi, *Sci. Hort.* 93,301–309.

36. Diatta, I.L.D., Kane, A., Agbangba, C.E., Sagna, M., Diouf, D., Aberlenc-Bertossi, F., Duval, Y., Borgel, A and Sane, D., 2014. Inoculation with arbuscular mycorrhizal fungi improves seedlings growth of two sahelian date palm cultivars (*Phoenix dactylifera* L., cv. Nakhla hamra and cv. Tijib) under salinity stresses. *Advances in Bioscience and Biotechnology.* 5, 64-72.

37. Douds, D.D., Nagahashi, G., Shenk, Jand Demchak, K., 2008. Inoculation of strawberries with AM fungi produced on-farm increased yield. *Biol. Agr. Hortic.* 26,209–219.

38. Dueck,T.A., P. Visser, W.H.O. Ernst and H. Schat, 1986.Vesicular-arbuscular mycorrhizae decrease zinc toxicity to grasses growing in zinc polluted soil, *Soil Biology and Biochemistry*,18, 331-333.

39. Dutra, P.V., Abad, M., Almela, V and Agustí, M., 1996. Auxin interaction with the vesicular-arbuscular mycorrhizal fungus *Glomus intraradices* Schenck and Smith improves vegetative growth of two citrus root stocks, *Sci. Hortic.* 66, 77–83.

40. Eftekhari, M., Alizadeh, M., Mashayekhi, K., Asghari, H and Kamkar, B., 2010. Integration of arbuscular mycorrhizal fungi to grape vine (*Vitis vinifera* L.) in nursery stage. *J. Adv. Lab. Res. Biol.* 1 (1), 102–111.

41. El-Fiki, A.I.I., El-Habaa, G., Hafez, M.E and Eid, K.H.E., 2007b. Studies on inocula of vesicular arbuscular mycorrhyzal fungi produced *in-vitro.* II: Physiological studies. Proc. 11th Cong. *Egypt. Phytopathol. Soc.,* Giza, Egypt.

42. El-Shamma, M.S., Zaied, N.S., Hafez, O.M., Saleh, M.A and El-habbaa, G.M., 2013. Efficiency of slow release Fertilizers and/or VA Mycorrhizal Fungi on Performance of Balady Guava Trees Grown in Calcareous Soil. *11th Arab International Conference on Polymer Science and Technology.* 1-20.

43. Entry, J.A.,P.T. Rygiewicz, L.S. Watrud and P.K. Donnelly, 2002.Influence of adverse soil conditions on the formation and function of arbuscular mycorrhizas, *Advances in Environmental Research,* 7, 123-138.

44. Estrada-Luna, A.A., Davies, Jr. F.T. and Egilla, J.N., 2000. Mycorrhizal fungi enhancement of growth and gas exchange of micropropagated guava plantlets (*Psidium guajava* L.) during *ex-vitro* acclimatization and plant establishment. *Mycorrhiza.* 10, 1–8.

45. Eswarappa, H., Sukhada, M., Gowda, K.N and Mohandas, S., 2002. Effect of VAM fungi on banana. *Curr. Res.* 31 (5- 6), 69-70.

46. Fidelibus, M.W., Martin, C.A., Wright, G.C and Stutz, J.C., 2000. Effect of arbuscular mycorrhizal (AM) fungal communities on growth of "Volkamer" lemon in continually moist or periodically dry soil. *Sci. Hortic.* 84, 127–140.

47. Fortuna, P., Citernesi, A.S., Morini, S., Vitagliano, C and Giovannetti, M., 1996. Influence of arbuscular mycorrhizae and phosphate fertilization on shoot apical growth of micropropagated apple and plum rootstocks. *Tree Physiology.* 16,757-763.

48. Game, B.C and Navale, A.M.,2006. Effect of VAM inoculation on nitrogen and phosphorus uptake by custard-apple seedlings. *Internat. J. Agric. Sci.* 2(2), 354-355.

49. Gai,J.P., G. Feng, X.B. Cai, P. Christie and X.L. Li., 2006. A preliminary survey of the arbuscular mycorrhizal status of grassland plants in South Tibet, *Mycorrhiza.* 16, 191-196.

50. Game, B.C., Navale, A.M and Pardeshi, S.R., 2009. Growth response of custard apple seedlings to arbuscular mycorrhiza. *Agricultural Science Digest.* 29(1), 72-74.

51. Gaidashovaa, S., Nsabimana, A., Karamurac, D., Astend, P.V and Declerck, S., 2012. Mycorrhizal colonization of major banana genotypes in six East African environments. Agriculture, *Ecosystems and Environment.* 157, 40–46.

52. Gildon .A and P.B. Tinker., 1983. Interaction of vesicular-arbuscular mycorrhizal infection and heavy metals in plants, *New Phytol.*, 95, 247-261.

53. Gil,M.I.,Tomas-Barberan,F.A and Hess-Pierce,B.,2000. Antioxidant activity of pomegranate juice and its relationship with phenolic composition and processing. *J. Agric. Food Chem.* 48, 4581-4589.

54. Gordon-Weeks, R., Tong, Y., Davies, T.G.E and Leggewie, G., 2003. Restricted spatial expression of a high-affinity phosphate transporter in potato roots. *Journal of Cell Science.* 116, 3135–3144.

55. Graham, J.H and Eissenstat, D.M., 1998. Field evidence for carbon cost of citrus mycorrhizas. *New Phytol.* 140, 103–110.

56. Grant,C., S. Bitman, M. Montreal, C. Plenchette and C. Morel., 2005. Soil and fertilizer phosphorus: Effects on plant supply and mycorrhizal development, *Canadian Journal of Plant Science*,85, 3-14.

57. Guillemin, J.P., Gianinazzi, S and Trouvelot, A., 1992. Screening of arbuscular endomycorrhizal fungi for establishment of micropropagated pineapple plants. *Agronomie.* 12, 831-836.

58. Harley,J.L and S.E. Smith, 1983. Mycorrhizal Symbiosis, *Academic Press,* London.

59. Hayman,D.S. and M. Tavares., 1985. Plant growth responses to vesicular-arbuscular mycorrhiza XV: Influence of soil pH on the symbiotic efficiency of different endophytes, *New Phytol.*100, 367-377.

60. Hawkins,H.J., A. Johansen and E. George., 2000. Uptake and transport of organic and inorganic nitrogen by arbuscular mycorrhizal fungi, *Plant and Soil.* 226, 275-285.

61. Hazarika, B.N., 2003. Acclimatization of tissue-cultured plants. *Curr. Sci.* 85, 12–25.

62. Harrier,L.A and C.A. Watson., 2003.The role of arbuscular mycorrhizal fungi in sustainable cropping systems, *Advances in Agronomy.* 79, 185-225.

63. Hamel,C., 2005. Impact of arbuscular mycorrhiza fungi on N and P cycling in the root zone,*Canadian Journal of Soil Science*, 84, 383-395.

64. Hare Krishna., R.K. Sairam., S.K. Singh., V.B. Patel., R.R. Sharma., Grover, M., Nain, L and Sachdev, A., 2008. Mango explants browning: Effect of ontogenic age, Mycorrhization and pre-treatments. *Scientia Horticulturae.* 118, 132–138.

65. Hernandez-Sebastia, C., Piche, Y and Desjardins, Y., 1999. Water relations of whole strawberry plantlets *in-vitro* inoculated with *Glomus intraradices* in a tripartite culture system. *Plant Science.* 143, 81–91.

66. İncesu,M., Turgut.Yeşiloğlu1, Berken Çimen, Bilge Yilmaz1, Çağdaş Akpınar, İbrahim Ortaş 2014. Effects on growth of persimmon (*Diospyros virginiana*) rootstock of arbuscular 1 mycorrhizal fungi species 2 *Turkish journal of Agri & Forestry* DOI: 10.3906/tar-1405-134

67. Ibrahim, H. I. M., Zaglol, M.M.A and Hammad, A.M.M., 2010. Response of balady guava trees cultivated in sandy calcareous soil to bio fertilization with phosphate dissolving bacteria and / or VAM Fungi. *Journal of American Science.* 6(9),399-404.

68. Jaizme-Vega, M.C and Azcón, R., 1995. Response of some tropical and subtropical cultures to endomycorrhizal fungi. *Mycorrhiza.* 5, 213–7.

69. Jaizme-Vega, M.C., Tenoury, P., Pinochet, J and Jaumot, M., 1997. Interactions between the root-knot nematode *Meloidogyne incognita* and *Glomus mosseae* in banana. *Plant and Soil.* 196, 27–35.

70. Janos, D.P., Schroeder, M.S., Schaffer, B and Crane, J.H., 2001. Inoculation with arbuscular mycorrhizal fungi enhances growth of *Litchi chinensis* Sonn. Trees after propagation by air-layering. *Plant Soil.* 233, 85-94.

71. Jefwa, J., Vanlauwe, B., Coyne, D., van, A.,sten, P., Gaidashova, S., Rurangwa, E., Mwashasha, M and Elsen, A., 2010. Benefits and potential use of Arbuscular Mycorrhizal Fungi (AMF) in banana and plantain *(Musa spp.)* systems in Africa Proc. IC. In: Banana and Plantain in Africa. Eds Dubois et al., 2010. *Acta Hort.* 879, 479-486.

72. Jefwaa, J.M., Okothb, S., Wachira, P., Karanja, N., Kahindi, J., Njuguini, S., Ichami, S., Mungatuc, J., Okotha, P and Huisinga, J., 2012. Impact of land use types and farming practices on occurrence of arbuscular Mycorrhizal fungi (AMF) Taita-Taveta district in Kenya. Agriculture, *Ecosystems and Environment.* 157, 32–39.

73. Johansen, A.,I. Jakobsen and E.S. Jensen., 1992. Hyphal transport of 15N-labelled nitrogen by a vesicular-arbuscular mycorrhizal fungus and its effect on depletion of inorganic soil N, *New Phytol.* 122(1992), 281-288.

74. Karagiannidis, N., Nikolaou, N and Mattheou, A.,1995. Influence of three VA-mycorrhiza species on the growth and nutrient uptake of three grapevine rootstocks and one table grape cultivar. *Vitis.* 34,85-89.

75. Kamble, S.R., Navale, A.M. and Sonawane, R.B., 2009. Response of mango seedlings to VA- mycorrhizal inoculation. *Inter. J. Plant Prot.* 2 (2), 161–164.

76. Khade, S.W. and Rodrigues, B.F., 2006. Arbuscular mycorrhizal fungi associated with varieties of *Carica papaya* L. In tropical agro-based ecosystem of goa, india. *Tropical and Subtropical Agroecosystems.* 10, 369 – 381.

77. Khanam,D., 2007. Assessment of arbuscular mycorrhizal association in some fruit plants in Bangladesh. *Bangladesh Journal of Microbiology.* 24, 34-37.

78. Koide,R.T. and B. Mosse., 2004. A history of research on Arbuscular mycorrhiza, *Mycorrhiza.* 14, 145-163.

79. Koffi, M.C., De la Providencia, E.I., Elsen, A and Declerck, S., 2009 Development of an *in vitro* culture system adapted to banana mycorrhization. *African Journal of Biotechnology.* 8,2750-2756.

80. Krikun, J and Levy, Y., 1980. Effect of vesicular arbuscular mycorrhiza on citrus growth and mineral composition. *Phytoparasitica.* 8,195–200.

81. Krishna, H., Singh, S.K., Patel, V.B., Minakshi, R.N., Khawale and P.S., Deshmukh., 2006. Arbuscular-mycorrhizal fungi alleviate transplantation shock in micropropagated grapevine *(Vitis vinifera* L.) *J. Hortic. Sci. Biotechnol.* 81, 259–263.

82. Kucey, R.M.N., and H.H. Jarzen., 1987. Effect of vesicular arbuscular mycorrhiza and reduced nutrient availability on growth and phosphorus and micronutrient uptake of wheat and field beans under green house conditions, *Plant and Soil.* 104, 71-79.

83. Kumaran, S. and Aziza, H.C., 1995. Influence of biological soil conditioner on mycorrhizal versus non – mycorrhizal guava seedlings. *Tropical Agriculture, St Augustine.* 72(1), 39-43.

84. Kuwada, K.L., Wamocho, S., Utemura, M., Matsushita, I. and Ishi, T., 2006. Effect of red green algae extra on hyphal growth of arbuscular mycorrhizal Fungi and on mycorrhizal development and growth of Papaya and Passion fruit .*Agronomy Journal*. 98 (5),1340 – 1344.

85. Lee, N. and Wetzstein, H.Y., 1988. Quantum flux density effects on the anatomy and surface morphology of *in-vitro* and *in-vivo* developed sweet gum leaves. *J. Am. Soc. Hortic. Sci.* 113, 167–171.

86. Li,T. and Z.W. Zhao., 2005. Arbuscular mycorrhizas in a hot and arid ecosystem in southwest China, *Applied Soil Ecology*. 29, 135-141.

87. Lingua, G., Bona, E., Manassero, P., Marsano, F., Todeschini, V., Cantamesssa, S., Copetta, A., D'Agostino, G., Gamalero, E. and Berta, G., 2013. Arbuscular mycorrhizal fungi and plant growth-promoting pseudomonads increases anthocyanin concentration in strawberry fruits (*Fragaria* × *ananassa* var. Selva) in conditions of reduced fertilzation. *Int J Mol Sci*. 14,16207–16225.

88. Lovato, P., Guillemin, J.P. and Gianinazzi, S., 1992. Application of Commercial arbuscular endomycorrhizal fungal inoculants to the establishment of micropropagated grapevine rootstocks and pineapple plants. *Agronomie*. 12,873-880.

89. Louro, R.P., Santos, A.V.D. and Machado, R.D., 1999. Ultrastructure of Eucalyptus grantum dis × Eucalyptus urophylla I. shoots cultivated *in vitro* in multiplication and elongation-rooting media. *Int. J. Plant Sci.* 160 (2), 217–227.

90. Martins,M.A., Gonclaves, G.F.D.E and Soares, A.C.F., 2000. Effects of arbuscular mycorrhizal fungi associated with phenolic compounds on the growth of papaya. *Pes. Agropec. Bra.* 35(7). 465-1471.

91. Mamatha, G., Bagyaraj, D.J. and Jaganath, S., 2002. Inoculation of field-established mulberry and papaya with arbuscular mycorrhizal fungi and a mycorrhiza helper bacterium. *Mycorrhiza*. 12, 313-316.

92. Manoharachary, C., Adholeya, A and Kunwar, I.K., 2009. Mycorrhiza: some glimpses. *Mycorrhiza News letter*. 20(4), 2- 6.

93. Matsubara, Y., Karikomi, T., Ikuta, M., Hori, H., Ishikawa, S and Harada, T., 2006. Effect of arbuscular mycorrhizal fungus inoculation on growth of apple (*Malus* spp.) seedlings. *Journal of the Japanese Society for Horticultural Science*. 65, 297-302.

94. Matsubara, Y., Ishigaki, T and Koshikawa, K., 2009. Changes in free amino acid concentrations in mycorrhizal strawberry plants, *Scientia Horticulturae* 119, 392–396.

95. Menge, J.A and Johnson, E.L.V., 1978. Commercial Am. Phytopathol. Soc. 4:173 (Abstr.). Production of mycorrhizal inoculum may benefit citrus. *Calif. Citrogr.* 63,139-143.

96. Menge, J.A., Munnecke, D.E., Johnson, E.L.V and Carnes, D.W., 1978a. Dosage response of the vesicular-arbuscular mycorrhizal fungi *Glomus fasciculatus* and *G. constrictus* to methyl bromide. *Phytopathology.* 68,1368-1372.

97. Menge, J.A., Davis, R.M., Johnson, E.L.V. and Zentmyer, G.A., 1978b. Mycorrhizal fungi increase growth and reduce transplant injury in avocados. *California-Agriculture.* 32, 6-7.

98. Menge, J.A., LaRue, J., Labanauskas, C.K. and Johnson, E.L.V., 1980. The effect of two mycorrhizas fungi upon growth and nutrition of avocado seedlings grow with six fertilizer treatments. *Journal of the American Society for Horticultural Science.* 105,400-404.

99. Menge, J.A., Raski, D.J., Lider, L.A., Johnson, E.L.V., Jones, N.O., Kissler, J.J and Hemstreet, C.L., 1983. Interactions between mycorrhizal fungi, soil fumigation, and growth of grapes in California. *Am. J. Enol. Vitic.* 34,117-121.

100. Marschner,P. and S. Timonen., 2004.Interactions between plant species and mycorrhizal colonization on the bacterial community composition in the rhizosphere, *Applied Soil Ecology.* 28, 23-36.

101. Manoharachary, C., Kunwar, I.K., Reddy, S.V. and Adholeya, A., 2009. Ecological implications and ectomycorrhiza. *Myco. News.* 21(1), 2-8.

102. Miller,R.L. and L.E. Jackson., 1998. Survey of vesicular- arbuscular mycorrhizae in lettuce production in relation to management and soil factor, *Journal of Agricultural Science (Cambridge).* 130, 173-182.

103. Morales, A., Castillo, C., Rubio, R., Godoy, R., Rouanet, J.L. and Borie, F., 2005. Niveles de glomalina en suelos de dos ecosistemas del sur de Chile. *R.C. Suelo Nutr. Veg.* 5, 37–45.

104. Motosugi, H., Yamamoto, Y., Naruo, T., Kitabyashi, H and Ishi, T., 2002. Comparison of the growth and leaf mineral concentrations between three grapevine rootstocks and their corresponding tetraploids inoculated with an arbuscular mycorrhizal fungus *Gigaspora margarita. Vitis.* 41,21-25.

105 Nagahashi, G.,D.D. Douds and G.D. Abney.,1996.Phosphorus amendment inhibits hyphal branching of VAM fungus *Gigaspora margarita* directly and indirectly through its effect on root exudation, *Mycorrhizae.* 6, 403-408.

106. Nagarajappa, A., Patil, C.P., Swamy, G.S.K. and Patil, B., 2003. Influence of VAM on Drought Tolerance of Papaya. *Karnataka J. Agril. Sci.* 16 (3),434-437.

107. Naher, U.A., Othman, R. and Panhwar, Q.A., 2013. Beneficial effects of mycorrhizal association for crop production in the tropics - a review. *Int J Agr Biol* .15,1021-1028.

108. Nemec, S., 1978. Response of six citrus rotsocks to the thre species of Glomus, amycorhizal fungus. *Procedings of the Florida State Horticultural Society.* 91,10 -14.

109. Nemec, S., 1986. VA mycorrhizae in horticultural systems. In: Safir GR (ed) Ecophysiology of VA mycorrhizal plants. *CRC Press*, Boca Raton, Fla, pp: 193-211.

110. Nikolaou, N., Angelopoulos, K and Karagiannidis, N., 2003. Effects of drought stress on mycorrhizal and non-mycorrhizal 'Cabernet Sauvignon' grapevine, grafted onto various rootstocks. *Experimental Agric.* 39, 241–252.

111. Ojha, S., Chakraborty, M.R., Chakrabarti, J. and Chatterjee, N.C., 2005. Fruit rot of custard-apple (*Annona squamosa*) – a new disease from Burdwan, West Bengal. *J Mycopath Res.* 43(1),143-144.

112. Ojha, S., Chakraborty, M.R., Dutta, S. and Chatterjee, N.C., 2008. Influence of VAM on nutrient uptake and growth of Custard-apple. *Asian Journal of Experimental Science.* 22(3),221-224.

113. Onkarayya, H and Mohandas, S., 1993. Studies on dependency of *Citrus* root stocks to VAM inoculation in Alfisol soil. *Adv. Hort. For.* 3, 81-91.

114. Ozdemir, G., Akpinar, C., Sabir, A., Bilir, H., Tangolar, S and Ortas, I., 2010. Effect of inoculation with mycorrhizalfungi on growth and nutrient uptake of grapevine genotypes (*Vitis* spp.). *Eur. J. Hort. Sci.* 75(3), 103-110.

115. Panneerselvam, P., Saritha, B., Sukhada Mohandas., Upreti, K.K., Poovarasan, S., Sulladmath, V.V and Venugopalan, R., 2013. Effect of mycorrhiza associated bacteria on enhancing colonization and sporulation of *Glomus mosseae* and growth promotion in sapota (*Manilkara Achras* (Mill.) Forsberg) seedlings. *Biological Agriculture and Horticulture.* 29(2),118–131.

116. Paul Schreiner R., 2007. Effects of native and nonnative arbuscular mycorrhizal fungi on growth and nutrient uptake of 'Pinot noir' (*Vitis vinifera* L.) in two soils with contrasting levels of phosphorus. *Applied soil ecology.* 36, 205–215.

117. Petgen, M.A., George, S.E and Romheld, V., 1998. Einfluss unterschiedlicher inokulationstiefen mit dem arbuskulären mykorrhizapilz *Glomus mosseae* auf die mykorrhizierung bei reben (*Vitis* sp.) in wurzelbeobachtungskästen. *Vitis.* 37,99-105.

118. Pfeffer,P., D. Douds, G. Becard and Y. Shachar-Hill., 1999. Carbon uptake and the metabolism and transport of lipids in an arbuscular mycorrhiza, *Plant Physiology.* 120, 587-598.

119. Plenchette, C.,J.A. Fortin and V. Furlan., 1983. Growth responses of several plant species to mycorrhizae in a soil of moderate P-fertlity, *Plant and Soil.* 110, 199-209.

120. Possingham, J.V and Groot-Obbink, J., 1971. Endotrophic mycorrhiza and the nutrition of grapevines. *Vitis.* 10,120-130.

121. Poovarasan, S., Sukhada Mohandas., Panneerselvam, P., Saritha, B. and Ajay, K.M., 2013. Mycorrhizae colonizing actinomycetes promote plant growth and control bacterial blight disease of pomegranate (*Punica granatum* L. cv Bhagwa). *Crop Protection.* 175–181.

122. Rawat,A., Mishra, N.K., Mishra, D.S., Kumar, P., Rai, R.B. and Damodaran, T., 2013. Concentration mediated effect of Arbuscular Mycorrhizal fungi (AMF) on Growth and nutrition of Air-layerd Litchi plants. *International Journal of current Research.* 5(7), 1730-1734.

123. Reddy, B.M.S., 1991. Selection of efficient VA mycorrhizal fungi for mango, acid lime, and papaya. *M.Sc. (Agriculture) Thesis, University of Agricultural Sciences*, Bangalore, India.

124. Reddy, B. and Bagyaraj, D.J., 1994. Selection of efficient vesicular arbuscular mycorrhizal fungi for inoculating the mango rootstock cultivar 'Nekkare'. *Sci. Horti.* 59 (1), 69–73.

125. Reddy, P.P., Rao, M.,S,Mohandas and Nagesh, M., 1995. Integrated management of citrus nematode. *Tylenchulus Semipenetrans* cobb using VA mycorrhiza,, *Glomus fasciculautm (Thaxt.)* Gerd&Trappe and oil cakes. *Pest Management in Horticultural Ecosystem.* 1,37-43.

126. Reddy, B., Bagyaraj, D. J. and Mallesha, B. C., 1996. Selection of efficient mycorrhizal fungi for papaya. *Biological Agriculture Horticulture.*13 (1),1-6.

127. Roux, N., Baurens, F.C., Dolezel, J., Hribová, E., Heslop-Harrison, P., Town, C., Sasaki, T., Matsumoto, T., Aert, R., Remy, S., Souza, M. and Lagoda, P., 2008. Genomics of banana and plantain (*Musa* spp.), major staple crops in the tropics. In: Moore PH, Ming R (eds) Genomics of tropical crop plants, Springer. pp. 83-111.

128. Rufyikiri,G., Declerck,S., Delvaux,B., Dufey, J.E., 2000a. Arbuscular mycorrhizal fungi might alleviate aluminium toxicity in banana plants. *New Phytologist.* 148, 343–52.

129. Rupnawar, B.S. and Navale, A.M., 2000. Effect of VA-mycorrhizal inoculation on growth of pomegranate layers. *J. Maharashtra Agric. Univ.* 25 (1), 44–46.

130. Salazar-García, S., 2002. Nutrición del Aguacate, Principios y Aplicaciones. Inifap Inpofos. *México.* pp. 165.

131. Sagar, P. and Roy, A.K., 2013. Efficiency of Arbuscular Mycorrhizal Fungi for Litchi [*Litchi Chinensis* (Gaertn.) Sonn] Marcots inoculation in nursery. *J. Mycopathol, Res* 51(1), 141-144.

132. Sbrana,C. and M. Giovannetti., 2005. Chemotropism in the arbuscular mycorrhizal fungus *Glomus mosseae, Mycorrhizae*, 15, 539-545.

133. Schubert, A., Mazitelli, M., Ariusso, O., and Eynard, I., 1990. Effect of vesicular-arbuscular mycorrhizal fungi and micropropagated grapevines: influence of endophyte strain, P fertilization and growth medium. *Vitis.* 29, 5–13.

134. Selvaraj, T., 1998. Studies on mycorrhizal and rhizobial symbioses on tolerance of tannery effluent treated Prosopis juliflora, *Ph.D. Thesis, University of Madras*, Chennai, India, p. 209.

135. Sedlacek, M., Pavlousek, P., Losak, T., Zatloukalova, A., Filipcik, R., Hlusek, J. and Vitezova, M., 2013. The effect of arbuscular mycorrhizal fungi on the content of macro and micro elements in grapevine (*Vitis vinifera*, L.) Leaves. *Acta Universitatis Agriculturae Et Silviculturae Mendelianae Brunensis.* 22, 187-192.

136. Shrestha, Y.H., Ishii, T., Matsumoto, I.and Kadoya, K., 1996. Effects of AM fungi on satsuma mandarin tree growth and water stress tolerance and fruit development and quality. *Journal of the Japanese Society for Horticultural Science.* 64, 801-807.

137. Shashikala, B.N., Reddy, B.J.D and Bagyaraj, D.J., 1999. Response of micropropagated banana plantlets to *Glomus mosseae* at varied levels of fertilizers of phosphorus. *Indian Journal of Experimental Biology.* 37, 499-502.

138. Shrivastava, A.K., Ahmed, R., Kumar, S and Sukhada, Mohandas., 2001. Role of VA Mycorrhiza in the management of wilt disease of guava (*Psidium guajava* L.) in the alfisols of Chotanagpur . *Indian Phytopathology.* 54(1), 78-81

139. Sharma, S.D., Kumar, P., Gautam, H.R. and Bhardwaj, S.K., 2009. Isolation of arbuscular mycorrhizal fungi and *Azotobacter chroococcum* form local litchi orchards and evaluate their activity in air-layers system *Scientia Hortic.* 123, 117–123.

140. Sheng, P.P., Liu, R.J and Li, M., 2012. Inoculation with an Arbuscular Mycorrhizal Fungus and Intercropping with Pepper Can Improve Soil Quality and Watermelon Crop Performance in a System Previously Managed by Monoculture. *American-Eurasian J. Agric. & Environ. Sci.* 12 (11), 1462-1468.

141. Singh, H.P. and Singh, T.A., 1993. The interaction of rock phosphate Bradyrhizobium, vesicular arbuscular mycorrhizae and phosphate solubilizing microbes on soybean grown in a sub-Himalyan mollisol, *Mycorrhiza.* 4, 37–43.

142. Singh, S., 2005. Fungal chitin and its use for estimation of mycorrhizal infection, *Mycorrhiza News,* 17, 2-9.

143. Sivaprasad, P., Ramesh, B., Mohankumaran, N., Rajmohan, K and Joseph, P.J., 1995. Vesicular-arbuscular mycorrhizae for the *ex-vitro* establishment of tissue culture plantlets. Mycorrhizae Biofertilizers for the future, Third Nat Conf *Mycorrhiza.* 281-283.

144. Silveira, S.V., Da-souza, P.V.D.and De-kolle, O.C., 2002. Influência de fungos micorrízicos arbusculares sobre o desenvolvimento vegetativo de porta-enxertos de abacateiro. *Pesquisa Agropecuária Brasileira.* 37, 303-309.

145. Smith, S.E and Read, D.J., 1997.Mycorrhizal Symbiosis (*Academic Press, San Diego*), Ed 2.

146. Smith, S.E.and Gianinazzi-Pearson, V., 1988.Physiological interactions between symbionts in vesicular-arbuscular mycorrhizal plants. *Annu Rev Plant Physiol Plant Mol Biol.* 39, 221–244.

147. Smith, S.E., Smith,F.A. and Jakobsen, I., 2003. Mycorrhizal fungi can dominate phosphate supply to plants irrespective of growth responses. *Plant Physiology.* 133, 16–20.

148. Soares, A.C.F., Martins, M.A., Mathias, L., Simone, M. and Freitas, M., 2005. Arbuscular Mycorrhizal Fungi and the Occurrence of Flavonoids in Roots of Passion Fruit Seedlings. *Sci. Agric. (Piracicaba, Braz.).* 62(4),331-336.

149. Sreeramulu, K.R., Gowda, V.N and Bagyara,. D.J., 1998. Influence of VA mycorrhiza on establishment and growth of khirni (*Manilkara hexandra* (Roxb.) Dub) seedlings Current Research, *University of Agricultural Sciences,* Bangalore. 27(3), 50–52.

150. Sukhada, Mohandas., 1989. Response of papaya (*Carica papaya* L.) to VAM fungal inoculation. In: Mycorrhizae for Green Asia (Jan 29-31, 1988), Eds. Mahadevan, A., Raman, N. and Natarajan, K ., Madras, *Alamer Printing works.* Royapettah, Madras, 260-261pp.

151. Sukhada, Mohandas., 1992. Effect of VAM inoculation on plant growth nutrient level and root phosphatase activity in papaya (*Carica papaya* cv. Coorg Honey Dew). *Fert. Res.* 31, 263–267.

152. Sukhada, Mohandas., 1995. 'Utilization of VAM Fungi in Banana Cultivation', Proceedings, *National Seminar on Mycorrhiza, Biofertilizers of the Future,* New Delhi, 2931 March 1995.

153. Sukhada Mohandas., Shivananda, T.N and Iyengar, B.R.V., 1995. Uptake of 32P labelled super phosphate by endomycorrhizal papaya (*Carica Papaya* cv Coorg Honey Dew). *J.Nuclear Agric.* 24(4), 30-31.

154. Sukhada, Mohandas., 1998. Inoculation response if mango to VAM fungi in glass house and field trials. In: National Symposium on Mango Production and Export, Lucknow, June 25–27, 1998.

155. Sukhada Mohandas., Kumar,B.P. and Manamohan, M., 2000. Response of pomegranate cv Ganesh to inoculation with beneficial microbes. In: *National Seminar on Hitech Horticulture,* Bangalore June 26-28.

156. Sukhada, Mohandas., Chandre Gowda, M.J. and Manamohan, M., 2004. Popularization of Arbuscular Mycorrhizal (AM) Inoculum Production and Application On-farm, *Acta Hort.* 638, 279-283.

157. Sukhada Mohandas., 2012a. Arbuscular mycorrhizal fungi benefit mango (*Mangifera indica* L.) plant growth in the field. *Scientia Horticulturae.* 143, 43–48.

158. Sukhada Mohandas., 2012b. Field response of papaya (*Carica papaya* L. cv. Coorg Honey Dew) to inoculation with arbuscular mycorrhizal fungi at different levels of phosphorus. *Journal of Horticultural Science and Biotechnology.* 87(5), 514- 518.

159. Syvertsen, J.P. and Graham, J.H., 1990. Influence of vesicular arbuscular mycorrhizae and leaf age on net gas exchange of Citrus leaves. *Plant Physiol.* 94,1424-1428.

160. Taylor, J. and Harrier, L.A., 2001. A comparison of development and mineral nutrition of micropropagated *Fragaria xananassa* cv. Elvira (strawberry) when colonized by nine species of arbuscular mycorrhizal fungi. *Appl.Soil Ecol.* 18,205-215.

161. Taylor, D.L., Hollingsworth, T.N., McFarland, J., Lennon, N.J., Nusbaum, C and Ruess, R.W., 2013. A first comprehensive census of fungi in soil reveals both hyperdiversity and fine-scale niche partitioning.

162. Trindade, A.V., Siqueira, J.O., and Stürmer, S.L., 2006. Arbuscular Mycorrhizal Fungi in Papaya Plantations of Espírito Santo and Bahia, Brazil. *Brazilian Journal of Microbiology.* 37,283-289.

163. Tuomi,J., Kytoviita M., and Hardling, R. 2001. Cost efficiency of nutrient acquisition of mycorrhizal symbiosis for the host plant, *Oikos*, 92, 62-70.

164. Vázquez-Hernández, M.V., Arévalo-Galarzaa, L., Jaen-Contreras, D., Escamilla-García, J.L., Mora-Aguileraa, A., Hernández-Castroc, E., Cibrián-Tovar, J., and Téliz-Ortiz, D.,2011. Effect of *Glomus mosseae* and Entrophospora colombiana on plant growth, production, and fruit quality of 'Maradol' papaya (*Carica papaya* L.). *Scientia Horticulturae.* 128,255–260.

165. Van Aarle,I.M., P.A. Olsson and B. Soderstrom., 2002. Arbuscular mycorrhizal fungi respond to the substrate pH of their extraradical mycelium by altered growth and root colonization, *New Phytol.* 155, 173-182.

166. Vinayak, K and Bagyaraj, D.J., 1990. Selection of efficient VA mycorrhizal fungi for Trifoliate orange. *Biol. Agri. Hort.* 6, 305-311.

167. Westphal, A., Snyder, N.L and Xing, L., 2008. Effects of Inoculations with Mycorrhizal Fungi of Soilless Potting Mixes during Transplant Production on Watermelon Growth and Early Fruit Yield. *Hortscience.* 43(2),354–360.

168. Williams, S.C.K., Vestberg, M., Uosukainen, M., Dodd, J.C and Jeffries, P., 1992. Effects of fertilizers and arbuscular mycorrhizal fungi on the post-vitro growth of micropropagated straw berry. *Agronomie.* 12,851–857.

169. Wu, Q. S and Zou, Y.N., 2009. Mycorhizal influence on nutrient uptake of citrus exposed to drought stres. Philp. *Agric. Scientist.* 92,3-38.

170. Wu, Q.S., Zou, Y.N., and Xh, H.E., 2010a. Contributions of arbuscular mycorrhizal fungi to growth, photosynthesis, root morphology and ionic balance of citrus seedlings under salt stress. *Acta Physiol Plant.* 32,297–304.

171. Wu, Q.S., Zou, Y.N., Liu, W., Ye, X.F., Zai, H.F., and Zhao, L.J., 2010b. Alleviation of salt stress in citrusseedlings inoculated with mycorrhiza: changes in leaf antioxidant defense systems. *Plant Soil Environ.* 56,470–5.

172. Wright, S.F.and Upadhyaya A., 1996. Extraction of an abundant and unusual protein from soil and comparison with hyphal protein of Arbuscular mycorrhizal fungi, *Soil Science.*161,575–586.

173. Wright,S.F. and A. Updhyaya., 1998. A survey of soils for aggregate stability and glomalin, a glycoprotein produced by hyphae of arbuscular mycorrhizal fungi, *Plant and Soil.* 198,97-107.

174. Xiutang, L., 1990. The effect of VA mycorrhiza on mango seedling absorbing nutrient and growth. *J. Guangxi Agric. Biol. Sci.* 4,65–68.

175. Yan Li, H., Yang, G.D., Shu, H.R.,Yang, Y.T., Ye, B.X., Nishida, I., and Zheng, C.C., 2006. Colonization by the Arbuscular Mycorrhizal Fungus *Glomus versiforme* induces a Defense Response against the Root-knot Nematode *Meloidogyne incognita* in the Grapevine (*Vitis amurensis Rupr.*), Which Includes Transcriptional Activation of the Class III Chitinase Gene VCH3. *Plant Cell Physiol.* 47(1), 154–163.

177. Zhang, M.Q.,Y.S. Wang, K.N. Wang and L.J. Xing., 1998.VA Mycorrhizal fungi of the south and east coasts of China, Seven new records of *Acaulospora, Mycosystema,* 17, 15-18, (In Chinese).

Chapter 4

Improvement in Plant Growth and Nutrition in Temperate Fruit Crops

Muthuraju, R[1]* and Lakshmipathy, R.[2] and Vijayalaksmi[3]

[1]Department of Agricultural Microbiology, University of
Agricultural Sciences, Bengaluru-560065,India

[2]Post Harvest Technology Centre, Bapatla-522101, India.

[3]Indian Institute of Horticultural Research, Hessaraghatta (ICAR), Bengaluru-560089,
India .

*Address for the correspondence: muthurajr@gmailcom

ABSTRACT

Arbuscular mycorrhizal fungi (AMF) represent an important component of the temperate agroecosystem. AMF enhance uptake of P and diffusion of limited nutrients like Zn, Fe and Cu, which also enhance tolerance or resistance to root pathogens and help the plants grow under abiotic stresses such as drought and metal toxicity. Among the different species of AMF, *Glomus sps.* are reported to be associated with most of the temperate fruits crops and found to be involved in enhancing plant growth, nutrient uptake and disease management. In this article the role of AMF on sustainable production of temperate fruit crops is discussed in detail.

Keywords: *Temperate Agroecosystem, Growth Improvement, Nutrient, Uptake, Glomus sps.*

1.0 Introduction

Arbuscular mycorrizal fungi (AMF) play a significant role in temperate fruits crops production. AMF form a key component of the microbial populations influencing plant growth and uptake of nutrients. AMF are major components of rhizosphere microflora in natural ecosystems and play a significant role in the re-establishment of nutrient cycling. They modify the structure and function of plant communities (Douds and Miller, 1999) and are useful indicators of ecosystem change (McGonigle and Miller, 1996). The benefits of AMF in agricultural ecosystems are now widely known. The increased growth of plants inoculated with AMF is not only attributed to improved phosphate uptake but also better availability of other diffusion limited nutrients like Zn and Cu (Krishna and Bagyaraj, 1991). Horticulture has emerged as one of the potential agricultural enterprise in accelerating the growth of economy. Its role in the country's nutritional security, poverty alleviation and employment generation programmes is becoming increasingly important. It offers not only a wide range of options to the farmers for crop diversification, but also provides ample scope for sustaining large number of agro industries which generate huge employment opportunities. At present, horticulture is contributing 24.5% of GDP from 8% land area. The temperate fruit crops includes pome fruits (apple and pear) and stone fruits (peach, plum, apricot and cherry) grown in relatively lower temperature. Apricot, avacado, grape, persimmon, pistachio are also grown in temperate regions of the world. In India these are mainly grown in the North-Western Indian States and hills. Due to introduction and adaptation of low chilling cultivars of crops like peach, plum and pear, they are also now being grown commercially in certain areas of the north Indian plains.

There are various species of AMF proved to have symbiotic association with temperate fruit crops (**Table 4.1).**

Table.4.1: AMF association in temperate fruit crops

S. No.	Temperate fruit crops	AMF associated/ suitable for the crop	Reference
1.	Apple	G. mosseae, G. macrocarpum	Miller et al., (1989)
		G.versiforme	Morin and Fortin, (1994)
2.	Apricot	Gigaspora margarita	Cruz, et al., 2014
3.	Avocado	G.fasiculatum G. deserticola and G.mosseae	Menge et al. (1980), Aguilar et al. (1992),
4.	Blueberry	Glomus sp.	Goulart,1992, (Golldack,et al.,2001).

Contd..

Table 4.1: Contd...

S. No.	Temperate fruit crops	AMF associated/ suitable for the crop	Reference
5.	Grape	G. etunicatum and G. clarum G. mosseae	Caglar and Bayram, 2006,Aguin, et al., 2004
6.	Peach	G. mosseae	Wu et al.,(2010)
7.	Plum	G. intraradices	Stewart et al., (2005)
8.	Persimmon	G. etunicatum	Incesu et al.(2014)
9.	Pistachio	G. caledonium	(Caglar and Akgun, 2006).
10.	Strawberry	G. intraradices, G.mosseae, and G.etunicatum	(Vilma,et al.,2010). Camprubi et al (2007)
11.	Sweet passion fruit	Scutellospora heterogama	Anjoset al., (2010)
12.	Tangerine	Glomus sp.	Wu et al., (2007)

As AMF supply increased quantity of nutrients, produce plant growth hormone and control root pathogens their symbiosis can be better exploited in cultivation of temperate fruit crops. This review article explains about the AMF and current knowledge concerning the interactions between AMF and temperate fruit crops.

2.0 The AMF symbiosis with temperate fruit crops

The root zone temperature is not only the factor influencing root activity but also the symbiotic relationships with beneficial soil microbes such as AMF. It also affects the growth, productivity and quality of many crops (Gosling *et al.*, 2006; Koide, 2000). The host response of mycorrhizal colonization varies significantly among and within species (Devi and Reddy, 2002; Fan *et al.*, 2008; Guo *et al.*, 2007). The host plants that are typically responsive to colonization by AMF possess poorly developed root systems whereas the root systems of "non-responsive" plants have the capacity to successfully mine the soil nutrient supply (Hata *et al.*, 2010).

The effects of AMF on the growth and development of horticultural plants have been studied and described in many research papers (Lovato *et al.*, 1995). In general, fruit crops have received more attention than vegetable crops. Obviously, the interest of horticulturists in AMF technology is due to the ability of AMF to increase the uptake of phosphorus and other nutrients, and to increase resistance to biotic and abiotic stress.

It is accepted that the first evidence of the positive influence of the AMF symbiosis on horticultural production was provided by Menge *et al.* (1977). They showed that AMF propagule inoculation was a prerequisite to get the establishment of citrus plants in biocide-treated nursery beds. Since that publication, a number of experiments have been carried out, as reviewed by Miller *et al.* (1986), Nemec (1986), Lovato *et al.* (1995). These reviews show the potentiality for the application of AMF inoculation in temperate fruit crops.

3.0 The role of AMF in following temperate fruit crops

3.1 Apple (*Malus domestica* Borkh)

Apple is an economically important fruit crop, which is predominantly a temperate crop though grown in subtropical regions of the world. Apple trees are very good symbiotic hosts for the AMF colonization. In apple (*Malus domestica*), *G. mosseae* singly and in combination with *G. macrocarpum* were more effective in increasing plant biomass than *G.macrocarpum* (Miller *et al.*, 1989). Colonization of apple roots by AMF has been reported to be associated with healthy trees (Trappe *et al.*, 1973), efficient nutrient uptake (Benson & Covey 1976; Covey *et al.*,1981) and enhanced plant growth (Plenchette *et al.*,1981).

Effect of two AMF species (*Glomus epigaeum* Daniels and Trappe and *G. macrocarpum* Tul. and Tul.) on the growth and mineral composition of shoots and roots of apple seedlings (*M.micromalus makino*) indicated that the two species differed in their effects on seedlings grown at different temperatures under sterile soil conditions. *G. epigaeum* was mutualistic at low and high temperature, but *G. macrocarpum* was ineffective at low temperature. In unsterilized soil, neither species was mutualistic at low temperature, but both increased plant growth at high temperature. In sterilized soil, mutualism by *G. macrocarpum* was related to colonization. Enhanced growth of seedlings was accompanied by improved uptake of immobile nutrient elements, mainly Cu, P, Zn. (Zq *et al.*, 1993). Studies on AMF (*Glomus etunicatum* (GE) and *Gigaspora margarila* (GM) inoculation on growth of apple seedlings, such as *M. pumila* Mill. var. domestica Schneid and *M. sieboldii Rehd* showed that plant height and dry weights of shoot and roots were greater in all AMF infected seedlings than in non-inoculated ones. With both fungal species, P concentrations in the top or roots were higher in infected plants than in non-inoculated plants. Consequently, it was confirmed that GE and GM infections and their plant growth enhancement through symbiosis occurred in the seedling stage in several apple cultivars (Matsubara., 1996).

A 12-week greenhouse experiment was undertaken to test the efficiency of inoculation of AMF on four apple (*M. domestica* Borkh) rootstock cultivars i.e. M.26, Ottawa 3 (Ott.3), P.16, and P.22. The rootstocks were treated with *G.aggregatum* Shenck and Smith emend.Koske, *G. intraradices* Shenck and Smith, and two isolates of *G. versiforme* (Karsten) Berch. AMF inoculated plants were taller, produced more biomass, and had a higher leaf P concentration than the uninoculated control plants. Mycorrhizal inoculation also significantly increased the leaf surface area of 'M.26' and 'Ott.3' compared to the control. *G.versiforme* (CAL)-inoculated plants generally had the best nutrient balance, the greatest final height and shoot biomass, and produced an extensive hyphal network. All the mycorrhizal plants had similar percentages of root colonization, but the size of the external hyphal network varied with fungal species. *G.versiforme* (OR) had a larger extramatrical phase than *G. aggregatum* and *G. intraradix*. Mycorrhizal efficiency was associated with a larger external hyphal network, but showed no relation with internal colonization. The growth enhancement due to mycorrhizal inoculation was attributed to improved P nutrition (Morin and Fortin, 1994).

AMF inoculation in apple trees improves the fruit quality and economical value. Morin *et al.* (1994) tested the beneficial effect of AMF inoculation on four varieties of apple seedlings planted in soil containing P (644 kg bray ha^{-1}). The mycorrhizal seedlings grew taller with increased plant dry mass and leaf P content. *G. versiforme* inoculated plants generally had the best nutrient balance, the greatest final height and shoot biomass, and produced an extensive hyphal network. Apple cv *Malus pumila* Mill.var.domestica Schneid. (cv McIntosh, American Summer, Pearmain, Jonathan, Goldern Deliicious, Starking Delicious, Fuji, Mutsu and Red gold) were inoculated with *G.etunicatum and Gigaspora margarita* to study the effect on growth enhancement. AMF colonization upto 50% was noticed in Starking Delicious and Jonathan. Increase in plant height and weight and P concentration was observed in colonized plants (Matsubara , 1996). Resendes *et al.* (2008) reported the efficiency of colonization of mycorrhizal and non-mycorrhizal fungi in the apple tree roots. The newly emerged roots were colonized with AMF which induced the longer root growth and enhanced hypal network in the root rhizosphere. Forge *et al.* (2001) studied the effect of AMF (6 types) colonization in apple (Ottawa) root stocks for growth and to control the nematode lesions under green house condition. The experimental results were not sufficient to differentiate the treated and control plants but the root stocks preinoculated with *G.mosseae* in field condition showed significant growth difference and enhanced leaf nutrient content (N, P, K, Mn, Cu) than the control and other five isolates. Schubert and Lubraco (2000) demonstrated AMF inoculation on apple seedlings at the time of transplantation in soil and three different peat based substrates (enriched with P fertilizer). After 112 days of inoculation the plants inoculated with *G.mosseae* showed significant level of P increases in plant tissue than the control plants. The depletion of P in the substrate was due to the intense root colonization and nutrient uptake, enhancement in plant growth.

In apple, lower incidence of root rot caused by a fungus *Dematophora necatrix* and better growth of apple seedlings upon inoculation with indigenous AMF was observed by Bharat and Bhardwaj (2001). Ridgway *et al.* (2008) studied Specific Apple replant disease (SARD) effects the growth and establishments of trees in replanted apple orchards. Apple roots are normally colonised by AMF. Four AMF inoculation treatments (three species of AMF, *Glomus mosseae, Acaulospora laevis* and *Scutellospora calospora,* and an uninoculated control) were applied to M26 apple rootstock seedlings in SARD and non-SARD soil. Of the fungi inoculated, *S. calospora* had greatest beneficial effect in improving shoot dry weight and shoot length in SARD soil. More disease symptoms occurred on main and feeder roots in SARD soil and none of the inoculated AMF reduced these. Both *A.laevis* and *S.calospora* significantly increased root length and gave higher percentage of AMF coloniszed roots in non-SARD soil. These results showed that AMF improve tolerance of apple to SARD and indicate that beneficial effect is species specific

Raj and Sharma (2009) have also observed reduction in root rot incidence and enhanced growth in apple seedlings after inoculation with AMF. The growth promotion and disease suppression was found more when AMF were inoculated along with growth promoting bacterium *Azotobacter* spp. The reduction in disease incidence was attributed to the increased growth and disease resistance in mycorrhizal inoculated seedlings than that in non-mycorrhizal ones. The increased growth of apple and peach seedlings in fumigated soil upon mycorrhizal inoculation was also observed by Bingyne and Shengrui (1998). They were of the opinion that the increased growth was because of increased activity and response of inoculated beneficial microflora including AMF in fumigated soil.

Derkowska, *et al.* (2013) conducted an experiment to study the effect of mycorrhization and mulching on the colonization of the roots of 'Gold Milennium' apple trees and 'Ojebyn' and 'Tiben' blackcurrant bushes by AMF. The highest percentage of mycorrhizal frequency and mycorrhizal intensity were observed in the roots of trees inoculated with the mycorrhizal compared to uninoculated control. Apple tree plants of clones Malling Merron 111 (MM.lll) and Malling '7 (M.7) grown in a greenhouse in pots containing calcined montmorillonite chips were inoculated with a AMF (*G. epigaeum* Daniels and Trappe) and fed with a nutrient solution. AMF inoculated roots of both clones were heavily colonized by the fungus. However, no growth differences were observed on the trees of MM.111 clone. The fungus significantly increased leaf Cu of both rootstocks and increased leaf P of M.7 but it had no effect on leaf Ca, Mg and Zn and it depressed leaf K of M.7 plants (Granger, *et al.*, 1982).

3.2 Avocado (*Persea americana* Mill)

By nature the avocado plants have hairless mangrove-like roots (Salazar-Garcia, 2002) and hence are highly dependent on the mycorrhizal symbiosis for the better growth (Alarcón *et al.*, 2001). Menge *et al.* (1978) stated that Avocado inoculated with beneficial mycorrhizal fungi showed upto 250 percent greater growth rate than non-mycorrhizal avocados in sterilized soil. Mycorrhizal avocados resist transplant shock because of better water absorption. Menge *et al.* (1980) studied the effect of AMF (*G.fasiculatum*) on the avocado cv. Topa Topa planted with AMF spores in the sterilized loamy sand soil in three different treatments (i) supplemented with all the nutrients (N, P, K, S, Ca, Mg, Cu, Zn, Mn, Fe, Mo, B) (ii) without any nutrients, (iii) containing only 10 x of P. The study revealed that the plants from the potting mix containing no nutrient had a significant growth difference than the plant that had all the nutrient and there was no difference found in P content between mycorrhizal and non-mycorrhizal plants grown in 10xP potting mix. Aguilar *et al.* (1992) studied the role of AMF inoculation in the growth establishment of micropropagated avocado plants in the sand-soil/ peat-perlite mix. The plants inoculated with *G. deserticola* and *G.mosseae* showed significant growth establishment in avocado plants planted in sand-soil mix. Silveria *et al.* (2002) studied the effect of 6 different species of AMF on the avocado rootstocks which were grafted with Carmen scions. After seven month of inoculation the grafted plants which received (*Scutellospora heterogama, A.scorbiculata, G.etunicatum, G.clarum*) inocula showed the enhanced nutrient uptake and higher accumulation of reserve substances, which favoured vegetative growth of plants. Micropropagated avocado plants had poor rooting which severely affected the field survival rate at the time of transplantation. Pre-inoculation of seedlings with AMF facilitated the establishment of rooting. Banuelos *et al.* (2013) studied the effect of 6 AMF (*Rhizophagus fasciculatum, Gigaspora margarita, Claroideoglomus etunicatum, Pacispora scintillans, Rhizophagus intraradices, A.laevis*) on avocado seedlings planted in the sterile and non sterile soil. The plants in sterile soil inoculated with *Rhizophagus fasciculatum* showed significant growth changes over plants in the non-sterile soil. The outcome of the study demonstrated the compatibility between AMF and host, the planting mixture (sterile/non-sterile) played a key role in the growth of avocado seedlings.

3.3 Apricot (*Prunus mume*)

Cruz, *et al.*(2014) evaluated the symbiotic effect of AMF *Gigaspora margarita*, and the bacteria *Paenibacillus rhizospherae* on the alleviation of white root rot of Japanese apricot seedlings caused by *Rosellinia necatrix*. Disease severity, root infection and AMF colonization were evaluated in this experiment. This study could show that the AMF and bacteria could be used as bicontrol agents against white root rot. Their presence in soils could be used as indicator of fruit tree tolerance to this soil-borne plant pathogen.

3.4 Blueberry (*Vaccinium corymbosum* L.)

In 1992, Goulart and his co-workers studied the interactive effects of cultural practices on mycorrhizal infection intensity level in field grown 'Bluecrop' highbush blueberry.This experiment was carried out in Spring season , a field planting of 'Bluecrop' highbush blueberry was established to evaluate the effects of various cultural practices on plant growth and mycorrhizal level. Treatments included mulch or no mulch, pre-plant amendment or no pre-plant amendment, and different levels of nitrogen fertilization. They found that the intensity of mycorrhizal infection level varied with location in the rhizosphere, and was dependent on whether or not mulch was employed, as well as on the mulch composition. Mulch and/or amendment increased plant growth and vigor.

The mycorrhizal concentration level decreased with increasing nitrogen. Mycorrhization of highbush blueberry on farmland in Germany revealed that mycorrhizal infection rates of the varieties 'Duke' and 'Reka' differed by fertilizer regimes. 'Duke' had a high mycorrhizal infection rate at all fertilizer regimes. The mycorrhizal infection rate of 'Reka' decreased with increasing amounts of inorganic fertilizer (Golldack, *et al.*, 2001). These facts suggest a different sensitivity of the mycorrizal symbiosis.

3.5 Grape vine (*Vitis vinifera* L).

Caglar and Bayram, (2006) studied the effects of AMF on the leaf nutritional status of four grapevine rootstocks (420 A, 41 B, 1103 P, and 'Rupestris du Lot') .The AMF species (*Glomus etunicatum, G. caledonium, G. clarum, G. mosseae*, and mixed inoculum) were used in this experiment. One-year-old grapevine rootstock cuttings were rooted in perlite and transferred to black polyethylene bags filled with fumigated growing medium. A 40 g mycorrhizal soil band was used for each inoculation. The percentage of AMF inoculation, leaf area, N, P, K, and total sucrose contents of the leaves were determined. Ten months after inoculation, grapevine rootstocks were colonized by the AMF at frequencies ranging from 47.0 to 64.1%. Inoculations with *G. etunicatum* and *G. clarum* significantly increased the leaf areas of 41 B, 420 A, and 1103 P rootstocks. The AMF increased leaf P, but not N and K concentrations. Leaf total sucrose concentrations of the grapevine rootstocks were increased two- to four-fold with certain inoculums compared with the control. Aguin, *et al.,* (2004) examined the effects of AMF inoculation on root morphology and nursery production of three grapevine rootstocks and results showed that AMF symbiosis in grapevine rootstocks enhanced plant growth and nutrition. They also indicated that inoculation with the AMF *Glomus aggregatum* in rooting beds of grapevine cuttings changed root morphology, increasing branching of first-order lateral roots. Similarly, the influence of *Glomus mosseae* on mycorrhizal colonization in grapevine rootstocks (*Vitis berlandieri x* V. *riparia*, cv. SO_4) indicated that *G.mosseae* could significantly increased shoot dry weight, leaf blade Zn and P concentrations as compared to nonmycorrhizal plants (Petgen *et al.,*1998).

3.6 Peach (*Prunus persica* L. Batsch)

Peach, belonging to a species of *Prunus* of the subfamily Prunoideae of the family Rosaceae, is widely distributed and produced all around the world. Inoculation with *G.fasciculatus* was equal or more effective in overcoming soil-fumigation nutrient deficiency effects in peach nursery seedlings than the standard nursery practice of side-dressing P and Zn at planting time (Larue *et al.,* 1975).

In a pot experiment, the growth performance, nutrient concentrations and nutrient efficiency of the peach seedlings inoculated with *G.mosseae, G. versiforme,* and *Paraglomus ocultum*, respectively was determined. The mycorrhizal role on nutrient uptake generally was the best in the *G. mosseae* treatment (Wu *et al,* 2010). The influence of AMF on the growth and yield of the Plum tree cultivar 'Čačanska Lepotica' and sour cherry tree cultivar –'Schattenmorelle' was estimated. Productivity of mycorrhized trees, calculated according to the cross-sectional area of the trunk, was higher than in the control (Stewart *et al.,* 2005).

Wu, *et al.* (2011) studied the growth performance, nutrient concentrations and mycorrhizal nutrient efficiency of the peach seedlings inoculated with *Glomus mosseae, G. versiforme,* and *Paraglomus occultum*, respectively. After 100 days of mycorrhizal inoculations, mycorrhizal colonization of one-year-old seedlings ranged from 23.4% to 54.9%, this symbiosis generally increases the plant growth performance, such as plant height, stem diameter, shoot, root or total dry weight and plant nutrient uptake (K, Ca, Mg, Fe, Cu and Mn), compared to non-mycorrhizal inoculated seedlings.

This experiment finally concluded that the AMF species *G.mosseae* showed great influence on the nutrient uptake of peach plants. It suggests that AMF could improve growth performance and part nutrient acquisition of peach, which were absolutely dependent on AMF species. In 2004, Calvet and his co-workers conducted an experiment with eighteen *Prunus* rootstock cultivars and they were inoculated with three AMF under greenhouse conditions in order to evaluate their affinity for mycorrhizal colonization. The rootstocks were peach–almond hybrids, peaches, plums and cherries of Spanish, French and Italian origin. Among different AMF, *G.intraradices* was found to be the most infective endophyte, achieving the highest mycorrhizal colonization rate in most of the rootstocks evaluated.

Sheng, *et al.*(2011) found that *Glomus mosseae, G. versiforme,* and *Paraglomus occultum* would alter root system architecture (RSA) of peach seedlings, and the alteration due to mycorrhization was related to allocation of glucose/sucrose to root. Inoculation with *G. mosseae and G. versiforme* significantly increased leaf, stem, root and total fresh weights, compared with non-AMF treatment. Mycorrhizal alterations of RSA in peach plants were dependent on AMF species, because only *G. mosseae* and *G. versiforme* but not *P. occultum* markedly increased root length, root projected area, root surface area and root volume.

Rutto *et al.* (2002) studied on effect of root-zone flooding on mycorrhizal and non-mycorrhizal peach seedlings growing in low P medium and the results revealed that mycorrhizal seedlings showed relatively faster development prior to flooding and recorded significantly higher concentrations of shoot P, K and Zn and biomass yield. Ethanol accumulation was significantly higher in the taproots of non-mycorrhizal as compared to mycorrhizal plants after 3 days of flooding. A more rapid decline in plant health was also observed in non-mycorrhizal as compared to mycorrhizal seedlings. The presence or absence of the fungal partner led to significant difference in the ratio of roots that remained viable after extended flooding. Therefore, it was clear that AMF infection confered limited tolerance to flooding on peach seedlings. This could be due to improved plant nutrition, the suppression of ethanol accumulation in roots and the extension of the duration of root activity in a flooded environment.

3.7 Persimmon (*Diospyros kaki* Thunb.),

Persimmion, is grown in many parts of the world displaying subtropical climate conditions including Turkey. Incesu *et al.* (2014) used two common rootstocks, *D. kaki* and *D.virginiana* Thunb in the study evaluate the effect of inoculation with AMF. In this study, the effects of five AMF species (*Glomus mosseae, G. clarium, G. etunicatum, G. caledonium* and *G. intraradices*) on plant growth, chlorophyll concentration and chlorophyll fluorescence (*Fv'/Fm'*) in *D. virginiana* were investigated at greenhouse conditions. Mycorrhizal inoculations increased shoot and root dry weights compared with the non-inoculated plants. Plants inoculated with *G. etunicatum* showed the highest total plant dry weight. Highest leaf chlorophyll concentration was measured in plans inoculated with. *G.caledonium*. Results of chlorophyll fluorescence were similar with all the AMF inoculations, however significantly differed from non-inoculated plants. The results demonstrated the benefit potential of the mycorrhizal inoculations in persimmon production.

3.8 Pistacia (*Pistacia vera* L.)

The efficiency of inoculation of AMF on the seedling growth of three Pistacia species rootstocks were evaluated.The highest rate of AMF colonization i.e *G. clarum* (96.7 %), *G.etunicatum*, (83.3 %) and *G. caledonium* (73.3%) in *P. khinjuck*, *P. vera* and *P.terebinthus* seedlings respectively were noticed. Mycorrhizal inoculations improved seedling height only in *P. terebinthus*. Certain mycorrhizal inoculations increased the leaf N, but not P and K contents. Seedlings inoculated with *G. caledonium* had higher reducing sugar contents. It was concluded that pre-inoculated *Pistacia* seedlings could have a better growth in the harsh field conditions (Caglar and Akgun, 2006).

3.9 Strawberry (*Fragaria×ananassa* Duchesne)

Camprubi *et al* .(2007) reported that the combination of *G. intraradices*, *Trichoderma aureoviride* and *Bacillus subtilis* was an alternatives technology for controlling pests and diseases in strawberry crop production. Application of *Glomus intraradices* altered the quality of strawberry fruits at different nitrogen levels. AMF also increased the uptake of minerals from the soil, thus improving the growth of the strawberry plans (Vilma*et al.*,2010). Vestberg*et al.*(1992) designed the experiment in pots with ten strawberry cultivars, four early maturing, three late maturing and three "special" cultivars, were inoculated with six strains of AMF. *Glomus macrocarpum* V3, *G. mosseae* Rothamsted and *Glomus*. spp. V4, were highly efficient, causing significant growth increases in most cultivars of strawberry. De Silva *et al.* (1996) found that endomycorrhizae enhanced the growth of strawberry (*F. xananassa* Duch. 'Sweetheart') seedlings inoculated with six spore levels ranging from 0 to 12,000 spores/plant of the AMF *G. intraradices*. Plant height, leaf area, and number of leaves increased significantly with inoculum spore densities ranging from 750 to 12,000 spores/plant in relation to control plants in the greenhouse and field. In the field study, control plants were infected with indigenous mycorrhizae, but inoculated plants produced more runners than the control plants, and foliar Cu and Ca increased linearly with increased spore density. Inoculated plants contained significantly more dry matter than the controls. From this experiment, they concluded that a minimum spore density of 750 spores/plant is sufficient for a positive growth response.

3.10 Sweet Passion fruit (*Passiflora edulis*)

The effects of inoculation of sweet passion fruit plants with *Scutellospora heterogama* on the symptoms produced by *Meloidogyne incognita* race 1 and its reproduction were evaluated in two greenhouse experiments. *M. incognita* (5000 eggs/plant) and *S. heterogama* (200 spores/plant) inoculations were carried out simultaneously in the first; in the 2nd, the nematodes were inoculated 120 days after the fungal inoculation. In both the experiments, 220 days after AMF inoculation, plant growth was stimulated by the fungus. In disinfested soil, control seedlings (without *S. heterogama*) were intolerant to parasitism of *M. incognita*, while the growth of mycorrhized seedlings was not affected. Sporulation of *S. heterogama* was negatively affected by the nematodes that did not impair the colonization.

M. incognita did not affect mycorrhizal seedling growth. The establishment of mycorrhiza prior to the nematode infection contributed to the reduction of symptoms severity and reproducti on of *M. incognita* in disinfested soil (Anjos, *et al* .,2010). Sedentary endoparasitic nematodes (*Meloidogyne* spp.) caused significant losses in a variety of crops.

Avalcante *et al.* (2002) showed the effect of spore density (200, 300 and 400 spores/plant) and AMF (*Gigaspora albida, G. margarita, Acaulospora longula, Glomus etunicatum* and *Scutellospora heterogama*) on production of yellow passion fruit seedlings. The experiment findings indicated that there was no significant interaction between density of inoculum and the AMF species in relation to host growth. However, shoot dry biomass and leaf area reached maximum values in the treatment with 300 spores/plant. Seedlings inoculated with *G. albida, G. margarita* and *G. etunicatum* recorded higher growth, colonization, and spores density in the rhizosphere than those associated with *A. longula* and *S. heterogama*, whose growth was similar to the control. Similarly, the inoculation with AMF increased the plant height, leaf number and stem girth compared to un-inoculated seedlings grown under equivalent P concentrations (Chebet*et al.*,2008).

3.11 Trifoliate orange

Commercially *in-vitro* propagated elite plants of five cultivars ('Chambly,' 'Glooscap,' 'Joliette,' 'Kent,' and 'SweetCharlie') were transplanted in non-inoculated growth substrate or in substrate inoculated with *G.intraradices* or with a mixture of species (*G. intraradices, G.mosseae,* and *G.etunicatum*) at the acclimatization stage and were grown for 6 weeks before transplantation in the field (Sha *et al* .,2010). It was reported that the symbiosis of AMF in trifoliate orange enhanced the soluble sugar and leaf chlorophyll content..

The effects of different phosphorus doses with mixed inoculums of AMF (*Acaulospora morrowae, Glomus clarum* and *Glomus etunicatum*) on growth and mineral nutrition of 'Cleopatra' mandarin revealed that the mixed AMF inocula promoted better plant growth, and increased the boron contents in the shoots. The inoculated plants recorded larger growth and reached average height of 12.30 cm at four months after sowing (Darocha, *et al.*, 1994). Jifon *et al* .(2002) studied the performance of AMF (*Glomus intraradices*)on gas exchange and growth responses of pot-grown sour orange (*Citrus aurantium*) and sweet orange (*C. sinensis*) at high soil-P under elevated and ambient CO_2. At a CO_2, growth of AMF treated sour orange was depressed (18%) compared with uninoculated seedlings, but at elevated CO_2, AMF sour orange plants were 15% larger than uninoculated plants.

4.0 Conclusion and future strategies

Symbiotic association with AMF is beneficial to temperate horticultural crops. Growth improvement to varying levels have been reported in different crops. It improves plant growth, nutrient mobilization, production of growth hormones and protection against pests and diseases. Application of AMF should necessarily be an important package of practice in sustainable development of temperate crops. The concentration of AMF spores, P concentration have been worked out in some crops. The effects of nutrients, other than phosphorus, on the abundance and distribution of AMF in the field needs to be studied. There is very little knowledge of functional attributes such as stress tolerance and nutrient uptake efficiency of crops. These aspects should be given more importance.

References

1. Benson, N.R and Covey Jr R.P., 1976. Response of apple seedlings to zinc fertilization and mycorrhizal inoculation. *Hort Science*. 11, 252-253.

2. Bharat, N., 2013. Effect of indigenous AM fungi and BCAs on health of apple seedlings grown in replant disease soil. *Indian Phytopath*. 66 (4) , 381-386.

3. Bharat, N.K. and Bhardwaj, L.N., 2001. Interactions between VA-mycorrhizal fungi and Dematophoranecatrix and their effect on health of apple seedlings. *Indian J. Plant Pathol.*19, 47-51.

4. Bingyne, X. and Shengrui, Y., 1998. Studies on replant problem of apple and peach. *Acta Hortic.*477, 83-88.

5. Calvet, C., Pinochet, J., Hernández-Dorrego, A., Estaún, V. and Camprubi, A., 2001, Field microplot performance of the peach-almond hybrid GF-677 after inoculation with arbuscular mycorrhizal fungi in a replant soil infested root-knot nematodes. *Mycorrhiza*. 10, 295-300.

6. Calvet, C., V. Estaun., A. Camprubi., A. Hernandez- Dorego., J. Pinochet, and M.A. Moreno., 2014.Aptiude for mycorrhizal root colonization in *Prunus rootstocks*. *Sci. Hort.* 10, 39-49.

7. Caruso, F.L., Neubauer, B.F., Begin,M.D., 1989. A histological study of apple roots affected by replant disease. *Canadian Journal of Botany*. 67, 742-749.

8. Catska, V., 1994. Interrelationships between vesicular-arbuscular mycorrhiza and rhizosphere microflora in apple replant disease. *Biologia Plantarum*. 36, 99-104.

9. Catska, V., Taube-Baab, H., 1994. Biological control of replant problems. *Acta Horti*. 363,115-120.

10. Chebet, D.K., Rutto, L.K., Wamocho, L.S. and Kariuki, W. 2008. Effect Of Arbuscular Mycorrhizal Inoculation On The Growth, Nutrient Uptake And Root Infectivity Of Tropical Fruit Seedlings. *Acta Hort.* (Ishs). 773,253-260.

11. Covey, R.P., Koch, B.L., Larsen, H.J., 1981. Influence of vesicular arbuscular mycorrhizae on the growth of apple and corn in low-phosphorus soil. *Phytopathology*. 71,712-715.

12. Devi, M. C. and M. N. Reddy. 2002. Phenolic Acid Metabolism of Groundnut (*Arachis hypogeal* L.) Plants Inoculated with AMF fungus and Rhizobium. *Plant Growth Regulation*. 37, 151-156.

13. Doss, D. D. and Bagyaraj, D. J., 2001. Ecosystem dynamics of mycorrhizae. In: *Innovative Approaches in Microbiology*. (Eds.) D.K. Maheswari and R.C. Dubey, Bishen Singh Mahendra Pal Singh Pub:, Dehra Dun, India. 115-129

14. Douds, D.D. and Millner, P., 1999. Biodiversity of arbuscular mycorrhizal fungi in agroecosystems.*Agric. Ecosyst. Environ.* 74, 77-93.

15. Fan, Y. Q., Y. S. Luan, L. An and K. Yu. 2008. Arbuscular Mycorrhizae Formed by *Penicillium pinophilum* Improve the Growth, Nurtient Uptake and Photosynthesis of Strawberry with Two Inoculum-Types. *Biotechnology Letters*. 39, 1489-1494.

16. Estaún, V., Calvet, C., Camprub´, A., 1994.Arbuscular mycorrhyzae and growth enhancement of micropropagated Prunus in different soilless potting mixes. *Agric. Sci. Finland.* 3, 263–267.

17. Forge, T., Muehlchen, A., Hackenberg, C., Neilsen, G and Vrain, T.,2001. Effects of preplant inoculation of apple (*Malus domestica* Borkh.) with arbuscular mycorrhizal fungi on population growth of the root-lesion nematode, *Pratylenchus penetrans Plant Soil*. 236,185–196.

18. Gilmore, A.E., 1971. The influence of endotrophic mycorrhizae on the growth of peach seedlings.*J. Am. Soc. Hort. Sci,.*96, 35–38.

19. Giovannetti, M. and Gianinnazzi-Pearson, V., 1994. Biodiversity in arbuscular mycorrhizal fungi.*Mycol. Res.*98,705-715.

20. Goulart,B.L.,M.Brittingham,J.Harper,P.Heinemann,W.Hock.E.Rajotte,J and RytterandJ.Travis.,1991-92. Small Fruit Production and Pest Management.*The Pennsylvania State University*,107pp.

21. Golldack, J., Schubert, R., Tauschke, M., Schwarzel, H., Hofflich, G., Lentzsch, P. and Munzenberger, B. 2001. Mycorrhization And Plant Growth Of Highbush Blueberry (*Vaccinium Corymbosum* L.) On Arable Land In Germany Http://Mycorrhiza.Ag.Utk.Edu/Latest/Icoms/ Icom3/Icom3.Htm (Accessed 14 June 2010).

22. Gosling, P., A. Hodge, G. Goodlass and G. D. Bending. 2006. Arbuscular Mycorrhizal Fungi and Organic Farming. *Agriculture Ecosystems and Environment*. 113, 17-35.

23. Granger, R.L., Plenchette, C., Fortin, J.A., 1983. Effect of vesicular–arbuscular (VA) endomycorrhizal fungus (*Glomus epigaeum*) on the growth and leaf mineral content of two apple clones propagated *in-vitro. Can. J. Plant Sci.* 63,551–555.

24. Guo, T. J. Zhang, P. Christie and X. Li. 2006. Influence of Nitrogen and Sulfur Fertilizers and Inoculation with Arbuscular Mycorrhizal Fungi on Yield and Pungency of Spring Onion. *Journal of Plant Nutrition*. 29,1767-1778.

25. Hata, S., Y. Kobae and M. Banba. 2010. Interactions between Plants and Arbuscular Mycorrhizal Fungi. *International Review of Cell and Molecular Biology*. 281,1-48.

26. Holevas, C.D., 1966. The effect of a vesucular–arbuscular mycorrhiza on the uptake of soil phosphorus by strawberry (*Fragaria sp. var.*Cambridge Favourite. *J. HortSci.* 41,57-64.

27. İncesu,M., Turgut Yeşiloğlu, Berken Çimen, Bilge Yilmaz, Çağdaş Akpınar, İbrahim.,2014 .Effects on growth of persimmon (*Diospyros virginiana*) rootstock of arbuscular mycorrhizal fungi species.

28. Koide, R. T. 2000. Functional Complementary in the Arbuscular Mycorrhizal Symbiosis. *New Phytologist.* 147, 233-235.

29. Krishna, K.R. and Bagyaraj, D.J., 1991. Effect of vesicular-arbuscular mycorrhiza and soluble phosphate on *Abelmoscus esculentus* (L.) Moench. *Plant Soil.* 64, 209-213.

30. Lovato, P.E., Gianinazzi-Pearson, V., Trouvelot, A. and Gianinazzi, S., 1996. The state of art of mycorrhizas and micropropagation. *Adv. Hortic. Sci.* 10, 46-52.

31. Y. and Hadara T., 1996. Effect of arbuscular mycorrhizal fungus infection on growth and mineral nutrient content of *Asparagus officinalis* L. seedlings. *J. Jap. Soc. Hortic. Sci.* 65,303-309.

32. McGonigle, T.P. and Miller, M.H., 1993. Mycorrhizal development and phosphorus adsorption in maize under conventional and reduced tillage. *Amer. J. Soil Sci. Soc.* 57, 1002-1006.

33. Menge, J.A., Lembright, H. and Johnson, E.L.V., 1977. Utilization of mycorrhizal fungi in citrus nurseries. *Hortic. Sci.* 21, 974-984.

34. Miller, D. D., M. Bodmer, and H. Schuep .,1989. Spread of endomycorhizal colonization and effects on growth of apple seedlings.*New Phytol.* 11,51-59.

35. Miller, J.C., Jr., Rajzpakse, S. and Garber, R.K., 1986.Vesicular arbuscular mycorrhizae in vegetable crops.

36. Mosse, B., 1957. Growth and chemical composition of mycorrhizal and non-mycorrhizal apples. *Nature* .179, 922-924.

37. Mosse, B., Stribley, D.P. and Le Tacon, F.,1981. Ecology of mycorrhizae and mycorrhizal fungi. *Adv. Microbial Ecol.* 5 ,137-209.

38. Morin, F., Fortin, J.A., Hamel, C., Granger, R.L and Smith, D.L.,1994. Apple rootstock response to vesicular-arbuscular mycorrhizal fungi in a high phosphorus soil. *J. Am. Soc. Hortic. Sci.* 119,578-583.

39. Nemec, S., 1980. Effects of eleven fungicides on endomycorrhizal development in sour oranges.*Can. J. Bot.*58,522-527.

40. Pirozynski, K.A. and Y. Dalphe., 1989. Geological history of the Glomaceae with particular reference to mycorrhizal symbiosis.*Symbiosis.* 7,1 – 36.

41. Plenchette, C.V., Furlan, V., Fortin, J.A., 1981. Growth stimulation of apple trees in unsterilized soil under field conditions with VA mycorrhiza inoculation. *Canadian Journal of Botany.* 59, 2003-2008.

42. Powell, L., Bagyaraj, D.,Clark, G. E. and Caldwell, K. L., 1985. Inoculation with vesicular-arbuscular mycorrhizal fungi in the greenhouse production of asparagus seedlings. *New Zealand Journal of Agricultural Research.* 28, 293-297.

43. Raj, H. and Sharma, S.D., 2009. Investigation on soil solarization and chemical sterilization with beneficial microorganisms for control of white root rot and growth of nursery apple. *Scientia Hortic.*119 , 126-131.

44. Resendes, M.L., Bryla, D.R and Eissenstat, D.M.,2008. Early events in the life of apple roots: variation in root growth rate is linked to mycorrhizal and non mycorrhizal fungal colonization. *Plant and Soil.* 313,175–186.

45. Ridgway,H.J., J. Kandula and A. Stewart (2008). Arbuscular mycorrhiza improve apple rootstock growth in soil conducive to specific apple replant disease. Disease Control in Horticultural Crops (2008) *New Zealand Plant Protection Society (Inc.).*

46. Schubert, A and Lubraco, G.,2000. Mycorrhizal inoculation enhances growth and nutrient uptake of micropropagated apple rootstocks during weaning in commercial substrates of high nutrient availability. *Applied Soil Ecology.* 15, 113–118.

47. Sha, W., Omi rświe, R., Czyńsk, I .and Aleksander Stachowia, K., 2010. The influence of mycorrhizal fungi on the growth and yielding of plum and sour cherry trees. *Journal of Fruit and Ornamental Plant Research.* 18(2), 71-77.

48. Stewart, L.I., Hamel, C., Hogue, R. and Moutoglis, P., 2005. Response of strawberry to inoculation with arbuscular mycorrhizal fungi under very high soil phosphorus conditions. *Mycorrhiza.*15(8),612-619.

49. Trappe, J. M., Stahly, E. A., Benson, N.R., Duff, D.M., 1973. Mycorrhizal deficiency of apple trees in high arsenic soils. *Hort. Science.* 8, 52-53.

50. Wacker, T. L., Safir, G. R. and Stephens, C. T., 1990. Effect of *Glomus fasciculatum* on the growth of asparagus and the incidence of *Fusarium* root rot. L Amer. *Soc. Hort. Sci.* 115, 550-554.

51. Wu, Q. S., Y. N. Zou, R. X. Xia, and M. Y. Wang., 2007. Five *Glomus* species affect water relations of Citrus tangerine during drought stress. *Bot. Stud.* 48,147-154.

52. Wu, Q.-S., Li, G.H., and Zou Y. N., 2010. Roles of arbuscular mycorrhizal fungi on growth and nutrient acquisition of peach (*Prunus persica* L. Batsch) seedlings. *The Journal of Animal & Plant Sciences.* 21(4), 746-750.

Improvement in Soil and Plant Nutrition by Arbuscular Mycorrhizal Fungi

Chapter 5

Cycling of Macro and Micronutrients

Balaji Seshadri[1, 2],* and Nanthi Bolan [1,2]

[1]*Centre for Environmental Risk Assessment and Remediation, Building–X, University of South Australia, Mawson Lakes, South Australia 5095, Australia.*

[2]*Cooperative Research Centre for Contamination Assessment and Remediation of the Environment, PO Box 486, Salisbury, South Australia 5106, Australia.*

*Address for the correspondence: Balaji.Seshadri@unisa.edu.au.

ABSTRACT

With their significant role in carbon sequestration and productivity, mycorrhizae's influence on mobilization and cycling of nutrients will be discussed in this chapter. Improvement of soil health depends on the availability of nutrients to plants, which will help in effective cycling of the nutrients thereby creating a balance between influx and efflux of nutrients. Microorganisms play an effective role in utilising the carbon as an energy source and also producing carbon based products including enzymes and organic acids, which helps in the mobilisation of macro and micronutrients responsible for plant growth and metabolism. The presence of mycorrhizae in the rhizosphere of plants can do the above mentioned functions pertaining to nutrient mobilization, in an effective way because of the ability of mycorrhizal hyphae to cover large area around the rhizosphere region. Hence, positionally unavailable nutrients such as phosphorus become accessible through exploration of soil volume. The challenge is not only accessing nutrients and delivering to the plant but also making the nutrients to be in available forms. Arbuscular Mycorrhizal fungi (AMF) have the ability to decompose complex organic materials (especially organic nitrogen and phosphorus) into readily available inorganic forms which are available to plants. The availability of potassium (third important macronutrient) and other micronutrients is largely influenced by indirect effects of mycorrhizae on the mobilization (mineralisation) of nitrogen and phosphorus. Interms of horticulture industry, the understanding of mycorrhizal phenomenon will be supportive during micropropagation which will help in improving the overall health of the plants. Mycorrhizal fungi can also be effective against root pathogens serving as a line of defence in the rhizosphere region.

Keywords: *Complex organic materials, Nitrogen, Phosphorus , Potassium, Mineralization, Arignine transportation,*

1.0 Introduction

All organisms require carbon (C) and other nutrients nitrogen (N), phosphorus (P) and potassium (K)) for metabolism and most soil microorganisms depend on plant roots for C source, where organic matter from plants release C through root exudates such as organic acids, phenolic compounds, enzymes, etc. With the consumed carbohydrates, the microorganisms not only sustain their life cycle but also play the important role of transforming the macro and micronutrients to plant available forms (Smith and Read, 1997). In the process, these microorganisms play vital role in enhancing soil productivity by transporting nutrients within the soil environment and also between plant roots in an ecosystem. While most soil microorganisms transform nutrients near the rhizosphere region, AMF can reach out of rhizosphere region and acquire the required nutrients for plant growth, using hyphae. This means that the mycorrhizal hyphae can always reach out to the nutrients by increasing length under nutrient limiting conditions.

The effect of mycorrhizae in increasing plant growth has been well documented by many workers for many plant species (Bolan, 1991; Marschner, 1995;Lambers *et al.*, 2006). The beneficial effect of mycorrhizae on plant growth has mostly been attributed to an increase in the uptake of nutrients, e.g. P. A growing body of circumstantial evidence suggests that mycorrhizal fungi or their accompanying microflora may improve soil P availability by solubilising inorganic forms of P or by mineralisation of organic P. The fungal hyphae absorb nutrients from soil by active metabolism and transports ions from assimilated metabolites back to the host plant via the rhizosphere. The absorption of poorly mobile nutrients such as inorganic P and K along with that bound to soil particulates (e.g. NH_4) depends on the above mentioned mechanism of 'hyphal foraging' or can be referred as hyphal prospecting of soil.

The characteristic structures of AMFs are the eponymous arbuscules: they are intracellular, highly branched hyphae where the majority of nutrient exchanges between the two partners are supposed to occur. Because AMF are obligate biotrophs, they depend on the carbohydrates supplied by the plant roots (Bago *et al.*,2003), whereas the plant improves its P and N nutrition via the fungal partner (Govindarajulu *et al.*,2005; Smith and Read, 2008). The uptake of inorganic P (Pi) is the key physiological process by which AMF improves plant growth (Bucher, 2007). AMF indeed express Pi transporters which acquire Pi from the soil and deliver it to the host plant (Harrison and van Buuren, 1995). On the other hand, the plant acquires the Pi from the interfacial apoplast via own mycorrhiza specific Pi transporters.

Besides P, N is the other macronutrient for which AMF play an important role. In soils, N is found in two major forms, organic and inorganic compounds. The organic compounds are simple molecules, as urea, amino acids (AA), amines and peptides, or complex ones as proteins, whereas inorganic N compounds are mainly represented by nitrate (NO_3^-) and ammonium (NH_4^+). Uptake of organic and inorganic N sources from the soil via specialised transporters has been described for ectomycorrhizal (ECM) and AMF (Gobert and Plassard 2008). Recent studies on AMFs provide evidence that the subsequent N-transfer from the fungus to the plant cells occurs in the form of ammonium (Govindarajulu *et al.*,2005, Jin *et al.*,2005). However, it has been proposed that depending on the nutritional and photosynthetic status of the plant there could be two ways of nitrogen release from the fungus to the plant (Chalot *et al.*,2006). The "traditional view" suggests an organic N-transfer under high C availability, whereas under C depletion the synthesis of organic N would be down regulated in the fungus, leading to the transfer of inorganic N.

A third important mineral element is given by sulphur (S). Interestingly, a study (Allen and Shachar-Hill, 2008) showed that AMF transfer also significant amounts of S to the plant. However, so far no plant transcripts coding for S-transporters have been shown to be regulated during the AMF interaction. Lastly, it can be questioned whether K nutrition might be also improved by AMF. Similar to S, so far no transcript data are available to support the K transfer during AMF symbiosis.

Nutrient-dependent regulation of AMF colonization provides an important feedback mechanism for plants to promote or limit fungal colonization according to their needs. A study by Nouri *et al.*(2014) shows that phosphate and nitrate can potentially exert negative regulation on AMF, whereas sulfate and the cations Mg_2^+, Ca_2^+, and Fe^{3+}have no effect. There have been a number of studies on the potential value of AMF in the complex nutritional feedback mechanisms in the rhizosphere and their colonization in the soil. This chapter will consolidate the overall information pertaining to the two major nutrients – N and P, by discussing the mechanisms of nutrient dynamics, micronutrient transformation and their importance in fruit crop production.

2.0 Nitrogen nutrition to plants

Nitrogen nutrition to plants is essential for growth and overall metabolism of plants. It is among the most limiting nutrients along with P and K. Plants take up N in the inorganic forms such as ammonia (NH_4) and nitrate (NO_3). Most of the current models (Bago *et al.*, 2008; Jin *et al.*, 2005; Govindarajalu *et al.*, 2005) of N transport in the AMF symbiosis involve:

i.   Uptake of inorganic N from the soil and N assimilation via the anabolic arm of the urea cycle into arginine (Arg) in the extra-radical mycelium (ERM),

ii. Arg translocation—likely in connection with polyphosphates (polyP)—to the intraradical mycelium (IRM), and

iii. Break down of Arg via the catabolic arm of the urea cycle into inorganic N, which is subsequently transferred across the mycorrhizal interface to the host (**Figure 5.1**).

Labeling experiments (Jin *et al.*, 2005; Cruz *et al.*, 2007; Johansen *et al.*, 1996) and enzymatic tests (Cruz *et al.*, 2007) has supported this model of N transport and has been further corroborated by molecular data demonstrating the expression of genes putatively involved in Arg biosynthesis in the ERM and Arg breakdown in the IRM (Tian *et al.*, 2010; Gomez *et al.*, 2009).

2.1 Utilization of Organic N

The breaking down of organic N into their less complex inorganic ones (bio-available) has been the challenge of many soil chemists and biologists. At higher altitudes and latitudes (e.g., heathlands), low temperatures and low soil pH usually restrain nitrification and (to a lesser extent) ammonification (Read, 1991). Studies of N relations in temperate and boreal ecosystems, where some of the fruit trees such as apple thrive, have demonstrated the importance of its organic forms for the plant nutrition (Read, 1991). It is a well-known fact that ECM and ericoid fungi use complex organic N, such as protein, and their host plants have access to peptides and proteins. Soluble amino acids are also a substantial source for all types of mycorrhizal associations in these ecosystems (Read, 1991; Botton and Chalot, 1995).

Interestingly, the AMF have also shown to increase decomposition and subsequent capture of inorganic N from organic material (Hodge *et al.*, 2001); they show therefore a kind of response which (for long time) has been considered characteristic of ECM fungi. Whiteside *et al.* (2012) stated that the breakdown of organic nitrogen in soil is a potential rate-limiting step in nitrogen cycling. There is clearly a need to understand the mechanism involved in such organic N mobilization by AMF and to detect the molecular basis of such events. A putative aminoacidpermease, like those characterised in yeast, has been recently detected in *Glomus mosseae* (Capellazzo *et al.*, 2008); the analysis of its expression, exclusively located in the extraradical hyphae and N dependent, might prove some insights into this "hidden" capacity of AMF.

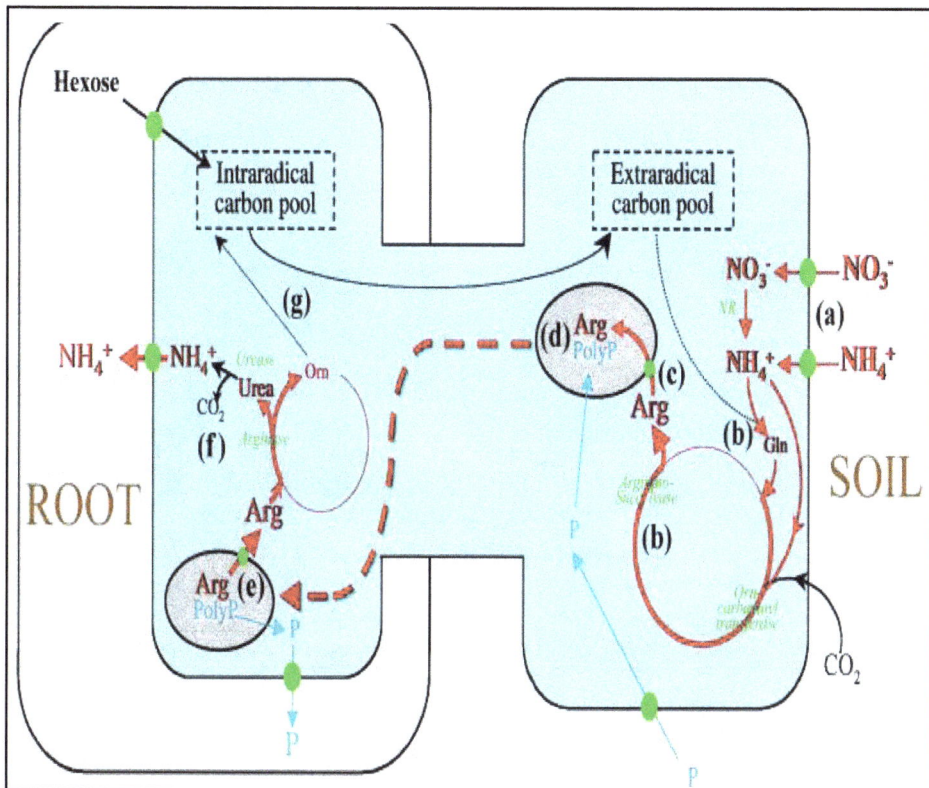

Fig. 5.1. A proposed model by which the urea cycle together with arginine translocation could act to move nitrogen in the arbuscular mycorrhizal symbiosis.

The flow of N is indicated in red, of phosphorus in blue, and of carbon in black; membrane transporters/channels and enzymes are shown in green. Individual steps in the Figureure labeled a–g. (a) is metabolized to Gln in the extraradical hyphae, then to Arg by the anabolic arm of the urea cycle (b). The Arg is imported into vacuoles by means of a specific Argtonoplast carrier (c). The cationic Arg binds to polyanionic polyphosphates within fungal vacuoles (d), and these move to the intraradical fungal mycelium by cytoplasmic streaming or a peristaltic tubular vacuolar system. Once within the root, Arg and P are released into the fungal cytoplasm (e). The N is further metabolized through the catabolic arm of the urea cycle producing urea and Orn, and finally ammonium which is then transferred to the host root (f). Resulting C skeletons in the intraradical fungus are re-incorporated into fungal C pools (g). Reproduced from Bago *et al.* (2008), with permission from John Wiley and Sons.

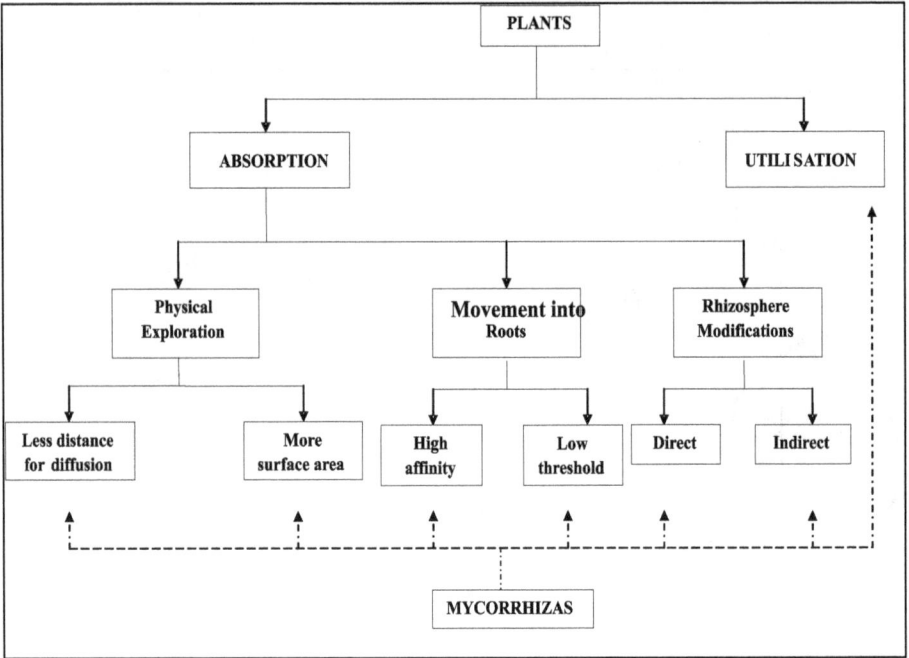

Fig. 5.2. Possible mechanisms by which mycorrhizal fungi increase the uptake of phosphorus by plants from soil (Reproduced from Bolan, 1991).

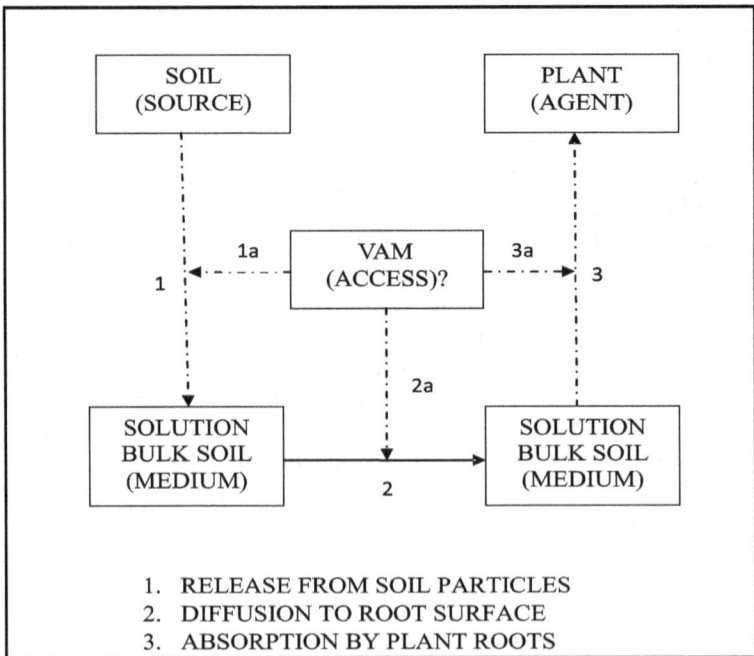

1. RELEASE FROM SOIL PARTICLES
2. DIFFUSION TO ROOT SURFACE
3. ABSORPTION BY PLANT ROOTS

Fig. 5.3. Rate-limiting steps in the uptake of phosphorus from soil by plants and the effect of AMF on these rate-limiting steps. (Reproduced fromBolan, 1991).

2.2. Mycorrhiza mediated uptake of NO_3^- and NH_4^+

AMF and ECM mycelia are extremely active scavengers of inorganic forms of N, such as either NH_4^+ or NO_3^- (Johansen *et al.*, 1996; Finlay *et al.*, 1988). Experiments performed on *Pisolithus* demonstrate that the fungus has access to NH_4^+ and Ca^{2+} ions trapped in between the vermiculite 2:1 layer. As vermiculite samples were separated from the mycelium by cellophane films, soluble fungal exudates may be considered responsible for phyllosilicate weathering. In the experiment, NH4+ ions usually considered as retrograded, were mobilized from the interlayer spaces and replaced by Mg^{2+}, Al^{3+}, Ca^{2+}, and Na^+ ions (Paris *et al.*, 1995).

NH_4^+ absorbed by mycelia, or derived from NO_3^- reduction, is rapidly assimilated into glutamate and glutamine, which are then used to synthesize other amino acids, such as alanine and γ-aminobutyrate, within the foraging hyphae.

Next, assimilated N is either incorporated into mycelia proteins or translocated to the host, glutamine being regarded as the main translocation form (Martin and Botton, 1993). NH_4^+ and NO_3^- are apparently assimilated a long way from the mycorrhizal roots when the hyphal web is permeating the rhizosphere and various soil horizons. In natural ecosystems, therefore, primary NH_4^+ assimilation is carried out by the fungus, then the conversion to glutamine and its transfer to the host occur (Botton and Chalot, 1995). In addition, both AMF and ECM fungi mediate N-transfer between plants and mycelial links (Read, 1997). In grassland and forest ecosystems, where plants are grown in very close association, these webs may be crucial in between-plant N cycling. Similarly, in fruit tree orchards, this relationship can be achieved if proper environmental conditions (similar to forest ecosystems) are provided.

3.0 Phosphorus nutrition to plants

3.1 Phosphorus transport

Nutrient ions reach the surface of roots by three processes. These include: (1) root interception; (2) mass flow and (3) diffusion. Root interception is the process by which the nutrients are absorbed when the root surface comes in direct contact with a nutrient ion. Mass flow is the movement of nutrient to the root surface in the convective flow of water during the absorption of water by plant. It is generally agreed that nutrient ions, such as nitrate, sulphate, calcium and magnesium which are less readily retained by soils than phosphate and potassium are supplied by mass flow. Diffusion is the movement of nutrient ions towards a root surface caused by concentration gradient. In the absence of mass flow, continued uptake of nutrient ions by roots reduces the concentration of available nutrients at the root surface causing a concentration gradient. When a concentration gradient exists, there is a net movement from the zone of higher concentration in the soil solution to the zone of lower concentration in the root surface. The amount of nutrient supplied by diffusion process depends on the rate of diffusion; surface adsorption reactions and the tortuous nature of the soil water film retard the rate of diffusion of various ions in soil.

For most plant species, the maximum amount of P supplied by root interception is less than 1% of the available P in soil and this amount is usually a small percentage of the plant requirement. Since the concentration of P in soil solution remains low (0.05-0.3 µg P mL^{-1}) the amount of P supplied by mass flow also accounts for only a small percentage of the P used by the plant. In many soils, diffusion is the main process by which most of the P is supplied for plant uptake. All factors which govern the rate of diffusion of P in soil and the extent of root growth determine the availability of P to plants growing in soil. These factors include both soil (moisture content, tortuosity, buffering capacity, and temperature) and plant factors (root length and mycorrhizal infection).In a short-term, the adsorption-desorption process mostly determines the concentration of P in soil solution which in turn determines the rate of diffusion of P from the surface of the soil particles to the surface of the roots. Other processes, such as solid-state diffusion, desorption, precipitation-dissolution and immobilisation-mineralisation operate over longer periods of time.

3.2 Mycorrhiza mediated plant uptake of phosphorus

The contribution of mycorrhizal associations in the efficient uptake of P by plants is illustrated in (**Figure 5. 2**). The increase in plant growth by mycorrhizal association is largely due to increased absorption of nutrients from soil solution. An increase in absorption of P by mycorrhizal plants could be brought about by: increased physical exploration of the soil; increased movement into mycorrhizal fungus hyphae; modification of the root environment; increased storage of absorbed P; efficient transfer of P to plant roots; and efficient utilisation of P within the plant.

AMF in association with plant roots seem likely to increase P uptake by more thorough exploration of soil volume thereby making 'positionally unavailable' nutrients 'available'. This is achieved by decreasing the distance for diffusion of P ions and by increasing the surface area for absorption. The extensive hyphal growth of mycorrhiza effectively 'short-circuits' the distance for diffusion and thereby increases the uptake. Further the diameter of hyphae of mycorrhizal fungi is much finer than that of root hairs. The fineness of hyphae has two-fold advantages. Firstly, it increases the surface area of hyphae for greater absorption of nutrients. Secondly, it enables the entry of hyphae into pores in soils and organic matter that cannot be entered by root hairs and thereby increases the area of exploration. It has been observed that both in soils and solution cultures, mycorrhizal roots absorbed P almost twice as fast as did non-mycorrhizal plants which has been attributed primarily to the higher affinity and lower threshold concentration of P for absorption by mycorrhizal than non mycorrhizal plants.

Differences between mycorrhizal and non-mycorrhizal plants in the absorption of anions and cations may lead to differences in rhizosphere pH. Such changes in rhizosphere pH may change the availability of P to plants. It has been observed that mycorrhizal plants utilise NH NH_4^+ more efficiently than non-mycorrhizal plants. Absorption and assimilation of ammonium by mycorrhizal plants could reduce the pH around an infected root and thereby affect the availability of P.

Inorganic P ions absorbed by hyphae is stored in three forms: soluble orthophosphate; soluble polyphosphate; and polyphosphate granules. The increase in P uptake by mycorrhizal plants can also be explained in terms of the rate of transfer of P between mycorrhizal fungus hyphae and plant root hairs. The concentration of inorganic P inside the hyphae is approximately 1000 times higher than that in soil solution, so P must be absorbed actively against an electrochemical potential. A high concentration of P in the fungus is maintained by the hydrolysis of polyphosphate. Translocation of P within the hyphae occurs passively down a concentration gradient between the P source in the external hyphae and a sink in the root.

The rate-limiting steps in the uptake of P by plants from soil are: (1) the diffusion of phosphate ions in the soil solution to the plant roots; (2) the concentration of phosphate at the root surface; and (3) the release of phosphate ions from the soil particles. Infection of plant roots with a mycorrhizal fungus could alter any of these rate-limiting steps and increase phosphate uptake by plants (**Figure 5. 3**).

4.0 Potassium uptake by AMF

The third important macronutrient, potassium (K) is relatively less mobile and the uptake of K in AMF plants has been studied by several researchers by analysing their plant tissue concentrations (Mosse, 1957; Holevas, 1966; Possingham and Obbink, 1971; Huang *et al.*, 1985). Interestingly was found to be at lower concentrations in the tissues of AMF than in those of non-mycorrhizal plants. Because of the simultaneous effects of P nutrition on growth, extrapolation of K data (similarly for other nutrients) from tissue concentrations can result in wrong determination. In an experiment conducted on P-deficient soils, AMF infected sub clover (*Trifolium subterraneum*), showed elevated concentrations of K in shoots rather than roots (Smith *et al.*,1981). Interestingly, on P treatment, K concentrations in both presence and absence of AMF were very similar, suggesting an indirect effect of AMF on K uptake in P deficient plants, as observed with sulphate (Rhodes and Gerdemann, 1978). Apart from P nutrition, accumulation of K is also strongly influenced by the form of N available (nitrate or ammonium), as well as by other cations, particularly Na^+. It might also be influenced by the synthesis and storage of polyP, so that carefully designed experiments to investigate the influence of AMF colonization on K nutrition need to take all these potentially difficult factors into account.

5.0 Role of AMF on micronutrient transformation

The transformation of micronutrients in mycorrhiza associated plants largely depends on the N and P concentrations in soil, where AMF associations play important roles in this nutrient cycling through their microbial activity and their involvement in plant nutrient acquisition (Barea, 1991; Bethlenfalvay and Schüepp, 1994).The extraradical mycelium connects plant roots to the surrounding soil microhabitat increasing the soil volume exploited by host plants. This allows plants to survive in nutrient and/or water depleted zones (Marschner and Dell,

1994). Mycorrhizal hyphae transport mineral nutrients over greater distances from depleted zones than do roots (Jakobsen, 1995). Thus, under low nutrient conditions AMF-colonized roots may have an enhanced uptake of relatively immobile macro and micronutrients (Kothari *et al.*, 1991; Leyval *et* al., 2002; Subramanian and Charest, 1999). In addition, mycorrhizal roots often have not only increased length but also modified root architecture (Berta *et al.*, 1995).Yet, under high nutrient conditions, a reduced accumulation of minerals has been observed in some AMF plants (Arines *et al.*, 1989; Azcon and Barea, 1992; Clark, 1997; Leyval *et al.*, 1991; Posta *et al.*, 1994; Rubio *et al.*, 2002; Weissenhorn *et al.*, 1995).

In most AMF associated plants, uptake of copper (Cu) and zinc (Zn) is more efficient compared to other micronutrients. Number of researchers worked on Cu uptake in plants and observed increased concentration of Cu in the roots, shoots and seedlings (Daft *et al.*, 1975; Gnekow and Marschner, 1989; Li *et al.*, 1991). Mosse (1957) observed increased concentration in AMF associated apple seedlings. Li *et al.* (1991) conducted a pot experiment on Cu uptake in fungal hyphae of *Trifolium repens* and observed that the fungi has taken up around 62 % of total Cu uptake, avoiding the effects of P nutrition.

Similarly, in the presence of P, the mobility of Zn in soils is very low and its uptake by organisms is also diffusion-limited. Strong P-Zn interactions have made the effects of AMF colonization on whole-plant uptake difficult. Gilmore (1971) observed through tracer studies that Zn deficiency symptoms in peach disappeared as AMF developed. Subsequently, a number of studies have shown unequivocally that Zn uptake via the AMF uptake pathway is important and can alleviate Zn deficiency in several species in both pot and field experiments (Thomson, 1990;Wellings *et al.*,1991). Jansa *et al.*(2003) demonstrated concurrent transfer of 32P and 65Zn to maize, via *G. Intraradices* from up to 14 cm, where Zn transfer was much lower than P transfer. However, there was a strong linear correlation between the amounts of P and Zn transported, proving the importance of hyphal length density in soil in the acquisition of both nutrients. Interestingly, AMF colonization is also implicated in reducing Zn accumulation and hence plant toxicity in soils with high Zn content (Burleigh *et al.*, 2003). Although the mechanism is unclear, it can be promoting Zn uptake under conditions of deficiency. Li and Christie (2001) found that the protective effect was independent of indirect effects due to increased growth and P uptake in AMF plants.

Interactions between P fertilization and deficiencies of trace elements are well known in several species of typically AMF plants (Wallace *et al.*, 1978). In general, when the availability of P is increased, P uptake and plant growth also increase. Concentrations of Cu and Zn in the tissues fall, sometimes to levels at which deficiency symptoms become apparent. Colonization of AMF has been shown to affect these interactions (Lambert *et al.*,1979; Timmer and Leyden, 1980), so that at moderate levels of P fertilization, deficiency symptoms are alleviated because AMF increase uptake of the trace elements and tissue concentrations rise. At very high levels of P, AMF colonization itself may be reduced with consequent reductions in AMF uptake and reappearance of the deficiency symptoms. Interactions such as these may be involved in some cases of alleviation of Zn toxicity in polluted sites

(Dueck *et al.*,1986). If the sites are P deficient, then AMF P uptake could result in increased growth and dilution of Zn in the tissues. However, as shown above, this cannot be the explanation in all cases.

Colonization of AMF can also cause physiochemical or microbiological changes to the rhizosphere thus affecting root hyphae uptake of some nutrients (Amora-Lazcano and Azcon, 1997; Amora-Lazcano *et al.*, 1998; Marschner, 1995). Nitrogen (N) and phosphorus (P) levels are considered among the most important factors affecting AMF association efficiency. Not only the level but also the balance between N and P fertilizer regimes is crucial in establishing an efficient symbiosis. Enhanced N acquisition in AMF plants had been explained by an increased N demand in AMF-colonized plants because of high P nutrition (George *et al.*, 1995).

Increased accessibility of soil nutrient pools to plants has been associated with AMF colonization (Clark and Zeto, 2000). Heavy fertilization of N and P decreased mycorrhizal colonization, affecting AMF functioning on mineral nutrient acquisition. Differences in the mycorrhizal behaviour in nutrient acquisition per root weight and colonized root weight are expected when the nutritive status in the growth medium changes.

The effects of mycorrhizal colonization on acquisition of relatively immobile nutrients by mycorrhizal plants are still unclear and inconsistent experimental results have been attributed to variations in soil conditions and/ or mycorrhizal development (Clark and Zeto, 2000). Root/AMF colonization often results in an enhanced uptake of relatively immobile metal-micronutrients such as Cu, Zn and Fe. However, previous studies show that these micronutrients were lower in the shoots of mycorrhizal maize plants but if soil treatments inhibit root colonization, the uptake of nutrients could also be affected (Liu *et al.*, 2000).

6.0 Horticultural significance of AMF

Horticultural fruit crop production is all about propagating high yielding varieties and custom based breeds. Inoculating AMF at the propagation stage has multiple benefits, including stronger roots, higher water conducting capacity, increased nutrient efficiency and protection from environmental stress including pathogenic effects (Kapoor *et al.*, 2008). Apart from the conventional propagation techniques, new technologies involving tissue culture has revolutionised the horticulture industry, where new disease-free and high yielding varieties have become possible over the past three decades (Raaman and Patharajan, 2006; Rajasekharan and Ganeshan, 2002).Plant tissue culture relies on growing plants on nutrient rich growth substrates devoid of microbes, which results in the production of plantlets without any mutualistic symbiosis (Dolcet-Sanjuan *et al.*, 1996). Several researchers have demonstrated the beneficial effects of AMF on improving the growth and development of several *in-vitro* raised horticultural plant species including fruit trees (**Table 5.1**).

The main benefit in the inoculation of AMF at the nursery level is their effect towards increasing the overall fitness of the host plant by increasing the uptake of minerals such as P that are relatively immobile in soils (Turk *et al.*, 2006). Before planting a fruit tree or any horticultural plant, AMF develop intensively inside roots and within the soil by forming an extensive extraradical mycelium, which helps the plant in not only exploiting mineral nutrients but also accessing water from the soil. In those plants, where the root system is weak or restricted, fungal hyphae act as a bridge between roots and nutrient sites in soil and facilitate efficient uptake of immobile nutrients by host plants (Azcon-Aguilar and Barea, 1996). Along with P nutrition, most AMF colonization also increases the acquisition of other mineral nutrients such as Ca, Cu, Mn and Zn. Mycorrhizal structures effectively take up these nutrients from lower concentrations in the soil at which normal plant roots fail (Jefferies *et al.*, 2003). This is achieved by increasing the surface area of absorptive system (roots) of plants and exploring soil by extraradical hyphae beyond the root hair and P-depletion zone. The absorbed P is then converted to polyphosphate granules in external hyphae and passed to the arbuscule for transfer to the host plant (Azcon-Aguilar and Barea, 1996).

Enhanced nutrient uptake and increased water conducting capacity as a result of AMF inoculation in young plants are also benefited by the development of superior root system (Ponton *et al.*, 1990). This can increase the survival of cuttings or seedlings of fruit tree species from mechanical stress, being raised on nutrient media. Apart from the mechanical stress, the AMF can also help the young plants to survive against some of the environmental stresses such as high metal concentration, adverse pH, salinity and drought through the above mentioned mechanisms. Soil borne pathogens can be reduced or suppressed under AMF inoculated system (Slezack *et al.*, 2000; Elsen 2001; Harrier and Watson, 2004).

For propagation (both conventional and tissue culture) of horticultural crops including fruit trees, soil fumigation or steam treatments are often used for high value crops, to reduce losses caused by plant pathogens. These measures eradicate both the pathogens and also the beneficial microbes including mycorrhizae, thereby resulting in 'stunting', poor yields and variable growth of many plant species including: *Citrus, Persea* sp. (avocado), Capsicum,

Cassava, *Vitis* sp.,*Malus* sp.(apple), *Prunus* sp.(peach) and a few other horticultural fruit trees. Although stunting can be reversed or reduced by fertilizer applications, most species that are highly AMF responsive, are unable to make effective use of P and other immobile nutrients unless their roots are colonized by AMF. This effect is most noticeable in P-fixing soils and varies with plant species (Yost and Fox, 1979; Haas *et al.*, 1987; Li *et al.*, 2005). Hence, micropropagation of horticultural fruit trees is best achieved in the presence of AMF and also helps in improving and maintaining the overall of health of the fruit trees, by supplying nutrients, water and serving as a defence against plant pathogens.

7.0 Conclusions

The interaction between AMF colonization and accumulation of heavy metals andother toxic elements is an area of considerable interest in relation both to production of safe food and bioremediation programmes. A number of different mechanisms may be involved, including tissue dilution of the toxic element due to interactions with P nutrition and growth, sequestration of the toxic metal in the fungus and development of tolerance by the fungi. It is known that the development of mycorrhizal colonization and its effectiveness on plant growth is enhanced in poor soils. Under agronomic practices, P limitation in soil is usually compensated by heavy P fertilization that reduces AMF colonization and effectiveness. The concept of effectiveness is defined as the ability of an AMF inoculum to increase plant nutrient uptake and plant growth. Thus, the effective use of AMF inocula and resulting colonization would results in similar plant growth and yield, and reduce the need for high levels of P application which historically results in soil and ground water contamination, under horticultural cropping systems. Conventional high input fertilizers tend to change by sustainable systems with a minimal synthetic inputs and environmental impact. The aim is to change to more sustainable forms of agriculture in which the results will depend on both the AMF inoculum efficiency and soil nutrient availability. However, in order to optimize management of AMF/host plants under field conditions, more knowledge on the effect of horticultural practices is required.

8.0 Future strategies

The complex system of nutrient accession, assimilation and absorption in a soil-mycorrhizae-plant system should be carried out at the molecular level focussing on metabolic profiling. Although the current methodologies on looking at the physiological aspects of this system is sufficient in terms of tracing the nutrient transport within the system, the use of genetic tools such as DNA fingerprinting and profiling will help to device specific strategies for the target plants. Development of mathematical models which involves the transcription and translation processes involved in the nutrient mobilizing metabolites will be helpful to determine the nutrient cycling. This also requires identification of various mycorrhizal taxa (not only in the AMF group, but also in the other groups) which can help in determining the dynamics of organic compounds in the soil and their actual uptake by mycorrhizal plants. The use of stable isotope can also help in identifying the nutrient transport in horticultural plants especially fruit trees.

References

1. Allen, J.W. and Shachar-Hill, Y., 2008.Sulfur transfer through an arbuscular mycorrhiza. *Plant Physiol.* 149, 549-560.

2. Amora-Lazcano, E.andAzcon, R., 1997. Response of sulphur cycling microorganisms to arbuscular mycorrhizal fungi in the rhizosphere of maize, *Appl. Soil Ecol.* 6, 217-222.

3. Amora-Lazcano, E., Vazquez, M.M. and Azcon, R., 1998.Response of nitrogen transforming microorganisms to arbuscular mycorrhizal fungi. *Biol. Fertil. Soil* .27, 65-70.

4. Arines, J., Vilarino, A. and Sainz, M., 1989.Effect of different inocula of vesicular arbuscular mycorrhizal fungi on manganese content and concentrations in red clover (*Trifolium Pratense* L.) plants, *New Phytol.* 122, 215-219.

5. Azcon, R. and Barea, J.M., 1992. The effect of vesicular arbuscular mycorrhizae in decreasing Ca acquisition by alfalfa in calcareous soils, *Biol. Fertil. Soil* .13,155-159.

6. Azcon-Aguilar, C., Barcelo, A., Vidal, M. and Vina, G., 1992.Further studies on the influence of mycorrhizae on growth and development of micropropagated avocado plants.*Agronomie* .12, 837-840.

7. Azcon-Aguilar, C. and Barea, J.M., 1996. Arbuscular mycorrhizas and biological control of soil-borne plant pathogens: an overview of the mechanisms involved. *Mycorrhiza* .6,457-464.

8. Barea, J.M. 1991. Vesicular/arbuscular mycorrhizae as modifiers of soil fertility. In: Stewart, B.A. (Ed.), Advances in Soil Science, Springer, New York. pp. 1 -40.

9. Bago, B., Pfeffer, P.E., Abubaker, J., Jun, J., Allen, J.W., Brouillette, J., Douds, D.D., Lammers, P.J. and Shachar-Hill, Y., 2003. Carbon export from arbuscular mycorrhizal roots involves the translocation of carbohydrate as well as lipid. *Plant Physiol.* 131, 1496-1507.

10. Bago, B., Pfeffer, P. and Shachar-Hill, Y., 2008. Could the urea cycle be translocating nitrogen in the arbuscular mycorrhizal symbiosis? *New Phytol.* 149(1), 4-8.

11. Bethlenfalvay, G.J., and Schüepp, H., 1994. Arbuscular mycorrhizas and agrosystem stability. In: Gianinazzi, S. and Schüepp, H. (Eds.). Impact of Arbuscular Mycorrhizas on Sustainable Agriculture and Natural Ecosystems, Birkhäuser, Basel. pp. 117-131.

12. Berta, G., Trotta, A., Fusconi, A., Hooker, J.E., Munro, M., Arkinson, D., Giovannetti, M., Morini, S., Fortuna, P., Tisserant, B., Gianinazzi-Pearson, V. and Gianinazzi, S., 1995.Arbuscular mycorrhizal induced changes to plant growth and root system morphology in *Prunus cerasifera, Tree Physiol.* 15, 281-293.

13. Bolan, N.S., 1991. A critical review on the role of mycorrhizal fungi in the uptake of phosphorus by plants. *Plant Soil* .134, 189-207.

14. Botton, B. and Chalot, M., 1995. Nitrogen assimilation: enzymology in ectomycorrhizas. In: Mycorrhiza: Structure, Function, Molecular Biology and Biotechnology (Eds. A. Varma & B. Hock),. Springer-Verlag, Berlin. pp. 325–363.

15. Branzanti, B., Gianinazzi-Pearson, V. and Gianinazzi, S., 1992. Influence of phosphate fertilisation on the growth and nutrient status of micropropagated apple infected with endomycorrhizal fungi during the weaning stage. *Agronomie*. 12, 841-845.

16. Bucher, M., 2007. Functional biology of plant phosphate uptake at root and mycorrhiza interfaces. *New Phytol.* 173,11-26.

17. Burleigh, S.H., Kristensen, B.K. and Bechmann, I.E, 2003. A plasma membrane zinc transporter from *Medicago truncatula* is upregulated in

roots by Zn fertilisation, yet down-regulated by arbuscular mycorrhizal colonization. *Plant Mol. Biol.* 52, 1077-1088.

18. Cassels, A.C., Mark, G.L. and Perappuram, C., 1996. Establishment of arbuscular mycorrhizal fungi in autotrophic strawberry cultures *in vitro* comparison with inoculation of microplants *in vivo. Agronomie.* 16, 625-632.

19. Capellazzo, G., Lanfranco, L., Fitz, M., Wipf, D. and Bonfante, P., 2008. Characterisation of an amino acid permease from the endomycorrhizzal fungus *Glomus mosseae. Plant Physiol.* 147(1), 429-437.

20. Chalot, M., Blaudez, D. and Brun, A., 2006. Ammonia: a candidate for nitrogen transfer at the mycorrhizal interface. *Trends Plant Sci.* 11, 263-266.

21. Clark, R.B., 1997. Arbuscular mycorrhizal adaptation, spore germination, root colonization, and host plant growth and mineral acquisition at low pH. *Plant Soil* .192, 15-22.

22. Clark, R.B. and Zeto, S.K., 2000. Mineral acquisition by arbuscular mycorrhizal plants , *J. Plant Nutr.*23,867-902.

23. Cruz, C., Egsgaard, H., Trujillo, C., Ambus, P., Requena, N., Martins-Loucao, M.A. and Jakobsen, I. 2007.Enzymatic evidence for the key role of arginine in nitrogen translocation by arbuscular mycorrhizal fungi. *Plant Physiol.* 144,782-792.

24. Daft, M.J., Hacksaylo, E. and Nicolson, T.H., 1975. Arbuscular mycorrhizas in plants colonising coal spoils in Scotland and Pennsylvania. In:Endomycorrhizas. Sanders, F.E., Mosse, B., Tinker, P.B (Eds.)pp.561-580.*Academic Press*, London, UK.

25. Dolcet-Sanjuan, R., Claveria, E., Camprubi, A., Estaun, V. and Calvet, C., 1996.Micropropagation of walnut trees (*Juglans regia* L.) and response to arbuscular mycorrhizal inoculation.*Agronomie* .16, 639-645.

26. Dueck, T.A., Visser, P., Ernst, W.H.O.andSchat, H., 1986. Vesicular-arbuscular mycorrhizae decrease zinc toxicity to grasses growing in zinc-polluted soil. *Soil Biol.Biochem.* 18,331-333.

27. Elsen, A., Declerck, S. and de Waele, D., 2001.Effects of *Glomus intraradices* on the reproduction of burrowing nematode (*Radopholussimilis*) in dixenic culture. *Mycorrhiza* .11, 49-51.

28. Finlay, R. D., Ek, H., Odham, G. and Söderström, B., 1988. Mycelial uptake, translocation and assimilation of nitrogen from 15N-labelled ammonium by *Pinus sylvestris* plants infected with four different ectomycorrhizal fungi. *New Phytolo.* 110,59-66.

29. George, E., Marschner, H. and Jakobsen, I., 1995. Role of arbuscular mycorrhizal fungi in uptake of phosphorus and nitrogen from soil, *Crit. Rev. Biotect.* 15, 257-270.

30. Gnekow, M.A. and Marschner, H., 1989. Role of VA-mycorrhiza in growth and mineral nutrition of apple *Malus pumila* var. *domestica* rootstock cuttings. *Plant Soil.* 119285-293.

31. Govindarajulu, M., Pfeffer, P.E., Jin, H.R., Abubaker, J., Douds, D.D., Allen, J.W., Bucking, H., Lammers, P.J. and Shachar-Hill, Y., 2005. Nitrogen transfer in the arbuscular mycorrhizal symbiosis. *Nature* .435,819-823.

32. Gobert, A. and Plassard, C., 2008. The beneficial effect of mycorrhizae on N utilization by the host-plant: myth or reality? In: Varma, A. (Ed), Mycorrhiza. Springer-Verlag, Berlin, pp. 209-240.

33. Gomez, S.K., Javot, H., Deewatthanawong, P., Torres-Jerez, I., Tang, Y.H., Blancaflor, E.B., Udvardi, M.K.and Harrison, M.J., 2009.*Medicagotruncatula* and *Glomus intraradices* gene expression in cortical cells harboring arbuscules in the arbuscular mycorrhizal symbiosis. *BMC Plant Biol.* 9, 10.

34. Haas, J.H., Bar-Yosef, B., Krikun, J., Barak, R., Markovitz, T. and Kramer, S., 1987.Vesicular-arbuscular mycorrhizal-fungus infestation and phosphorus fertigation to overcome pepper stunting after methyl bromide fumigation.*Agron. J.*79, 905-910.

35. Harrison, M.J. and van Buuren, M.L., 1995. A phosphate transporter from the mycorrhizal fungus *Glomus versiforme*. *Nature* .378,626-629.

36. Harrier, L.A. and Watson, C.A., 2004.The potential role of arbuscular mycorrhizal (AM) fungi in the bioprotection of plants against soil-borne pathogens in organic and/or sustainable farming systems.*Pest Manag. Sci.* 60, 149-157.

37. Holevas, C.D., 1966. The effect of vesicular-arbuscular mycorrhiza on the uptake of soil phosphorus by strawberry *Fragariasp*.var.Cambridge Favourite. *The Journal of Horticultural Science* .41,57-64.

38. Hodge. A., Campbell, C.D. and Fitter, A.H., 2001. An arbuscular mycorrhizal fungus accelerates decomposition and acquires nitrogen directly from organic material. *Nature* .413, 297-299.

39. Huang, R.S., Smith, W.K.and Yost, R.S., 1985. Influence of vesicular-arbuscular mycorrhizal on growth water relations, and leaf orientation in *Leucaenaleucocephala* Lam. DeWit. *New Phytol.* 99,229-243.

40. Jakobsen, I., 1995. Transport of phosphorus and carbon in VA mycorrhizas. In: Varma, A. and Hock B. (Eds.). Mycorrhizas: Structure Function, Molecular Biology and Biotechnology, Springer, Berlin, Germany. pp. 297-324.

41. Jansa, J., Mozafar, A. and Frossard, E. 2003.Long-distance transport of P ad Zn through the hyphae of an arbuscular mycorrhizal fungus in symbiosis with maize.*Agronomie* .23, 481-488.

42. Jefferies, P., Gianinazzi, S., Perotto, S., Turnau, K. and Barea, J.M., 2003.The contribution of arbuscular mycorrhizal fungi in sustainable maintenance of plant health and soil fertility. *Biol. Fert. Soils* .37, 1-16.

43. Jin, H., Pfeffer, P.E., Douds, D.D., Piotrowski, E., Lammers, P.J.and Shachar-Hill, Y., 2005. The uptake, metabolism, transport and transfer of nitrogen in an arbuscular mycorrhizal symbiosis.*New Phytol.* 168, 687-696.

44. Johansen, A., Finlay, R.D. and Olsson, P.A. 1996. Nitrogen metabolism of external hyphae of the arbuscular mycorrhizal fungus *Glomus intraradices*. *New Phytol.* 133, 705-712.

45. Kapoor, R., Sharma, D. and Bhatnagar, A.K., 2008. Arbuscular mycorrhizae in micropropagation systems and their potential applications. *Scientia Horticulturae.* 116, 227-239.

46. Kothari, S.K., Marschner, H.and Romheld, V., 1991. Contribution of the VA mycorrhizal hyphae in acquisition of phosphorus and zinc by maize frown in a calcareous soil. *Plant Soil* .131, 177-185.

47. Krishna, H., Singh, S.K., Minakshi, Patel, V.B., Khawale, R.N., Deshmukh, P.S. and Jindal, P.C., 2006. Arbuscular-mycorrhizal fungi alleviate transplantation shock in micropropagated grapevine (*Vitis vinifera* L.). *J. Horticult. Sci. Biotech.* 81, 259-263.

48. Lambert, D.H., Baker, D.E. and Cole, H. Jr., 1979. The role of mycorrhizae in the interactions of phosphorus with zinc, copper, and other elements. *Soil Sci. Soc. Am. J.* 43,976-980.

49. Lambers, H., Shane, M.W., Cramer, M.D., Pearse, S.J. and Veneklaas, J., 2006. Root structure and functioning for efficient acquisition of phosphorus: Matching morphological and physiological traits. *Ann. Botany* .98, 693-713.

50. Leyval, C., Berthelin, J., Schontz, D., Weissenhorn, I. and Morel, J.L., 1991. Influence of endomycorrhizas on maize uptake of Pb, Cu, Zn and Cd applied as Mineral Salts or sewage sludges, In: Farmer, J.G.(Ed.), Heavy Metals in the Environment, CEP, Edinburgh, pp. 204-207.

51. Leyval, C., Joner, E.J., del Val C. and Haselwandter, K., 2002. Potential of Arbuscular Mycorrhizal Fungi for Bioremediation. In: Mycorrhizal Technology in Agriculture: From Genes to Bioproducts, Gianinazzi, S. (Ed.). BirkhauserVerlag, Basel, Switzerland, ISBN: 9783764364854, pp: 175-186.

52. Li, X.L., Marschner, H. and George, E., 1991. Acquisition of phosphorus and copper by VA-mycorrhizal hyphae and root-to-shoot transport in white clover. *Plant Soil.* 136, 49-57.

53. Liu, A., Hamel, C., Hamilton, R.I., Ma, B.L. and Smith, D.L., 2000. Acquisition of Cu, Zn, Mn and Fe by mycorrhizal maize (*Zea mays* L.) grown in soil at different P and micronutrient levels, *Mycorrhiza.* 9, 331-336.

54. Li, X.L and Christie, P., 2001. Changes in soil solution Zn and pH and uptake of Zn by arbuscular mycorrhizal red clover in Zn contaminated soil. *Chemosphere,* 42,201-207.

55. Li, H.Y., Zhu, Y.G., Marschner, P., Smith, F.A. and Smith, S.E., 2005. Wheat responses to arbuscular mycorrhizal fungi in a highly calcareous soil differ from those of clover, and change with plant development and P supply. *Plant Soil* 277, 221-232.

56. Martin, F. and Botton, B. 1993.Nitrogen metabolism of ectomycorrhizal fungi and ectomycorrhiza.*Adv. Plant Pathol.* 9, 83-102.

57. Marschner, H. and Dell, B., 1994. Nutrient uptake in mycorrhizal symbiosis, *Plant Soil* .159, 89-102.

58. Marschner, H. 1995. Mineral nutrition of higher plants. *Academic Press*, London.

59. Marin, M., Mari, A., Ibarra, M. and Garcia-Ferriz, L., 2003.Arbuscular mycorrhizal inoculation of micropropagated persimmon plantlets. *J. Horticult. Sci. Biotech.* 78,734-738.

60. Mosse, B., 1957. Growth and chemical composition of mycorrhizal and non-mycorrhizal apples. *Nature* .179,922-924.

61. Nouri, E., Breuillin-Sessoms, F., Feller, U. and Reinhardt, D., 2014. Phosphorus and nitrogen regulate arbuscular mycorrhizal symbiosis in *Petunia hybrida*. PLoS ONE 9(3): e90841. doi:10.1371/journal.pone.0090841.

62. Paris, F., Bonnaud, P., Ranger, J., Robert, M. and Lapeyrie, F., 1995. Weathering of ammonium- or calcium-saturated 2:1 phyllosilicates by ectomycorrhizal fungi *in vitro*. *Soil Biol. Biochem.* 27, 12371244.

63. Pinior, A., Grunewaldt-Stöcker, G., Alten, H.V. and Strasser, R.J., 2005. Mycorrhizal impact on drought stress tolerance of rose plants probed by chlorophyll a fluorescence, proline content and visual scoring. *Mycorrhiza* .15, 596-605.

64. Possingham, J.V. and Groot Obbink, J., 1971. Endotrophic mycorrhiza and the nutrition of grape vines.*Vitis* .10,120-130.

65. Ponton, F., Piche, Y., Parent, S. and Caron, M., 1990. The use of vesicular arbuscular mycorrhiza in boston fern production, I, Effect of peat-based mixes. *Hort. Sci.* 25, 183-189.

66. Posta, K., Marschner, H. and Romheld, V., 1994. Manganese reduction in the rhizosphere of mycorrhizal and nonmycorrhizal maize, *Mycorrhizal* .5,119-124.

67. Quatrini, P., Gentile, M., Carimi, F., De Pasquale, F. and Puglia, A.M., 2003. Effect of native arbuscular mycorrhizal fungi and *Glomus mosseae* on acclimatization and development of micropropagated *Citrus limon* (L.). Burm. *J. Horticult. Sci. Biotech.* 78, 39-45.

68. Rajasekharan, P.E. and Ganeshan, S., 2002. Conservation of medicinal plant biodiversity an Indian perspective. *J. Medi. Aroma. Plant Sci.* 24, 132-147.

69. Raaman, N., and Patharajan, S., 2006. Integration of arbuscular mycorrhizal fungi with micropropagated plants. In: Mukerji, K.G. and Manoharachary, C. (Eds.), *Current Concepts in Botany*. I.K. International Publishing House, India.

70. Read, D.J., 1991. Mycorrhizas in ecosystems.*Experientia* .47, 376-3 91.

71. Read, D.J., 1997. Mycorrhizal fungi: The ties that bind. *Nature,*.388, 517-518.

72. Rhodes, L.H. and Gerdemann, J.W., 1978. Translocation of calcium and phosphate by external hyphae of vesicular-arbuscular mycorrhizae. *Soil Science* .126,125-126

73. Rodríguez-Romero, A.S., Piñero-Guerra, M.S. and Jaizme-Vega, M.C., 2005.Effect of arbuscular mycorrhizal fungi and rhizobacteria on banana growth and nutrition.*Agron. Sustain. Dev.* 25, 395-399.

74. Rubio, R., Borie, F., Schalchli, C., Castillo, C. and Azcon, R., 2002.Plant growth responses in a natural acidic soil as affected by arbuscular mycorrhizal inoculation and phosphorus sources, *J. Plant Nutr.*25, 1389-1405.

75. Sbrana, C., Vitagliano, C., Avio, L. and Giovanneti, M., 1992. Influence of vesicular-arbuscular mycorrhizae on transplant stress of micropropagated apple and peach rootstocks. In: Micropropagation, Root Regeneration and Mycorrhizas. Joint Meeting between COST 87 and COST 8.10, Dijon, France, p. 51.

76. Schubert, A., Bodrino, C. and Gribaudo, I., 1992., Vesicular arbuscular mycorrhizal inoculation of Kiwifruit (*Actinidia deliciosa*) micropropagated plants. *Agronomie* .12, 847-850.

77. Singh, S.K., Minakshi, G., Khawale, R.N., Patel, V.B., Krishna, H. and Saxena, A.K., 2004. Mycorrhization as an aid for biohardening of *in vitro* raised Grape (*Vitis vinifera* L.) plantlets. In: ISHS *Acta Horticulturae* 662: VII International Symposium on Temperate Zone Fruits in the Tropics and Subtropics.

78. Slezack, S., Dumas-Gaudot, E., Paynot, M. and Gianinazzi, S., 2000. Is a fully established arbuscular mycorrhizal symbiosis required for bioprotection of *Pisum sativum* roots against *Aphanomy ceseuteiches*? MPMI 13, 238–241.

79. Smith, S.E. and Read, D.J., 1997. Mycorrhizal Symbiosis, 2nd edition. *Academic Press*, London, UK.

80. Smith, S.E. and Read, D.J., 2008. Mycorrhizal Symbiosis, 3rd edition. *Academic Press*, New York, USA.

81. Smith, S.E., Smith, F.A. and Nicholas, D.J.D., 1981. Effects of endomycorrhizal infection on phosphate and cation uptake by Trifolium subterraneum. *Plant Soil.* 63,57-64.

82. Subramanian, K.S. and Charest, C., 1999. Acquisition of N by external hyphae of arbuscular mycorrhizal fungus and its impact on physiological response in maize under drought-stressed and wellwatered conditions. *Mycorrhiza.* 9, 69-75.

83. Thomson, J.P., 1990. Soil sterilization methods to show VA-mycorrhizae aid P and Zn nutrition of wheat in vertisols. *Soil Biol. Biochem.* 22, 229-240.

84. Timmer, L.W. and Leydan, R.F., 1980. The relationship of mycorrhizal infection to phosphorus-induced copper deficiency in sour orange seedlings.*New Phytol.* 85,15-23.

85. Tisserant, B. and Gianinazzi-Pearson, V., 1992. Micropropagation of *Platanus* X *acerifolia* (Wild) and post vitro VA endomycorrhization: problems and progress. In: Micropropagation, Root Regeneration and Mycorrhizas. Joint Meeting between COST 87 and COST 8.10, Dijon, France, p. 48.

86. Tian, C., Kasiborski, B, Koul, R, Lammers, P.J., B cking, H. and Sachar-Hill, Y., 2010. Regulation of the nitrogen transfer pathway in the arbuscular mycorrhizal symbiosis: Gene characterisation and the coordination of expression with nitrogen flux. *Plant Physiol.* 153, 1175-1187.

87. Turk, M.A., Assaf, T.A., Hameed, K.M. and Al-Tawaha., 2006. Significance of mycorrhizae. *World J. Agric. Sci.* 2, 16-20.

88. Vidal, M.T., Azcon-Aguilar, C., Barea, J.M. and Pliego-Alfaro, F., 1992. Mycorrhizal inoculation enhances growth and development of micropropagated plants of avocado. *Hort. Sci.* 27, 785-787.

89. Wallace, A., Mueller, R.T. and Alexander, G.V., 1978. Influence of phosphorus on zinc, iron, manganese, and copper uptake in plants. *Soil Sci.* 126,336-341.

90. Wellings, N.P., Wearing, A.H. and Thompson, J.P., 1991. Vesicular-arbuscular mycorrhizae (VAM) improve phosphorus and zinc nutrition andgrowth of pigeonpea in a Vertisol. *Aust. J. Agr. Res.* 42(5), 835-845.

91. Weissenhor, I., Leyval, C., Belgy, G. and Berthelin, J., 1995. Arbuscular mycorrhizal contribution to heavy metal uptake by maize *(Zea mays* L.) in pot culture with contaminated soil, *Mycorrhiza.* 5, 245-251.

Chapter 6

Nutrient Dynamics in Fruit Crops

K.S. Subramanian[1*], M. Thangaraju[2], N. Balakrishnan[1] and Vijayalakshmi[3]

[1]Department of Nano Science & Technology,
[2]Department of Agricultural Microbiology,
Tamil Nadu Agricultural University, Coimbatore 641 003

[3]Indian Institute of Horticultural Research,(ICAR) Bangalore 560089

*Address for the correspondence: kssubra2001@rediffmail.com

ABSTRACT

Arbuscular mycorrhizal fungi (AMF) play a significant role in symbiotic association with plant root in over 80% of the terrestrial plant species and facilitate biogeocycling of macro and micronutrients. Since AMF influence microorganisms which mineralize organic matter and nitrification in the soil, the availability or organic source of P and N increases in soil which are efficiently absorbed by the plant through the hyphae of mycorrhizal fungi. The improved nutritional status assist the host plant to produce higher biomass which is primarily attributed to the uptake of immobile nutrients especially phosphorus. Improved intake of highly immobile phosphate ions is particularly important in low-input organic farming, as P is most limiting under these situations. In addition to P, AMF symbiosis also facilitates N transport by the external mycelium which sets in motion host metabolic changes that provides greater ability for the plants to sustain water deficit conditions where NO_3^- ion mobility is impended by restricted mass flow and diffusion. The external mycelium of AMF also has the ability to transport immobile micronutrients such as

Zn and Cu. Mycorrhizae colonized soils retain stable aggregates, crumb structure and water holding capacity. Thus, mycorrhizal symbiosis offers a wholesome package of nutritional benefits and assists in maintenance of soil health. Further, mycorrhizal colonization provides protection against root feedingpathogens and nematodes by eliciting phytoalexins or serve as physical barrier. The AMF is capable of accumulating heavy metals and therefore prevent these toxicants enter into the aerial parts besides keeping the rhizosphere free from contaminants. Impact of AMF on nutrient cycling in few fruit crops is presented here.

Keywords: Mycorrhiza, Fruit Crops, Phosphorus, Nitrogen, Growth,

1.0 Introduction

Large amount of mineral fertilizers are used to enhance fruit yield. AMF are an alternative to maintain good soil health and enhance fruit production. They also help in preserving soil and water resources for future generations. Mycorrhizal plants accumulate P, K, Ca, Cu and Mn in the leaf in higher concentrations than non-mycorrhizal plants (Ross and Harper 1970; Nopamombodi *et al.*, 1987). Mycorrhizal plant remains in close contact with the soil for a longer period of time than a non-mycorrhizal. In agriculture soil-plant systems, the main impacts of AMF may be different from those in natural systems where N and P availability limit plant productivity (Hamel, 2004). Under well irrigated conditions, the major impact of AMF which are extensions of root and regulators of photosynthesis-derived C input to soil could be through microbial processes and soil quality. AMF due to their influence on plant and soil also influence soil N and P fertility.

Peach, belonging to a species of *Prunus* of the subfamily *Prunoideae* of the family *Rosaceae,* is widely distributed and produced all around the world. Inoculation with *G. fasciculatum* overcame soil-fumigation nutrient deficiency effects very effectively in peach nursery seedlings than the standard nursery practice of side-dressing with P and Zn at planting time (Larue *et al.*, 1975). AMF increased 25-75% growth in peach compared to the control plants (McGraw and Schenck 1980). Wu *et al.* (2007) has revealed that five *Glomus* species differed in their ability to improve water relations of red tangerine (*Citrus tangerine*) seedlings. Mycorrhizal symbiosis helped peach seedlings to overcome soil-fumigation nutrient-deficiency effects in nursery, uptake nutrients and alleviate flooded stress, but not overcome peach replant problem (Rutto *et al.*, 2002; Rutto and Mizutani, 2006). Knowing the happenings at the root zone is very crucial in understanding how AMF influence soil N and P dynamics.

2.0 Improved nutrition

2.1 Root system

The mycorrhizae benefit the rhizosphere by the mycelia they form around the roots. The internal and external fungal hyphae make contact with ten entry points per cm of root surface (Ocampo *et al.*, 1980). The external mycelium considerably increases the contact of the root with the medium in which it grows. One cm of root can generally explore about 1-2 cm^3 of soil using root hairs. When colonized by AMF assuming radial growth of mycelium around the root, the hyphae can increase this area from 5 to 200 times. A rhizospheric soil volume of 200 cm^3 may be an exception, but 12-15 cm^3 per infected root is common. The AMF mycelium appears to be more resistant to abiotic stresses such as drought, toxic elements, and soil acidity than the root itself (Sylvia and Williams 1992). The principal function of mycorrhiza is to increase the soil volume explored for nutrient uptake and to enhance the efficiency of nutrient absorption. It has been shown that mycorrhizal plants can absorb and accumulate several times more phosphate from the soil or solution than non-mycorrhizal plants (Mosse *et al.* 1981; Smith and Dowd,1981). In plant mycorrhizal association, mycorrhizosphere play an important role in nutrient transformation, the term "mycorrhizosphere" is referred to the combined rhizosphere and hyphosphere zones of mycorrhizal plant (Linderman 1988). The hyphosphere is the thin layer of soil surrounding arbuscular mycorrhizal hyphae, and the bulk soil, which includes senescing roots.AMF influence nutrient cycling in the root zone. At 15 cm depth, growing roots are known to create travelling waves of bacteria, (Zelenev *et al.*, 2000) which contribute to nutrient mobilization in the soil which can be reabsorbed by roots and mycorrhizae (Hamel, 2004). There is great amount of nutrient efflux from plant roots to the rhizosphere which is reabsorbed by mycorrhizal hyphae. This is an important step in nutrient cycling. The few AMF hyphae also absorb amino acids as they are situated along the plant roots (Friese and Allen,1991). Hawkins *et al.* (2000) demonstrated the ability of AMF to absorb a limited amount of amino acids. By supplying C source to rhizosophere microorganisms, plants also stimulate a large number of microorganisms, which enhance the P solubilization or mineralization process (Richardson 2001). The mycorrhizal hyphae by virtue of their capacity to colonize root hairs and also spread as extraradical hyphae extending away from the roots complement nutrient uptake and minimize the impact of root exudation losses.

2.2 Phosphorus

Soils of the tropics are very low in P avaliability and fix much of the phosphorus fertilizer added (> 80%). The available portion of total soil P is commonly less than 1% and is mainly controlled by chemical reactions and to a lesser extent by biological processes. Furthermore, the rate of diffusion of PO$^-$ ions in soil solution is less than 1mm / day. It has been proved beyond doubt that AMF symbiosis enhances host plant P nutrition which is highly limiting in agricultural production systems (Smith and Read, 1997). There are four possible hypothesis associated with the improved host plant P nutrition. The external mycelium of mycorrhizal fungus can take up P in the form of trehalose phosphate more effectively than roots at low concentrations. The external mycelium can proliferate far beyond the rhizosphere and increases the soil volume which is exploited for P uptake. The hyphal transport of P has been estimated to be 20-90% and likely to fulfill the entire requirement of fertilizer P. Rapid absorption of soluble form of P by the external hyphae leads to a shift in the equilibrium towards the release of bound P from the soil reserves (Smith and Read, 1997).

Atkinson (1986) and Neilsen *et al.* (1990) reported that P can limit fruit tree growth. As the plant needs a high P inflow rate and as P has a low mobility in soil, P depletion zones are likely to develop in the soil around apple roots because P is absorbed faster than it can diffuse toward the root surface. Bolan *et al.* (1987) emphasize that the P nutrition of apple trees is very dependent on the mycorrhizal fungi that extend the absorbing surface of plant root systems. Therefore, responsiveness to mycorrhizal colonization is a characteristic that could be considered in the evaluation of apple rootstocks. Shokri and Maadi (2009) observed that as salinity of soil increased, the concentration of P decreased in soils of non-mycorrhizal plants in comparison to mycorrhizal plants. Improved P uptake by AMF in plants grown under saline conditions may also reduces the negative effects of Na^+ and Cl^- ions by maintaining vacuolar membrane integrity, which facilitates compartmentalization within vacuoles and selective ion intake thereby preventing ions from interfering in metabolic pathways of growth (Cantrell and Lindermann, 2001). Bucher (2007) indicated that arbuscular mycorrhizal exhibited improved P acquisition efficiency and thus require lower inputs of P fertilizer for optimal growth.

AMF influences the release of phosphatase enzymes and organic anions in the soil (Trolove *et al.*, 2003) and also the size of the soil microbial biomass (He *et al.*, 1997) in addition to physically enhancing plant P extraction capabilities by producing extracellular phosphatases (Smith and Read 1997, Koide and Kabir 2000). In a compartmentalized *in-vitro* system, AMF also shown to take up 32P from AMP in the absence of heterotrophic organisms and to translocate limited amounts of this hydrolyzed P to a host root (Joner *et al.*, 2000a). This demonstrates that AMF have the potential to provide their host plants with some amounts of P from organic sources. Although AMF can contribute to P mineralization, the contribution of AMF-derived phosphatase to P mineralization may be insignificant (Joner *et al.*, 2000b). Feng *et al.* (2003) reported that the ability of AMF to improve plant P uptake is largest when the P source is organic. Mycorrhizal plants compete better than non-mycorrhizal for P liberated from decomposing organic matter. Soil microbial biomass is important to plant nutrition because these microbes constantly release the immobilized plant available P in the soil and preventing P from being fixed by soil minerals.

2.3 Nitrogen

Nitrogen is considered "kingpin" among the essential nutrients which is considered indispensable in any organic or conventional agricultural production systems. In tropical soils, the available N status is very low due to excessive loss in leaching, gaseous loss through volatilization and intense microbial activity. Under such situation mycorrhizal fungi play a vital role in improving plant nutrition through N acquisition and assimilation mechanisms. The external mycelium plays a crucial role in direct N acquisition and transport to the root cells thereby contributing to plant growth and nutrition. AMF have the potential to increase the availability of organic N. AMF may directly take up the organic molecules or enhance mineralization. Tracer studies have revealed that the extraradical mycelium in AMF can derive 15N from the soil (Frey and Schüepp, 1993; Tobar *et al.*, 1994). In another compartmental box experiment, Subramanian and Charest (1999) have shown that the amount of NO_3^- ions being transported by the external hyphae was about 30-35% under water deficit conditions. The extraradical AMF hyphae can uptake NH_4^+ as NO_3^- from the soil organic residues (Johansen, 1999; Bago *et al.*, 2001; Hawkins and George, 2001) in addition to significant amounts of N-organic and amino acids (Hawkins *et al.*, 2000). AMF enchanced NH_4^+ uptake in plants by increasing the soil nitrification rate (Hamel, 2004). AMF have the potential to improve plant competitive ability for N uptake in N-limited systems. However, only a high affinity NH_4^+ transporter expressed in extraradical mycorrhizal hyphae has been characterized up to now (López-Pedrosa *et al.*, 2006). Additionally, functional AMF can generate important pH changes in the mycorrhizosphere, as it has been detected in various assays using *in-vitro* culture systems (Bago *et al.*, 1996; Bago and Azcón-Aguilar, 1997).

It has been observed that the development of mycelium in the presence of NO_3^- is associated with a pH increase, and, the opposite occurs in NH_4^+ presence. Despite this, there is no concordance in respect to the effect of fertilization with different N-sources on mycorrhizal plants. Some studies suggest the preference for NO_3^- (Azcón *et al.*, 1992), while others recommend the preference for NH_4^+ (Cuenca and Azcón, 1994), with genotypical variations within the same plant species, mainly in limited N environments (Nakamura *et al.*, 2002). AMF have the potential to improve plant. AMF activity in acidic soils is of particular relevance, since it favours root absorption of typically limited elements in this type of soil, especially P (as previously mentioned), N and some microelements (Clark *et al.*, 1999; Clark and Zeto, 2000; Borie *et al.*, 2002). Several studies showed that different agronomic practices can affect AMF functionality as well as spore density, mycelia and colonized roots present in the soil. Among others, crop system, rotation design and amount and type of the used fertilizer are of particular relevance (Jeffries and Barea, 2001).

Giri and Mukerji (2004) also recorded highest accumulation of N in shoots of mycorrhizal *Sesbania grandiflora* and *S. aegyptiaca* than non mycorrhizal plants at all salinity levels. Increased N uptake in mycorrhizal plants may be due to a change in N metabolism brought about by changes in the enzymes associated with N metabolism (Mathur and Vyas, 1996).

Several workers have reported that improved N nutrition may also help to reduce the toxic effects of Na ion by reducing its active uptake and this may indirectly help in maintaining the chlorophyll content of the plant (Evelin *et al.*, 2009). Usha *et al.* (2004) studied effects of AMF (*Glomus deserticola*) inoculation on the rhizosphere dynamics of Kinnow mandarin and found that *G. deserticola* modifies rhizosphere favourably to improve soil nitrogen availability and consequent uptake by plant and thus results in better growth, fruit yield, and quality of Kinnow.

2.4 Micronutrients

The micronutrient status of tropical soils is low due to the continuous fertilization of cultivated crops without the inclusion of micronutrients. Colonization of the roots by AMF has been shown to improve the productivity of such soils by enhancing the uptake of slowly diffusing micronutrients such as Cu and Zn (Sylvia *et al.*, 1993). Li *et al.* (1991) demonstrated hyphal uptake and translocation of Cu to *Trifolium repens* L. This contributed about 62% of the total Cu uptake and the mycorrhizal response was independent of the effects of P nutrition. A number of studies have clearly shown that Zn uptake via mycorrhizae is important for the alleviation of Zn deficiency in several plant species (Evans and Miller, 1988; Sylvia *et al.*, 1993).

Wang and Xia (2009) reported that in trifoliate orange (*Poncirus trifoliata*) seedlings inoculated with *G. versiforme* at different pH levels of nutrient solution mycorrhizal efficiencies of Cu and Mn uptake were absolutely dependent on mycorrhizal fungal species and plant tissues and mycorrhizal hyphae could take part in Cu and Mn uptake (Li *et al.*, 1991; Giovannetti, 2008). Therefore, mycorrhizal symbiosis significantly increased Cu and Mn concentrations of roots. However, compared with those of non-mycorrhizal leaves, lower Cu and Mn concentrations of mycorrhizal leaves might ascribe to the reason that mycorrhizal hyphae are only concerned with Cu and Mn uptake but not with Cu and Mn translocation. It has been proved that mycorrhizal symbiosis can improve Zn nutrition as a secondary consequence of P nutrition (Subramanian *et al.*, 2009). In order to get sufficient Zn from the soil, plant roots need to be infected by mycorrhizal fungi which is also a good indication of mycorrhizal dependency. Also it has been shown that excess P can reduce Zn uptake (Cakmak and Marschner, 1987). Mycorrhizal-inoculation also significantly increased Cu concentration in shoot but not in roots.Timmer and Leyden (1980) and Graham and Syvertsen (1989) reported that mycorrhizal inoculated plants had higher Cu concentration than non-mycorrhizal plants.Mycorrhizae-inoculated plants had less Fe and Mn concentration than non-inoculated plants. Treeby (1992) used two citrus rootstocks differing in mycorrhizal dependency and lime tolerance, rough lemon (*Citrus jambhiri*) and trifoliate orange (*Poncirus trifoliata* Raf.) in two soils with different pH and, it was concluded that mycorrhizal fungi increased the supply of Fe to the host plant in an acid soil, but not in an alkaline soil.

2.5 Improvement of soil physical fertility

Mycorrhizal impacts on soil aggregate stability can also influence soil physical conditions mycorrhizae assist in development of water stable aggregates that are formed by the binding of soil particles that are important in the retention of organic matter and habitable pore space in these aggregates. The external mycelium of mycorrhizal fungus assists in physical entanglement but also binding with amorphous substances that are present on the surface of the hyphae (Oades, 1993). These processes collectively help in soil conservation and to minimize erosional hazard.

3.0 AMF mediated nutrient cycling in some selected fruit crops

3.1 Apple (*Malus pumila* L.)

Mycorrhizal inoculation to apple 'M.26' rootstock was attributed to improved P nutrition in P-rich soil (644 kg·ha⁻¹) (Plenchette *et al.*, 1983). *G. versiforme*, especially the California isolate, was best for growth promotion and nutrition of all apple rootstocks. The external AMF hyphae reached beyond the depletion zone around the root hairs, absorbed soil P and translocated it, perhaps in the form of polyphosphate granules, to the arbuscules where P is transferred to the plant cell in exchange of carbon (Mago and Mukerji, 2003). Apple inoculated with *G. versiforme* and *G. macrocarpum* increased the rate of recovery from water stress, improved plant growth (leaf number, stem diameter and DW) and enhanced the absorption of most minerals especially Zn and Cu (Liu, 1989). An *et al.* (1993) studied the effects of two mycorrhizal fungi (*G. epigaeum* Daniels and Trappe and *G. macrocarpum* Tul.and Tul.) on the growth and mineral composition of shoots and roots of apple seedlings (*M. micromalus* Makino). The AMF inoculated apple seedlings enhanced growth and improved uptake of immobile nutrient elements, mainly Cu, P, Zn.

The beneficial roles of mycorrhizal colonization in fruit crops depend on both fungal and fruit tree species. In apple (*M. domestica*), *G. mosseae* singly and in combination with *G.macrocarpum* was more effective in increasing plant biomass than *G. macrocarpum* (Miller *et al.*, 1989). In eighteen *Prunus* rootstocks inoculated with three AMF, *G. etunicatum, G.intraradices*, and *G. mosseae*, only *G. intraradices* proved to be the most infective endophyte, achieving the highest root colonization in most of the rootstocks evaluated (Calvet *et al.*, 2004).

3.2 Banana (*Musa* spp.)

Romero (2005) found the capacity of AMF to improve plant growth and nutrient absorption in banana (*Musa* AAA subgroup Cavendish clone Valery) and their relationship to the nematode *Radopholus similis*. AMF treatment produced a better P absorption. Lin. (1992) evaluated the comparative agronomic effectiveness of three rock phosphate (Idaho, Florida and North Carolina) and superphosphate uptake by mycorrhizal fungi *Glomus aggregatum* using small banana (*M. paradisiaca* L.) corms as planting material in pots. A concentration of 0.2 mg P L⁻¹ in solution was maintained in the soil. Plant dry weight, P percentage in the 3rd leaf, and total P uptake were increased when plants fertilized with insoluble rock phosphates were inoculated with AMF. Phosphorus uptake by plants fertilized with Idaho, Florida, and North Carolina rock phosphates was 0.18, 0.42, and 0.97% as much as by plants fertilized with superphosphate. The beneficial effect of mycorrhiza on phosphate uptake was 136, 30, 2 and 24 % for plants fertilized with Idaho, Florida and North Carolina rock phosphate, and superphosphate, respectively.

3.3 Citrus (*Citrus* spp.)

Shamshiri *et al.* (2012) reported that Kinnow trees budded on Jatti Katti rootstock inoculated with three different AMF viz., *Glomus manihotis* ,

G. mosseae, and *Gigaspora gigantia* separately or in combination could increase plant phosphorous. Ten -year-old Kinnow mandarin plants inoculated with AMF (*G. deserticola*) and *Azotobacter chroococcum* in different combinations with organic-farm-yard manure (FYM) and inorganic fertilizers recorded higher release of organic acids such as malic, citric, shikimic, and fumaric acids as compared to un-inoculated control. This study indicated that *G. deserticola*, when compared with *A. chroococcum*, modified the rhizosphere favourably to improve soil nutrient availability and consequent uptake by plants and thus result in better growth, fruit yield, and quality of Kinnow.

The effect of phosphorus doses and mycorrhizal fungi on mineral-nutrition and growth of Cleopatra Mandarine (*Citrus*-Reshni Hort Ex Tan) was studied by Darocha *et al.* (1994). This study was carried out to verify the effects of phosphorus doses on inoculation with a mixed population of AMF on growth and mineral nutrition of 'Cleopatra' mandarin until transplanting. The plants were grown in trays containing commercial substrate, inoculated with a mixed population of *Acaulospora morrowae*, *G. clarum* and *G. etunicatum* and non-inoculated substrate which received six phosphorus doses (0, 320, 640, 1280, 2000 and 2560 g P_2O_5/m^3 of substrate) supplied as simple superphosphate. The inoculation promoted better plant growth, and increased the boron contents in the shoots. Ishii *et al.* (1995) studied the effect of the reduction of phosphorus fertilizer in *Citrus iyo* orchards on the development of arbuscular mycorrhizae and the quality of fruit. Shrestha et. al. (1996) reported that satsuma mandarin trees which were inoculated with AMF grew larger and had better fruit quality as compared with non AMF control trees under low concentrations of applied phosphorus (P) condition. The formation of AMF was inhibited by a great amount of fertilizers, especially P.

3.4 Grape (*Vitis vinifera* L.)

Almaliotis *et al.* (2008) studied the mycorrhizal colonization, soil fertility and P nutrition in 53 vineyards of cv. Victoria and the experimental results indicated that soil P concentration was negatively correlated with the number of mycorrhiza spores in soil, root colonization, but positively correlated with leaf P. The grapevine root stocks inoculated with *G. etunicatum* and *G.clarum* significantly increased the leaf areas and leaf P, but not N and K concentrations. Similarly *Gigaspora margarita* increased the growth and leaf mineral concentrations of the tetraploid grapevine rootstocks (Motosugi *et al.*, 2002). Mortimer *et al.* (2005) observed in grapevine that the AMF symbiosis increased P uptake and AMF roots were more efficient at P utilization.

Cheng *et al.* (2004) grew grapevines and cover crops in specially designed containers in the greenhouse that restricted their root systems to separate compartments, but allowed AMFi to colonize both root systems. In this experiment, the cover crops i.e. grass (*Bromus hordeaceus*) and a legume (*Medicago polymorpha*), were labeled with 99 atom% N^{-15} solution and results indicated AMF-mediated N^{-15} transfer from cover crops to grapevines after 5th and 10th days of labeling. Karagiannidis *et al.* (1995) studied the influence of three AMF species, *Glomus fasciculatum*, *G. mosseae* and *G. macrocarpum* in a pot experiment on the shoot and root dry weight, shoot nutrient concentration of P, K, Ca, Mg, Zn, Mn and Fe as well as the total P uptake in three grapevine rootstocks (41 B, 110 R, 5 BB) and cv. *Razaki* table grape vines (syn. *Dattier de Beyrouth)*. Mycorrhizal colonization increased the shoot P concentration but did not affect the levels of K, Ca, Mg and Zn.

Caglar and Bayram (2006) evaluated the effects of AMF species (*Glomus etunicatum, G. caledonium, G. clarum, G. mosseae,* and mixed inoculum) on the leaf nutritional status of four grapevine rootstocks (420 A, 41 B, 1103 P, and 'Rupestris du Lot'). Ten months after inoculation, grapevine rootstocks were colonized by the AMF at frequencies ranging from 47.0 to 64.1%. Inoculations with *G. etunicatum* and *G. clarum* significantly increased the leaf areas of 41 B, 420 A, and 1103 P rootstocks. The AMF increased leaf P, but not N and K concentrations. Total leaf sucrose concentrations of the grapevine rootstocks were increased two- to four-fold with certain inocula compared to control. The effects of *Gigaspora margarita* BECKER and HALL on growth and leaf mineral concentrations of the tetraploid grapevine rootstocks Gloire de Montpellier (Gloire, *Vitis. riparia* MICHX.), Rupestris St. George (St. George, *V .rupestris* SCHEELE), and Couderc 3309 (3309, *V. riparia x V. rupestris)* were compared with those of their corresponding diploids by Motosuget *al.*(2002). The percentage of AMF infection in the inoculated tetraploid grapevines of each rootstock was as high (above 90 %) as in the inoculated diploids. Tetraploid and diploid rootstocks with AMF-inoculation had significantly higher P concentrations in the leaves than the non-inoculated grapevines, but tetraploid grapevines with AMF-inoculation had lower Ca and Mg concentrations. The tetraploid grapevines with thicker roots and more compact root systems were considered to depend more on arbuscular mycorrhizae than the original diploid rootstock cultivars. Hydroxyapatite was used as phosphate fertilizer. Considerable responses to mycorrhizal inoculation were observed in both grapevine rootstocks and Razaki vines, according to vine and mycorrhiza species. Mycorrhizal colonization increased the shoot P concentration but did not affect the levels of K, Ca, Mg and Zn. On the contrary the mycorrhizal vines showed a low concentration of Mn and Fe (M/NM down to 0.2).

Grapevine mineral nutrition is an important factor influencing wine quality. For this, the effect of rootstocks on the nutrient content in different tissues of 'Cabernet Sauvignon' grapevines grown in the Serra Gaúcha region was evaluated by Miele *et al.* (2009). There was a significant effect of the rootstock on the contents of N, P, K, Ca, and Mg in limb, petiole, rachis and berry of the 'Cabernet Sauvignon' grapevine. This effect varied as a function of the analyzed nutrient and tissue. But there was no significant effect on the nutrients present in the shoot. Besides this, nutrient content varied according to the kind of tissue analyzed. In this way, contents of N and Ca were higher in limb; contents of P and K, in rachis; and contents of Mg, in petiole.

Bavaresco *et al.* (1992) experimented on the woody cuttings from three ungrafted rootstocks, with decreasing resistance to lime-induced chlorosis. The plants were treated with *Pseudomonas fluorescens* cells and *Glomus mosseae* inoculum in unsterile calcareous soil to test the effects of these organisms on some physiological parameters involved in chlorosis occurrence. They found *P. fluorescens* and *G. mosseae* treatments increasing ferrous iron and leaf chlorophyll, mycorrhizal colonization, N, P, Mn and Cu concentration in leaf fresh matter in the rootstock more susceptible to lime-induced chlorosis.

3.5 Mango (*Mangifera indica* L.)

Sukhada *et al.* (2012) studied the benefits of AMF inoculation to nutrient uptake in field grown mango plants. In AMF treated plants available soil P, leaf P, Zn and Cu were significantly higher compared to uninoculated plants. The seedlings (Arka Puneeth grafted on AMF colonized Totapuri rootstock) inoculated with AMF also increased root acid and alkaline phosphatase activity (compared to un-inoculated control).

3.6 Papaya (*Carica papaya* L.)

Sukhada (1992) studied the effect of AMF inoculation on plant-growth, nutrient level and root phosphatase-activity in Papaya (*C. papaya* Cv Coorg Honey Dew) found *G.mosseae* and *G. fasciculatum* highly beneficial in improving P, N and Zn concentrations in the leaf. The root phosphatase activity was higher in AMF inoculated plants.

Minhoni *et al.*(2003) studied the effect of phosphorus, soil fumigation and mycorrhizal colonization on papaya cv Sunrise Solo. Separate treatments with AMF inoculation (*Glomus macrocarpum)* and increasing levels (60, 120, 240, and 480 mg kg^{-1} of P in soil influenced these parameters significantly, while the factor substratum fumigation had no significant effect. The inoculation effect was most expressive when 60 mg kg^{-1} of P in soil was added. Such was the reduction in phosphorus requirement of papaya by *Glomus macrocarpum* inoculation, that the studied parameters between inoculated plants without phosphate addition and un-inoculated ones in substratum enriched with over 240 mg kg^{-1} of P in soil did not differ.

Khade *et al.* (2009) conducted the experiment to study the effects of AMF on mineral nutrition of *C. papaya* var. Surya.This study revealed that total potassium and total phosphorus content of mycorrhizal leaf petiole was higher in inoculated plants as compared to controls and varied significantly within the treatments. *Glomus mosseae* was the most effective species of AMF, in influencing mineral nutrition of papaya followed by mixed inoculum (G.I +G.M) and *G. intraradices* respectively.

3.7 Pomegranate (*Punica granatum* L.)

Aseri *et al.* (2008) studied the positive response of biofertilizers in pomegranate nursery seedlings followed by their transplantation in harsh field conditions of Indian Thar Desert. The experimental results showed that *G. mosseae* was found to be the most effective for better establishment of pomegranate plants under field conditions. A significant improvement in the plant height, plant canopy, pruned material and fruit yield was evident in 5-year-old pomegranate plants under field conditions.

3.8 Peach (*Prunus* spp.)

Rutto *et al.* (2006) studied the effect of AMF inoculation on growth and nutrition in peach seedlings. The results showed that mycorrhizal seedlings demonstrated better growth and biomass yield. In general, mycorrhizal seedlings had better P and Ca nutrition.

3.9 Strawberry (*Fragaria×ananassa* Duchesne)

Vestberg, (1992) studied the effect of arbuscular mycorrhizal inoculation on the growth and root colonization of ten Strawberry cultivars. Out of six AMF strains, *G. macrocarpum* V3, *G.mosseae* Rothamsted and *G.* sp. V4, found efficient in increasing all the growth parameters, mycorrhizal dependency and sporulation as compared to other treatmnents. Vilma *et al.* (2010) evaluated the influence of three N treatments in combination with inoculation with the *G. intraradices* on the quality of strawberry fruits and the results indicated that mycorrhization significantly affected some colour parameters and the concentrations of most phenolic compounds. There was a significant difference in fruit yield between mycorrhizal and non-mycorrhizal plants.

3.10 Sweet passion fruit (*Passiflora edulis* Sims. f. *flavicarpa* Deg.)

Dos Anjos *et al.* (2005) who studied required P dose and efficient AMF to promote growth of seedlings of sweet passion fruit observed that *Scutellospora heterogama* and *Gigaspora albida* increased plant height at P values estimated between 15.40 and 16.07 mg dm^{-3} in disinfected soil and 14.85 and 15.60 mg dm^{-3} in non disinfected soil. Chebet *et al.* (2008) reported that inoculation of AMF could increase the uptake of P in leaf tissue of passion fruit (*Passiflora edulis* var. Edulis), Rough lemon (*Citrus limon*) and papaya (*Carica papaya* var. solo).

4.0 Strategies to exploit AMF symbiosis in fruit crop production

In order to derive full benefits from indigenous mycorrhizal population or to sustain the inocula potential, following strategies can be adopted.

a. **Minimal soil disturbance:** Heavy tillage has a major effect on the symbiosis, reducing root colonization primarily by disrupting the hyphal network. No till system favours mycorrhizal symbiosis and therefore their adoption would enhance the functions of mycorrhiza and aid the sustainability of the system

b. **Fertilizers:** The literature on fertilizer impact on colonization of mycorrhiza is contradictory. Application of soluble P appears to reduce the germination of mycorrhizal spores and extent of colonization. However, some fungal species (*G. intraradices*) are insensitive to fertilizer application

c. **Crop rotations:** Mycorrhizas are obligate symbionts and thus highly dependent on living host plants. Fallowing also results in reduction of inoculum potential but the effect is smaller than the inclusion of non-mycotrophic plants in the cropping sequence.

d. **Pesticides:** Mycorrhizae are highly sensitive to pesticides such as methyl bromide and if used, it impedes the development of mycorrhizal symbiosis in the crop. Most fungicides have shown to be detrimental to either colonization or functions of mycorrhizas. Rationalization of their use by selection of fungicides that do not interact negatively with mycorrhiza could bring significant benefits.

e. **Crop Breeding:** Some crop varieties are more responsive to mycorrhizal colonization and can be exploited for low-input production system.

5.0 Conclusions and future strategies

Fruit crops being perennials, mycorrhizae play a pivotal role in nutrient cycling and associated fertilizer management. Mycorrhizal symbiosis is known to improve the nutritional status or fruit crops particularly P. Fruit trees which possesses very poor root systems and the are highly benefited from mycorrhizal symbiosis. Apart from P, micronutrients such as Zn and B are reported to be enhanced as a result of immobile nature of nutrient ions in the soil. There are wide range of nutritional benefits from mycorrhizae and therefore deserves due consideration for organic fruit production. The use of AMF as inoculants in crops and environmental rehabilitation is becoming more widely accepted as being the key to maintaining soil health and vitality because of the intimate link they form between the host plant and the soil environment. Mycorrhizae inoculation generally increased growth of fruit crops under the various climatic conditions, and also significantly elevated P, N and micronutrient concentrations of soil and plants. Fruit crops inoculation with mycorrhizal inoculation could reduce the amounts of fertilizers and pesticides being used, and could represent a sustainable technique to improve crop yield and profitability.

There is a need to work out the exact fertilizer requirement of fruit crops when AMF inoculation is given. Using stable isotopes the requirement of P and N can be worked out. There is a need to continue the work on phosphate transporter genes which may be cloned and utilized further for scavenging more nutrients from soil.

References

1. Akhtar, M.S and Siddiqui, Z.A., 2007. Biocontrol of a chickpea root-rot disease complex with *Glomus intraradices, Pseudomonas putida* and *Paenibacillus polymyxa. Austra. Plant Path.* 36(2),175-180.

2. Al-Karaki, G.N and Clark, R.B., 1998. Growth, mineral acquisition and water use by mycorrhizal wheat grown under water stress. *J. Plant. Nutr.* 21,263–276.

3. Almaliotis, D., Karagiannidis, N., Chatzissavvidis, C., Sotiropou-Los, T.and Bladenopoulou, S. 2008. Mycorrhizal colonization of table grapevines (cv. Victoria) and its relationship with certain soil parameters and plant nutrition. *Agrochimica.* 52(3),129-136.

4. Al-Agely, A. and Sylvia, D., 2008. Compatible host/ mycorrhizal fungi combinations for micropropagated Sea oats: II field evaluation. *Mycorrhiza.* 18(5), 257-261.

5. Alberto Miele., Luiz Antenor Rizzon .and Eduardo Giovannini., 2009. Efeito Do Porta-Enxerto No Teor De Nutrientes Em Tecidos Da Videira 'Cabernet Sauvignon'. *Rev. Bras. Frutic., Jaboticabal - SP,* v. 31, n. 4, p. 1141-1149.

6. An Z.Q., Shen T. and Wang H.G . 1993. Mycorrhizal Fungi In Relation To Growth And Mineral-Nutrition Of Apple Seedlings . *Sientia Horticulturae .* 54 (4),275-285 .

7. Asimi,S.,Gianinazzi-Pearson,V and Gianinazzi,S.,1980.Influence of increasing soil phosphorus levels on interactions between vesicular arbuscular mycorrhizae and *Rhizobium* in soybeans. *Can. J. Bot,* 58, 2200–2205.

8. Aseri, G.K., Jain, N., Panwar, J., Rao, A.V.and Meghwal, P.R., 2008. Biofertilizers improve plant growth, fruit yield, nutrition, metabolism and rhizosphere enzyme activities of Pomegranate (*Punica granatum* L.)in Indian Thar Desert. *Scientia Horticulturae.* 117(2),130-135.

9. Atkinson, D., 1986. The needs of the fruit tree for mineral nutrients. *Adv. Plant Nutr.* 2,93–128.

10. Azcón,R., Gómez,M.and Tobar, R., 1992. Effects of nitrogen-source on growth, nutrition, photosynthetic rate and nitrogen-metabolism of mycorrhizal and phosphorus-fertilized plants of *Lactuca-sativa* L. *New Phytol.* 121(2), 227-234.

11. Azcon-Aguilar,C and Barea,J.M.,1997. Applying mycorrhiza biotechnology to horticulture: significance and potentials. *Sci. Hort,* 68, 1-24.

12. Bavaresco, L., and Fogher, C., 1992. Effect Of Root Infection With *Pseudomonas-fluorescens* and *Glomus-mosseae* In Improving Fe-Efficiency Of Grapevine Ungrafted Rootstocks. *Vitis.* 31(3),163-168.

13. Bago, B., Vierheiling, H., Piche, Y and Azcón-Aguilar, C. 1996. Nitrate depletion and pH changes induced by the extraradical mycelium of the arbuscular mycorrhizal fungus *Glomus intraradices* grown in monoxenic culture. *New Phytol.* 133(2), 273-280.

14. Bago, B and Azcón-Aguilar, C. 1997. Changes in the rhizospheric pH induced by arbuscular mycorrhiza formation in onion (*Allium cepa* L.). Z. Pflanz. Bodenk. 160(3), 333-339.

15. Bâ, A.M., Plenchette, C., Danthu, P., Duponnois, R and Guissou, T., 2000. Functional compatibility of two arbuscular mycorrhizae with thirteen fruit trees in Senegal. *Agrofor. Syst.* 50, 95-105.

16. Bago, B., Pfeffer, P and Shachar-Hill, Y. 2001. Could the urea cycle be translocating nitrogen in the arbuscular mycorrhizal symbiosis. *New Phytol.* 149, 4-8.

17. Bolan, N.S., Robson, A.D and Barrow, N.J. 1987. Effects of vesicular arbuscular mycorrhiza on the availability of iron phosphates to plants. *Plant Soil*. 99, 401–410.

18. Borie, F., Redel, Y., Rubio, R., Rouanet, J.L.and Barea, J.M. 2002. Interactions between crop residues applications and mycorrhizal developments and some soil-root interface properties and mineral acquisition by plants in an acidic soil. *Biol. Fértil. Soils* .36(2), 151-160.

19. Bucher, M., 2007. Functional biology of plant phosphate uptake at root and mycorrhizal interfaces. *New Phytol.* 173(1), 11-26.

20. Cakmak, I and Marschner, H., 1987. Mechanism of Phosphorus–Induced Zinc Deficiency in Cotton. III. Changes in Physiological Availability of Zinc in Plants. *Physiol. Plant.* 70, 13–20.

21. Caris, C., Hordt, W., Hawkins, H.J., Romheld, V and George, E., 1998. Studies of iron transport by arbuscular mycorrhizal hyphae from soil to peanut and sorghum plants. *Mycorrhiza.* 8, 35-39.

22. Cantrell, I.C and Linderman, R.G. 2001. Pre-inoculation of lettuce and onion with VA mycorrhizal fungi reduces deleterious effects of soil salinity. *Plant Soil.* 233, 269-281.

23. Calvet, C., Estaun, V., Camprubi, A., Hernandez- Dorrego, A., Pinochet, J and Moreno M.A. 2004. Aptitude for mycorrhizal root colonization in *Prunus* rootstocks. *Sci. Hort.* 100, 39-49.

24. Caglar, S. and Bayram, A., 2006. Effects of Vesicular-Arbuscular Mycorrhizal (VAM) fungi on the leaf nutritional status of four grapevine rootstocks. *European Journal Of Horticultural Science.* 71(3),109-113.

25. Cartmill, A.D., Alarcon, A and Valdez, A., 2008. Arbuscular mycorrhizal fungi enhance tolerance of *Rosa multiflra* cv. Burrto bicarbonate in irrigation water. *J. Plant Nutri.* 30(7-9), 1517-1540.

26. Cavagnaro, T.R., Sokolow, S and Jakson, L., 2009. Mycorrhizal effects on growth and nutrition of tomato under elevated atmospheric carbon dioxide. *Func. Plant Biol.* 34(8), 730-736.

27. Cheng, X.M. and Baumgartner, K., 2004. Survey of arbuscular mycorrhizal fungal communities in Northen California vineyards and mycorrhizal colonization potential of grapvine nursery stock. *Hortscience.* 39(7),1702-1706.

28. Chebet,D.K., L.K. Rutto, L.S. Wamocho and W. Kariuki ., 2008. Effect of Arbuscular Mycorrhizal Inoculation on the Growth, Nutrient Uptake and Root Infectivity of Tropical Fruit Seedling. Proc. XXVII IHC-S16 Citrus and Other Trop.and Subtrop. Fruit Crops Ed.-in-Chief: *Dae-Geun Oh Acta Hort.* 773.

29. Clark, R.B., Zobel, R.W and Zeto, S.K., 1999. Arbuscular mycorrhizal fungal isolate effectiveness on growth and root colonization of *Punicum virgatum* in acid soil. *Soil Biol. Biochem.* 31, 1757-1763.

30. Clark, R.B and Zeto, S.K., 2000. Mineral acquisition by arbuscular mycorrhizal plants. *J. Plant Nutr.* 23(7), 867-902.

31. Cuenca, G and Azcón, R., 1994. Effects of ammonium and nitrate on the growth of VA-mycorrhizal *Erythrina poeppigiana* O.I, cook seedlings. *Biol. Fert. Soil.* 18, 249-254.

32. Darocha .M.R., Deoliveira. E.and Correa, G.D., 1994. Effect Of Phosphorus Doses And Mycorrhizal Fungi On Mineral-Nutrition And Growth Of Cleopatra Mandarine (Citrus-Reshni Hort Ex Tan). *Pesquisa Agropecuaria Brasileira.* 29(5),725-731.

33. Deguchi, S., Shimazaki, Y., Vozumi. S., Tawaraya, K., Kawamoto, H and Tanaka, O., 2007. White clover living mulch increases the yield of silage corn via arbuscular mycorrhizal fungi colonization. *Plant Soil.* 29(1-2), 291-299.

34. dos Anjos, E.C.T., Cavalcante, U.M.T., dos Santos, V.F.and Maia, L.C., 2005. Production of mycorrhized sweet passion fruit seedlings in disinfected and phosphorus fertilized soil. *Pesquisa Agropecuaria Brasileira.* 40(4),345-351.

35. Evans, D.G and M.H. Miller., 1988. Vesicular-arbuscular mycorrhizas and the soil-induced reduction of nutrient absorption in maize. I. Casual relations. *New Phytol.* 110, 67-74.

36. Evelin, H., Kapoor, R and Giri, B., 2009. Arbuscular mycorrhizal fungi in alleviation of salt stress: *a review. Ann Bot.* 104, 1263-1280.

37. Feng,G.,Song,Y.C., Li, X.L and Christle,P.,2003. Contribution of arbuscular mycorrhizal fungi to utilization of organic sources of phosphorus by red clover in a calcarcous soli. *Appl.Soil Ecol.*22,139-148.

38. Friese, C.F and M.F .Allen., 1991. The spread of VA mycorrhizal fungal hyphae in the soil: inoculum types and external hyphal architecture. *Mycologia* 83, 409–418.

39. Frey, B and Schüepp, H., 1993. Acquisition of N by external hyphae of AM fungi associated with maize. *New Phytol.* 124, 221-230.

40. Gange, A.C., Brown, V.K and Sinclair, G.S., 1993. Vesicular-arbuscular mycorrhizal fungi: a determinant of plant community structure in early successino. *Func. Ecol.* 7, 616-622.

41. Garg, N and Manchanda, G., 2008. Effect of arbuscular mycorrhizal inoculation of salt-induced nodule senescence in *Cajanus cajan* (pigeonpea). *J. Plant Growth Regul.* 27, 115-124.

42. Genre, A and Bonfante, P., 2010. The making of symbiotic cells in arbuscular mycorrhizal roots. In: Koltai, H., and Kapulnik, Y. (Eds.) Arbuscular Mycorrhizas: Physiology and Function. *Springer-Verlag,* Berlin Heidelberg, pp 57-71.

43. Giovannetti, M., 2008. Structure, extend and functional significance of belowground arbuscular mycorrhizal networks. In: Varma, A. (Ed.)

Mycorrhiza: State of the Art, Genetics and Molecular Biology, Eco-Function, Biotechnology, Eco-Physiology, Structure and Systematics. Third edition. *Springer-Verlag*, Berlin Heidelberg, pp 59-72.

44. Giri, B and Mukerji, K.G., 2004. Mycorrhizal inoculant alleviates salt stress in *Sesbania aegyptiaca* and *Sesbania grandiflora* under field conditions: Evidence for reduced sodium and improved magnesium uptake. *Mycorrhiza*. 14, 307-312.

45. Gnekow, M.A and H. Marschner., 1989. Role of VA-mycorrhiza in growth and mineral nutrition of apple (*Malus pulmila* var. *domestica*) rootstock cuttings. *Plant Soil*. 119, 285–293.

46. Graham, J.H and Syvertsen, J.P., 1989. Vesicular–Arbuscular Mycorrhizas Increase Chloride Concentration in Citrus Seedlings. *New Phytol*. 113, 29–36.

47. Hawkins, H.J., Johansen, A and George, E. 2000. Uptake and transport of organic and inorganic nitrogen by arbuscular mycorrhizal fungi. *Plant Soil*. 226(2), 275-285.

48. Hawkins, H.J and George, E., 2001. Reduced N-15-nitrogen transport through arbuscular mycorrhizal hyphae to *Triticum aestivum* L. supplied with ammonium vs. nitrate nutrition. *Ann. Bot*. 87(3), 303-311.

49. Hartwig, U.A., Wittmann, P., Braun, R., Hartwig-Raz, B., Jansa, J., Mozafar, A., Lüscher, A., Leuchtmann, A., Frossard, E and Nösberger, J., 2002. Arbuscular mycorrhiza infection enhances the growth response of *Lolium perenne* to elevated atmospheric pCO_2. *J. Exp. Bot*. 53(371), 1207-1213.

50. Hamel, C., 2004. Impact of arbuscular mycorrhizal fungi on N and P cycling in the root zone. *Can J Soil Sci* 84, 383–395.

51. He,Z.L.,Wu,J.,O'Donnell,A.G. and Syers,J.k. 1997. Seasonal responses in microbial biomass carbon,phosphorous and sulphur in soils under pasture.*Biol.Fertil.Soils*, 24, 421-428.

52. Ishii, Takaaki, Sunao Kirino, Ming Zeng, Jiro Aikawa, Isao Matsumoto and Kazuomi Kadoya., 1995. Effect of the reduction of phosphorus fertilizer for *Citrus iyo* orchards on the development of vesicular-arbuscular mycorrhizae and the quality of fruit.

53. Jeffries, P and Barea, J.M., 2001. Arbuscular Mycorrhiza: a key component of sustainable plant-soil ecosystems. In: B. HOCK. The Mycota vol. IX Fungal Associations. Ed. Springer-Verlag, Berlin, Heildelberg. 95-113.

54. Johansen, A., 1999. Depletion of soil mineral N by roots of *Cucumis sativus* L-colonized or not by arbuscular mycorrhizal fungi. *Plant Soil* .209(1), 119-127.

55. Joner, E.J.,2000. The effect of long term fertilization with organic or inorganic fertilizers on mycorrhizal mediated phosphorus uptake in subterranean clover. *Biol. Fertil. Soil*. 32(5), 435-440.

56. Joner, E.J., Raynskov, S. and Jakobsen. I. 2000a. Arbascular mycorrhizal phosphate transport under monoxenic conditions using radio-labelled inorganic and organic phosphate.*Biotechnol.Let*, 22, 1705-1708.

57. Joner,E.J.,Van Aarle,I.M.and Vosatka,M.200b. Phosphatase activity of extra-radical arbuscular mycorrhizal hyphae:*A review. Plant soil* .226,199-210.

58. Karagiannidis, N., Nikolaou, N.and Mattheou, A., 1995. Influence Of 3 Va-Mycorrhiza Species On The Growth And Nutrient-Uptake Of 3 Grapevine Rootstocks And One Table Grape Cultivar. *Vitis* .34, 85-89.

59. Koide, R. T. and Kabir, Z., 2000. Extraradical hyphae of the mycorrhizal fungus *Glomus intraradices* can hydrolyse organic phosphate. *New Phytologist, 148*, 511-517.

60. Khade,W., Bernard, F., Rodrigues.,Prabhat.and K. Sharma., 2009. Arbuscular mycorrhizal status and root phosphatase activities in vegetative *Carica papaya* L. varieties Franciszek Go ´rski Institute of Plant Physiology, Polish Academy of Sciences, Krako ´w.

61. Larue, J.H and Clellan, W.D., and Peacock, W.I., 1975. Mycorrhizal fungi and peach nursery nutrition. *California Agric.* 5, 6-7.

62. Lampkin, N., 1990. Organic Farming, Farming Press Books, IPS WICH, UK.

63. Linderman, R.G., 1988. Mycorrhizal interactions with the rhizosphere microflora: The mycorrhizosphere effect. *Phytopathology* .78,366-371.

64. Liu, R.J., 1989. Effects of vesicular-arbuscular mycorrhizas and phosphorus on water status and growth of apple. *Journal of Plant Nutrition.* 12(8), 997-1017.

65. Li, X.L., Marschner, H and George, E., 1991. Acquisition of phosphorus and copper by VA mycorrhizal hyphae and root-to-shoot transport in white clover. *Plant Soil.* 136, 49-57.

66. Lin M.L. and Fox, R.L., 1992. The Comparative Agronomic Effectiveness Of Rock Phosphate And Superphosphate For Banana. *Fertilizer Research.* 31(2),131-135.

67. Linderman, R.G., 1994. Role of AM fungi in biocontrol. In: Pfleger FL, Linderman RG (eds). Mycorrhizae and plant health. *The American Society of Phytopathology,* St. Paul, Minn. USA.

68. López-Pedrosa, A., González-Guerrero, M., Valderas, A., Azcón-Aguilar, C and Ferrol, N., 2006. Gint AMTl encodes a functional highaffinity ammonium transporter that is expressed in the extraradical mycelium of *Glomus intraradices. Fungal Genet. Biol.* 43(2), 102-110.

69. Mathur, N and Vyas, A., 1996. Biochemical changes in *Ziziphus xyloropus* by VA mycorrhizae. *Bot. Bull. Acad. Sinica.* 37, 209-212.

70. Mago, P and Mukerji K.G., 2003. Mycorrhizal technology in forestry and agriculture. In:Compendium of Mycorrhizal Research Vol II: Role of Mycorrhiza in biotechnology, (Eds.) Mukerji, K.G. and. Chamola, B.P), *A.P.H. Publishing Corporation,* New Delhi. 129- 146.

71. McGraw, A.C.and Schenck, N.C. 1980. Growth stimulation of fungi. Proc. *Fla, State Hort. Soc* .93, 201-205.

72. Mena-Violante, H.G., Ocampo-Jimenez, O., Dendooven, L., Martinez-Soto, G., Gonzalez- Castaneda, J., Davies, F.T and Olalde-Portugal, V., 2006. Arbuscular mycorrhizal fungi enhanced fruit growth and quality of chile ancho (*Capsicum annuum* L. cv San Luis) plants exposed to drought. *Mycorrhiza.* 16, 261-267.

73. Miller, D. D., Bodmer, M and Schuepp, H., 1989. Spread of endomycorrhizal colonization and effects on growth of apple seedlings. *New Phytol.* 111, 51-59.

74. Minhoni, MTA; Auler, PAM. 2003. Effect of phosphorus, soil fumigation and mycorrhizal colonization on papaya growth. Revista Brasileira De Ciencia Do Solo. 27(5):841-847.

75. Miransari, M., 2010. Contribution of arbuscular mycorrhizal symbiosis to plant growth under different types of soil stress. *Plant Biol.* 12, 563- 569.

76. Mosse, B., Warner, A. and Clarke, C.A. 1981. Plant growth responses to *Vesicular- Arbuscular Mycorrhiza.* Xlll. Spread and introduced VA endophyte in the fie ld and residual growth effect in the second year. *New Phytologist,* 90,521-528.

77. Mohandas, S., 1992. Effect Of VAM Inoculation On Plant-Growth, Nutrient Level And Root Phosphatase-Activity In Papaya (*Carica-Papaya* Cv Coorg Honey Dew). *Fertilizer Research.* 31(3),263-267.

78. Motosugi, H., Yamamoto, Y., Naruo,T., Kitabayashi, H. and Ishii, T., 2002. Comparison of the growth and leaf mineral concentrations between three grapevine rootstocks and their corresponding tetraploids inoculated with an arbuscular mycorrhizal fungus *Gigaspora margarita. Vitis.* 41(1),21-25.

79. Mortimer, P.E., Archer, E. and Valentine, A.J., 2005. Mycorrhizal C costs and nutritional benefits in developing grapevines. *Mycorrhiza.* 15(3),159-165

80. Mohandas Sukhada , R. Manjula . and R.D. Rawal., 2011. Evaluation of arbuscular mycorrhizal and other biocontrol agents against Phytophthora parasitica var. nicotianae infecting papaya (*Carica papaya* cv. Surya) and enumeration of pathogen population using immunotechniques; *Biological Control* .58 ,22–29.

81. Nakamura, T., Adu-Gyamfi , J.J., Yamamoto, A., Ishikawa, S., Nakano, H and Ito, O., 2002. Varietal differences in root growth as related to nitrogen uptake by sorghum plants in low-nitrogen environment. *Plant Soil* . 245(1), 17-24.

82. Nasim, G., 2010. The role of arbuscular mycorrhizae in inducing resistance to drought and salinity stress in crops. In: Ashraf, M., Ozturk, M., and Ahmad M.S.A. (Eds.). Plant Adaptation and Phytoremediation. Springer-*Verlag,* Berlin Heidelberg, 119-141.

83. Neilsen, G.H., Neilsen, D and Atkinson. D., 1990. Top and root growth and nutrient absorption of *Prunus avium* L. at two soil pH and P levels. *Plant Soil.* 121, 137–144.

84. Nopamornbodi, O., Thamsurakul, S. and Charoensook, P., 1987. Selection of efficient VA mycorrhizal species for increasing soybean mungbean and peanut growth. Annual Report, 1987. *Soil Sci. Div. Bangkok.* 16 pp.

85. Oades, J.M., 1993. The role of biology in formation, stabilization and degradation of soil structure. *Geoderma.* 56, 377-400.

86. Plenchette, C., Furlan, V and Fortin. J.A., 1981. Growth stimulation of apple trees in unsterilized soil under field conditions with VA mycorrhiza inoculation. *Can. J. Bot.* 59, 2003–2008.

87. Plenchette, C., Furlan, V and Fortin. J.A., 1983. Responses of endomycorrhizal plants grown in a calcined montmorillonite clay to different levels of soluble phosphorus. II. Effect on nutrient uptake. *Can. J. Bot.* 61, 1384–1391.

88. Pelletier, S and Dionne, J., 2004. Inoculation rate of arbuscular-mycorrhizal fungi *Glomus intraradices* and *Glomus etunicatum* affects establishment of landscape turf with no irrigation or fertilizer inputs. *Crop Sci.* 44, 335-338.

89. Richardron,A.E.,2001.Prospects for using soil microorganisms to improve the acquisition of phosphrous by plants.*Aust.J.Plant physiol,*28,897-906.

90. Ross, J.P. and Harper, J.A. 1970. Effect of *Endogone mycorrhiza* on soybean yields.

91. Rodriguez-Romero, A.S.,Guerra, M.S.P.and Jaizme-Vega, M.D., 2005. Effect of arbuscular mycorrhizal fungi and rhizobacteria on banana growth and nutrition. *Agronomy For Sustainable Development.* 25(3), 395-399.

92. Rutto, K.L., Mizutani, F and Kadoya, K., 2002. Effect of root-zone flooding on mycorrhizal and nonmycorrhizal peach (*Prunus persica* Batsch) seedlings. *Sci. Hort.* 94, 285-295.

93. Rutto, K. L and Mizutani, F., 2006. Peach seedlings growth in replant and non-replant soils after inoculation with arbuscular mycorrhizal fungi. *Soil Biol. Biochem.* 38, 2536-2542.

94. Rutto, K.L.and Mizutani, F., 2006. Effect of mycorrhizal inoculation and activated charcoal on growth and nutrition in peach (*Prunus persica* Batsch) seedlings treated with peach root-bark extracts. *Journal Of The Japanese Society For Horticultural Science.* 75(6),463-468.

95. Shrestha, Y.H., Ishii, T., Matsumoto, I.and Kadoya K., 1996. Effects of vesicular-arbuscular mycorrhizal fungi on Ssatsuma mandarin tree growth and water stress tolerance and on fruit development and quality. *Journal Of The Japanese Society For Horticultural Science.* 64(4),801-807.

96. Shokri, S and Maadi, B., 2009. Effects of arbuscular mycorrhizal fungus on the mineral nutrition and yield of *Trifolium alexandrium* plants under salinity stress. *J Agrono.* 8, 79-83.

97. Shamshiri,M.H., K. Usha. and Bhupinder Singh., 2012. Growth and Nutrient Uptake Responses of Kinnow to Vesicular Arbuscular

Mycorrhizae International Scholarly Research Network ISRN Agronomy Volume 2012, Article ID 535846, 7 pages doi:10.5402/2012/535846 ,Smith, W.H. & Dowd, M.L. 1981. Biomass production of mycorrhiza in Florida. Forestry, 79,508-511.

98. Smith, S.E and Read, D., 1997. Vesicular-arbuscular mycorrhizas. In: Mycorrhizal symbiosis. IInd Edition, *Academic Press, Harcourt Brace & Company,Publishers,* New York. 9-160.

99. Subramanian, K.S., Charest, C., Dwyer, L.M and Hamilton. R.I., 1997. Effects of mycorrhizas on leaf water potential, sugar and P contents during and after recovery of maize. *Can. J. Bot.* 75, 1582-1591.

100. Subramanian, K.S and Charest, C., 1998. Arbuscular mycorrhizae and nitrogen assimilation in maize after drought and recovery. *Physiol. Plant.* 102, 285-296.

101. Subramanian, K.S and Charest, C., 1999. Acquisition of N by external hyphae of an arbuscular mycorrhizal fungus and its impact on physiological responses in maizeunder drought-stressed and well-watered conditions. *Mycorrhiza.* 9, 69-75.

102. Subramanian, K.S and Santhana krishnan, P., 2003. Responses of field grown mycorrhizal tomato plants under varying intensities of drought. Paper presented at the Fourth International Conference on Mycorrhizas, August 10-15. 737.

103. Subramanian, K.S., Tenshia, V., Jayalakshmi, K and Ramachandran, V., 2009. Role of arbuscular mycorrhizal fungus (*Glomus intraradices*) – (fungus aided) in zinc nutrition of maize. *J. Agric. Biotech. Sustain. Dev.* 1, 29-38.

104. Sukhada Mohandas., 1992. Effect of VAM inoculation on plant growth nutrient level and root phosphatase activity in papaya (*Carica papaya* cv. Coorg Honey Dew). *Fert. Res.* 31, 263–267.

105. Sukhada Mohandas., R. Manjula., R.D. Rawal., H.C. Lakshmikantha., Saikat Chakraborty and Y.L. Ramachandra ., 2010 Evaluation of arbuscular mycorrhiza and other biocontrol agents in managing *Fusarium oxysporum* f. sp. *Cubense* infection in banana cv. Neypoovan, *Biocontrol Science and Technology.* 20,2, 165-181.

106. Sukhada Mohandas., 2012. Arbuscular mycorrhizal fungi benefit mango (*Mangifera indica* L.) plant growth in the field, Indian Institute of Horticultural Research, Hessaraghatta, Bangalore, India. *Scientia Horticulturae,* 143, 43-48p.

107. Sylvia, D.M. and Williams , S.E. 1992. Vesicular-Arbuscular Mycorrhizae and Enviromental Stresses. Pages 101-123. In: Mycorrhizae in Sustainable Agriculture. Edited by G.J. Bothlenfalvay and R.G. Linderman. *American Society of Agronomy,* Inc. Madison, Wisconsin, USA.

108. Sylvia, D.M., Hammond, L.C., Bennet, J.M., Hass J.H and Linda. S.B., 1993. Field response of maize to a VAM fungus and water management. *Agron. J.* 85, 193-198.

109. Timmer, L.W and Leyden, F.F., 1980. The Relationship of Mycorrhizal Infection to Phosphorus-Induced Copper Deficiency in Sour Orange Seedlings. *New Phytol.* 85, 15–23.

110. Tobar, R.M., Azcón, R and Barea, J.M., 1994. Improved nitrogen uptake and transport from 15_N-labelled nitrate by external hyphae of arbuscular mycorrhizae under water-stressed conditions. *New Phytol.* 126, 119-122.

111. Treeby, M.T., 1992. The Role of Mycorrhizal Fungi and Non-Mycorrhizal Microorganisms in Iron Nutrition of Citrus. *Soil Biol. Biochem.* 24, 857–864.

112. Trolove, S. N., Hedley, M. J., Kirk, G. J., Bolan, N. S. and Loganathan, P.,2003. Progress in selected areas of rhizosphere research on P aquisition. *Australian Journal of Soil Research.* 41, 471-499.

113. Usha, K., Saxena, A and Singh, B., 2004. Rhizosphere dynamics influenced by arbuscular mycorrhizal fungus (*Glomus deserticola*) and related changes in leaf nutrient status and yield of Kinnow mandarin {King (*Citrus nobilis*) x Willow Leaf (*Citrus deliciosa*)}," *Austr. J. Agric. Res.* 55(5), 571–576.

114. Vestberg M. 1992. Arbuscular Mycorrhizal Inoculation Of Micropropagated Strawberry And Field Observations In Finland. *Agronomie.* 12(10),865-867.

115. Vilma Castellanos-Morales,a Javier Villegas,b Silvia Wendelin,c Horst Vierheilig,d Reinhard Ederc and Ra´ul C´ardenas-Navarroa., 2010. Root colonisation by the arbuscular mycorrhizal fungus *Glomus intraradices* alters the quality of strawberry fruits (*Fragaria ananassa* Duch.) at different nitrogen levels. *J Sci Food Agric.* 90, 1774–1782.

116. Wang, M.Y and Xia R.X., 2009. Effects of arbuscular mycorrhizal fungi on growth and iron uptake of *Poncirus trifoliata* under different pH. *Acta Microbiol. Sin.* 49,1374-1379.

117. Wu, Q.S., Zou, Y.N., Xia, R.X and Wang, M.Y., 2007. Five *Glomus* species affect water relations of *Citrus tangerine* during drought stress. *Bot. Stud.* 48, 147-154.

118. Zelenev,V.V.,Van Bruggen,A.H.C.and Semenov,A.M.2000."BACWAVE" a spatial temporal model for traveling waves of bacterial populations in response to a moving carbon source in soil .*Microb.Ecol.* 40,260-272.

Part IV

Stress Management by Arbuscular Mycorrhizal Fungi

Chapter 7

Management of Moisture Stress

Amar Bahadur[1], Sukhen Chandra Das[1]and Vijayalakshmi[2]

1College of Agriculture, Tripura, Lembucherra,

Agartala (W.T.) 79921, India.

2Indian Institute of Horticultural Research (ICAR),
Bengaluru 560089, India.

*Address for the correspondence: sukhenchandra@rediffmail.com

ABSTRACT

Moisture stress is one of the most important factors limiting growth and yield in fruit crops. Arbuscular mycorrhizal fungi (AMF) symbiosis plays significant role in protecting the plants against osmotic stresses by altering rates of water movement into, through and out of host plants, hydrating plant tissue and there by its physiology. The positive impact of AMF on different factors are nutritional factors increasing leaf conductance and transpiration including P and K uptake, physiological factors directly enhancing root water uptake and providing adequate water to preserve physiological activity in plants, physical factors increasing nutrient acquisition by the plant and cellular factors affecting stomatal conductance, photosynthesis and proline accumulation. This chapter discusses the effect of AMF on management of moisture stress in fruit crops, focusing on the probable mechanism by which AMF changes water relations of their host plant under ample moisture supply and water stress conditions.

Keywords: *AMF , Glomus mosseae, Symbiosis, ROS, Anti oxidant enzymes,*
Aquaporines, LEA Proteins,

1.0 Introduction

Mycorrhiza is a symbiotic (mutualistic) relationship between a fungus and a plant, in which the fungus obtains at least some of its sugars from the plant, while the plant benefits from the efficient uptake of mineral nutrients (or water) by the fungal hyphae (Sylvia *et al.*, 2005). The hyphae are narrower than root hairs, extend further and explore a greater volume of soil, thus having a much higher surface to volume ratio for absorption of nutrients and water. Additionally, the fungi secrete enzymes which break down organic matter and tightly bound micronutrients, such as phosphorus, zinc and copper, enabling the plant to absorb the required minerals from the surrounding soil. Mycorrhizal associations vary widely in structure, form and function, involving different combinations of plant and fungus, which allow plants to use a range of survival strategies. The fungi produce enzymes which release nitrogen from organic matter which can be a critical limiting factor for growth of these plants. It is reported that under water stress condition, the photosynthetic activity of plant suffers affecting the productivity of the plant (Boyer, 1968; Cailloux, 1972). Consequently, AMF are important in sustainable production of horticultural crops, because they improve nutrient uptake (Bolan, 1991), photosynthesis (Johnson, 1984) and water relations and thus increase the drought resistance of host plants (Ruiz-Lazano *et al.*, 1995). Grassland plants are colonized by fungi which form highly branched structures both within root cells and outside the root, and this significantly increases rates of phosphorus-uptake which is frequently a limiting factor for growth. Most AMF are not host-specific and several plants belonging to different species may be linked by a single hyphal network, just as individual plants may form AMF associations. In some cases, the combined associations act synergistically.

Both biotic and abiotic stresses are major constraints to agricultural production. A plant under stress suffers from hormonal and nutritional imbalance, ion toxicity, physiological disorders and susceptibility to diseases. In a changing climate, horticulturalists and landscape professionals are looking for ways to ensure the establishment and healthy growth of plants, whilst minimizing the use of both water and chemical fertilizers. The demands on potable water have never been greater than they are today and with rising populations demands are likely to increase further.

2.0 Moisture stress alleviation by AMF

Moisture stress or drought stress is mostly experienced by plants when they lack water. Lack of sufficient moisture often results in poor or irregular fruit development, reduced yield, plant height, plant growth; smaller fruit size and improper fruit set problems in many of the fruit crops. AMF plays significant role in plant growth under different levels of drought stress conditions. It protects the plants against osmotic stresses by altering rates of water movement into, through and out of host plants, with consequent effects on tissue hydration, plant physiology and also indirectly improving P nutrition (Auge, 2001).

Moreover, AMF increase drought tolerance under nutrient limiting conditions, particularly in low P, N, S, Cu and Zn soils. AMF colonization affects the host plant positively by improving growth, nutrient content, flower quality and thereby alleviating the stress imposed by water deficit, which is not observed in non-mycorrhizal plant (Abdul-Wasea *et al.*, 2010). Plant drought resistance is the result of accumulative physical, nutritional, physiological and cellular effects (Aliasgharzad *et al.*, 2006). The symbiotic association of AMF in different fruit crops under moisture stress tolerance conditions have been studied in citrus (Fidelibus *et al.*,1997) and (Wu and Zou, 2009; Wu *et al.*, 2010a, b), banana (Barea *et al.*, 1997), apple (Runji, 1989), grape (Nikolaou, *et al.*,2003), litchi (Pandey and Misra, 1971; Koske *et al.*, 1989), Peach (Calvet,*et al.*,2004), Pomegranate (Josfina, *et al.*, 2014), Plum (Świerczyński and Stachowiak, 2010), papaya (Boyer, 1968; Cailloux, 1972) and (Ruiz-Lazano *et al.*, 1995), Strawberry (Taylor and Harrier, 2001; Vestberg *et al.* (2004) and Jamun (Devachandra *et al.*, 2009).

3.0 Role of AMF hyphae in water transfer

The AMF hyphae with a diameter of 2–5 mm can penetrate soil pores inaccessible to root hairs (10–20 mm diameter) and absorbs water which is not accessable to non-mycorrhizal plants (Allen ,1982; Hardie ,1985). The rate of transport of water by extraradical hyphae to the root is 0.28 ngs^{-1} per entry point, which is sufficient to modify plant water relations (Allen, 1991). Faber *et al.*, (1991) estimated the rates of water transport by extraradical hyphae to be ranging from 375 to 760 nl H_2O h^{-1}. Fitter, (1985), George *et al.* (1992) and Koide, (1993) showed that rates of water uptake by hyphae on the basis of hyphal entry points per unit of root length, hyphal cross sectional area or water potential gradients suggested that hyphal water transport rates are negligible. Ruiz- Lozano and Azconz (1995) through an experiment with mycorrhizal lettuce plants grown in containers which had three vertical compartments, the upper root compartment containing the root system of the plants separated from the next compartment by a 50-mm nylon screen which allowed penetration by AMF hyphae but not by roots showed that when water was applied in lower compartment, plant fresh weight increased by 150% in plants colonized by AMF *Glomus fasciculatum* compared to the plants which had their roots only immersed in water. Similarly Walker and Kaoske conducted the same with plants colonized by *G. deserticola* and obtained an increase of 215% in plant fresh weight as compared to a non-inoculated P fertilized treatment. This experiment also resulted in increased leaf water content and gas exchange in mycorrhizal plants with water applied to the hyphal compartment II. By this they concluded that the positive effects of the AMF on plant growth and water uptake were enhanced by addition of water to the hyphal compartment.

3.1 Nutritional Effect

AMF improved host plant growth, water relations and acquisition of nutrients especially P from soil (Maronek *et al.*, 1981). AMF association is a biological strategy (Cekic *et al.*, 2012), this association allows plants to explore larger volumes of soil, improves water-use efficiency, absorb more water and nutrients especially absorption of immobile mineral elements such as phosphorus (P) and other micronutrients and provides resistance to soil pathogens ,salinity and drought (Al-Karaki, 2000., Beltrano *et al.*, 2003). AMF symbiosis increases leaf area, delays senescence (Beltrano *et al*, 2003., Beltrano and Ronco, 2008). The protection of mycorrhizal plants against water stress is related to the effect that the endophyte has on increasing leaf conductance and transpiration as well as P and K uptake. AMF symbiosis declines in leaf water potential (Davies *et al.*, 1992, Subramanian and Charest 1995, 1997, El-Tohamy *et al.*, 1999). AMF plants are also reported to return to the control level more quickly than non-mycorrhizal plants after relief of drought stress (Subramanian and Charest 1997). *Glomus versiforme* notably increased the P, K and Fe contents in leaves of tangerine subjected to normal watering and water stress. Higher P, K and Fe contents in leaves of mycorrhizal plants helped host plants to enhance their drought resistance.

During soil drying, mycorrhizal plants often maintain higher gas exchange rates than non-mycorrhizal plants of similar size and nutrient status (Auge *et al.* 1987; Bethlenfalvay *et al.* 1987; Auge 1989; Auge, *et al.* 1992; Sanchez-Deaz and Honrubia 1994; Ruiz-Lozano *et al.* 1995a, 1995b; Goicoechea *et al.* 1997). Duan *et al.* (1996) reported that mycorrhizal plants maintained higher stomatal conductance, transpiration rate and shoot water than non-mycorrhizal plants. They suggested that AMF probably increase the ability of the root system to scavenge water in dried soil, resulting in less strain on foliage, and hence higher stomatal conductance and shoot water content at particular soil water potential. The physical, chemical and biological actions of AMF hyphae and hyphal exudates on soils effect soil structure (Jastrow and Miller 1991; Oades and Waters 1991), which in turn affects soil moisture retention properties (Hamblin 1985).

3.2 Physiological Effect

The mycorrhizal symbiosis occurs in almost all fruit tree species grown in nursery or field (Calvet *et al.*, 2004). Experiments have indicated that AMF is able to alter water relations and play a great role in the growth of host plant during drought stress. AMF increase both their dehydration avoidance and dehydration tolerance capacities (Auge,*et.al.*,1987). Dehydration is avoided to a larger extent in AMF plants, as compared to non-mycorrhizal plants of similar size, through increased accumulation of solutes, which lower bulk leaf symplastic and osmotic potential. Mycorrhizal inoculation may directly enhance root water uptake providing adequate water to preserve physiological activity in plants, especially under severe drought conditions (Faber *et al.*, 1991; Smith and Read, 1997). This stressful condition can trigger an increase in reactive oxygen species (ROS) production that, in turn, can induce cellular, anatomical, and morphological changes that improve drought tolerance. AMF protect photochemical systems against water deficiency (Borkowska, 2002). Gaurav *et al.*(2010) tested the hypothesis that growth and water-use characteristic of AMF plants would differ from those of non-AMF plants that were well supplied with P. Mycorrhizal lettuce plants subjected to drought had increased SOD activity compared to non-mycorrhizal controls (Ruiz-Lozano *et al.* 1996) and *G. mosseae* possessed CuZn-SOD activity and that mycorrhizal clover roots exhibited two additional SOD isoforms as compared to non-mycorrhizal roots: a myc CuZn-SOD and a Mn-SOD (Palma *et al.* 1993). Molecular analysis have confirmed this response at the transcriptional level (Ruiz-Lozano *et al.* 2001a).

3.2.1 Aquaporins

To cope up with water limitation and damage from water deficit, plants have adopted certain mechanisms to increase water permeability in their tissues. Aquaporins are water channel proteins that facilitate the passive movement of water molecules down a water potential gradient. They belong to the major protein (MIP) family of transmembrane channels that are represented in all kingdoms (Chrispeels and Agre, 1994). There are two classes of plant aquaporins present in plasma membrane and tonoplast. The vacuolar and plasma membrane aquaporins, acting together, are responsible for the cytosolic osmoregulation that is necessary for maintaining normal metabolic processes (Kjelbom *et al.*, 1999). The inhibition studies of aquaporins *ex-vitro* and antisense mutant studies have indicated that aquaporins are also important for the bulk flow of water in plants (Kjelbom *et al.*, 1999). The genes encoding for aquaporin in plant cells shows high expression of aquaporins in tissues involved in water transport in transcellular water flow through living cells (Barrieu *et al.*, 1998). AMF symbiosis development induces the expression of genes encoding aquaporins (Roussel, *et al.*, 1997; Krajinski *et al.*, 2000).

3.2.2 Late embryogenesis abundant proteins

Late embryogenesis abundant (LEA) proteins are another important group of proteins involved in drought tolerance in mycorrhizal plants. LEA proteins also accumulate in vegetative plant organs during periods of water deficit, and it has been proposed that LEA proteins play an important role in maintenance of the structure of other proteins and membranes, in sequestration of ions, in binding of water, and in functioning as molecular chaperones (Close, 1996). These are most soluble proteins induced by a dehydration stress and have been observed in over 100 independent studies of drought stress, cold acclimatization, salinity stress, embryo development and responses to ABA (Close, 1996). It would be of interest to determine whether the AMF symbiosis alters the pattern of LEA protein accumulation under osmotic stress and whether such alteration functions in the protection of the host plants against drought (Ruiz-Lozano, 2003).

3.3 Cellular effect

Ruiz-Lozano (2003) reported that AMF symbiosis might increase the drought resistance of higher plants by promoting antioxidant enzymes. In fact, the symbiosis with AMF has been proposed as one of the mechanisms of water stress avoidance (Augé, 2004; Ruiz- Lozano and Azcón, 1996; Ruíz-Lozano *et al.*, 1995). Plants can respond to drought stress at morphological, metabolic and cellular levels with modifications that allow the plants to avoid the stress or to increase its tolerance (Bray, 1997). Higher levels of sugar may improve the ability of the plants to withstand drought stress and recover after the condition is restored (Kameli and Lösel, 1996). The AMF *Glomus intraradices* (Schenck and Smith) colonized rough lemon seedlings (*Citrus jambhiri* Lush), showed increased root growth and transpiration rate and reduced leaf water potentials relative to non-infected control plants (Levy *et al.*, 1983). Recovery from water stress was studied on mycorrhizal and non-mycorrhizal rough lemon seedlings (*C. jambhiri* Lush) of similar age. Mycorrhiza affected stomatal conductance, photosynthesis and proline accumulation, but not leaf water potential, suggesting that most of the effect of the mycorrhizal association is on stomatal regulation rather than on root resistance.

3.4 Physical effect

Mycorrhizal plants grown in the amended soil reached the highest proline and sugar contents and were the least-damaged (in terms of plant growth) by drought, the effects achieved by the improved physical characteristics of composted soil and the ability of these fungi to acquire water (Porcel *et al.*, 2006; Ruíz-Lozano and Azcón, 1996). *G. deserticola* was the most adapted and aggressive colonizer as well as the most effective species for increasing drought tolerance of the host plant both in terms of maintaining growth under stress conditions and in permitting more efficient use of water (Ruiz-Lozano, *et al.*,1995).

4.0 Moisture Stress alleviation in selected fruit crops due to AMF inoculation

4.1 Apple (*Malus pumila* L.)

Among the temperate fruits, apple is the premier fruit of the world. AMF influence the water status, mineral uptake, and growth of the seedlings of apple and establish the probable mechanism by which AMF change water relations of their host plant under ample moisture supply and water stress conditions. Runjin, (1989) found that sterilized soil inoculated with *Glomus versiforme* and *G. macrocarpum* enhanced element uptake, improved water status, drought tolerance and growth of the plants. The mycorrhizal plants were more water-use efficient than non-mycorrhizal plants. Shoot dry matter differences between mycorrhizal and non-mycorrhizal plants represented the benefit derived by plants from AMF root associations. The mycorrhizal plants used less water to produce one unit of shoot dry matter (WUE-Water Use Efficiency) than non-mycorrhizal plants, but water-stressed and well-watered plants did not differ in WUE (Al-Karaki, 1998). Plants treated with *Glomus* isolates differed in colonization level, leaf P concentration, root length, transpiration flux and leaf conductance (Fidelibus *et al.*, 2001). The mycorrhizal plants fully recovered their photosynthetic activity when watering was restored. In apple (*M. domestica*), *G. mosseae* singly and in combination with *G. macrocarpum* were more effective in increasing plant biomass than *G. macrocarpum* (Miller *et al.*, 1989).

4.2 Citrus (*Citrus* spp.)

Citrus is the world's leading fruit crop and in India, it is the third important fruit crop next to Mango and Banana. However, it is grown in regions that are frequently subjected to water deficiency, which restricts the yield and quality of the crop. *Citrus* has very few and short root hair's and is highly dependent on arbuscular mycorrhizae, since the mutualistic symbiosis replaces some of the root hairs' functions. In field, most citrus plants lack root hairs and are strongly dependent on AMF symbiosis, because AMF symbiosis replaces partly the function of root hairs to uptake nutritients and water. Grafted citrus trees are usually used in orchard, but little is known about biochemical responses of grafted citrus trees colonized by AMF to drought stress. Wu *et al.* (2007) reported the efficacy of five *Glomus* species, *Glomus mosseae*, *G. geosporum*, *G. versiforme*, *G. etunicatum* and *G. diaphanum* for the ability to improve water relations of *Citrus tangerine* Hort. ex Tanaka under well-watered and drought stress conditions. Various glasshouse and field experiments proved that inoculation with AMF enhanced the growth and ion uptake in citrus plants, improved tolerance to drought and salt stress and also the quality of fruit (Wu and Zou, 2009; Wu *et al.*, 2010a, b). They concluded that AMF symbiosis might increase the drought tolerance of citrus plants by promoting the antioxidant content and also the activities of antioxidant enzymes (Wu *et al.*, 2006 a, b).

Cardon and Whitbeck, (2007) inoculated citrus seedling with AMF *Glomus* isolates from different geographic origins arid, semi-arid or mesic areas and found that plants had different patterns of plant growth and water use with different isolates. Fidelibus *et al.* (2001) showed that four *Glomus* species isolated from arid, semiarid and mesic areas stimulated the root growth (dry weight and length) of *C. volkameriana*. The leaf P concentrations were 12-56% higher in AMF plants than in non-AMF plants under well-watered conditions. AMF plants had greater whole-plant transpiration than non-AMF plants under well-watered conditions, under mild water stress and during recovery from moderate and severe soil drying. AMF plants had lower leaf conductance than non-AMF plants when exposed to severe soil drying. Also, they suggested that *Glomus* isolates that increased root growth and whole-plant transpiration might improve the field performance of young citrus rootstock and mitigate against desiccation after soil drying by amplifying the potential for root exploration of soil for water. Plant colonized by AMF could tolerate and recover more rapidly from soil water deficits than plants without AMF (Allen and Boosalis, 1983, Bildus *et al.*, 1986, Henderson and Davies, 1990). Johnson and Hummel (1985) reported increased resistance to drought and transplant stress by Carrizo citrange seedlings inoculated with *G. intraradices* as compared to un-inoculated ones.

4.3 Grapes *(Vitis vinifera L.)*

Nikolaou *et al.* (2003) studied the effects of drought stress on leaf photosynthesis and water relations of Cabernet Sauvignon grapevine scion grafted onto eight different rootstocks. Foliar growth, leaf phosphorus concentrations and drought tolerance were greater in the inoculated than in the non-inoculated plants. Some drought-sensitive rootstocks colonized with mycorrhizal fungi and subjected to drought for eight days showed much-improved drought resistance compared with non-infected rootstocks of the same varieties. They concluded that mycorrhizal colonization improved the water status of non-irrigated vines. The root infection by the mycorrhiza increases active absorptive surface area and stimulate nutrient and water uptake even in water stress condition. Nikolaou *et al.* (2003a) studied the effect of AMF colonization and water stress and isopentenyladenine/ isopentenyladenosine or zeatin/zeatin riboside content in (grapevine L. 'Cabernet Sauvignon'). Cytokinin production of mycorrhizal plants was greater than that of non-mycorrhizal plants. In both well-watered and stressed plants, concentrations of Z/ ZR in shoot tips ranged from 2.75 to 87.7 ng g^{-1} dry weight being higher than those of iAde/iAdo in most cases, ranging from 4.01 to 25.2 ng g^{-1} dry wt. Significantly lower concentrations for both types of cytokinins were found in stressed plants. Mycorrhizal stressed vines had higher predawn leaf water potential, stomatal conductance and CO_2 assimilation rates than non-mycorrhizal stressed ones. Rootstock had a considerable effect on shoot growth, leaf P content, and cytokinin production.

4.4 Jamun (*Syzygium cuminii* Skeel)

Devachandra *et al.* (2009) studied AMF inoculated Jamun seedlings for drought tolerance as it is a dry land crop. Seeds of jamun were sown in pots inoculated with 9 different mycorrhizal strains.Low moisture stress was imposed on six months old plants by withholding the application of water after irrigating to the field capacity of the soil. When the plants were subjected to drought situation, the proline content of leaves increased. The un-inoculated control recorded higher proline content before and after imposition of stress. The AMF inoculated seedlings recorded lower proline content, indicating that the seedlings were less affected by low moisture stress. After imposing drought, the percent soil moisture of AMF inoculated seedlings were lower than those of un-inoculated control. Significantly least soil moisture content was noted in *G. fasciculatum*, followed by *Sclerocystis dussii* .This phenomenon is associated with profused root morphology as influenced by AMF inoculation, which enhances the water uptake by seedlings. The relative water content in the leaves was higher in seedlings inoculated with AMF as compared to the un-inoculated control seedlings. *G. fasciculatum* inoculated seedlings registered highest relative water content, which might have resulted in lowered percent of leaf curled. The percent leaf curled was maximum in un-inoculated control seedlings. In contrast to non-mycorrhizal controls, AMF inoculated seedlings showed higher nitrogen assimilation. Better nitrogen nutrition characterized by higher soluble proteins during drought has been recorded by other researchers.

4.5 Litchi (*Litchi chinensis* L.)

It is well known that arbuscular mycorrhizae can enhance the growth of seedlings of many tropical tree species (Janos 1980), including species of commercial fruit trees (Read, 1993), in low phosphorus soil. It is less certain, however, that AMF will enhance the growth of trees propagated from large stem cuttings or air-layers (girdled, rooted branches) that involve considerable woody tissue containing abundant carbohydrate reserves (Menzel *et al.*, 1995) and possibely substantial mineral nutrients. Litchi is commercial tropical fruit tree, forms arbuscular mycorrhizae (Pandey and Misra, 1971; Koske *et al.*, 1989) and root hairs. It sometimes produces 'tuberculate' ultimate rootlets. As litchi does not produce true-to-type seedlings and grafting and budding are not reliable (Pandey and Misra, 1975), propagation is typically by air-layering. Pandey and Misra (1975) examined the effect of AMF inoculation on litchi in pots of sterilized soil, but all control plants without mycorrhiza died.

4.6 Papaya (*Carica papaya* L.)

Cruz *et al.* (2000) studied the effect of AMF on tree growth, leaf water potential, and levels of 1-aminocyclopropane-1-carboxylic acid and ethylene in the roots of papaya under water-stress conditions. LWP decreased during water-stress treatment which was more severe in the non-AMF plants. Plant fresh weight was higher for AMF than non-AMF plants under both conditions. Under well-irrigated conditions, the ethylene concentration in the roots was increased by the presence of AMF, although there was no significant difference between AMF and non-AMF roots in ACC levels. ACC increased in both AMF and non-AMF roots under water-stress conditions. The water-stress treatment resulted in a marked increase in ethylene concentration in non-AMF roots but the concentration in AMF roots was slightly lower than normal conditions.

Forty five days old papaya seedlings inoculated with respective AMF were transplanted to the earthen pots of size 30 cm x 40 cm and drought was imposed at different levels. Effect of AMF on chlorophyll 'a', chlorophyll 'b' and total chlorophyll content before and after drought imposition was found to be significant. Significant differences in the proline content were noticed both at 10 and 20 days after drought (Nagarajappa, *et al.*, 2003).

4.7 Pomegranate (*Punica granatum* L.)

Josfina *et al.* (2014) reported that AMF alleviate oxidative stress in pomegranate plants growing under drought stress conditions. They evaluated the effect of inoculation of two strains of the AMF *Rhizophagus intraradices* (Schenck and Smith), Walker and Schüßler (GA5 and GC2) on pomegranate plants under two irrigation conditions. The study showed that, in most cases, mycorrhizal plants increased antioxidant defenses, such as the ROS-scavenging enzymes superoxide dismutase (SOD), catalase (CAT), and ascorbate peroxidase (APX) in shoots under both irrigation levels, whereas the response for roots was ambiguous. AMF inoculation maintained the levels of malondialdehyde (MDA), probably by rapidly increasing antioxidant defenses and preventing lipid damage. Early AMF inoculation (particularly with the GC2 strain) in pomegranate propagation protects plants against abiotic stress according to the authors.

4.8 Peach (*Prunus persica* L. Batsch)

Qiang-Sheng *et al.*(2011) studied Root system architecture (RSA) and spatial conFigureuration of mycorrizal peach (*Prunus persica* L. Batsch) root system inoculated with the AMF species *Glomus mosseae*, *G. versiforme*, and *Paraglomus occultum* in the soil, which substantially determines the capacity of a plant to take up nutrients and water. The inoculated seedling showed 0-1 cm and 3-4 cm increase in root length and also altered RSA seedlings, and the alteration due to mycorrhization was related to allocation of glucose/sucrose to root (Aglucose/ sucrose). The inoculation with *G. mosseae* and *G. versiforme* significantly increased leaf, stem, root and total fresh weights, compared with non-AMF treatment. Alteration of RSA depended on the AMF species, because only *G. mosseae* and *G. versiforme* but not *P. occultum* markedly increased root length, root projected area, root surface area and root volume for scavenging more moisture from soil.

Rutto *et al.* (2002) studied the effect of root-zone flooding on mycorrhizal and non-mycorrhizal peach seedlings growing in a low P medium .The flooding improved the mycorrhizal seedling development and also significantly increased concentrations of shoot P, K and Zn and biomass yield. The accumulation of ethanol was significantly higher in the taproots of non-mycorrhizal as compared to mycorrhizal plants after 3 days of flooding. The rapid decline in plant health was observed in non-mycorrhizal as compared to mycorrhizal seedlings. The presence or absence of the fungal partner led to significant difference in the ratio of roots that remained viable after extended flooding. Therefore, it showed that infection of AMF confers limited tolerance to flooding on peach seedlings, because due to improved plant nutrition, the suppression of ethanol accumulation in roots and the extension of the duration of root activity in a flooded environment.

Mycorrhizal symbiosis helped peach seedlings overcome soil-fumigation nutrient-deficiency effects in nursery, alleviate flooded stress, but not overcome peach replant problem (Rutto *et al.*, 2002; Rutto and Mizutani, 2006).

4.9 Strawberry (*Fragaria virginiana* Duchesne)

Borkowska, (2002) found that mycorrhization strongly affected growth and tolerance to water deficiency of the strawberry plants cultivated in greenhouse. Mycorrhizal plants showed higher biomass accumulation (crowns and roots) and larger leaf area. Positive role of AMF in protecting photochemical systems against water deficiency was evident. Mycorrhizal plants fully recovered their photosynthetic activity when watering was restored. Borowiicz, (2010) studied AMF effect on three genotypes of wild strawberry, on water relations. Increased root colonization by these root symbionts increased host's tolerance to drought.

5.0 Conclusion and future strategies

The role of AMF in alleviating moisture stress is very much evident from the results of large amount of work conducted on the subject. AMF protects the plants against osmotic stress by altering the rate of water movement into the plants through its hyphal network which acts as extended root system and helping the plants absorb many times more water compared to nonmycorrhizal plants thereby altering the physiology of the plant. The mechanism of moisture stress tolerance induced by AMF includes nutritional factors increasing leaf conductance and transpirational factor including P, K ,and Fe uptake and cellular factors affecting stomatal conductance, photosynthesis and proline accumulation. AMF increases the dehydration avoidance and tolerance capacities of plants. AMF symbiosis is also known to induce expression of genes encoding aquaporins, the water channel proteins that facilitate the passive movement of water molecules down a water potential gradient. AMF also induce production of LEA proteins under water deficit conditions which help to maintain the structure of other proteins and membranes in sequestration of ions , in binding of water and in functioning as molecular chaperons.

It is essential to concentrate on osmotic-stress regulated gene which play a role in the response of AMF symbiosis by creating cDNA libraries from AMF plants grown under osmotic stress. It is also required to check if the symbiosis alters the pattern of LES protein accumulation under stress and whether this alteration protects the plants from stress.

References

1. Abdul-Wasea, A.A. and K.M. Elhindi., 2010. Alleviation of drought stress of marigold (*Tagetes erecta*) plants by using arbuscular mycorrhizal fungi. *Saudi Journal of Biological Sciences*. 18, 93–98.

2. Allen, M.F., 1982. Influence of vesicular-arbuscular mycorrhizae on water movement through Bouteloua gracilis Lag ex Steud. *New Phytol* . 91,191–196.

3. Allen, M.F., Boosalis, M.G., 1983. Effects of two species of VA mycorrhizal fungi on drought tolerance of winter wheat. *New Phytologist*. 93, 67-76.

4. Allen, M.F., 1991. The ecology of mycorrhizae. Cambridge University.

5. Al-Karaki, G.N., 1998. Benefit, cost and water-use efficiency of arbuscular mycorrhizal durum wheat grown under drought stress.*Mycorrhiza*. 8,41-45.

6. Al-Karaki, G.N., 2000. Growth of mycorrhizal tomato and mineral acquisition under salt stress. *Mycorrhiza*. 10, 51-54.

7. Allen, M., W. Swenson., J.I. Querejeta., L.M. Egerton-Warburton, and K.K. Treseder., 2003. "Ecology of Mycorrhizae: A Conceptual Framework for Complex Interactions Among Plants and Fungi." *Phytopathology*. 41, 271-300.

8. Aleksandrowicz-Trzcińska M., 2004. Kolonizacja mikoryzowa i wzrost sosny zwyczajnej (*Pinus silvestris* L.) w uprawie założonej z sadzonek wróżnym stopniu zmikoryzowanych. *Acta Sci.pol.Silv.Colender.Rat.Ind. Lignar*. 3(1), 5-15.

9. Aliasgharzad N., Neyshabouri M.R. and Salimi G., 2006. Effects of arbuscular mycorrhizal fungi and *Bradyrhizobium japonicum* on drought stress of soybean. *Biologia*. 61, 324–328.

10. Auge, R .M., Schekel, K. A and R .L. Wample., 1987 Leaf water and carbohydrate status of VA mycorrhizal Rose exposed to drought stress. *Plant and Soil* . 99, 291–302.

11. Auge, R.M.,1989. Do VA mycorrhiza enhance transpiration by influencing host phosphorus status, *J Plant Nutr*. 12,743–753.

12. Auge, R.M., Stodola, A.J.W., Brown, M.S. and Bethlenfalvay, G.J., 1992a. Stomatal responses of mycorrhizal cowpea and soybean to short-term osmotic stress. *New Phytol* .120,117–125.

13. Augé, R.M., 2001. Water relation, drought and vesiculararbuscular mycorrhizal symbiosis. *Mycorrhiza*. 11, 3-42.

14. Augé, R. M., 2004. Arbuscular mycorrhizae and soil/plant water relations. *Can. J. Soil Sci.* 84, 373-381.

15. Azcon-Aguilar, C and J. M. Barea., 1997. Applying mycorrhiza biotechnology to horticulture: significance and potentials. *Sci. Hort.* 68,1-24.

16. Baylis, G. T. S., McNabb, R. F. R and T. M. Morrison., 1963. The mycorrhizal nodules of podocarps. *rans. Br. Mycol. Soc.* 46, 378–384.

17. Barea, J.M., 1991. Vesicular-arbuscular mycorrhizae as modifiers of soil fertility. *Adv. Soil Sci.* 15 , 1 - 40.

18. Barea, J. M and P. Jeffries., 1995. Arbuscular mycorrhizas in sustainable soil plant systems. In: A. Varma, B. Hock (eds). Mycorrhiza: Structure, Function, Molecular Biology and Biotechnology. Springer-Verlag, Heidelberg. 521-559.

19. Barea J.M., Calvet C., Estaun V and A. Camprubi., 1996. Biological control as key component in sustainable agriculturae. *Plant Soil*.185,171-172.

20. Barea J.M., Azcón-Aguilar C and R. Azcón ., 1997. Interactions between mycorrhizal fungi and rhizosphere microorganisms within the context of sustainable soil-plant systems, in: Gange A.C., Brown V.K. (Eds.), Multitrophic interactions in terrestrial systems, *Blackwell Science*, Oxford. 65–77.

21. Barrieu, F., Chaumont, F.and Chrispeels, M.J., 1998. High expression of the tonoplast aquaporin ZmTIP1 in epidermal and conducting tissues of maize. *Plant Physiol.*,117,1153–1163.

22. Barea, J. M., Azcón, R and C. Azcón-Aguilar., 2002a. Mycorrhizosphere interactions to improve plant fitness and soil quality. *Anton. Leeuw. Int. J. G.* 81, 343-351.

23. Barea, J. M., Gryndler, M., Lemanceau, P., Schüepp, H.and Azcón, R., 2002b. The rhizosphere of mycorrhizal plants. In: S. Gianinazzi, H. Schüepp, J. M. Barea, K. Haselwandter (eds). *Mycorrhiza Technology in Agriculture: from Genes to Bioproducts*. Birkhäuser Verlag, Basel, Switzerland, 1-18.

24. Barea, J. M., Palenzuela, J., Azcón, R., Ferrol, N.and Azcón-Aguilar, C., 2002c. Micorrizas yrestauración de la cubierta vegetal en ambientes mediterráneos. In: J. M. Barea-Azcón, E. Ballesteros, J. M. Luzón, M. Moleón, J. M. Tierno, R. Travesí, R. (eds). Biodiversidad y Conservación de Fauna y Flora en Ambientes Mediterráneos, Granada, España, 83-105.

25. Barea, J. M., Toro, M., Orozco, M. O., Campos, E and R. Azcón., 2002d. The application of isotopic (P^{32} and N^{15}) dilution techniques to evaluate the interactive effect of phosphate solubilizing rhizobacteria, mycorrhizal fungi and *Rhizobium* to improve the agronomic efficiency of rock phosphate for legume crops. *Nutr. Cycl. Agroecosys.* 63, 35-42.

26. Bethlenfalvay, G.J., Brown, M.S., Mihara, K.L. and Stafford, A.E., 1987. The *Glycine-Glomus-Bradyrhizobium* simbiosis. V. Effects of mycorrhizal on nodule activity and transpiration in soybean under drought stress. *Plant Physiol.* 85,115–119.

27. Beltrano, J., Ronco, M.G., Salerno, M.I., Ruscitti, M. and O Peluso., 2003. Respuesta de planta de trigo (*Triticum aestivum* L.) micorrizadas en situaciones de déficit hídrico y de rehidratación del suelo. *Rev. Cien. Tec.* 8, 1-7.

28. Beltrano, J and M. Ronco., 2008. Improved tolerance of wheat plants (*Triticum aestivum* L.) to drought stress and rewatering by the arbuscular mycorrhizal fungus *Glomus claroideum*: Effect on growth and cell membrane stability. *Braz. J. Plant Physiol.* 20, 29-37.

29. Biermann, B and R.G. Linderman., 1983. Increased geranium growth using pretransplant inoculation with mycorrhizal fungus. *J. Am. Soc. Hort. Sci.* 108 , 972 - 976.

30. Bildusas, I.J., Dixon, R.K., Pfleger, F.L and E.L. Steward., 1986. Growth nutrition and gas exchange of *Bromus inermis* inoculated with *Glomus fasciculatum*. *New Phytologist.* 102, 303-311.

31. Boyer, J.S., 1968. Relationship of water potential to growth of leaves. *Plant Physiology.* 43,1056-1062.

32. Bolan, N.S., 1991. A critical review on the role of mycorrhizal fungi in the uptake of phosphorus by plants. *Plant and Soil.* 134, 189-207.

33. Borkowska, B., 2002. Growth and photosynthetic activity of mikropropagated strawberry plants inoculated with endomicorrhizal fungi (AMF) and growing under drought strees. *Acta Physiol. Plantarum.* 24 (4), 365-370.

34. Brundrett, M., 1991. Mycorrhizas in natural ecosystem. In: Begon, M., Fitter, A.H., Macfadyen, A. (eds.). *Advances in Ecological Research. Academic Press Limited, London.*171-313.

35. Bray, D.E., 1997. Plant responses to water deficit. *Trends Plant Sci.* 2, 48-54.

36. Cailloux, M., 1972. Metabolism and absorption of water by root hairs. *Canadian Journal of Botany.*50,557-573.

37. Caron. M., 1989. Potential use of mycorrhizae in control of soil borne diseases. *Can. J. Plant Pathol.* 11 , 177 - 179.

38. Calvet, C., V. Estaun., A. Camprubi., A. Hernandez- Dorrego., J. Pinochet and M.A. Moreno 2004. Aptitude for mycorrhizal root colonization in *Prunus* rootstocks. *Sci. Hort.* 100,39-49.

39. Cardon, Z.G and J.L. Whitbeck., 2007. The Rhizosphere, Elsevier Academic Press. 235.

40. Cekic, F.O., Unyayar, S and I. Ortas., 2012. Effects of arbuscular mycorrhizal inoculation on biochemical parameters in *Capsicum annuum* grown under long term salt stress. *Turk J .Bot*. 36, 63-72.

41. Chrispeels,M.J. and Agre, P., 1994. Aquaporins: water channel proteins Of plant and animal cells. *Trends Biochem Sci*.19,421–425.

42. Close, T., 1996. Dehydrins: Emergence of a biochemical role of a family of plant dehydration proteins. *Physiol Plant* .97,795–803.

43. Cruz, A.F., Ishii, T. and Kadoya, K., 2000. Effects of arbuscular mycorrhizal fungi on tree growth, leaf water potential, and levels of 1-aminocyclopropane-1-carboxylic acid and ethylene in the roots of papaya under water-stress conditions. *Mycorrhiza*. 10,115-119.

44. Davies, J.F.T., Potter, J.R and R.G. Linderman., 1992. Mycorrhiza and repeated drought exposure affect, drought resistance and extraradical hyphae development of pepper plant independent of plant size and nutrient content. *J. Plant Physiol*. 139, 289 - 294.

45. Dehne H.W., 1982. Interactions between vesicular-abuscular mycorrhizal fungi and plant pathogens. *Phytopathology*.72,1115- 1119.

46. Declerck, S., B. Devos., B. Delvaux and C. Plenchette., 1994. Growth response of micropropagated banana plants to VAM inoculation. *Fruits*. 49, 103–109.

47. Devachandra, N.,C. P. Patil., P .B. Patil., G.S.K. Swamy and M.P. Durgannavar., 2009. Mycorrhizal inoculation in combating low moisture stress in jamun (*Syzygium cuminii Skeel*). *Mycorrhiza News* .20(4), 7-10.

48. Dixon, R.K., Rao M.V and V.K. Garg., 1994. Water relations and gas exchange of mycorrhizal Leucaena leucocephala seedlings. *J Trop For Sci*. 6, 542-552.

49. Duan, X., Newman, D.S., Reiber, J.M., Green, C.D., Saxton, A.M.and Auge, R.M., 1996. Mycorrhizal influence on hydraulic and hormonal factors implicated in the control of stomatal conductance during drought. *J Exp Bot* .47,1541–1550.

50. El-Tohamy W., Schnitzler, W.H and U. El-Behairy. 1999. Effect of VA mycorrhiza on improving drought and chilling tolerance of bean plants. *J Appl Bot*,.73, 178-183.

51. Faber, B. A., Zasoski, R. J., Munns D. N and K. Shackel., 1991. A method for measuring hyphal nutrient and water uptake in mycorrhizal plants. *Can. J. Bot*. 69, 87–94.

52. Fitter, A.H .,1985. Functioning of vesicular-arbuscular mycorrhizas under field conditions. *New Phytol* .99,257–265.

53. Fidelibus, M.W., C.A. Martin and J.C. Stutz., 2001. Geographic isolates of *Glomus* increase root growth and wholeplant transpiration of *Citrus* seedlings grown with high phosphorus. *Mycorrhiza.* 6, 119-127.

54. Gadkar, V., David-Schwartz R., Kunik, T and Y. Kapulnik Y., 2001. Arbuscular mycorrhizal fungal colonization. Factors involved in host recognition. *Plant Physiology.* 127, 1493–1499.

55. Gaurav, S.S., S.P.S. Sirohi., B. Singh and S. Pradeep., 2010. Effect of Mycorrhiza on Growth, yield and tuber deformity in potato (*Solanum tuberosum* L.) grown under water stress conditions. *Prog. Agric.* 10, 35–41.

56. George E., Hssuser, K.U., Vetterlein, D., Gorgus, E.and Marschner ,H.,1992. Water and nutrient translocation by hyphae of *Glomus mosseae. Can J Bot* .70,2130–2137.

57. Gianinazzi, S., Trouvelot, A and V. Gianinazzi-Pearson., 1990. Role and use of mycorrhizas in horticultural crop production. *Proceedings of the Internaitonal Society for Horticultural Science,* August 1990, Firenze.

58. Goicoechea, N., Antolin, M.C and Sanchez-Diaz, M., 1997. Gas exchange is related to the hormone balance in mycorrhizal or nitrogenfixing alfalfa subjected to drought. *Physiol Plant* .100,989–997.

59. Grant, C., S. Bittman., M. Montreal., C. Plenchette and C. Morel., 2005. Soil and fertilizer phosphorus: Effects on plant P supply and mycorrhizal development. *Can. J. Plant Sci.* 85,3–14.

60. Hardie, K.,1985. The effect of removal of extraradical hyphae on water uptake by vesicular-arbuscular mycorrhizal plants. *New Phytol.* 101,677–684.

61. Henderson, J.C and F.T. Davies., 1990. Drought acclimation and the morphology of mycorrhizal *Rosa hybrida* L. cv. *'Fredy'* is independent of leaf elemental content. *New Phytologist* .115, 503-510.

62. Janos, D. P., 1980. Vesicular-arbuscular mycorrhizae affect lowland tropical rain forest plant growth. *Ecology.* 61, 151–162.

63. Jastrow, J.D.and Miller, R.M., 1991. Methods for assessing the effects of biota on soil structure. *Agric Ecosyst Environ.* 34,279–303.

64. Johnson, C. R., 1984. Phosphorus nutrition on mycorrhizal colonization, photosynthesis, growth and nutrient composition of *Citrus aurantium. Plant and Soil.*80, 35-42.

65. Johnson, C.R., Hummel, R.L., 1985. Influence of mycorrhizae and drought stress on growth of *Poncirus* x *Citrus* seedlings. *Hort Science.* 20, 754-755.

66. Kameli, A.and Lösel, D.M., 1996. Growth and sugar accumulation in durum wheat plants under water stress. *New Phytologist.* 132, 57-62.

67. Kapoor, R., Sharama, D and A.K. Bhatnagar. 2008. Arbuscular mycorrhizae in micro propagation systems and their potential applications. *Sci Hortic.* 116,227-239.

68. Khanam , D., 2007. Assessment of Arbuscular Mycorrhizal Association in Some Fruit Plants in Bangladesh. *Bangladesh J Microbiol.* 24(1),34-37.

69. Kjellbom, P., Larsson, C., Johansson, I., Karlsson, M.and Johanson, U., 1999. Aquaporins and water homeostasis in plants. *Trends Plant Sci* . 4,308–314.

70. Koske, R.E., Gemma, J .N and W.C. Mueller., 1989. Observations on 'sporocarps' of the VA mycorrhizal fungus *Rhizophagus litchii*. *Mycol. Res.* 92, 488–490.

71. Koide, R., 1993. Physiology of the mycorrhizal plant. *Adv Plant Pathol* . 9,33–54.

72. Krajinski, F., Biela, A., Schubert, D., Gianinazzi-Pearson, V., Kaldenhoff, R.and Franken, P.,2000. Arbuscular mycorrhiza development regulates the mRNA abundance of Mtaqp1 encoding a mercury- insensitive aquaporin of *Medicago truncatula*. *Planta.* 211, 85–90.

73. Księżniak, A., 2007. Endomikoryzyi ichzastosowaniew urządzaniu ogrodów. Mat. Kon. "Mikoryza w architekturze krajobrazu", Poznań. 11-12.

74. Larue, J. H., W. D. Clellan and W. I. Peacock., 1975. Mycorrhizal fungi and peach nursery nutrition. *California Agric.* 5,6-7.

75. Levy,Y., Syvertsen, J.P.and Nemec, S., 1983. Effect of drought stress and vesicular-arbusuclar mycorrhiza on citrus transpiration and hydraulic conductivity of roots. *New Phytol.* 93,61.

76. Maronek, D.M., J.W. Hendrix and J. Kiernan., 1981. Mycorrhizal fungi and their importance in horticultural crop production. *Hort. Rev.* 3, 172-213.

77. María Josefina Bompadre., Vanesa Analía Silvani., Laura Fernández Bidondo., María Del Carmen Ríos de Molina., Roxana Paula Colombo., Alejandro Guillermo Pardo. and Alicia Margarita Godeas.,2014. Arbuscular mycorrhizal fungi alleviate oxidative stress in pomegranate plants growing under different irrigation conditions, *Botany, e-First Article* : pp. 187-193.

78. Menzel, C. M., Rasmussen, T. S and Simpson, D. R., 1995. Carbohydrate reserves in litchitrees (*Litchi chinensis* Sonn.). *J. Hort. Sci.* 70, 245–255.

79. Mena-Violante, H. G., O. Ocampo-Jimenez., L. Dendooven., G. Martinez-Soto., J. Gonzalez- Castaneda., F. T. Davies and V. Olalde-Portugal., 2006. Arbuscular mycorrhizal fungi enhanced fruit growth and quality of chile ancho (*Capsicum annuum* L. cv San Luis) plants exposed to drought. *Mycorrhiza.* 16,261-267.

80. Miller, D. D., M. Bodmer and H. Schuepp., 1989. Spread of endomycorrhizal colonization and effects on growth of apple seedlings. *New Phytol.* 111,51-59.

81. Miransari, M., 2010. Contribution of arbuscular mycorrhizal symbiosis to plant growth under different types of soil stress. *Plant Biol.* 12,563- 569.

82. Murkute1, A. A., Sharma. S and S. K. Singh., 2006. Studies on salt stress tolerance of citrus rootstock , genotypes with arbuscular mycorrhizal fungi. *Hort. Sci. (Prague).* 33 (2), 70–76.

83. Nagarajappa, A., Patil , C.P., Swamy, G.S.K and P. B. Patil., 2003. Influence of VAM on Drought Tolerance of Papaya. *Karnataka J. Agril. Sci.* 16 (3), 434-437.

84. Nikolaou, N., K. Angelopoulos and N. Karagiannidis., 2003. Effects of drought stress on mycorrhizal and non-mycorrhizal cabernet sauvignon grapevine, grafted onto various rootstocks. *Expl. Agric.* 39,241-252.

85. Oades, J.M.and Waters, A.G., 1991. Aggregate hierarchy in soils. *Aust J Soil Res* 29,815–828.

86. Oehl, F., Sieverding, E., Ineichen, K., Mäder, P., Boller, T and A. Wiemken., 2003. Impact of Land Use Intensity on the species diversity of arbuscular mycorrhizal fungi in agroecosystems of Central Europe. *Applied Environmental Microbiology.* 69, 2816–2824.

87. Parke J.L., Linderman R.G and C.H. Black., 1983. The role of ectomycorhizas in drought tolerance of Douglas-fir seedlings. *Newphytol.* 95, 83-95.

88. Pandey, S.P and A.P. Misra., 1971.*Rhizophagus* in mycorrhizal association with *Litchi chinensis* Sonn. Mycopathol. *Mycol. Appl.* 45, 337–354.

89. Pandey, S. P and A.P. Misra.,1975. Mycorrhiza in relation to growth and fruiting of *Litchi chinensis* Sonn. *J. Indian Bot. Soc.* 54, 280– 293.

90. Palma, J.M., Longa, M.A., del Rio, L.A. and Arines, J .,1993. Superoxide dismutase in vesicular-arbuscular red clover plants. *Physiol Plant.* 87,77–83.

91. Pearson, J.N and I. Jakobsen., 1993. Symbiotic exchange of carbon and phosphorous between cucumber and three arbuscular mycorrhizal fungi. *New Phytol.* 124 , 481 - 488.

92. Phillips, J. M and D. S. Hayman., 1970. Improved procedures for clearing roots and staining parasitic and vesicular-arbusular mycorrhizal fungi for rapid assessment of infection. *Trans. Br. Mycol. Soc.* 55,158-161.

93. Porcel, R., Aroca, R., Azcón, R and J.M. Ruíz-Lozano., 2006. PIP aquaporin gene expression in arbuscular mycorrhizal *Glycine max* and *Lactuca sativa* plants in relation to drought stress tolerance. *Plant Mol. Biol.* 60, 389-404.

94. Qiang-Sheng W. U., Ying-Ning Z.O .U., Ren-Xue X.I.A and W.A.N.G. Ming-Yuan., 2007. Five *Glomus* species affect water relations of *Citrus tangerine* during drought stress . *Botanical Studies.* 48, 147-154.

95. Qiang-Sheng, W.U., Guo-Huai, L. and Ying-Ning, Z.O.U., 2011. Improvement of RootSystem Architecture in Peach (*Prunus persica*) Seedlings by Arbuscular Mycorrhizal Fungi, Related to Allocation of Glucose/Sucrose to Root, *Not Bot Horti Agrobo*, 39(2),232-236.

96. Read, D., 1993. Mycorrhizas. Appendix C. *In* Tropical Soil Biology and Fertility. Eds. *J M Anderson and J S I Ingram.* CAB International, Wallingford, Oxon, UK. 121–131.

97. Roussel, H., Bruns, S., Gianinazzi-Pearson, V., Hahlbrock, K.and Franken, P., 1997. Induction of a membrane intrinsic protein-encoding mRNA in arbuscular mycorrhiza and elicitor-stimulated cell suspension cultures of parsley. *Plant Sci* .126,203–210.

98. Runjin, L., 1989. Effects of vesicular-arbuscular mycorrhizas and phosphorus on water status and growth of apple. *Journal of Plant Nutrition.* 12,997-1017.

99. Ruiz-Lozano, J.M and Azcon, R.,1995. Hyphal contribution to water uptake in mycorrhizal plants as affected by the fungal.

Chapter 8

Management of Salt Stress

Nisha Kadian[1], Kuldeep Yadav[1], Ashok Aggarwal[1], Sowmya H.D[2]

[1]*Department of Botany, Kurukshetra University,*

Kurukshetra -136 119, India

[2]*Division of Biotechnology, Indian Institute of Horticultural Research (ICAR)*

Bengaluru-560089, India

* *Address for the correspondence: nishakadian04@gmail.com*

ABSTRACT

Salinization of soil is a serious problem and has received great attention from agriculturalists all over the world due to its adverse effects on plant growth and yield. The dominant sources of salt are rainfall and rock weathering. Overall, salinity affects all the major processes, such as growth, photosynthesis, protein synthesis and plant metabolisms leading to many deleterious effects on plants at different life stages. In the recent years, the use of soil microorganisms have generated great interests in agriculture, by reducing the damage from salt stress and providing sustainable solutions for crop production in such climates. Arbuscular mycorrhizal fungi (AMF) play an important role by providing multifaceted benefits in terms of plant growth, nutrient uptake, stress tolerance, disease tolerance and also help plants to become established in new areas. The present review provides an insight into the beneficial effects of AMF in alleviating salt stress in important fruit crops around the world.

Keywords: *AMF, Proline, aquaporines, antioxidants, phosphatase activity, growth enhancement, yield,*

1.0 Introduction

Throughout evolutionary time, plants have been confronted with changing environmental conditions, among which water stress is considered as the most important abiotic factors limiting plant growth and yield in many areas. Soil salinity has received great attention from agriculturalists all over the world due to its adverse effects on plant growth and yield. More than 800 million hectares (around 6%) of the total surface land of earth is affected by salinity (Arzani, 2008). Among the 1.5 billion hectares of the total cultivable land present in the world, about 5% is (approximately 77 million hectares) affected by salinity (Evelin *et al.*, 2009); particularly arid and semi arid areas facing major ecological and agronomical problems (Ruiz-Lozano *et al.*, 2012) Approximately 20% of the irrigated cultivable land is severely affected by soil salinity (Yamaguchi and Blumwald, 2005; Wu *et al.*, 2010). The percentage is increasing day by day and is expected to have devastating global effects, resulting in 30% land loss within the next 25 years and upto 50% by the middle of twenty-first century (Wang *et al.*, 2003) due to improper irrigation practices (Jun-li and Yue-hu, 2009; Bothe, 2012). The response of plants to various stresses depends on several factors such as developmental stage, severity of stress, duration, and genotype.

The salt affected soils were overlaid with irrigation and drainage layer that shows a good correlation between the area under irrigation and the extent of Salt Affected Soil (SAS). Salinity is the concentration of dissolved mineral salts present in the soils (soil solution) and waters. The dissolved mineral salts consist of the electrolytes of cations and anions. The major cations in saline soil solutions consist of Na^+, Ca^+, Mg^{2+} and K^+ and the major anions are Cl^-, SO_4^{2-}, HCO_3^-, CO_3^{2-} and NO_3^-. Other constituents contributing to salinity in hyper saline soils and water include B, Sr^{2+}, SiO^{2+}, Mo, Ba^{2+} and Al^{3+} (Hu and Schmidhalter, 2002). Water soluble salts accumulate in the soil solum (the upper part of the soil profile, including the A and B horizons) or regolith (the layer or mantle of fragmental and unconsolidated rock material, whether residual or transported) to a level that impacts on agricultural production, environmental health and economic welfare (Rengasamy, 2006). The dominant sources of salt are rainfall and rock weathering. Rainfall contains low amounts of salt, but over the time, salt deposited by rain can accumulate in the landscape. Wind-transported (aeolian) material from soil or lake surfaces is another source of salt. Poor quality irrigation water also contributes to salt accumulation in irrigated soils. Sea water intrusion onto land, as occurred in recent tsunami-affected regions can deposit huge amounts of salts in soils of coastal lands.

This particular process contributing salt, combined with the influence of other climatic and landscape features and the effects of human activities, determine where salt is likely to widely accumulate in the landscape (Rengasamy, 2006). The most common causes are (1) land clearing and the replacement of perennial vegetation with annual crops, and (2) irrigation schemes using salt-rich irrigation water or having insufficient drainage. High salt concentration (Na^+) in particular which deposit in the soil can alter the basic texture of the soil resulting in decreased soil porosity and consequently reduced soil aeration and water conductance. High salt depositions in the soil generate a low water potential zone in the soil, making it increasingly difficult for the plant to acquire both water as well as nutrients. Therefore, salt stress essentially affects the establishment, growth and development of plants leading to huge losses in productivity (Evelin *et al.*, 2009).

Plants growing in saline soil are subjected to three distinct physiological stresses. (1) The toxic effect of specific ions such as sodium and chlorides, prevalent in saline soils, disrupt the structure of enzymes and other macromolecules, damage cell organelles, disrupt photosynthesis and respiration, inhibit protein synthesis, and induce ion deficiencies (Juniper and Abbott, 1993). (2) Plants exposed to the low osmotic potentials of saline soil are at risk of physiological drought because they must maintain lower internal osmotic potentials to prevent water moving from the roots into the soil. (3) Finally, salinity also produces nutrient imbalance in the plant caused by decreased nutrient uptake and/or transport to the shoot (Marschner 1995; Adiku *et al.*, 2001). Overall, salinity affects all the major processes, such as growth, photosynthesis, protein synthesis and plant metabolisms (Ramoliya *et al.*, 2004) and leads to many deleterious effects on plants and at different life stages (Figure **8. 1)**.

To counteract this problem, many strategies were proposed to overcome salt detrimental effects such as searching for new salt-tolerant crops, genetically engineering plants, removing excessive salt accumulation in groundwater and desalinizing water for irrigation (Ashraf and Harris, 2004; Flowers, 2004; Zhang and Blumwald, 2001). Although these strategies appear efficient, yet they are costly and out of reach for developing countries that are the most affected.

The modern agriculture is getting more and more dependent on the steady supply of synthetic inputs i.e. chemical fertilizers. Indiscriminate and injudicious use of chemical fertilizers for the crop production has compounded the problem of environmental pollution, deterioration of soil health and residue problems. Adverse effects of the chemical fertilizers have compelled the scientific fraternity to look for alternatives in the form of Biofertilizers. The term biofertilizer refers to all organisms, which add, conserve and mobilize the plant nutrients in the soil. Such microorganisms have come to be called as "biofertilizers" a term, which is a misnomer, compared to commercial fertilizers manufactured on large scales in factories. Biofertilizers are based on renewable energy sources and ecofriendly compared to commercial fertilizers (Yawalkar *et al.*, 1996). Biofertilizers play a very significant role in improving soil fertility by fixing atmospheric nitrogen, both symbiotic and non-symbiotic, solubilise insoluble soil phosphate and produce plant growth substances in the soil.

Symbioses with beneficial microorganisms constitute the universal and ecologically highly effective strategy of adaptation of plants towards nearly all types of environmental challenges. Representatives of many groups of fungi AMF and bacteria participate in plant-microbial symbioses (PMS) wherein they can colonize the plant surfaces, tissues or intra-cellular compartments using two basic adaptive strategies: nutritional and defensive. Construction of niches for hosting the symbiotic microbes involves the complicated developmental programs implemented under the joint control by plant and microbial partners and based on the cross-regulation of their genes.

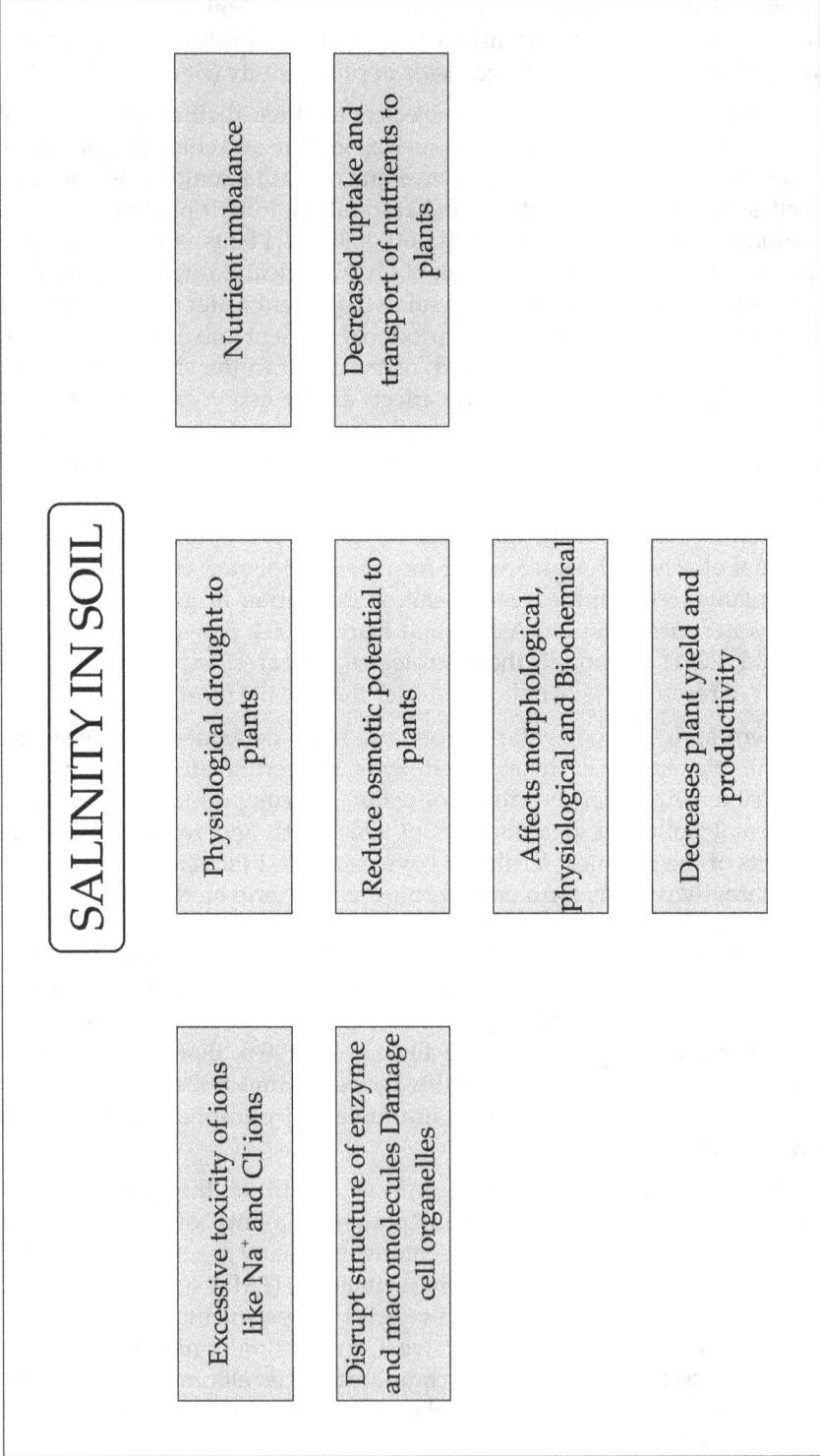

Fig. 8.1: Influence Of AMF On Growth Performance, Physiological Parameters, Mineral Nutrients And Yield Of Various Crops Under Salinity Stress. (Source: Marschner 1995; Adiku *et al.* 2001).

2.0 AMF and Salt stress

Plants, in their natural environment are colonized both by external and internal microorganisms. To cope with this stress, AMF are ubiquitous among a wide array of soil microorganisms inhabiting the rhizosphere and play a key role in alleviating the toxicity induced by salt stress. These fungi constitute an important integral component of the natural ecosystem and are known to exist in saline environments (Giri *et al.*, 2003). The proportion of vascular plant species forming AMF is commonly over estimated (Trappe, 1987), probably as a result of the low proportion of species and environments surveyed (Brundett and Abbott, 1991).

Salinization of soil is a serious problem and is increasing steadily in many parts of the world, in particular in arid and semi-arid areas (Giri *et al.*, 2003; Al-Karaki, 2006). AMF can help to overcome the problem of salinity stress. Plants growing in saline soils are subjected to physiological stresses. Although AMF exist in saline soils, the growth and colonization of plants may be affected by the excess of salinity, which can inhibit the growth of microbes due to osmotic and/or toxic effect of salts (Juniper and Abbott, 2006). Although it is clear that AMF mitigate growth reduction caused by salinity. So far, studies on salt stress tolerance in mycorrhizal plants have suggested that AMF plants grow better due to improved mineral nutrition and physiological processes like photosynthesis or water use efficiency, the production of osmoregulators, higher K^+/Na^+ ratios and compartmentalization of sodium within some plant tissues (Ruiz-Lozano *et al.* 1996; Giri *et al.*, 2003; Al-Karaki, 2006). Ecological consequences of the interactions between plants and the AMF for plant nutrition, growth, competition, stress tolerance and fitness, as well as for soil structuring have been often addressed. The key effects of AMF symbiosis can be summarized as follows: (1) enhancing uptake of low mobile ions, (2) improving quality of soil structure, (3) enhancing plant community diversity, (4) improving rooting and plant establishment, (5) improving soil nutrient cycling, and (6) enhancing plant tolerance to (biotic and abiotic) stress (Smith and Read, 2008).Thus, AMF plays a major role in the plant tolerance to salinity with the regulation of their root hydraulic properties (Aroca *et al.*, 2007; Ruiz-Lozano and Aroca, 2010) (**Figure 8. 2**).

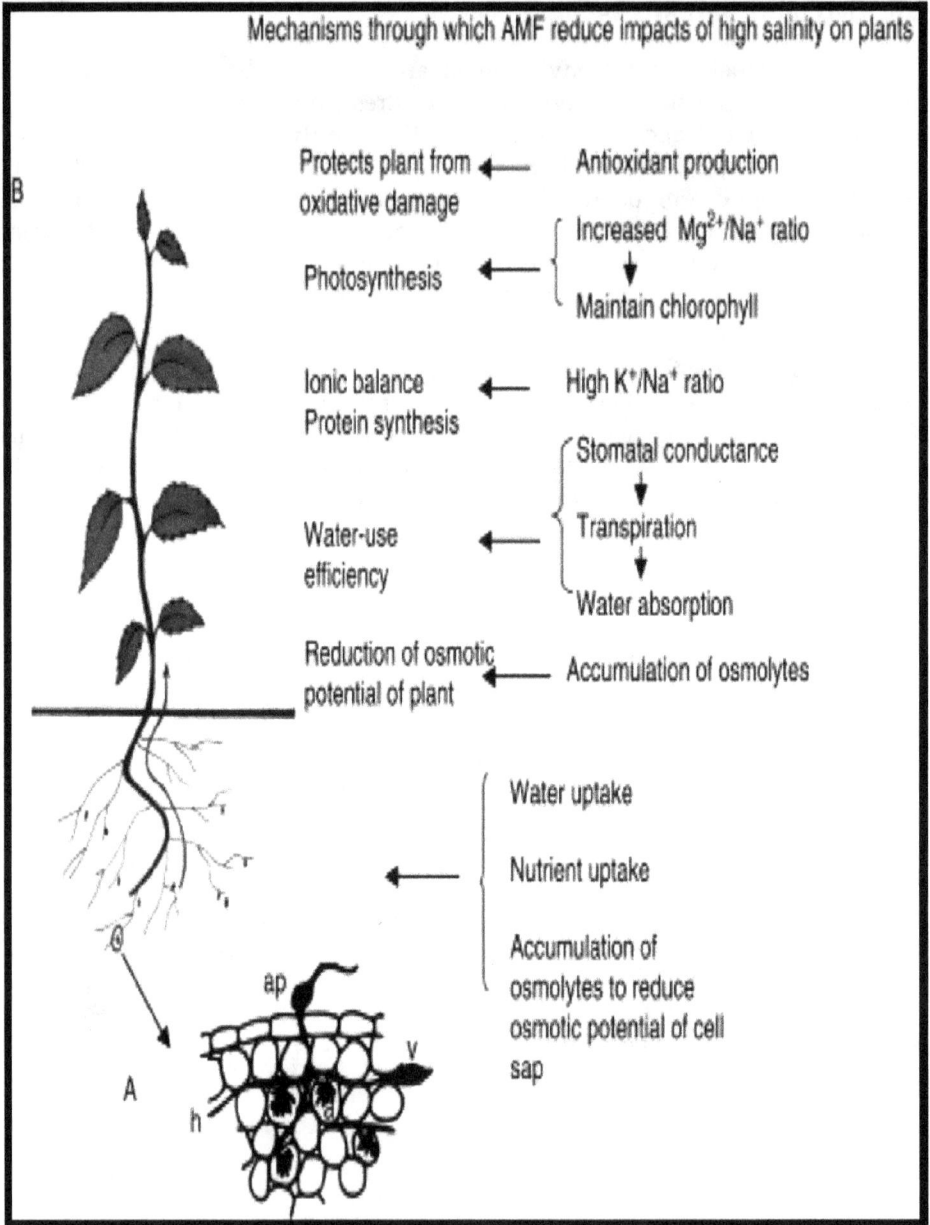

Fig. 8. 2. Intricate functioning of AMF in ameliorating salt stress in plants. In AMF symbiosis, the fungus forms an appressorium (ap) on the root surface and enters the root cortex by extending its hyphae (h). The hyphae form arbuscules (a) and vesicles (v) in the cortex. Salinity deprives plants of the basic requirements of water and nutrients, causing physiological drought and a decrease in osmotic potential accompanied by nutrient deficiency, rendering plants weak and unproductive. Arbuscular mycorrhiza help plants in salt stress by improving water and nutrient uptake (Source: Evelin *et al.*, 2009).

3.0 Influence of AMF on various plant parameters under salinity stress.

3.1 Plant height

The salt levels present in the soil increases osmotic stress, decreases nutrient uptake and also affect the physiological and biochemical process leading to decreased plant height. Under saline soil, salinity lowered the water potential of the roots, and this caused quick reductions in growth rate along with a suite of metabolic changes (Munns, 2002). According to Niknam and McComb (2000) one of the causes of reduction in growth rate under stress was inadequate photosynthesis owing to stomatal closure and consequently limited carbon dioxide uptake. More importantly, however, stress inhibited cell division and expansion directly. The four reasons that were usually introduced as solely responsible for reduction of plant growth under salt-stressed conditions, were reviewed by Al-Yassin (2004) as: osmotic stress caused by lowering the availability of external water, specific ion toxicity effects caused by metabolic processes in the cell, nutritional imbalances caused by these ion-toxicity effects, and combination of any two of the above-mentioned factors. Salt-induced water deficiency, osmotic dehydration, high transpiration and ion toxicity had been suggested as the main causes of salt damage in plants. AMF counteract the toxic effects of salts by the better acquisition of nutrition especially phosphorus and other elements by extraradical mycorrhizal hyphae, and transferring them to the root tissues (Wu *et al.*, 2010). Colla *et al.* (2008) also reported improved growth of *Cucurbita pepo* colonized with *Glomus intraradices* under salinity stress. Roldan *et al.* (2008) reported that mycorrhizal inoculation significantly increased growth of the plants at different salinity level in comparison to control plants. It might be due to the reason that mycorrhization increased the fitness of the host plant through better acquisition of mineral nutrition especially P under salt stress.

3.2 Plant Biomass

Under saline soil, greater CO_2 assimilation could adequately provide carbohydrates for the fungal partner and result in more benefits to plants from AMF association. These AMF isolates from saline soils have a better ability to improve the survival, growth and ultimately biomass of host plants (Tain *et al.*, 2004). Younesi *et al.* (2013) also reported the increase in biomass in salt-stressed soybean plant. The increment in root and shoot biomass may be due the reason that there was no indigenous microbe to compete with the inoculated strain in pure culture,the effect of inoculation was more obvious. Al-Karaki (2000) also observed a higher shoot and root dry weight in mycorrhizal tomato plant than in non-mycorrhizal tomato plant. This increment may be due to more absorption of nutrients especially P via an increase in root surface area through AMF (Prakash *et al.*, 2011) and ability of plants for replacement of K by Na (Haijiboland and Joudmand, 2009).

3.3 Root length

Under salt stress, reduction in root length is caused due to possible diversion of photosynthates to the tissues damaged by the salts for respiration and to reduce further damage. Thus nutrients fall short to meet the demand for new growth which remains retarded (Munns and Termaat,1986).The reduction in root growth, due to salinity, ultimately affect the root volume. This might be due to reduction in uptake of phosphorus affecting root growth because of less production of energy carrying ATP and reduction in protein synthesis rate due to low nitrogen content. It is also probable that salinity caused a decrease in plant hormone content (Auxin and Gibberellin) and increased the inhibitors level which restricted the growth and subsequent rooting under saline soil. AMF inoculated plants experienced increased root elongation under salinity condition, as AMF absorb more of the nutrients especially away from the P depletion zone and resulted in the better growth of the plant. The root growth is increased by increasing mycorrhizal inoculation of plant root and providing a desirable limitation of phosphorus. It may be due to space and nutrition for its multiplication and survival in sterilized soil resulting in absorption of more nutrients from the soil. Moradi *et al.* (2013) reported an increment in root length with the chickpea plants are inoculated with G. *mosseae* and G. *intraradices* in calcareous soil. Correlation of root length with mycorrhizal inoculation amount of root is probably related to suitable ventilation of soil, that is the result of hypha network of mycorrhizal fungi that connects particles of soil and as a result the root spreads into deep soil (Turk *et al.*, 2006). Similar results are observed by Enteshari and Hajbagheri, 2011 in *Ocimum basilicum* when inoculated with G. *mosseae* and G. *intraradices*.

3.4 Mycorrhizal spore number and root colonization

A significant promoting effect on mycorrhizal colonization and spore number was observed in *Vitis* spp. when inoculated with *Glomus fasciculatum* at different levels of salinity (Belew *et al.*, 2010). The suppressed spore number and colonization of AMF under different salinity level may be attributed to the reduced hyphal extension of AMF. Salinity hampers colonization capacity, spore germination (Hirrel, 1981), growth of hyphae of AMF (Mc Millen *et al.*, 1998) and reducing number of arbuscules (Tian *et al.*, 2004). Similar results were also reported by Mirdhe and Lakshman, (2011), in *Phaseolus vulgaris* L. (Kidney bean) and *Psophocarpus tetragonolobus* (L.) D.C (Winged bean) when inoculated with *Glomus fasciculatum*.

3.5 Effect on chlorophyll

Under saline conditions, mycorrhization helps in better absorption of Mg in plants and the antagonistic effect of Na^+ on Mg^+ uptake is counter balanced and suppressed resulting in increased chlorophyll synthesis (Giri and Mukerji, 2003). In *Glomus etunicatum* inoculated maize plant, increase in photosynthesis rate, transpiration and chlorophyll a, b density was reported under stress (Zhu *et al.*, 2010). Al- Khaliel (2010) also observed that inoculation of peanut with G. *mosseae* exhibited triggering effect on chlorophyll content under different salinity levels.

According to Parida and Das (2005) the total chlorophyll content were found to be decreased in tomato plants under higher salt stress. Similarly, in *Capsicum annuum* mycorrhizal treated plants showed higher chlorophyll over control under salt stress (Demir, 2004).

3.6 Electrolyte leakage

Higher salt concentration increased membrane permeability and mycorrhizal plants has lower root plasma membrane permeability and stability (Kaya *et al.*, 2009) due to mycorrhizal mediated enhanced P uptake and antioxidant production. Mycorrhizal treated plants have lower electrolyte leakage as compared to non-mycorrhizal plants by maintaining improved integrity and stability of membrane (Zhongoun *et al.*, 2007; Garg and Manchanda, 2008). Mycorrhizal plants had much lower root plasma membrane electrolyte permeability than the non-mycorrhizal plants (Kaya *et al.*, 2009). The increased membrane stability has been attributed to mycorrhizal mediated enhanced P uptake and increased antioxidant production (Feng *et al.*, 2002).

3.7 Proline

One of the main consequences of salinity stress is the loss of intracellular water. Plants accumulate many metabolites that are also known as "compatible (organic) solutes" in the cytoplasm to increase their tolerance against salt stress-induced water loss from the cells. Proline acts as major osmoprotectant osmolytes, which are synthesized by plants in response to stress, including salinity stress, and thereby help in maintaining the osmotic stress of the cell to ameliorate the abiotic stress effect. Proline also plays role in scavenging free radicals, stabilizing subcellular structure and buffering cellular redox potential under stresses. Under saline conditions, plants accumulate proline as a non-toxic and protective osmolyte to maintain osmotic balance under low water potentials (Ashraf and Foolad, 2007; Parida *et al.*, 2002). It also acts as a reservoir of energy and nitrogen for utilization during salt stress conditions (Goas *et al.*, 1982). Sharifi *et al.* (2007) also reported a higher proline concentration in AMF soybean than non-AMF plants at different salinity level. This increment in proline could be due to the induction of proline biosynthesis enzymes and/or to the reduction of oxidation to glutamate (Stewart 1981). Several roles have been attributed to this supra optimal level of proline; for instance, osmoregulation and detoxification of free radicals (Kaul *et al.*, 2008). Marked increase in proline content occurs during salt stress; this accumulation, mainly as a result of increased proline biosynthesis, is usually the most outstanding change among the free amino acids (Hurkman *et al.*, 1989) and the colonization of plant roots by AMF further stimulated the salinity-induced accumulation of proline by inhibiting proline dehydrogenase in mycorrhizal plants.

3.8 Protein

The higher protein content in treated plants could be attributed due to higher osmotic regulation, synthesis of osmolytes, membrane stability, and better mineral acquisition and ultimately leads to higher protein synthesis. Mycorrhizal and nodule symbioses often act synergistically on infection rate, mineral nutrition and plant growth (Patreze and Cordeiro, 2004) which support the need for both N and P and increased tolerance of plants to salinity stress (Rabie and Almadini, 2005).

3.9 Aquaporins

Under salinity condition plants encounter negative water potential condition due to insufficient amount of water (Ouziad *et al.*, 2006), a process where aquaporins participate (Luu and Maurel 2005). Aquaporins are the membrane water channel proteins plays a crucial role to facilitate and regulate the passive movement of water molecules down a water potential gradient in the cells and tissues (Kirch *et al.*, 2000; Tyerman *et al.*, 2002; Kruse *et al.*, 2006). Basically plant aquaporins are classified into four groups based on their sequence homology and stereotypical intron positions within each group includes plasma membrane intrinsic proteins (PIPs), tonoplast intrinsic proteins (TIPs), nodulins like intrinsic proteins (NIPs) and small and basic intrinsic proteins (SIPs) (Maurel *et al.*, 2008). Additionally these aquaporins are also involved in diverse biological functions including cell elongation, reproductive growth, seed germination, stomatal movement, organ movement and cell division (Suleyman *et al.*, 2000; Morillon *et al.*, 2001; Moshelion *et al.*, 2002; Eisenbarth *et al.*, 2005). The expression of aquaporins genes in AMF treated *P. vulgaris* plants under different osmotic stressed condition like drought, cold or salinity was studied by Aroca *et al.* (2007). The results showed that three of the PIP genes showed differential regulation by AMF under different stress condition applied. Additionally they have also reported that expression of these genes was more in the AMF treated plants under salt stress condition. The role of AMF on differential expression of aquaporins isoforms under salinization condition was studied by Ouziad *et al.* (2006). Under salinity condition the transcription level of TIP and PIP was found to be up-regulated in the leaves and down regulated in roots of *Lycopersicon esculentum*. Similarly Jahromi *et al.* (2008) reported that the expression level of PIP1 and PIP 2 was down regulated under saline free condition in *L. sativa* AMF treated plants, whereas under saline condition in AMF treated plants showed the up regulation of only LsPIP2 gene and at high salinity level (100 mM NaCl) expression level of LsPIP1 gene was found to be more.

3.10 Phosphatase Activity

Phosphatase enzymes produced by the extraradical hyphae of AMF has the ability to hydrolyze extracellular phosphate ester bonds and ultimately make P available to the plants (Joner and Johansen, 2000) and thus the increase in P content may have been due to the enhanced activities of phosphatases (Feng *et al.*, 2002). The P cycle enzymes activities are actually inversely related to the available P in the soil. Among both the phosphatases enzymes, acid phosphatase activity was found higher than alkaline phosphatase activity in all the inoculated plants. This is due to the slight acidic pH of the soil which stimulates the activity of acidic phosphatases, which has a major role in mineralization of organic P. The increase in phosphatase activity could be due to active mycorrhizal infection and the presence of Mycorrhizal-specific phosphatase (MSPase) enzyme in root extract of mycorrhizal inoculated plants. Such increase in these activities was related to the degree of active mycorrhizal infection of each fungal species. Gianinazzi-Pearson and Gianinazzi (1978) and Ezawa and Yoshida (1994) detected mycorrhizal-specific Phosphatase (MSPase) only in the mycorrhizal root extract, and it was of fungal origin. The close relation between mycorrhizal growth responses and the active arbuscular phase of the infection supports the hypothesis that the phosphatase enzyme is somehow involved in assimilation of phosphorus by AMF (Abdel-Fattah, 2001).

3.11 Phosphorus (P) uptake

It has been reported that external hyphae of AMF deliver P to the plants probably due to the extended network of AMF hyphae allowing them to explore more soil volume (Ruiz-Lozano and Azcon, 2000). Indeed, mycorrhizal hyphae extend beyond the depletion zone around roots and capture nutrients that are several centimeters away from root surfaces and thus suppress the adverse effect of different salinity stress (Matamoros *et al.*, 1999). Improved P nutrition in AMF inoculated plants may improve plant growth, enhanced nodulation and nitrogen fixation in legumes (Garg and Manchanda, 2008).

3.12 Nitrogen (N) uptake

AMF can function as a facilitator for N uptake through activation of a plant ammonium transporter (Guether *et al.*, 2009). Application of AMF can help in better assimilation of nitrogen in host plant through extra radical mycelia take up inorganic nitrogen from the soil in the form of nitrate and assimilated it via nitrate reductase, located in the arbuscule-containing cells (Kaldorf *et al.*, 1998) and the GS-GOGAT cycle leading to the formation of arginine. Arginine, so formed, is transported from the extra-radical to the intra-radical mycelia where it is catabolized again producing amongst other substances ammonia, which equilibrates with ammonium according to the pH. These processes are consistent with increased expression of enzymes involved in primary nitrogen fixation in the extra-radical mycelia, whereas enzymes involved in arginine catabolism are up-regulated in the intra-radical mycelia. Thus, improved uptake of N in mycorrhizal plants under salt stress may be due to better nutrient uptake and maintenance of ionic balance and better acquisition of N (both nitrate and ammonium ions) from the soil. Giri and Mukerji (2004) also recorded highest accumulation of N in shoots of mycorrhizal *Sesbania grandiflora* and *S. aegyptiaca* than non mycorrhizal plants at all salinity levels.

3.13 Sodium (Na+) uptake

Under saline soil, plants tend to take up more Na+ ions resulting in decreased K+ uptake. But mycorrhizal treated plants prevent the Na+ translocation and lower the level of Na+ to the shoot tissue (Dixon *et al.*, 1993; Zuccarini and Okurowska, 2008). The concentration of Na increased in AMF plants with increasing salinity levels upto a certain level and subsequently decreased at higher salinity. According to Hammer *et al.* (2011), AMF exclude Na+ by discriminating its uptake from soil or during its transfer to plants. The accumulation of Na is strongly influenced by the available form of N (NO_3 or NH_4) and it may also be influenced by the synthesis and storage of polyphosphate as well as by other cations, particularly K (Giri *et al.*,

2003). High concentration of Na creates various osmotic and metabolic problems like reduced photosynthesis and protein synthesis for plants (Weissenhorn *et al.*, 1995). It appears that the role of AMF in alleviating salt stress is partly to prevent Na absorption to root. These findings suggest that AMF induce a regulatory effect on the translocation of Na^+ to the aerial parts, thus maintain favourable $K^+: Na^+$ and $Ca^+: Na^+$ ratio in shoots over non-mycorrhizal plants and thus, AMF induce as a buffering agent on the uptake of Na^+ when the content of Na^+ is within the permissible limit.

3.14 Potassium (K^+) uptake

Potassium plays a key role in plant metabolism. It activates a range of enzymes, and plays an important role in stomatal movements and protein synthesis. Mycorrhizal colonization can enhance K^+ absorption under saline conditions as reported by various workers (Alguacil *et al.*, 2003; Giri *et al.*, 2007; Sharifi *et al.*, 2007) while preventing Na^+ translocation to shoot tissues. Shokri and Maadi (2009), Porras-Soriano *et al.* (2009) also reported the efficacy of *G. intraradices* in maintaining favourable $K^+: Na^+$ ratio.

3.15 Antioxidant activity

Adverse environmental conditions such as salinity may alter the metabolic pathways and results in the generation of reactive oxygen species (ROS) such as hydrogen peroxidase (H_2O_2), hydroxyl radicals (OH^-), super oxides radicals (O_2^-) and singlet oxygen (Jaffel *et al.*, 2011). These compounds alter the metabolic function by oxidative damage to protein, lipid and nucleic acid. Besides, it also affects various physiological related functions such as accumulation of amino acids, amides, polyamines, quaternary ammonium compounds, proteins in the cytoplasm. These nitrogen containing compounds play an important role in osmotic potential balance, stabilization of macromolecules, cellular pH, scavenging free radicals (Siddique *et al.*, 2010). Plants develop efficient scavenging system for ROS removal and to protect cells (Foyer *et al.*, 1994). It includes non-enzymatic molecules that act as ROS scavengers such as Glutathionate, ascorbate, flavonoids and carotenoids and specific ROS scavenging antioxidative enzymes such as superoxide dismutase (SOD), catalase (CAT), glutathione reductase (GR), and peroxidase (POX). There is a correlation between antioxidant capacity and salinity tolerance, which has been reported in several plant species (Tu"rkan and Demiral, 2009). Several studies have reported that application of AMF alleviates salt stress condition by enhancing the accumulation of antioxidant compounds or by activating antioxidant enzymes (Garg and Manchanda, 2009; Kumar *et al* .,2013).

3.16 Effect on yield

Mycorrhizal colonization significantly improved yield of salt stressed plants due to production of growth stimulating substances by the mycorrhizal hypae. The stimulated energy rich carbon compounds were derived from the host assimilates, transported via fungal hyphae to the rhizosphere, and changes the pH of rhizosphere making conditions more conducive for secretion of some stimulatory substances (Johansson *et al.*, 2004). Another possible mechanism for increase in yield could be the stimulation of nutrient transfer from soil to roots, or through production of growth promoting compounds such as cytokinin like molecules, possibly kinetin (Tsavkelova *et al.*, 2006). The effects of salinity in yield for a range of the agricultural crops were reported by Kaya *et al.* (2009) on *Capsicum annum* when inoculated with *G. mosseae*. The improved yield might be due to enhanced photosynthesis associated with increased P uptake (Dietz and Foyer, 1986) and high amounts of assimilates were produced to support both symbiosis and pod yield.

4.0 Alleviation of salt stress in fruit crops

Role played by AMF in alleviating salt stress has been studied in many fruit crops and the results are encouraging enough to use AMF inoculum as a package of practice in fruit crop cultivation.

4.1 Citrus (*Citrus* spp.)

A mixture of AMF (*Rhizophagus irregularis* and *Funneliformis mosseae*) colonized seedlings of citrus root stocks Cleopatra mandarin (*Citrus* reshni Hort. ex Tan.) and Alemow (*Citrus macrophylla* Wester) showed improved growth under salinity condition (50 mM) when compared to control plants. AMF-inoculated Cleopatra mandarin seedlings showed better response under salinity than Alemow seedlings under salinization. The colonization of AMF helped the uptake of more P, K, Fe ,Cu and Mg^{++} ion concentration in roots of Cleopatra mandarin than in Alemow plants. (Navarro *et al.*, 2013).

Zou and Wu (2010) studied three month old Trifoliate orange treated with 5 different types of AMF *Diversispora spurca, Glomus etunicatum, G. mosseae, G. versiforme* and *Paraglomus occultum* under saline condition. Mycorrhizal inoculated plants markedly enhanced the root architecture, including root length, surface area, projected area, number of root tips and volume excepte *G. etunicatum*. Studies have also shown that under saline conditions all these four AMF except *G. etunicatum* stimulated the sucrose concentration in roots, glucose content in leaves and roots and allocation of sucrose in roots.

The relative water content also increased in the leaves than in uninoculated control plants. Besides, root colonization, number of entry points, vesicles and arbuscules were more in the plants treated with *G. versiforme* followed by *D. spurca, G. mosseae, P. occultum, G. etunicatum*. Wu *et al.* (2010) studied the response of antioxidant defense system in three month old trifoliate orange plants under saline condition by inoculating two different *Glomus* strains *G. mosseae* or *G. versiforme*. Mycorrhizal seedling showed elevated level catalase enzyme activity and antioxidant compound ascorbate and glutathione and reduced level of hydrogen peroxidase and malondialdehyde activity in leaves under saline condition. They also reported that *G. mosseae* colonized seedlings showed marked increase in antioxidant defense system than plants colonized by *G. versiforme* under saline condition.

Similarly, trifoliate orange leaves treated with two different mycorrhizal *Glomus mosseae* and *G. versiforme* improved the osmotic adjustment under saline conditions. Mycorrhization of seedlings enhanced the concentration of K^+ and also increased K^+/Na^+, Mg^{2+}/Na^+ and Ca^{2+}/Na^+ ratios and reduces Na^+ and Ca^{2+}concentrations when exposed to 100 mM NaCl concentration. Besides, colonization also reduced the concentration of sucrose in leaves but ameliorate the concentration of glucose, fructose and proline content. AMF treated plants also improved the plant biomass also under stress condition (Zou and Wu 2011). Khalil *et al.* (2011) investigated the effect of different salinity levels (1500, 3000 and 4500 ppm) on the growth and chemical composition of Sour orange and Volkamer lemon root stocks in presence of AMF. The AMF colonized root stocks showed improved leaf and root dry weight, leaf area, trunk cross section area. The colonization of AMF improved the uptake of P, K, Mg and Zn and reduced the leaf N, Ca and Na however; there was no difference in leaf Cl, Fe, Mn and Cu content in the AMF colonized root stocks under saline condition as compared to control.

Wu and Zou *et al.* (2013) also studied the effect of root H^+ efflux and morphological changes in root architecture in AMF inoculated trifoliate orange under different salinity inducing chemical compounds like NaCl, KCl and $NaNO_3$ at 100mM. The AMF inoculated plants showed significant increase in growth performance. Besides, AMF colonization also markedly enhanced the root architecture (length, projected area, surface area, volume, and average diameter) under saline condition. They have also observed more acidic environment in rhizosphere of mycorrhizal seedlings than in non-mycorrhizal seedlings due to H^+ efflux and it was found to be 14.3%, 31.7%, 10.3%, and 16.7% under the NaCl, KCl, $NaNO_3$ and non-salt conditions, respectively.

4.2 Date palm (*Phoenix dactylifera* L.)

Diatta *et al.* (2014) studied the effect of different AMF on the growth of date cultivar Nakhla Hamra (NHH) and Tijib under salt stress condition. The seedling grown under green house were treated with different concentration (0 to 16 gl^{-1}) of NaCl and then incubated with 5 different *Glomus* strains including *G. aggregatum*, *G. intraradices*, *G. verriculosum*, *G. mosseae* and *G. fasciculatum* and the plants without irrigating NaCl used as control. The experimental results showed that under saline condition (8 gl^{-1}) *G. intraradices* treated NHH cultivar showed better stem (37.3 cm) and root growth (77 cm) and Tijib cultivar in presence of *G. fasciculatum* with respectively 31.9 cm and 51.27 cm for stem and root. Biochemical analysis also showed that the enhancement of proline level depended mainly on the type of cultivar used, level of stress and also on the type of strains used.

4.3 Grapes (*Vitis vinifera* L.)

Wahab *et al.* (2011) found that mycorrhization of grapevines plants under low and medium saline condition (1000 and 2000 ppm) improved vegetative growth parameters likes shoot length, shoot diameter, number of leaves/plant, average leaf area, total leaf area/plant, coefficient of wood ripening, shoot and root biomass, total biomass and root/shoot ratio but also improved nutritional acquisition in the plants as compared to non-mycorrhizal plants. They also found that increase in the level of salinity decreased the total microbial count, spore number, percentage of AMF infection and dehydrogenase activity at rhizosphere region. At higher salinity level (3000 ppm) the vegetative growth parameters of the superior grape rootings were severely affected and AMF failed to support the improvement of plant growth.

An influence of AMF on micropropagated rootstocks salt Creek (*Vitis champini*) and Male hybrid (*V. vinifera*) was studied by Ramajayam *et al.* (2013) during acclimatization and also under salt stress condition. The colonization of AMF under saline condition was more in Male hybrid (19 %) than in salt Creek (12 %) root stock. Colonization of Mycorriza enhanced the shoot and root biomass 20 % and 40 % in both root stocks respectively as compared to control plants. The rootstocks showed enhanced level of total chlorophyll content in Salt Creek (24%) and Male hybrid (17%) and P, K, Ca and Mg and reduction in the Na$^+$/ K$^+$ ratio. Under saline conditions, Male hybrid showed a less degree of dependence on mycorrhizae than Salt Creek. The increased absorption of P, K, Ca and Mg reduction in Na$^+$/ K$^+$, high chlorophyll and proline contents in mycorrhizal plantlets influenced lowering the level of stress induced by salinity. Khalil *et al.*(2013) treated fifteen months old grape root stocks Dogridge (*V. champini*, 1103 Paulsen (*V. berlandieri* X *V. rupestris*) and harmony (*V. champini* X 1613) with *Glomus intraradices* under different salt stress condition (0.65, 1.56 and 4.68 dS m^{-1}). Salinity reduced the colonization of *G. intraradices* where as colonization of mycorrhiza potentially enhanced plant growth parameters (plant height, stem diameter, leaf area, number of leaves per plant, total dry weight) under saline condition. The plants also showed improved level of chlorophyll contents, proline , total carbohydrates and also increased level of P and K. In contrary the concentration of Na and Cl was reduced compared to control plants.

Belew *et al.* (2010) reported that *Glomus fasciculatum* symbiosis in grape rootstocks (Salt Creek, St. George, Dogridge and 1613) showed positive growth performance under saline condition. Among these rootstocks mycorrhizal plant 1613 showed better vegetative growth response in terms of improved shoot length, leaf number, internode, length and total dry weight while, Dogridge plants showed least response under saline condition. Based on overall growth and total dry matter accumulation, they ranked the salt tolerance of the four rootstocks, in decreasing order, as Dogridge, Salt Creek, 1613 and St. George.

4.4 Pistachio (*Pistacia vera* L.)

Shamshiri *et al.* (2014) investigated the impact of *Glomus mosseae* on alleviation of salt stress on seedling of pistachio (*Pistacia vera* cv. Badami-Riz- Zarand) in green house condition. Mycorrizal treated plants perpetuated increased shoot and root biomass at different salinity condition (0.5, 3.0, 6.0 m and 9.0 dSm^{-1}) within a period of three to five weeks. The leaf dry weight did not affect even at different saline condition. They also reported the decrease in the level of Na $^+$ and Cl$^-$ uptake and root to shoot transport which inturn reduced the accumulation of these ion in plants and alleviated the salt stress.

4.5 Strawberry (*Fragaria* × *ananassa* Duch.)

Fan *et al.* (2011) investigated the influence of AMF on biomass and root morphology in three elite strawberry cultivars (Kent, Jewel, and Saint-Pierre) in glass house. The plants were treated with three NaCl concentration (0, 30, and 60 mmol/L) and were inoculated with AMF *Glomus irregulare*. The presence of AMF significantly changed root morphology and increased root-length percentages, shoot and root tissue biomass, and specific root length (SRL) in all cultivars. In contrast, salt alone changed root morphology and decreased shoot and root tissue biomass, R/S ratio, and SRL. The AMF colonization rates were reduced linearly and significantly with increasing salinity levels. Cultivars responded differently to AMF than to salt stress. 'Saint-Pierre' seemed to be the most tolerant cultivar to salinity, while 'Kent' was the most sensitive. Consequently, AMF symbiosis highly enhanced salt tolerance of strawberry plants, which confirmed the potential use of mycorrhizal biotechnology in sustainable horticulture in arid areas.

Sinclair *et al.* (2014) studied three day–neutral strawberry cultivars (Albion, Charlotte and Seascape) under greenhouse condition, treated with three different type of AMF *Funneliformis caledonius*, *F. mosseae* and *Rhizophagus irregularis* and mixture of *R. irregularis* and *F.mosseae* on Seascape cultivar under 4 different salinity conditions (0-200 mM). All the cultivars behaved differently for both salinity and AMF and Seascape showed more tolerance to salinity as compared to other cultivars. AMF treated plants showed enhanced growth and also coarse diameter of the roots. Stimulation in accumulation of root and shoot biomass at higher salinity was more in plants treated with mixture of AMF than plants treated with individual AMF. It was reported that mixture of AMF inoculated plants showed higher degree of root and shoot biomass but failed to improve the fruit quality than individual AMF treated plants.

5.0 Conclusion and future strategies

AMF strains showing consistant response to inoculation should be selected for adoption in different salt susceptible lines in crop production (Herrera-Peraza *et al.*, 2011). Farmers can inoculate crops with selected mycorrhizal inoculants during the nursery stage, as it cannot be predicted from the soil conditions whether the native AMF community is sufficient to sustain a stable production of crops (Castillo *et al.*, 2010).

More work needs to be done to elucidate the mechanisms of salinity tolerance induced by AMF symbiosis and to distinguish the salinity-induced processes of the protective mechanisms regulated by AMF symbiosis.

References

1. Abdel-Fattah Gamal, M., 2001. Measurement of the viability of arbuscular-mycorrhizal fungi using three different stains; relation to growth and metabolic activities of soybean plants. *Microbiol. Res.* 156, 359–367.

2. Abd El-Wahab, M.A., El-Helw, H. A. and Tolba, H. I., 2011. Physiological Studies on the Effect of Inoculation with Arbuscular Mycorrhizae Fungi on Superior Grape Rootings under Salt Stress Conditions. *Nature and Science.* 9 (1),85-100.

3. Adiku, G., Renger, M., Wessolek, G., Facklam, M. and Hech-Bischoltz, C., 2001. Simulation of dry matter production and seed yield of common beans under varying soil water and salinity conditions. *Agricul. Water Manage.*47, 55-68.

4. Al-Karaki, G.N., 2000. Growth and mineral acquisition by mycorrhizal tomato grown under salt stress. *Mycorrhiza.* 10, 51-54.

5. Alguacil, M.M., Hernandez, J.A., Caravaca, F., Portillo, B.and Roldan, A., 2003. Antioxidant enzyme activities in shoots from three mycorrhizal shrub species afforested in a degraded semi-arid soil. *Physiol Plant.* 118, 562-570.

6. Al-Yassin, A., 2004. *Influence of salinity* on citrus: a review paper. *J. Cent. Eur. Agric.* 5, 263- 272.

7. Al-Karaki, G.M., 2006. Nursery inoculation of tomato with arbuscular mycorrhizal fungi and subsequent performance under irrigation with saline water. *Sci. Hort.* 109, 1-7.

8. Al-Khaliel, A.S., 2010. Effect of salinity stress on mycorrhizal association and growth response of peanut infected by *Glomus mosseae. Plant Soil Environ.* 56, 318–324.

9. Aroca, R., Porcel, R.and Ruiz-Lozano, J.M., 2007. How does arbuscular mycorrhizal symbiosis regulate root hydraulic properties and plasma membrane aquaporin in *Phaseolus vulgaris* under drought, cold or salinity stresses? *New Phytol.* 173, 808–816.

10. Arzani, A., 2008. Improving salinity tolerance in crop plants: a biotechnological view. *In vitro Cell. Dev. Biol. Plant.* 44, 373–383.

11. Ashraf, M.and Harris, P.J.C., 2004. Potential biochemical indicators of salinity tolerance in plants. *Plant Sci.* 166, 3-16.

12. Ashraf, M.and Fooland, M.R., 2007. Roles of glycine betaine and Proline in improving plant abiotic stress resistance. *Environm. Experi. Bot.* 59, 206-216.

13. Belew, D., Astatki, T., Mokashi, M.N., Getachew, Y.and Patil, C.P., 2010. Effects of salinity and mycorrhizal inoculation (*Glomus fasciculatum*) on growth responses of Grape Rootstocks (*Vitis* spp.). *S. Afr. J. Enol. Vitic.* 31, 82-88.

14. Bothe, H., 2012. Arbuscular mycorrhiza and salt tolerance of plants. *Symbiosis* .58, 7–16.

15. Brundett, M.C.and Abbott, L.K., 1991. Roots of jarrah forest plants. 1. Mycorrhizal associations of shrubs and herbaceous plants. *Aust. J. Bot.* 39, 445–457.

16. Brundrett, M.C., 2004. Diversity and classification of mycorrhizal associations. *Biol. Rev. Camb. Philos. Soc.* 79, 473–495.

17. Castillo, J.P., Verdu,´ M.and Valiente-Banuet, A., 2010. Neighnorhood phylodiversity affects plant performance. *Ecology.* 91, 3656–3663.

18. Colla, G., Rouphael, Y., Cardarelli, M., Tullio, M., Rivera, C.M.and Rea, E., 2008. Alleviation of salt stress by arbuscular mycorrhizal in zucchini plants grown at low and high phosphorus concentration. *Biol. Fert. Soil.* 44,501-509.

19. Demir, S., 2004. Influence of arbuscular mycorrhiza on some physiological growth parameters of pepper. *Turk. J. Biol.* 28, 85-90.

20. Dietz, K.J and Foyer, C., 1986. The Relationship between Phosphate and Photo-synthesis in Leaves: Reversibility of the effects of phosphate deficiency on photosynthesis. *Planta.* 167, 376-381.

21. Dixon, R.K., Garg, V.K.and Rao, M.V., 1993. Inoculation of *Leucaena* and *Prosopis* seedlings with *Glomus* and *Rhizobium* species in saline soil: rhizosphere relations and seedlings growth. *Arid Soil Res. Rehab.* 7, 133–144.

22. Diatta, I. L. D., Kane, A., Agbangba, C. E., Sagna, M., Diouf,, D., Aberlenc-Bertossi,, F., Duval, Y., Borgel,, A. and Sane, D., 2014. Inoculation with arbuscular mycorrhizal fungi improves seedlings growth of two sahelian date palm cultivars (*Phoenix dactylifera* L., cv. Nakhla hamra and cv. Tijib) under salinity stresses. *Advances in Bioscience and Biotechnology.* 5 , 64-72.

23. Eisenbarth, D.A.and Weig, A.R., 2005. Dynamics of aquaporins and water relations during hypocotyl elongation in *Ricinus communis* L. seedlings. *J Exp Bot*, 56,1831-1842.

24. Enteshari,S.Hajbagheri ,S.,2011. Effects of mycorrhizal fungi on some physiological characteristics of salt stressed *Ocimum basilicum* L. *Iranian Journal of Plant Physiology*, 1, 215-222.

25. Evelin, H., Kapoor, R. and Giri, B., 2009. Arbuscular mycorrhizal fungi in alleviation of salt stress: *A review. Ann. Bot.* 104, 1263-1280.

26. Ezawa, T., Saito, M., Yoshida, T., 1994. Comparison of phosphatase localization in the intraradical hyphae of arbuscular mycorrhizal fungi *Glomus* spp, and *Gigaspora* spp. *Plant Soil.* 176, 57-63.

27. Fan, L., Dalpé,Y., Fang, C., Dubé, C. and Khanizadeh, S., 2011. Influence of arbuscular mycorrhizae on biomass and root morphology of selected strawberry cultivars under salt stress. *Botany.* 89, 397–403.

28. Feng, G., Zhang, F.S., Li, X.L., Tian, C.Y., Tang, C.and Rengel, Z., 2002. Improved tolerance of maize plants to salt stress by arbuscular mycorrhiza is related to higher accumulation of soluble sugars in roots. *Mycorrhiza*. 12, 185-190.

29. Flowers, T.J., Torke, P.F. and Yeo, A.R., 1977. The mechanism of salt tolerance in halophytes. *Annual Review of Plant Physiol*. 28, 89-121.

30. Flowers, T.J.2004. Improving crop salt tolerance. *Journal of Experimental Botany*, 55: 307-319.

31. Foyer, C.H., P. Descourvières, and K.J. Kunert. 1994. Protection against oxygen radicals: An important defence mechanism studied in transgenic plants. *Plant. Cell. Environ*. 17,507–523.

32. Frank, A.B., 1885. Ueber die auf wurzelsymbiose beruhende Ernahrung gewisser Baume durcha Unterirdische pilze. *Ber. Dtsch. Bot. Ges*. 3, 128-145.

33. Friberg, S., 2001. Distribution and diversity of arbuscular mycorrhizal fungi in traditional agriculture on Niger Inland delta, Mali, West Africa. *CBN Skriftserie* .3, 53-80.

34. Gallaud, I., 1905. Etudes sur les mycorrhizes endotrophes. *Rev. Ge´n. Bot*. 17,5–48, 66–83, 123–136, 223–239, 313–325, 425–433, 479–500.

35. Garg, N., Manchanda, G., 2008. Effect of arbuscular mycorrhizal inoculation of salt induced nodule senescence in *Cajanus cajan* (pigeonpea). *J. Plant Growth Regul*. 27, 115–124.

36. Garg, N. and Manchanda, G. 2009. Role of arbuscular mycorrhizae in the alleviation of ionic, osmotic and oxidative stresses induced by salinity in *Cajanus cajan* (L.) Mill sp. *J. Agron. Crop Sci.*, 195, 110-126.

37. Gianinazzi-Pearson ,Gianinazzi, S.V. and Dexheimer, J., 1978. Enzymatic studies on the metabolism of vesicular-arbuscular mycorrhizal. III. Ultrastructural localization of acid and alkaline phosphatase in onion roots infected by *Glomus mosseae* (Nicol. and Gerd.) *New Phytol*. 84, 489-500.

38. Giri, B., Kapoor, R. and Mukerji, K.G., 2003. Influence of arbuscular mycorrhizal fungi and salinity on growth, biomass, and mineral nutrition of *Acacia auriculiformis*. *Biol. Fert. Soil*. 38, 170–175.

39. Giri, B. and Mukerji, K.G., 2004. Mycorrhizal inoculants alleviates salt stress in *Sesbania aegyptiaca* and *Sesbania grandiflora* under field conditions: evidence for reduced sodium and improved magnesium uptake. *Mycorrhiza*. 14, 307-312.

40. Giri, C., Pengra, B., Zhu, Z., Singh, A. and Tieszen, L., 2007. Monitoring mangrove forest dynamics of the Sundarbans in Bangladesh and India using multi-temporal satellite data from 1973–2000. Estuarine, *Coastal and Shelf Science*, 73, 91–100.

41. Goas, G., Goas, M. and Larher, F., 1982. Accumulation of free proline and glycine betaine in Aster tripolium subjected to a saline shock: a kinetic study related to light period. *Physiol. Plant.* 55, 383-388.

42. Guether, M., Neuhauser, B., Balestrini, R., Dynowski, M., Ludewig, U. and Bonfante, P., 2009. A mycorrhizal-specific ammonium transporter from *Lotus japonicas* acquires nitrogen released by arbuscular mycorrhizal fungi. *Plant Physiol.* 150, 73-83.

43. Hafeez, F.Y., Hameed, S.and Malik, K.A., 1999. *Frankia* and *Rhizobium* strains as inoculum for fast growing trees in saline environment. *Pak. J. Bot.* 31, 173–182.

44. Hajiboland, R.and Joudmand, A., 2009. The K/Na replacement and function of antioxidant defense system in sugar beet (*Beta vulgaris* L.) cultivaris. *Soil Plant Sci.* 59, 246-259.

45. Hammer, E.C., Nasr, H., Pallon, J., Olsson, P.A.and Wallander, H., 2011. Elemental composition of arbuscular mycorrhizal fungi at high salinity. *Mycorrhiza.* 21, 117-129.

46. Herrera-Pereza, R.A., Hamel, C., Fernández, F., Ferrer, R.L.and Furrazola, E., 2011. Soil-strain compatibility: the key to effective use of arbuscular mycorrhizal inoculants? *Mycorrhiza.* 21, 183–193.

47. Hirrel, M.C., 1981. The effect of sodium and chloride salts on the germination of *Gigaspora margarita. Mycologia.* 73, 610-617.

48. Hurkman, W.J., Fornari, C.S.and Tanaka, C.K., 1989. A comparison of the effect of salt on polypeptides and translatable mRNAs in roots of a salt-tolerant and a salt-senestive cultivar of Barley. *Plant Physiol.* 90, 1444-1456.

49. Hu, Y. and Schmidhalter, U., 2002. Limitation of salt stress to plant growth. In: B.Hock and C.F. Elstner (Ed.), Marcel Dekker Inc., New York. *Plant toxicology.* 91-224.

50. Jahromi, F., Aroca, R., Porcel, R. and Ruiz-Lozano, J.M., 2008. Influence of salinity on the *in vitro* development of *Glomus intraradices* and on the in vivo physiological and molecular responses of mycorrhizal lettuce plants. *Microb Ecol* .55,45–53.

51. Jaffel, K., Sai, S., Bouraoui, N.K., Ammar, R.B., Legendre, L., Lacha^ al, M. and Marzouk, B., 2011. Influence of salt stress on growth, lipid peroxidation and antioxidative enzyme activity in borage (*Borago officinalis* L.). *Plant Biosyst.* 145, 362–369.

52. Joner, E.J. and Johnsen, A., 2000. Phosphatase activity of external hyphae of two arbuscular mycorrhizal fungi. *Mycol. Res.* 104, 81-86.

53. Johansson, J.F., Paul, L.R. and Finlay, R.D., 2004. Microbial interactions in the mycorrhizosphere and their significance for sustainable agriculture. *FEMS Microbiol Ecol.* 48, 1-13.

54. Juniper, S. and Abbott, L., 1993. Vesicular-arbuscular mycorrhizas and soil salinity. *Mycorrhiza.* 4, 45-58.

55. Juniper, S. and Abbott, L.K., 2006. Soil salinity delays germination and limits growth of hyphae from propagules of arbuscular mycorrhizal fungi. *Mycorrhiza*. 16, 371-379.

56. Jun-li, T. and Yue-Hu, K., 2009. Changes in soil properties under the influences of cropping and drip irrigation during the reclamation of severe salt-affected soils. *Agric. Sci. China*. 8, 1228–1237.

57. Kaldorf, M., Schemelzer, E. and Bothe, H., 1998. Expression of maize and fungal nitrate reductase in arbuscular mycorhiza. *Mol. Plant–Microbe In.* 11, 439–448.

58. Kaul, S., Sharma, S.S. and Mehta, I.K., 2008. Free radical scavenging potential of L-proline evidence from *in-vitro* assays. *Amino Acids*. 34, 315–320.

59. Kaya, C., Ashraf, M., Sonmez, O., Aydemir, S., Tuna, A.L. and Cullu, M.A., 2009. The influence of arbuscular mycorrhizal colonization on key growth parameters and fruit yield of pepper plants grown at high salinity. *Sci. Hort*. 121, 1–6.

60. Khalil, H. A., Eissa,A .M., El-Shazly, S. M., Aboul Nasr, A.M., 2011. Improved growth of salinity-stressed citrus after inoculation with mycorrhizal fungi. *Scientia Horticulturae*. 130,624–632.

61. Khalil H.A., 2013. Influence of Vesicular- arbuscula Mycorrhizal Fungi (*Glomus* spp.) on the response of Grapevines Rootstocks to salt stress. *Asian Journal of Crop Science*.5(4), 393-404.

62. Kirch, H.H., Vera-Estrella, R., Golldack, D., Quigley, F., Michalowski, C.B., Barkla, B.J., and Bohnert, H.J., 2000. Expression of water channel proteins in Mesembryanthemum crystallinum. *Plant Physiol*. 123, 111–124.

63. Kumar, B.N., Rajput, S., Dey, K.K., Parekh, A., Das, S., Mazumdar, A., Mandal ,M.,2013. Celecoxib alleviates tamoxifen-instigated angiogenic effects by ROS-dependent VEGF/VEGFR2 autocrine signaling. *BMC Cancer*. 2013 Jun 3,13,273.

64. Luu ,D.T. and Maurel, C., 2005. Aquaporins in a challenging environment: molecular gears for adjusting plant water status. *Plant Cell Environ* . 28,85–96.

65. Marschner, H., 1995. Mineral Nutrition of Higher Plants. *Academic Press, London*. 889.

66. Matamoros, M.A., Baird, L.M. and Escuredo, P.R., 1999. Stress-induced legume root nodule senescence: physiological, biochemical and structural alterations. *Plant Physiol*. 121, 97-111.

67. Maurel, C., Verdoucq ,L., Luu, D.T. and Santoni, V., 2008. Plant aquaporins: membrane channels with multiple integrated functions. *Annu Rev Plant Biol*. 59,595-624.

68. Mirdhe, R.M. and Lakshman, H.C., 2011. Seasonal variation in three leguminous tree seedlings associated with AM fungi. *Asian J. Bio Science* .6, 82-86.

69. Morillon, R., Lienard, D., Chrispeels, M.J., and Lassalles, J.P., 2001. Rapid movements of plant organs require solute-water cotransporters or contractile proteins. *Plant Physiol.* 127, 720–723.

70. Moshelion, M., Becker, D., Biela, A., Uehlein, N., Hedrich, R., Otto, B., Levi, H., Moran, N., and Kaldenhoff, R., 2002. Plasma membrane aquaporins in the motor cells of Samanea saman: Diurnal and circadian regulation. *Plant Cell* .14, 727–739.

71. Munns, R. and Termaat, A. 1986. Whole-plant responses to salinity. *Australian Journal of Plant Physiology*,13, 143-160.

72. Munns, R., 2002. Comparative physiology of salt and water stress. Plant, *Cell Environ.* 25, 239-250.

73. Navarro, J.M., Pérez-Tornero, O and Morte, A., 2013. Alleviation of salt stress in citrus seedlings inoculated with arbuscular mycorrhizal fungi depends on the rootstock salt tolerance. *Journal of Plant Physiology.* 171,76–85.

74. *Niknam*, S.R. and *McComb*, J., *2000*. Salt tolerance screening of selected Australian woody species - *a review, Forest Ecology and Management*, 139, 1-19.

75. Ouziad, F., Wilde, P., Schmelzer, E., Hildebrandt, U. and Bothe, H., 2006. Analysis of expression of aquaporins and Na^+/H^+ transporters in tomato colonized by arbuscular mycorrhizal fungi and affected by salt stress. *Environ Exp Bot* .57,177–186.

76. Parida, A., Das, A.B. and Das, P., 2002. NaCl stress causes changes in photosynthetic pigments, proteins and other metabolic components in the leaves of a tree mangrove, *Bruguiera parviflora*, in hydroponic cultures. *J. Plant Biol.* 45, 28–36.

77. Patreze, C.M. and Cordeiro, L., 2004. Nitrogen fixing and vesicular arbuscular mycorrhizal symbioses in some tropical legume trees of tribe Mimosaceae. *Forest Ecol. Manag.* 196, 275–285.

78. Parida, S.K. and Das, A.B., 2005. Salt tolerance and salinity effects on plants. *Ecotox. Environ. Safe.* 60, 324–349.

79. Parkash, V., Sharma, S. and Aggarwal, A., 2011. Symbiotic and synergistic efficacy of endomycorrhizae with *Dendrocalamus strictus* L. *Plant Soil Environ.* 57, 447– 452.

80. Porras-Soriano, A., Soriano-Martín, M.L., Porras-Piedra, A. and Azcón, R. 2009. Arbuscular mycorrhizal fungi increased growth, nutrient uptake and tolerance to salinity in olive trees under nursery conditions. *J. Plant Physiol.* 166, 1350-1359.

81. Ramoliya, P.J. and Patel, H.M., 2004. Effect of salinization of soil on growth and macronutrient accumulation in seedlings of *Salvadora persica* (Salvadoraceae). *Forest Eco. Manage.* 2002, 181-193.

82. Rabie, G.H. and Almadini, A.M., 2005. Role of bioinoculants in development to salt-tolerance of *Vicia faba* plants under salinity stress. *Afr. J. Biotechnol.* 4, 210-222.

83. Ramajayam, D, Singh S.K., Singh A.K., Patel,V.B. and Alizadeh M. 2013. Mycorrhization alleviates salt stress in grape rootstocks during *in vitro* acclimatization *Indian J. Hort.* 70(1), ,26-32.

84. Rengasamy, P., 2006. World salinization with emphasis on Australia. *J. Exp. Bot.* 57, 1017-1023.

85. Roldan, A., Diaz-Vivancos, P., Hernandez, J.A., Carrasco, L. and Caravaca, F., 2008. Superoxide dismutase and total peroxidase activities in relation to drought recovery performance of mycorrhizal shrubs seedlings grown in an amended semiarid soil. *J. Plant Physiol.*165, 715–722.

86. Ruiz-Lozano, J.M., Azcon, R. and Gomez, M., 1996. Alleviation of salt stress by arbuscular mycorrhizal *Glomus* species in *Lactuca sativa* plants. *Plant Physiol.* 98, 767-722.

87. Ruiz-Loznano, J.M. and Azcon, R., 2000. Symbiotic efficiency and infectivity of an autochthonous arbuscular mycorrhizal *Glomus* sp. From saline soils and *Glomus destricola* under salinity. *Mycorrhiza.* 10, 137-143.

88. Ruiz-Sanchez, M., Aroca, R., Munoz, Y., Polon, R and Ruı´z-Lozano, J.M., 2010. The arbuscular mycorrhizal symbiosis enhances the photosynthetic efficiency and the antioxidative response of rice plants subjected to drought stress. *J. Plant Physiol.* 167, 862–869.

89. Ruiz-Lozano, J.M. and Aroca, R.,2010. Host response to osmatic stresses: Stomatal beharviour and water use effeciency of arbuscular mycorrhizal plants. In: Koltai H,Kaputink Y, editors.Arbascular mycorrhizas: Physiology and Function.2 nd ed Dordrecht. *The Netherland:* Springer *Science and Business Media B.V.*Pp 239-256.

90. Ruiz-Lozano, J.M., Porcel, R., Azcon , C. and Aroca R., 2012. Regulation by arbuscular mycorrhizae of the integrated physiological response to salinity in plants: new challenges in physiological and molecular studies. *J. Experimn. Bot.* 126, 1-12.

91. Sharifi, M., Ghorbanli, M. and Ebrahimzadeh, H., 2007. Improved growth of salinity-stressed soybean after inoculation with salt pre-treated mycorrhizal fungi. *J. Plant Physiol.* 164, 1144-1151.

92. Shokri, S. and Maadi, B., 2009. Effects of arbuscular mycorrhizal fungus on the mineral nutrition and yield of *Trifolium alexandrinum* plants under salinity stress. *J. Agrono.* 8, 79-83.

93. Shamshiri, M.H., Pourizadi, F. and Karimi, H.R., 2014. Role of Mycorrhizal Symbiosis in Growth and Salt Avoidance of Pistachio Plants. *Journal of Stress Physiology & Biochemistry.*10, 155-167.

94. Siddique, N.A., Mujeeb, M., Najmi, A.K. and Akram, M., 2010. Evaluation of antioxidant activity, quantitative estimation of phenols and flavonoids in different parts of *Aegle marmelos, Afr. J. Plant. Sci.* 4(1),001-5.

95. Sinclair, G., Charest, C., Dalpe, Y. and Khanizadeh, S., 2014. Influence of colonization by arbuscular mycorrhizal fungi on three strawberry cultivars under salty condition.Agricultural and food science. 23, 146-158.

96. Smith, S.E. and Read, D.J., 2008. Mycorrhizal symbiosis. *Academic Press,* New York.

97. Stewart , C.R. 1981. Proline accumulation: Biochemical aspects. In: Paleg LG, Aspinal LD (eds.), Physiology and biochemistry of drought resistance in plants. *Academic Press, Sydney.* 243-259.

98. Suleyman, I.A., Atsushi, A., Yoshitaka, N., and Murata, N., 2000. Inactivation of photosystems I and II in response to osmotic stress in Synechococcus: Contribution of water channels. *Plant Physiol.* 122, 1201–1208.

99. Tian, C. Y., Feng, G., Li, X. L. and Zhang, F. S., 2004. Different effects of arbuscular mycorrhizal fungal isolates from saline or non-saline soil on salinity tolerance of Plants. *Appl. Soil. Ecol.* 26, 143-148.

100. Trappe, J. M., 1987.Phylogenic and ecological aspects of mycotrophy in the angiosperms from an evolutionary stand point. In: Ecophysiology of Vesicular Arbuscular Mycorrhizal Plants. *Ed. Safir G., C. R. C. Press,* Boca, Raton.5-25.

101. Tsavkelova, E.A., Klimova, S.Y., Cherdyntseva, T.A. and Netrusov, A.I., 2006. Microbial producers of plant growth stimulators and their practical use: a review. *Appl. Biochem. Micro.* 42, 117-126.

102. Turk, M.A., Assaf, T.A., Hameed, K.M. and Tawaha, Al-A.M., 2006. 'Significance of mycorrhizae'. *World J. Agri. Sci.* 2, 16-20.

103. Türkan, I. and Demiral, T. 2009. Recent developments in understanding salinity tolerance . *Environmental and Experimental Botany,* 67, 2-9.

104. yerman, S.D., Niemietz, C.M. and Bramley, H., 2002. Plant aquaporins: Multifunctional water and solute channels with expanding roles. *Plant Cell Environ.* 25, 173–194.

105. Wang, W., Vinocur, B. and Altman, A., 2003. Plant responses to drought, salinity and extreme temperatures: toward genetic engineering for stress tolerance. *Planta.* 218, 1–14.

106. Weissenhorn, I., Leyval, C., Belgy, G. and Berthelin, J., 1995. Arbuscular mycorrhizal contribution to heavy metal uptake by maize (*Zea mays* L.) in pot culture with contaminated soil. *Mycorrhiza.* 5, 245–252.

107. Wu, Q. S., Zou, Y. N., Liu, W., Ye, X. F., Zai, H. F. and Zhao, L. J., 2010. Alleviation of salt stress in citrus seedlings inoculated with mycorrhiza: changes in leaf antioxidant defense systems. *Plant Soil Environ.* 56,470-475.

108. Wu, Q.S., Zou, Y.N. and He, X.H., 2010. Contributions of arbuscular mycorrhizal fungi to growth, photosynthesis, root morphology and ionic balance of citrus seedlings under salt stress. *Acta Physiol. Plant.* 32, 297-304.

109. Wu, Q.S., and Zou,Y.N., 2013. Mycorrhizal symbiosis alters root h+ effluxes and root system architecture of trifoliate orange seedlings under salt stress. *The Journal of Animal & Plant Sciences,* 23(1),143-148.

110. Yawalkar, K.S. Aggarwal, J.P. and Bokde, S. 1996. Manures and fertilizers. Ed. VIII, *Agri-Horticultural Publishing House*, Nagpur, pp. 297-299.

111. Yamaguchi, T. and Blumwald, E., 2005. Developing salt-tolerant crop plants: challenges and opportunities. *Trends Plant Sci.* 10, 615-620.

112. Younesi, O., Moradi, A. and Namdari, A., 2013. Influence of arbuscular mycorrhiza on osmotic adjustment compounds and antioxidant enzyme activity in nodules of salt stressed soybean (*Glycine max*). *Acta agriculturae Slovenica.* 101, 219-230.

113. Zhang, H. and Blumwald, E. 2001. Transgenic salt-tolerant tomato plants accumulate salt in foliage but not in fruit. *Nature Biotechnology* .19, 765-768.

114. Zhongoun, H., Chaoxing, H., Zhibin Zhirong, Z . and Huaisong, W.,2007. Changes of antioxidative enzymes and cell membrane osmosis in tomato colonized by arbuscular mycorrhizae under NaCl stress. *Colloids Surf, B: Biointerfaces* 59, 128-133.

115. Zhu, X.C., Song, F.B. and Xu, H.W., 2010. Arbuscular mycorrhizae improves low temp stress in maize via alterations in host water status and photosynthesis. *Plant Soil.* 331, 129-137.

116. Zou, Y.N. and Wu, Q.S., 2011a. Efficiencies of five arbuscular mycorrhizal fungi in alleviating salt stress of trifoliate orange. *International Journal Agricultural and Biology.* 13, 991–995

117. Zou, Y.N. and Wu, Q.S., 2011b. Sodium Chloride Stress Induced Changes in Leaf Osmotic Adjustment of Trifoliate Orange (*Poncirus trifoliata*) Seedlings Inoculated with Mycorrhizal Fungi. *Not Bot Horti Agrobo.* 39,64-69.

118. Zuccarini, P. and Okurowska, P., 2008. Effects of mycorrhizal colonization and fertilization on growth and photosynthesis of sweet basil under salt stress. *J. Plant Nutr.* 31, 497-513.

Chapter 9

Management of Plant Pathogens

R. Ramesh

ICAR–Research Complex for Goa, Old Goa 403 402, Goa, India

Author for the correspondence: r.ramesh@icar.gov.in

ABSTRACT

Arbuscular Mycorrhizal fungi (AMF) are ubiquitous organisms that form symbiotic relationships with roots of most terrestrial plants. This symbiotic association benefits plant's survival, nutrition and growth due to AMF's ability to exploit soil nutrients, influence soil micro flora and host plants. These fungi play a key role in nutrients cycling and protect plants against environmental and other stresses. This beneficial nature of AMF could be exploited to improve plant growth and health. Several evidences of AMF inoculation as a means of biological control against soil borne diseases have been provided. The earlier the AMF establish symbiosis with host plants, the sooner the host plants get benefited from this mutualistic relationship in terms of improved growth and reduced incidence of diseases. This article discusses about AMF as a bio-protectant against plant pathogens and plant parasitic nematodes, their proposed mechanisms of disease control. Besides mechanisms such as improved plant nutrition and competition, experimental evidence supports a major role of induced plant defenses in the mycorrhizal protection. Studies indicate that pathogen suppression by AMF involves changes in mycorrhizosphere microbial dominance and populations. Interaction of AMF with plant pathogens and other microflora in mycorrhizosphere region conclude that the interactions vary with the host plant and the cultural system. Further, the beneficial effect of AMF inoculation as systemic and localized on disease suppression is reported based on new scientific evidences. Biotic and abiotic factors important for determination of efficiency of AMF as a disease control agent are discussed throughout the paper. Though biological control of plant diseases using AMF has been reported mostly in annual crops, their use in fruit crops is mostly restricted to growth improvement. However, bio-protectant effect of AMF in fruit crops is being realized nowadays which is evident from the growing number of reports and it is clear that AMF have great potential in managing diseases in fruit crops.

Keywords: *AMF, plant pathogen, immune response activation,biocontrol, mycorrhizosphere, Jasmonic acid regulation, defence related gene.*

1.0 Introduction

Fruit crops are infected by pathogenic microorganisms, nematodes which cause significant loss in crop growth, yield and sometimes complete crop failure. There are many issues associated with controlling plant pathogens due to the difficulties in reducing pathogen inoculum and lack of good sources of plant resistance. Amongst the diseases, soil borne diseases are severe and are generally managed by several agricultural practices, such as use of disease resistant cultivars, chemical fungicides, crop rotation and soil fumigation etc. However, soil borne pathogens are difficult to control due to their high degree of persistence in the soil, their resistance to majority of the fungicides/ bactericides and lack of host resistance. Many researchers reported the use of alternate approaches based on either manipulating or adding microorganisms to enhance plant protection against pathogens. The beneficial microorganisms (antagonistic bacteria and fungi) inhibit the pathogen growth by competing for nutrients and space, by producing antibiotics, by parasitizing pathogens, or by inducing resistance in the host plants. These microbes have been used since decades for biocontrol of pathogens.

Most of the higher plants are known to form one of the most intricate fungal-root associations with a special group of microorganisms known as AMF. AMF are one of the most dominant and important organisms in the soil, comprising 5–50% of the total microbial biomass in soils (Olsson *et al.*, 1999), and are obligatory associated with the majority of vascular plants (Brundrett, 2002; Millner and Wright, 2002). It is probably the oldest and most widespread plant symbiosis on earth. Fossil records and phylogenetic evidence showed the symbiosis is more than 450 million years (Smith and Read, 2008), which indicates a considerable selective advantage for both partners. The fungus colonizes the root cortex and forms intracellular structures called 'arbuscules' where the exchange of nutrients between the partners takes place. The extracellular hyphal network spreads widely into the surrounding soil, thereby reaching out of the nutrient depletion zone and improving the supply of inorganic nutrients, especially phosphate and nitrate (Smith *et al.*, 2011). In return, the heterotrophic fungal partner receives photosynthates from the host plant (Smith and Smith, 2011).

AMF receive carbon (12–27%) from the plant host in the form of simple hexose sugars, which are used for fungal growth and exuded into the mycorrhizosphere

(Tinker *et al.*, 1994). Root and mycorrhizal exudates attract other soil organisms, which use these exudates and transform organic matter, soil minerals into plant available nutrients. A major part of ecosystem carbon flux is the transfer of carbon from plants to fungi and from the fungi to the soil (Johnson *et al.*, 2005). The result of this carbon transfer was the formation of a zone of intense microbial growth and activity in the area surrounding AMF colonized roots and extra-radical hyphae (i.e. the mycorrhizosphere). Abiotic and biotic factors such as soil pH, soil chemical composition, root and hyphal exudation and soil microflora change in the mycorrhizosphere according to the interactions of the plant, fungus, soil, and soil microbes (Johnson *et al.*, 2005; Rillig and Mummey, 2006). In the mycorrhizosphere, AMF influence bacterial, fungal, and micro arthropod communities by providing them substrates in the forms of decomposing fine, ephemeral hyphae and the deposition of hyphal biomolecules, and by influencing soil structure (Andrade *et al.*, 1998; Rillig and Mummey, 2006). AMF interact with almost all organisms in the mycorrhizosphere including beneficial, plant pathogenic, saprophytic and even predatory microfauna (Bagyaraj, 1984).

2.0 AMF and plant disease control

Mutual benefits of AMF-plant symbiosis are the basis of the evolutionary success of the interaction, ensured through a tight bidirectional control of the mutualism (Kiers *et al.*, 2011). In plants, this regulation results in important changes in the primary and secondary metabolism and regulation of the defense mechanisms (Harrison, 1999; Hause and Fester, 2005). These changes usually have a deep impact on plant physiology, altering the plant's ability to cope with stresses. Early studies on mycorrhiza showed an improved growth and/or yield of mycorrhizal plants, attributed primarily to the improved nutritional status of the plant (reviewed in Linderman, 1994). Later, several authors reported a higher tolerance of mycorrhizal plants to abiotic stresses, such as drought, salinity, or presence of heavy metals (Miransari, 2010; Smith *et al.*, 2010). Studies also revealed the higher resistance of mycorrhizal plants to a wide range of soil borne fungal and bacterial pathogens, nematodes, or root-attacking insects (Azcón-Aguilar and Barea, 1997; Whipps, 2004).

With the high cost of pesticides, environmental and public health hazards associated with these pesticides and development of pesticide resistant pathogens, AMF may provide a more suitable and environmentally acceptable alternative for sustainable agriculture. With regard to bio-protective properties of AMF, the mycorrhizal symbiosis has become a focal point of research as an alternative to pesticides in sustainable agriculture (Harrier and Watson, 2004; Mukerji and Ciancio, 2007; Fester and Sawers, 2011). AMF are a major component of the rhizosphere of plants and may affect the incidence and severity of root diseases (Linderman, 1992). Comprehensive reviews on the possibilities of AMF in the biocontrol of plant diseases were published as early as by Schonbeck (1979) and many reviews were published since then.

AMF and their associated interactions with plants reduce the damage caused by plant pathogens (Harrier and Watson, 2004). These interactions have been documented for many plant species. The observations of various studies on the use of AMF for plant disease management indicated that (1) AMF associations reduce the damage caused by plant pathogens, especially those caused by fungi and nematodes (2) AMF symbiosis enhances resistance or tolerance in roots but is not equal in different crops (3) Protection is not effective against all pathogens (4) Disease protection is modulated by soil and other environmental conditions (Akhtar and Siddiqui, 2008). Therefore, the interactions between different AMF and plant pathogens vary with the host plant and the cultural system. Further, the beneficial effect of AMF inoculation may be both systemic and localized. Several biotic and abiotic factors are very important for determination of efficiency of AMF as a disease control agent. The most important factors are, soil moisture, soil contents, host genotype, mycorrhizal level inocula, inoculation time of mycorrhiza, mycorrhizal fungi species virulence, inocula potential of pathogen and soil microflora (Singh *et al.*, 2000). Before going to discuss about the bio-protectant nature of AMF, it is important to understand mycorrhizosphere, where all the interaction/ influence takes place.

3.0 Mycorrhizosphere

Disease suppressive soils occur naturally or due to specific management practices, and are thought to involve soil specific bacteria, fungi, or actinomycetes. Among the rhizosphere microbial populations with the greatest influence, AMF are the most important, but only in combination with bacterial associates. AMF formation causes an increase in levels of antagonistic bacteria, provided the background soil contains effective antagonists to be selectively increased (Linderman, 1988). The symbiotic association of AMF with the roots of plants has existed for millions of years and includes associations with other functional groups of soil microbes *viz.* rhizobacteria, other rhizosphere fungi, and diverse fauna. The combination of these organisms in an undisturbed and natural ecosystem contributes to the successful growth and health of plants. Plant diseases particularly root diseases are rare in undisturbed ecosystems compared to disturbed agro-ecosystems due to microbial support systems in the rhizosphere soil associated with plant roots.

The 'rhizosphere' phenomenon, as described by Hiltner (1904), was induced initially by nutrients released from plant roots. The altered microflora in the rhizosphere due to mycorrhizae led to the expanded concept of the 'mycorrhizosphere' (Linderman, 1988) in which mycorrhizae significantly influence the microflora both in qualitative and quantitative terms due to altered root physiology and exudation (Bagyaraj, 1984; Gryndler, 2000). This dynamic process is not only influenced by the initial enrichment of plant specific root exudates but also primarily by the microbial interactions that occur therein. Other factors that influence the process are the age of the plant, the nature and treatment of the soil, foliar applications, environmental factors, fertilizer applications and host nutrition. Because AMF establish a persistent interface between the host root and the soil, it becomes perhaps the only stable microbial system in the rhizosphere. Thus, 'rhizosphere soil' is soil adjacent to roots and influenced by root exudates, while 'mycorrhizosphere soil' is soil adjacent to mycorrhizae and influenced by exudates from both the root tissue and the fungal hyphae.

4.0 Mechanisms of disease control by AMF

Different hypotheses have been proposed to explain bio-protection by AMF against plant pathogens during different interactions. These include,

1. Competition for colonization sites and host photosynthates

2. Anatomical or architectural changes in the root system and damage compensation

3. Modification of microbial community changes: The mycorrhizosphere effect

4. Improvement of plant nutrition uptake

5. Physiological and biochemical changes

6. Activation of plant defense/ resistance mechanisms

However, very little is known of the physiological, cellular, or molecular mechanisms that are really active. Of the various mechanisms proposed for biocontrol of plant diseases, effective bio-protection is a cumulative result of all mechanisms working either separately or together.

4.1 Competition for colonization sites and host photosynthates

Physical competition between AMF and rhizosphere microorganisms to occupy more space in the root is the first mechanism to explain the interaction between AMF and soil microorganisms (Bansal and Mukerji, 1996). AMF and soil borne plant pathogens are colonizing the same host tissues and there may be direct competition for space if colonization is occurring at the same time (Smith and Read, 2008) because both usually develop within different cortical cells of roots. Davies and Menge (1980) observed localized competition between AMF and *Phytophthora*. They observed reduced development of *Phytophthora* in AMF colonized and adjacent uncolonized root systems, and pathogens never penetrated arbuscule-containing cells (Cordier *et al.*, 1996). Similarly *Aphanomyces* was suppressed on pea roots by AMF only when the two organisms were present on the same root (Rosendahl, 1985). Vigo *et al.* (2000) observed that the number of infection sites was reduced within mycorrhizal root systems and colonization by the AMF had no effect on the spread of necrosis.

AMF depend on the plant host photosynthates and a competition for the same is possible between AMF and other microbes, especially with microbial symbionts and plant pathogens for photosynthetically assimilated carbon (Smith and Read, 1997). When AMF have primary access to the photosynthates, the higher carbon demand may inhibit the pathogen growth (Linderman, 1994). Though approximately 4-20% net photosynthates of host are transferred to the fungus; there is only a limited data to support this mechanism (Smith and Read, 2008). This phenomenon may have an important role in interactions with endoparasitic nematodes because of the obligate nature of both organisms for host-derived compounds (Azcon-Aguilar and Barea, 1996).

4.2.Anatomical or architectural changes in the root system and damage compensation

Roots offer structural support to the plants and function in absorption of water and supply mineral nutrients for a wide range of microorganisms. The colonization by AMF results in morphological changes to the root, leading to an increased surface area of root and these changes will ultimately affect the plant's responses to other organisms. AMF colonized root system contains shorter, more branched, adventitious roots of larger diameters and lower specific root lengths (Schellenbaum *et al.*, 1991; Berta *et al.*, 1993). The AMF inoculated plants possess a strong vascular system, which provides greater mechanical strength to reduce the effects of pathogens (Schonbeck, 1979). Dehne *et al.* (1978) observed increased lignifications in the endodermal cells of tomato and cucumber plants colonized by AMF and attributed that such responses may account for reduced incidence of *Fusarium* wilt. Becker (1976) reported a similar effect on pink root of onion (*Pyrenochaeta terrestris*). Mycorrhizal plants produced wound barriers at a faster rate than non-mycorrhizal plants and increased wound barrier formation inhibited *Thielaviopsis* black root rot of mycorrhizal holly plants (Wick and Moore, 1984).

It is also suggested that AMF reduce the effect of pathogen attack by compensating for the loss of functional root and biomass caused by soil borne fungal and nematode pathogens (Linderman, 1994; Cordier *et al.*, 1996). This suggests an indirect contribution to the biological control through the conservation of root system and increasing the root absorbing surface area by AMF hyphae growing out into the soil and maintenance of root cell activity through arbuscules formation (Gianinazzi-Pearson *et al.*, 1995).

4.3 Modification of microbial community changes: The mycorrhizosphere effect

The growth and health of plants influenced by the microbial shifts occur in the mycorrhizosphere (Azcon-Aguilar *et al.*, 2002). This effect has not been specifically evaluated as mechanisms for AMF associated biocontrol, but there are indications that such a mechanism does operate (Linderman, 1994). Some reports suggest that AMF alter the composition of functional groups of microbes in the mycorrhizosphere, including the numbers and/or activity of pathogens and antagonists (Meyer and Linderman, 1986; Linderman, 1994). Numbers of pathogen antagonistic actinomycetes were greater in the rhizosphere of AMF plants than in non-mycorrhizal controls (Secilia and Bagyaraj, 1987) and *G. fasciculatum* harbored actinomycetes antagonistic to *F. solani* than those of non-mycorrhizal plants. Plant root systems colonized by AMF differ in their effects on the bacterial community composition within the rhizosphere and rhizoplane (Burke *et al.*, 2002). The number of facultative anaerobic bacteria, fluorescent *pseudomonas*, *Streptomyces* species and chitinase producing actinomycetes differ depending on the host plant and the isolate of AMF (Harrier and Watson, 2004).

The role of AMF in improving plant nutrition and their interactions with other soil biota has been reported with reference to the host plant growth. Plants colonized by AMF differ from non-mycorrhizal plant in rhizosphere microbial community (Marschner *et al.*, 2001). These differences have been attributed to alterations in root respiration rate and quality and quantity of exudates. AMF symbiosis can also cause qualitative and quantitative changes in rhizospheric microbial populations; the resulting microbial equilibria could influence the growth and health of plants (Bansal and Mukerji, 1994; Azaizeh *et al.*, 1995; Marschner *et al.*, 2001; Pivato *et al.*, 2009). These changes impact the microbial community of the mycorrhizosphere and, among other effects, may lead to a shift in its composition favouring specific microbiota with the capacity to antagonize possible root pathogens (Barea *et al.*, 2005; Badri and Vivanco, 2009). In certain cases, altered root exudates directly impact microbial pathogens and nematodes. Exudates from mycorrhizal tomatoes transiently paralyzed nematodes, and subsequently their penetration into mycorrhizal tomato roots was decreased (Vos *et al.*, 2011). Similarly, sporulation of the oomycete *Phytophthora fragrariae* was severely reduced in the presence of mycorrhizal strawberry root exudates (Norman and Hooker, 2000), and concentrated exudates from mycorrhizal tomato roots were repulsive to *Phytophthora nicotianae* zoospores (Lioussanne *et al.*, 2008). Caron *et al.* (1985) showed a reduction in *Fusarium* populations in mycorrhizosphere soil of tomatoes and a corresponding reduction in root rot in AMF plants compared with non-AMF plants, probably due to the increased antagonism in the AMF mycorrhizosphere. Different factors (e.g. altered exudation patterns, putative direct AMF effects, different root size and architecture, altered physiology) may contribute to quantitative and qualitative microbial community changes in the mycorrhizosphere caused by AMF (Andrade *et al.*, 1998; Artursson *et al.*, 2006; Toljander *et al.*, 2007; Finlay, 2008).

4.4 Improvement of plant nutrition uptake

Improvement of plant growth due to root colonization by AMF occurs as a result of enhancement of the mineral nutrient uptake by the plants. Mycorrhizal plants contain higher concentrations of phosphorus than non-mycorrhizal plants (Hayman, 1978; Bowen, 1980). Improvement of phosphorus nutrition following AMF colonization of phosphorus-deficient roots results in a decrease in membrane permeability and reduction in root exudation (Graham *et al.*, 1981). In general, AMF infection in phosphorus deficient plants affects membrane permeability and exudation patterns in a fashion similar to that caused by phosphorus fertilization in non-mycorrhizal plants. AMF-induced decreases in root exudation have been correlated with reduction of soil borne disease (Graham and Menge, 1982), while improved nutritional status of the host brought about by AMF-root colonization may affect quantitative changes in root exudates (Linderman, 1985; Reid, 1984). Declerck *et al.* (2002) suggested a similar effect whereby AMF or added phosphorus reduced root rot of bananas caused by *Cylindrocladium spathiphyylii*. Studies indicated that phosphorus induced changes in root exudation and thereby reduced the germinations of pathogen spores (Graham, 1982; Sharma *et al.*, 2007). Changes in exudation due to phosphorus nutrition alter the chemotaxic attraction of the nematodes to the roots and affects exclusion of nematode species that require a hatching stimulus (Baker and Cook, 1982).

The increasing nutrient uptake resulted in more vigorous plants; thus, the plant itself may be more resistant or tolerant to pathogen attack (Linderman, 1994). Davis (1980) found this type of response on *Thielaviopsis* root rot of citrus, where AMF plants were larger than non-mycorrhizal plants until the latter were fertilized with additional phosphorus. The AMF spores germinate and thick-walled hyphae penetrate the host root causing internal infection. After penetrating into the root, the hyphae spread inter- and/or intra-cellularly in the root cortex without damaging the integrity of the cells (Strack *et al.*, 2003). AMF may increase host tolerance to pathogen by increasing the uptake of essential nutrients rather than phosphorus alone which are otherwise deficient in the non-mycorrizal plants (Gosling *et al.*, 2006). In addition to phosphorus, AMF can enhance the uptake of Ca, Cu, Mn, S and Zn (Pacovsky *et al.*, 1986; Smith and Gianinazzi- Pearson, 1988). Depending on the host plant and AMF isolate, colonization of the root system can increase phosphorus nutrition and other mineral nutrients (Clark and Zeto, 2000). However, in some cases enhanced mineral nutrition of mycorrhizal plants has no affect against pathogens (Graham and Egel, 1988). Therefore, enhanced mineral nutrition of AMF plants does not account for all protection conferred by AMF to host plant (Caron *et al.*, 1986).

4.5 Physiological and biochemical changes

The physiological and biochemical changes caused by AMF in the host plant generally reduce the severity of nematode diseases (Dehne, 1982). Phenolic compounds have been shown to be formed after mycorrhizal colonization (Sylvia and Sinclair, 1983) and are thought to play a role in disease resistance (Goodman *et al.*, 1967). Production of phytoalexin (isoflavonoid compounds) was greater on mycorrrhizal roots of soybean than on non-mycorrhizal roots of soybean (Morandi, 1987) and attributed that these compounds are responsible for the increased resistance to fungal pathogens in AMF plants (Kaplan *et al.*, 1980). An increase in lignin and phenols in mycorrhizal plants was observed and was associated with reduced nematode reproduction (Sikora, 1978; Umesh *et al.*, 1988; Singh *et al.*, 1990). Increased phenylalanine and serine concentrations in tomato roots due to inoculation with AMF have been observed (Suresh, 1980). These two amino acids are known to be inhibitory to root-knot nematodes (Prasad, 1971; Reddy, 1974). Krishna and Bagyaraj (1986) reported that higher amounts of catechols inhibit *Sclerotium rolfsii* growth *in-vitro*. Suresh and Bagyaraj (1984) reported that AMF inoculation increased the quantities of sugars and amino acids in plant tissue which may be responsible for the reduction of nematode infestation. However, inferences based on the absence of galling on segments of roots and split root experiments revealed a more localized effect (Tylka *et al.*, 1991; Fitter and Garbaye, 1994). Various evidences indicate structural and biochemical changes in the cell walls of plants colonized by AMF. Dehne and Schonbeck (1979) and Becker (1976) reported enhanced lignification of endosperm cell walls and vascular tissues. Dehne *et al.* (1978) also demonstrated an increased concentration of antifungal chitinase in AMF roots. They suggested that increased arginine accumulations in AMF roots suppress sporulation of *Thielaviopsis*, as reported earlier (Baltruschat and Schonbeck, 1975).

4.6 Activation of plant defense/ resistance mechanisms

Besides mechanisms such as improved plant nutrition and competition, experimental evidences support a major role of plant defenses in the mycorrhizal protection. The use of experimental split-root systems has confirmed that the protection by mycorrhiza is manifested in non-colonized areas of the root system. For example, systemic protection in the root has been confirmed against oomycetes and bacterial pathogens in tomato (Cordier *et al.*, 1998; Pozo *et al.*, 2002; Zhu and Yao, 2004; Khaosaad *et al.*, 2007), against fungal pathogens in barley (Khaosaad *et al.*, 2007), and in banana and grapevine against nematodes (Hao *et al.*, 2012). Observations of systemic protection against pathogens in non-colonized root portions from mycorrhizal plants and enhanced resistance of the aerial parts indicated the involvement of plant defense mechanisms (Cordier *et al.*, 1998; Pozo *et al.*, 2002; Pozo and Azcón-Aguilar, 2007).

Defense mechanisms are coordinated by the plant immune system, strikingly similar in some aspects to the innate immune system in animals (Ausubel, 2005). This system allows the plant to distinguish non-self-alien organisms by recognizing structurally conserved microbe-associated molecules, such as flagellin, lipopolysaccharides, or peptidoglycans, which are collectively, termed microbe-associated molecular patterns (MAMPs, or PAMPs in the case of pathogens). PAMPs are recognized by transmembrane pattern recognition receptors (PRRs), which leads to the induction of the appropriate responses in the host and the PAMP-triggered immunity (PTI) (Ausubel, 2005; Jones and Dangl, 2006; Thomma *et al.*, 2011). In the evolution, microbes have evolved effector proteins that are secreted into the host to suppress PTI, and allow successful host colonization by the pathogen, thus causing effector-triggered susceptibility of the plant. In some cases, intracellular proteins of the plant recognize pathogen effectors or their modified target proteins and activate immune responses that are quicker, more prolonged, and more robust than those in PTI, resulting in effector-triggered immunity (ETI) (Jones and Dangl, 2006; Thomma *et al.*, 2011).

The activation of specific plant defense mechanisms as a response to AMF colonization is an obvious basis for the protective behavior of AMF. The elicitation of specific plant defense reactions, could predispose the plant to an early response to attack by a root pathogen leading to accumulation of phytoalexins (Morandi *et al.*, 1984), enzymes of the phenylpropanoid pathway (Lambais and Mehdy, 1993; Volpin *et al.*, 1994), pathogenesis-related (PR) proteins (Balestrini *et al.*, 1994), callose, and phenolics. This suggests that the AMF are able to elicit a defense response, but that symbiosis-specific genes somehow control the expression of the genes related to plant defense during AMF establishment (Gianinazzi- Pearson *et al.*, 1995, 1996). These observations suggest a systemic suppression of the defense reaction during the establishment of the AMF association.

5.0 Effects of AMF interactions on plant pathogens and other microorganisms

The occurrence of AMF and plant pathogenic fungi/ plant parasitic nematodes/ other plant pathogens in the roots of different crops and their dependence for nutrition on the host may result in a common association of AMF, the pathogens and host plant. Association of these two groups of organisms generally exert opposite effects on the host plant. Thus, it is of utmost importance to determine the effect of interaction of these organisms on plant growth and yield including their mutual effects.

5.1 Mycorrhiza effects on below ground interactions

Many studies showed the protective effect of colonization by AMF against infections by microbial pathogens in different plant systems. The majority of these reports focus on soil borne pathogens such as fungi from the genera *Fusarium*, *Rhizoctonia*, *Macrophomina*, or *Verticillium*; bacteria such as *Erwinia carotovora*; or oomycetes like *Phytophthora*, *Pythium*, and *Aphanomyces*. In most cases, the protective effect is not only related to damage compensation or tolerance, but frequently the reduced damage also correlates with a decrease of the pathogen content within plant tissues (reviewed by Whipps, 2004). Similarly, studies that showed a clear reduction of the detrimental effects by endoparasitic nematodes such as *Pratylenchus* and *Meloidogyne* in mycorrhizal plants (Pinochet *et al.*, 1996; Elsen *et al.*, 2008; Vos *et al.*, 2011). Recently, a decrease on the development of ectoparasitic nematodes also has been described (Hao *et al.*, 2012).

Beneficial organisms also are influenced by the presence of mycorrhiza. A positive effect of mycorrhiza on beneficial plant-microbe interactions has been reported. For example, synergistic effects have been described with regard to plant associations with nitrogen-fixing bacteria (Niranjan *et al.*, 2007; Larimer *et al.*, 2010); phosphate solubilizing bacteria (Kohler *et al.*, 2007); biocontrol agents (Martínez-Medina *et al.*, 2010; Saldajeno and Hyakumachi, 2011); and plant growth promoting microorganisms (Chandanie *et al.*, 2006, 2009). These effects on below-ground interactions may result from a combination of diverse mechanisms as described in the earlier sections.

5.2 Mycorrhiza effects on above ground interactions

Systemic protection by a mycorrhizal association can even be observed in the aerial parts of a colonized plant. However, reports on AMF effects on pests and pathogens attacking shoots are less studied compared to below ground interactions and the results are variable. Early studies described a higher susceptibility of AMF plants to viruses, and biotrophic pathogens survived better on mycorrhizal plants, although an increased tolerance has been observed in terms of plant mass and yield (Gernns *et al.*, 2001; Whipps, 2004). Effect of the symbiosis on hemibiotrophs varies from no effect to reduction of the disease, for example, against *Colletotrichum orbiculare* in cucumber (Lee *et al.*, 2005; Chandanie *et al.*, 2006). However, necrotrophic pathogens are negatively affected in their proliferation, and symptom development is less severe on mycorrhizal plants. Examples are the incidence of *Alternaria solani* in tomato (Fritz *et al.*, 2006; de la Noval *et al.*, 2007), *Magnaporthe grisea* in rice (Campos-Soriano *et al.*, 2012), and *Botrytis cinerea* in roses and tomato (Møller *et al.*, 2009; Pozo *et al.*, 2010). In case of phytoplasmas, specialized obligate parasites of phloem tissue, most reports show a reduction of disease incidence (Kamińska *et al.*, 2010, Batlle *et al.*, 2011). However, it should be noted that the analysis of mycorrhizal effects on phytoplasmas and viruses is difficult because of the potential impact of mycorrhization on the insect vector and the complexity of studies on multi-trophic interactions.

5.3 Influence of AMF on plant interactions with pathogens

The effects of the AMF symbiosis on plant interactions with other organisms, and, in particular, the induction of resistance against deleterious organisms seem to result from the combination of multiple mechanisms (as described in the previous sections) that may operate simultaneously. Apart from that, a major transcriptional reprogramming takes place upon mycorrhizal colonization of the roots (Liu *et al.*, 2007; Güimil *et al.*, 2005). This reprogramming happens due to alterations in the primary and secondary metabolism in mycorrhizal plants (Hause *et al.*, 2007; Toussaint, 2007). The majority of the changes affects the host's secondary metabolism, and has far-reaching consequences for the plant. One example is the alteration in root exudates, composed of various secondary metabolites such as phenolic compounds, strigolactones, and allelopathic compounds that regulate multiple interactions in the rhizosphere (López-Ráez *et al.*, 2010, 2011). Accordingly, all those changes may influence mycorrhizal effects on plant interactions below ground. The symbiosis also increases the rate of photosynthesis and influences the carbon assimilation and allocation, thereby possibly affecting the source-sink relations that may influence the suitability of the plant for above ground attackers (Wright *et al.*, 1998).

6.0 Modulation of the host plant's immune system by AMF

During AMF establishment, modulation of plant defense responses occurs thus achieving a functional symbiosis. As a consequence of this modulation, a mild, but effective activation of the plant immune responses seems to occur, not only locally but also systemically. This activation leads to a primed state of the plant that allows a more efficient activation of defense mechanisms in response to attack by pathogens.

As biotrophs, AMF share some similarities with biotrophic pathogens, and are able to trigger plant defense responses during the early stages of the interaction (Paszkowski, 2006), the responses that are subsequently suppressed (García-Garrido and Ocampo, 2002; Liu *et al.*, 2003). Thus, for a successful colonization, the fungus has to cope with these reactions and actively modulate plant responses. Jung *et al.* (2012) have proposed that this modulation may result in pre-conditioning of the tissues for efficient activation of plant defenses upon a challenger attack, a phenomenon that is called 'priming' (Pozo and Azcón-Aguilar, 2007). Priming sets the plant in an 'alert' state in which defenses are not actively expressed but in which the response to an attack occurs faster and/ or stronger compared to non-primed plants, leads to increased plant resistance. Thus, priming confers important fitness benefits (Conrath *et al.*, 2006; Walters and Heil, 2007). Moreover, priming seems to be the mechanism underlying the Induced Systemic Resistance (ISR) observed in plants interacting with beneficial microorganisms (Conrath *et al.*,

2006; Van Wees *et al.*, 2008) and the plant immune responses by beneficial microbes is often dependent on a functional jasmonic acid (JA) signaling pathway, as has been described for rhizobacteria and AMF (Verhagen *et al.*, 2004; Pozo *et al.*, 2010).

In response to colonization by AMF, a quick but transient increase of endogenous salicylic acid (SA) occurs in the roots with a concurrent accumulation of defensive compounds, such as reactive oxygen species, specific isoforms of hydrolytic enzymes, and the activation of the phenylpropanoid pathway (Pozo *et al.*, 1998; Fester and Hause, 2005; de Román *et al.*, 2011). These reactions are temporally and spatially limited compared to the reaction during plant-pathogen interactions, suggesting a role in the establishment or control of the symbiosis (García-Garrido and Ocampo, 2002). Indeed, SA signaling seems to have a negative effect on AMF colonization (de Román *et al.*, 2011), and AMF establishment requires inhibition of certain SA-regulated responses as described for other mutualistic symbiosis (Soto *et al.*, 2009). Despite our lack of knowledge on how the AMF evade and manipulate the host's innate immune system, recent studies support that AMF can actively suppress SA-dependent defense reactions by secreting effector proteins that interfere with the host's immune system (Kloppholz *et al.*, 2011). Not only SA, but also the level of other phytohormones related to defense, such as JA, ABA, and ET, is altered during the plant interaction with the AMF (Hause *et al.*, 2007; Ludwig-Müller, 2010).

Primed defense responses in mycorrhizal plants were first observed in root tissues. Mycorrhizal transformed carrot roots displayed stronger defense reactions at sites challenged by *Fusarium* (Benhamou *et al.*, 1994). AMF colonization systemically protected tomato roots against *Phytophthora parasitica* infection. Only mycorrhizal plants formed papilla-like structures around the sites of pathogen infection through deposition of pectins and callose, preventing the pathogen spread, and they accumulated significantly higher PR-proteins than non-mycorrhizal plants upon *Phytophthora* attack (Cordier *et al.*, 1998; Pozo *et al.*, 2002). Priming for callose deposition also was reported to underlie protection against *Colletotrichum* in cucumber (Lee *et al.*, 2005). Similarly, mycorrhizal potatoes showed amplified accumulation of the phytoalexins rishitin and solavetivone upon *Rhizoctonia* infection, whereas AMF alone did not affect the levels of these compounds (Yao *et al.*, 2003). Primed accumulation of phenolic compounds in AMF date palm trees also has been related to protection against *F. oxysporum* (Jaiti *et al.*, 2007), and priming has been involved in mycorrhizal induction of resistance against nematodes (Li *et al.*, 2006; Hao *et al.*, 2012).

However, the primed response is not restricted to the root system alone. Recently, it has been shown that priming of defenses also in shoots of mycorrhizal plants (Pozo *et al.*, 2010). AMF symbiosis induced systemic resistance in tomato plants against the necrotrophic foliar pathogen *Botrytis cinerea*. While the amount of pathogen in leaves of mycorrhizal plants was significantly lower, the expression of some jasmonate-regulated, defense related genes was higher in those plants (Pozo *et al.*, 2010). In addition to this priming effect on above-ground tissues, the AMF hyphal network may even extend the induction of resistance to neighboring plants, acting as a plant to plant underground communication system (Song *et al.*, 2010). They observed that healthy 'receiver' plants activate JA-regulated, defense-related genes when neighboring 'donor' plants, connected via a common mycorrhizal network in the soil, were infected by the foliar pathogen *Alternaria solani*.

7.0 Synergy of AMF with other endophytic bacteria/microorganisms

7.1 The mycorrhizosphere effect

A relatively small volume of soil around plant roots is under the direct influence of root exudates, termed the 'rhizosphere'. This zone is characterized by increased levels of microbial activity. However, 80% of all plant species form symbiotic relationships with AMF. Consequently, the volume of soil influenced by plant-derived carbon via AMF can be extended to encompass the 'mycorrhizosphere'. AMF have a selective influence on microbial communities in the mycorrhizosphere as discussed earlier in this paper. The enhanced microbial activity surrounding mycorrhizal roots compared with non-mycorrhizal roots is called the 'mycorrhizosphere effect' (Linderman, 1988).

With the development of metagenomics technologies and DNA sequencing methods, the true extent of quantitative and qualitative changes in the microbial community due to AMF is beginning to emerge. The chemical basis driving mycorrhizosphere development is less well resolved, although there is indication that carbon exudation by AMF in the form of the glycoprotein glomalin plays a role (Rillig *et al.*, 2003). The consequences of the mycorrhizosphere effect, including recruitment of PGPRs, may not only boost nutrient mobilisation by AMF but could also provide non nutritional benefits, such as disease suppression through antibiosis and/or competitive exclusion as discussed in the previous sections. Importantly, increased densities of selected rhizobacteria in the mycorrhizosphere have the potential to suppress pests and diseases in systemic plant tissues through priming of inducible defenses (Cameron *et al.*, 2013). They proposed that AMF induced resistance is a cumulative effect of direct plant responses to mycorrhizal infection and indirect immune responses to ISR-eliciting rhizobacteria in the mycorrhizosphere.

7.2 Mycorrhizal helper bacteria (MHB)

It is clear from the various research findings that extramatrical hyphae from AMF have access to resources from a vast volume of soil. Later it was discovered that some mycorrhizosphere-inhabiting bacteria, called 'mycorrhiza helper bacteria' (MHB), can stimulate mycorrhizal symbioses (Frey-Klett *et al.*, 2007). These helper bacteria play vital role in growth and spore germination of the AMF and the absence of the MHB is highly affecting the ecological fitness of AMF (Lumini *et al.*, 2007). The concept of mutualism between AMF and soil bacteria is not new. Mosse (1962) first proposed the idea that AMF and bacteria interact directly in the soil, showing that mycorrhizal roots can enhance the survival of soil bacterium, *P. fluorescens*. Since then, multiple studies have demonstrated that MHB can promote mycorrhizal infection and symbiosis through stimulation of mycelial extension and reducing the impact of adverse environmental conditions (Frey-Klett *et al.*, 2007). Whether increased AMF growth and survival by MHB are due to production of growth factors, detoxification of soil allelochemicals, or antagonism of competitors and/or parasites remains unresolved (Frey-Klett *et al.*, 2007).

The tripartite association between plant-mycorrhizal-endophytic bacteria is beneficial in plant nutrient dynamics and plant growth. Bonfante and Anca (2009) reviewed the interaction between the three organisms and pointed out that bacteria represent the third component of mycorrhize and bacterial communities are influenced by mycorrhizal fungi. Naumann *et al.* (2010) described a vertically inherited, monophyletic and globally distributed lineage of endobacteria living in arbuscular mycorrhizal cytoplasm also identified the association of bacteria. Studies showed that actinomycetes were able to improve plant growth and nutrition, and benefit root colonization by AMF. Co-inoculation with both types of microorganisms showed synergistic effects at enhancing plant growth and nutrient acquisition (Franco-Correa *et al.*, 2010). Studies supported the use of actinomycetes as plant growth promoting, biocontrol organism as well as mycorrhiza helper bacteria (Sukhada *et al.*, 2013; Poovarasan *et al.*, 2013).

8.0 Status of AMF as bio-protectant in fruit crops

Biological control of plant diseases using AMF has been reported mostly in annual crops for many years. Though association of AMF with fruit crops has been reported widely, its bio-protectant application is restricted to only few crops, probably due to the difficulty in assessing the effect of AMF in reducing the disease in fruit crops and in understanding the mechanisms of disease control through AMF. Nevertheless, AMF constitute an important partner in the management of diseases in fruit crops as most of higher plants are naturally colonized by AMF. Improvement of plant growth parameters by AMF and other microorganism in fruit crops has been well documented by many workers. There are numerous studies evaluating biocontrol of plant diseases by using the antagonistic effects of either AMF or rhizobacteria or endophytic bacteria or antagonistic fungi, but usually independently. Fewer studies, although growing in number, report on evaluating the effectiveness of concurrent fungi- bacteria/ fungi inoculation in combating plant diseases in fruit crops. Some of the studies of disease control involving AMF in fruit crops are discussed in this section.

8.1 Apple (*Malus* spp.)

Inoculation of AMF has been reported to reduce the incidence of white root rot (*Dematophora necatrix*) of apple and *Phytophthora* root rot of citrus (Davis and Menge, 1980; Bharat and Bhardwaj, 2001). Apple seeds coated with two native isolates of *Azotobacter chrococcum* and inoculated with AMF (*Glomus fesiculatum*, *Glomus macrocarpum*, *Glomus mosseae* and *Gigaspora* sp.) in solarized plots reduced the incidence of white root rot and improved the growth parameters (Raj and Sharma, 2009). Soil inoculation of AMF to apple seedlings reduced the percentage of powdery mildew infection similar to those treated by fungicide Flint

(Yousefi, 2011). Inoculation of AMF to apple roots suppressed the *Botryosphaeria* canker severity and enhanced plant growth (Krishna *et al.*, 2010). The timing of inoculation also had a significant effect on disease development and plant survival as plants pre-inoculated with mycorrhiza performed better over those inoculated simultaneously with *Botryosphaeria ribis* and AMF. The most possible mechanism involved with resistance imparted by AMF inoculation against canker development on aerial parts of apple could be the accumulation of secondary compounds in the aerial parts of plants colonized by AMF. The production of secondary metabolites in aerial plants of mycorrhizal plants has been reported by Moraes *et al.* (2004) and Krishna *et al.* (2005). The application of AMF for improving growth, stress alleviation and disease management in apple has been reported by various authors in the past (Ridgway *et al.*, 2008; Raj and Sharma, 2009).

8.2. Banana (*Musa* spp.)

AMF *Glomus mosseae* and fungal biocontrol agent *Trichoderma harzianum* combination was more effective than individual treatment in reducing the impact of *Fusarium oxysporum* f. sp. *cubense* infection under field conditions (**Plate 9.1**) Sukhada *et al.*, 2010 using FITC conjugated antibodies it was shown that the pathogen population reduced over a period of time in *G. mosseae* and *T. harzianum* treatment compared to higher population in control. AMF and *Trichoderma* treated banana roots had developed cell wall thickenings in their cortex which probably resisted the entry of the pathogens (**Plate 9.4**). Similar synergistic effect of *G. mosseae* fungi and *T. harzianum* has been reported by Filion *et al.* (1999). Application of AMF *G. clarum* reduced the incidence and severity of *Fusarium* wilt of banana and increased growth and yield (Lin *et al.*, 2012). Declerck *et al.* (2002) reported that application of AMF reduced root rot of bananas crop caused by *Cylindrocladium spathiphyylii*. Poovarasan *et al*, (2013b) evaluvated the effect on *Streptomyces* spp isolated from *Glomus* spores on *Fusrium Oxysporum* f.sp. *cubense* infecting banana and found that infection was substantially reduced in the inoculated suckers (**Plate 9.2**).

8.3 Grape (*Vitis vinifera* L.)

Yan Li *et al.* (2006) tested the effect of AMF inoculation in grape vine seedlings for resistance against root knot nematodes. The seedlings inoculated with *G.versiformea* increased the plant height and induced a resistance response, which resulted in the growth restriction of nematode in the mycorrhizal grapevine roots, and which involved transcriptional activation of defense genes like the class III Chitinase gene and VCH3.

8.4 Guava (*Psidium guajava* L.)

Root inoculation AMF is reported to be effective in reduction of wilt caused by *Fusarium oxysporum* f.sp. *psidii* (Srivastava *et al.*, 2001). Actinomycetes isolated from *Glomus mosseae* spores predominant in the rhizosphere of guava improved the plant growth (Sukhada *et al.* 2013) and indicated the possibilities of using mycorrhizae associated actinomycetes as bioinoculant for growth promotion, nutrient mobilization and biocontrol agent in guava seedling production.

8.5 Pomegranate (*Punica granatum* L.)

Role of mycorrhizae associated actinomycetes in promoting plant growth and control of *Xanthomonas axonopodis* pv. *punicae* causing blight disease in pomegranate was studied (Poovarasan *et al.*, 2013). The findings from this study clearly indicated the possibilities of using mycorrhizae associated actinomycetes as bio inoculant for growth promotion and for control of bacterial blight in pomegranate.

8.6 Papaya (*Carica papaya* L.)

Effect of AMF on *Pythium aphanidermatum*, the papaya foot rot pathogen indicated that *Glomus mosseae* and *Glomus deserticola* significantly suppressed the pathogen and also enhanced the seedling growth when applied alone (Olawuyi *et al.*, 2013). Further they reported that *G. deserticola* was superior in protecting the plant. Both the AMF were good bio-fertilizers which was evident from the growth promotion activities. Hernández-Montiel *et al.* (2013) reported that application of AMF complex consisting of *Glomus intraradices, Glomus mosseae, Glomus etunicatum* and *Gigaspora albida* reduced root rot in papaya caused by *Fusarium oxysporum*, individually and in combination with *Pseudomonas* sp. Further they reported the reduced colonization of *F. oxysporum* in papaya seedlings.

They postulated that rhizobacteria and AMF together formed a mutualistic relationship that enhanced disease control against *F. oxysporum* and stimulates growth in papaya. Sukhada *et al.*, (2011) evaluated the effect of AMF and other biocontrol agents against *Phytophthora parasitica* var. nicotianae infecting papaya (C. *papaya* cv. Surya) and enumerated the pathogen population using immunotechniques. *Glomus mosseae, Trichoderma harzianum* and *Pseudomonas fluorescens* were inoculated at the time of planting in the nursery and at the time of transplanting in single, dual and tripartite combinations allowing colonization up to 90 days. Plants were challenged thereafter with *Phytophthora* inoculum multiplied on specialized *Phytophthora* medium. Uninoculated plants and those inoculated with pathogen only were controls. All the BCAs in general improved plant growth and reduced severity of disease compared to uninoculated control in both pot experiments and under field conditions. Plants preinoculated with *G. mosseae + T. harzianum*, provided the best results (**Plate 9.3**) when challenged with *Phytophthora*, with increased plant height, girth and yield and also reduced disease severity over plants not inoculated with BCAs. Studies with *Phytophthora* antibodies conjugated with fluorescein isothiocyanate (FITC) and enzyme-linked immunosorbent assay (ELISA) further substantiated the reduction in infection and population of *Phytophthora* in BCA pretreated plants.

8.7 Strawberry (*Fragaria×ananassa* Duchesne)

Camprubi *et al* (2007) concluded through their experimental findings that a combined application of *G. intraradices, Trichoderma aureoviride* and *Bacillus subtilis* was an alternative technology for controlling pests and diseases in strawberry crop production.

8.8 Replant diseases of fruit crops

Replant diseases are one of the important constraints in developing new orchards and are difficult to manage. Non equilibrium of soil microbe community structures due to replanting results in reducing crop yield and soil quality. Under replant conditions, crop root secretion and litter decayed materials provide rich nutrients for the pathogen. AMF have positive effects on replant diseases in grape or apple orchard (Camprubí *et al.*, 2008; Raj and Sharma, 2009). Inoculation with *Glomus etunicatum* was successful during the first 6 months of growth only when apple seedlings were grown in a sterile substrate for the first three weeks. When young grafted apple trees were inoculated directly in orchard soil with apple replant disease, the effect of AMF inoculation was negligible after 6 months. Application of a potent indigenous AM F culture (*Glomus fasciculatum*) to newly planted apple seedlings fumigated or non-fumigated soil proved as a good management practice for apple replant problem (Mehta and Bharat, 2013).

AMF inoculation reduced watermelon (*Citrullus lanatus*) replant problems through effectively modifying the soil microbe population, community structure, and increasing the soil enzyme activities (Zhao *et al.*, 2010). The inoculation with AMF *Glomus versiforme* enhanced soil bacteria and actinomycete population, and decreased the fungal numbers, and the fungi/total microbe ratio in watermelon replanting soils, and improved soil proteinase, polyphenoloxidase, urease, and saccharase activities.

8.9 Nematode control

AMF reduce the damage caused by sedentary endoparasitic nematodes in some annual crops, sometimes by reducing nematode population growth (reviewed in Ingham, 1988). A microplot approach used to demonstrate the inoculation with *Glomus mosseae* or *G. intraradices* improved the growth of plum (Camprubi *et al.*, 1993), peach and cherry (Pinochet *et al.*, 1995), quince (Calvet *et al.*, 1995) and apple (Pinochet *et al.*, 1993) rootstocks in soil infested with *Pratylenchus vulnus*. Effects of pre-plant inoculation of apple with AMF on population growth of the *Pratylenchus penetrans* was studied (Forge *et al.*, 2001). Greenhouse experiments did not provide evidence for reduced nematode reproduction in AMF inoculated plants. However, results from the field study suggested suppression of *P. penetrans* population growth, especially in fumigated plots recolonized by the nematode. After two growth seasons, populations of *P. penetrans* were significantly fewer in roots and root zone soil of *G. mosseae* inoculated plants than in non-inoculated plants. Similar results were reported in field microplot studies in suppression of *P. vulnus* by *G. mossae* on peach (Pinochet *et al.*, 1995) and apple rootstocks (Pinochet *et al.*, 1993). Reduced leakage of carbohydrate from root cells has been proposed as a mechanism to explain suppression of some fungal pathogens by AMF (Graham, 2001). Alterations in carbohydrate leakage could affect the attraction of *P. penetrans* to roots, and changes in the nutritional quality of cortical tissue induced by AMF could affect the fecundity of *P. penetrans*.

9.0 Conclusion

AMF are obligate biotrophs which establish a persistent interface between the plant root and the soil; it becomes perhaps the only stable microbial system in the rhizosphere. AMF-plant symbiosis cause qualitative and quantitative changes in rhizospheric microbial populations; the resulting microbial equilibria could influence the growth and health of plants. Effects of AMF interactions with plant pathogens and other microorganisms vary depending on several biotic and abiotic factors and in general, early AMF establishment have been shown to reduce damage caused by soil borne plant pathogens. This prophylactic ability of AMF could be exploited in cooperation with other rhizospheric and endophytic microbial angatonists to improve plant growth and health. Direct and indirect interactions have been suggested as mechanisms by which AMF can reduce the abundance of pathogenic fungi or nematode in plant roots and rhizosphere. These have generally been proposed in response to observations of negative correlations in the abundance of AMF structures and pathogenic microorganisms in roots and soil (Bodker *et al.*, 2002; Filion *et al.*, 2003). Recent research suggests that if competition between AMF and pathogenic organisms is occurring, it is for resources other than just the occupation of space within the root system. Presumably, pathogenic organisms and AMF exploit common resources within the root, including infection sites, space, and photosynthate within the root. Among the potential mechanisms involved in the resistance of mycorrhizal systems, the induction of plant defenses is the most controversial. The regulatory mechanisms of AMF on host plant defense response have not been elucidated, regardless whether this defense responses is of a specific activation, or this response play a unique role on the mycorrhizal symbionts formation process. Recent advances regarding signaling processes in mutualistic and pathogenic associations are expanding our understanding of plant interactions with their environment. During mycorrhiza formation, modulation of plant defense responses occurs, potentially through crosstalk between SA and JA-dependent signaling pathways. AMF induced resistance in plants correlates with induction of JA-dependent plant defenses associated with priming that activates defense mechanisms upon attack. This type of natural defense may be among the reasons to explain why root associations with AMF have been conserved during evolution and are widespread among plant species. AMF and their biocontrol efficiency was very well studied in annual crops and based on the knowledge gained over decades, consistent efforts may be made to utilize them in control of plant pathogens in fruit crops.

10.0 Future strategies

The challenges to achieve biocontrol of plant pathogens through use of AMF include the obligate nature of AMF, limited understanding of the mechanisms involved, and the role of environmental factors in these interactions. AMF, being obligate biotrophs, show a genetic, cellular and physiological complexity that makes the study of their biology as well as their effective agronomical exploitation rather difficult. With the improvement in research techniques, especially the ongoing development of plant and fungi genome sequencing technology and the raise of functional genomics, it will be possible to study in detail the relationship between a certain symbiosis related gene and plant defense response regulation. Latest developments in 'Omics' tools constitute a powerful means of describing the complexity of plants and soil borne microorganisms. The increasing availability of 'omics' data on mycorrhiza and of computational tools that allow systems biology approaches represents a step forward in the understanding of AMF symbiosis. Many numbers of genes that are differentially regulated upon AMF inoculation was observed (Salvioli and Bonfonte, 2013).

An important advance in AMF research is the identification of new 'myc-' mutants in different plant species will uncover key steps in symbiosis regulation. Application of mutants will help in finding many key signaling molecules that work during the symbiotic interaction between fungi and plant in general and understanding plant defense response upon mycorrhiza formation process in particular. This will provide an unprecedented opportunity for research into mycorrhizal molecular biology, the interaction between symbiotic partners, and allowing the underlying mechanisms to be gradually uncovered (Song *et al.*, 2011). Exciting challenges remain ahead, such as the identification of key regulators in defense modulation during AMF symbiosis. Future advances should allow identification of marker genes or metabolites associated with induced resistance, as well as generation of predictive models concerning the outcome of particular interactions (Pozo and Azcón-Aguilar, 2007). Since mycorrhizal establishment alters the volatile emission, and volatiles have been shown to prime distal plant parts or even neighbouring plants for a faster induction of defense responses (Heil and Ton, 2008), it remains to be determined if changes in volatiles in mycorrhizal plants also prime neighbouring plants for efficient activation of defense against attackers.

In spite of the nature of obligate biotrophs, methods for the efficient production of large scale AMF are now available (Ijdo *et al.*, 2011). Further, the advances that have been made in AMF detection and quantification now lead to a better inocula characterization and standardization, as well as a more reliable estimation of the presence of AMF in both the soil and mycorrhizal roots (Stockinger *et al.*, 2010). These technologies along with the knowledge gained over decades of use of AMF as bio-protectant in annual crops will aid in the better deployment of AMF to control plant diseases in fruit crops.

Plate 9.1. Field grown Banana Cv Neypoovan plants preinoculated with AMF and *Trichoderma* and challenged with the pathogen show resistance to *Fusarium oxysporum* f. sp. cubense **(left) compared to plants treated with pathogen alone onthe right (Sukhada Mohandas, NATP report 2004).**

A   B   C

Plate 9.2. Longitudinal sections of the rhizome of banana inoculated with *Fusarium oxysporum f.sp.*cubense after 40 days of inoculation with acthnomycetes. A. Control uninoculated plant B- *G.mosseae* colonizing *Streptomyces avermitilis* inoculated plant C -*G.mosseae* colonizing *Streptomyces cinnamonensis* inoculated plant (Source: Poovarasan, *et al.,*2013b)

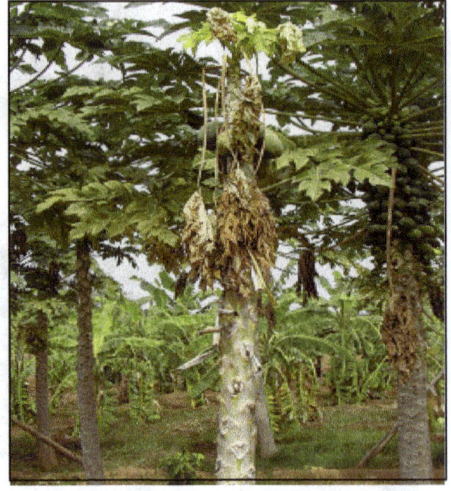

Plate 9. 3: Papaya Plants precolonized with AMF +*Trichderma* and challenged with *Phytophthora parasitica*, (Only Pathogen treated control on the right) (Source: Sukhada Mohandas, NATP Report 2005).

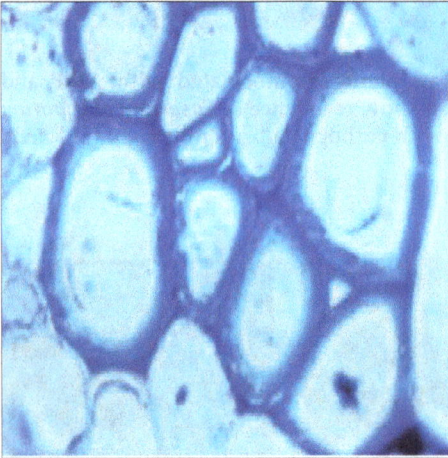

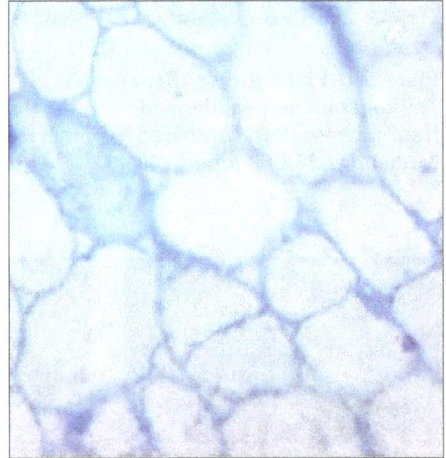

Plate 9.4. Anatomical changes observed in banana roots treated with AMF and Trichoderma. (A) Ultra thin (90 nm) transverse section (TS) (400) of uninoculated control banana roots and ultra thin TS of banana roots inoculated with *G. mosseae and T. harzianum* showing densely stained amorphous material (AM) resulting in wall thickenings (B) . Control on the right (B) (Source: Sukhada Mohandas (2005), NATP report 2005.

References

1. Akhtar, M.S and Siddiqui, Z.A., 2008.Arbuscular mycorrhizal fungi as potential bioprotectants against plant pathogens. In: Mycorrhizae: Sustainable Agriculture and Forestry, Edited by Siddiqui, Z.A., Akhtar, M.S and Futai ,K., Springer, Dordrecht, The Netherlands. pp. 61-98.

2. Andrade, G., Mihara, K.L., Linderman, R.G and Bethlenfalvay, G.J., 1998. Soil aggregation status and rhizobacteria in the mycorrhizosphere. *Plant and Soil.* 202, 89-96.

3. Artursson, V., Finlay, R. D and Jansson, J. K., 2006. Interactions between arbuscular mycorrhizal fungi and bacteria and their potential for stimulating plant growth. *Environmental Microbiology.* 8, 1-10.

4. Ausubel, F.M., 2005. Are innate immune signaling pathways in plants and animals conserved?.*Nature Immunology.* 6, 973-979.

5. Azaizeh, H.A., Marschner, H.,Romheld, V and Wittenmayer, L., 1995. Effects of a vesicular-arbuscular mycorrhizal fungus and other soil microorganisms on growth, mineral nutrient acquisition and root exudation of soil grown maize plants. *Mycorrhiza* .5, 321-327.

6. Azcón-Aguilar, C and Barea, J.M., 1997. Arbuscular mycorrhizas and biological control of soil-borne plant pathogens - An overview of the mechanisms involved. *Mycorrhiza.* 6, 57-464.

7. Azcon-Aguilar, C., Jaizme-Vega, M.C and Calvet, C., 2002. The Contribution of Arbuscular Mycorrhizal Fungi for Bioremediation. In: Gianinazzi, S., Schuepp, H., Barea, J.M., and Haselwandter, K. (Eds.), Mycorrhizal Technology in Agriculture-From Genes to Bioproducts. Birkhauser Verlag, Berlin, ISBN-10: 0-89054-245-71, pp. 187-197.

8. Baltruschat, H and Schonbeck, F., 1975. Studies on the influence of endotrophicmycorrhiza on the infection of tobacco by *Thielaviopsis basicola. Journal of Phytopathology.* 84, 172-188.

9. Baker, K. F and Cook, R. J., 1982. Biocontrol of Pathogens. *The American Phytopathological Society, St. Paul, MN.*

10. Bagyaraj, D.J., 1984. Biological interaction with VA mycorrhizal fungi. In: Powell, C.L. Bagyaraj, D.J. (Eds.), VA Mycorrhiza, *CRC Press*, Boca Raton, Florida. 131-153.

11. Balestrini, R., Romera, C., Puigdomenech, P and Bonfonte, P., 1994, Location of a cell-wall hydroxyproline-rich glycoprotein, cellulose and -1,3 glucans in apical and differentiated regions of maize mycorrhizal roots. *Planta.*195, 201-209.

12. Bansal, M and Mukerji, K.G., 1994. Positive correlation between AM-induced changes in root exudation and mycorrhizosphere mycoflora. *Mycorrhiza.* 5, 39-44.

13. Bansal, M and Mukerji, K. G., 1996. Root Exudates and its Rhizosphere Biology. In: Concepts in Applied Microbiology and Biotechnology, Mukerji, K.G., V.P. Singh and S. Dwivedi (Eds.). *Adita Books Private Ltd.,* New Delhi, pp: 79-119.

14. Barea, J. M., Pozo, M. J., Azcon, R and Azcon-Aguilar, C., 2005.Microbial co-operation in the rhizosphere. *Journal of Experimental Botany.*56, 1761-1778.

15. Badri, D.V and Vivanco, J.M., 2009. Regulation and function of root exudates. *Plant, Cell & Environment.* 32, 666-681.

16. Batlle, A., Laviña, A., Sabaté, J., Camprubí, A., Estaún, V and Calvet, C., 2011. Tolerance increase to *Candidatus phytoplasma prunorum* in mycorrhizal plums fruit trees. *Bulletin of Insectology.*64, 125-126.

17. Becker, W. N., 1976. Quantification of onion vesicular-arbuscular mycorrhizae and their resistance to *Pyrenochae taterrestris. Ph.D. dissertataion,* University of Illinois, Urbana, IL.

18. Berta, G., Fusconi, A and Trotta, A., 1993. VA mycorrhizal infection and the morphology of root systems. *Environmental and Experimental Botany.* 33, 159-173.

19. Benhamou, N., Fortin, J. A., Hamel, C., St Arnaud, M and Shatilla, A., 1994. Resistance responses of mycorrhizal Ri TDNA- transformed carrot roots to infection by *Fusarium oxysporum* f. sp. *chrysanthemi. Phytopathology.* 84, 958-968.

20. Bharat, N.K and Bhardwaj, L.N., 2001. Interactions between VA-mycorrhiza and *Dematophora necatrix* and their effect on apple seedlings. *Indian Journal of Plant Pathology.* 19, 47-51.

21. Bowen, G. D., 1980. Misconceptions, concepts and approaches in rhizospheric biology, In: *Contemporary Microbial Ecology,* eds., Ellwood, D. C., Hedger, J. N., Lathem, M. J., Lynch, J. M., and Slater, J. H., Academic, London, pp. 283-304.

22. Bodker, L., Kjoller, R., Kristensen, K and Rosendahl, S., 2002. Interactions between indigenous arbuscular mycorrhizal fungi and *Aphanomyces euteiches* in field grown pea. *Mycorrhiza.* 12, 7-12.

23. Bonfante, P and Anca, I.A., 2009. Plants, mycorrhizal fungi, and bacteria: a network of interactions. *Annual Review of Microbiology.* 63, 363–383.

24. Brundrett, M.C., 2002. Co-evolution of roots and mycorrhizas of land plants. *New Phytologist.* 154, 275-304.

25. Burke, D.J., Hamerlynck, E.P and Hahn, D., 2002. Interactions among plant species and microorganisms in salt marsh sediments. *Applied and environmental microbiology.* 68, 1157-1164.

26. Caron, M., Fortin, J. A and Richard, C., 1985.Influence of substrate on the interaction of *Glomus intraradices* and *Fusarium oxysporum* f. sp. *radicis-lycopersicio*n tomatoes. *Plant and Soil .*87, 233-239.

27. Caron, M., Fortin, J. A and Richard, C., 1986. Effect of phosphorus concentration and *Glomus intraradices* on *Fusarium* crown and root-rot of tomatoes. *Phytopathology*.76, 942-946.

28. Camprubi, A., Pinochet, J., Calvet, C and Estaún, V., 1993. Effects of the root lesion nematode *Pratylenchus vulnus* and the vesicular arbuscular mycorrhizal fungus *Glomus mosseae* on the growth of three plum rootstocks. *Plant and Soil*. 153, 223-229.

29. Calvet, C., Pinochet, J., Camprubí, A and Fernandez, C., 1995. Increased tolerance to the root-lesion nematode *Pratylenchus vulnus* in mycorrhizal micropropagated BA-29 quince rootstock. *Mycorrhiza*. 5, 253–258.

30. Camprubí, A., Estaún, V., Nogales, A., García-Figureueres, F., Pitet, M and Calvet, C., 2008. Response of the grapevine rootstock Richter 110 to inoculation with native and selected arbuscular mycorrhizal fungi and growth performance in a replant vineyard. *Mycorrhiza*. 18, 211-216.

31. Camprubi, A., Estaun, V., El- Bakali, M.A., Garcia-Figureueres, F.and Calvet, C.,2007. Alternative strawberry production using solarization, metham sodium and beneficial soil microbes as plant protection methods. *Agron. Sustain. Dev.* 27, 179-184.

32. Campos-Soriano, L., García-Martínez, J and Segundo, B. S., 2012. The arbuscular mycorrhizal symbiosis promotes the systemic induction of regulatory defense-related genes in rice leaves and confers resistance to pathogen infection. *Molecular Plant Pathology*. doi:10.1111/j.1364-3703.2011.00773.x.

33. Cameron, D.D., Neal, A.L., van Wees, S.C and Ton, J., 2013. Mycorrhiza-induced resistance: more than the sum of its parts. *Trends in Plant Science*. 18, 539-545.

34. Chandanie, W., Kubota, M and Hyakumachi, M., 2006. Interactions between plant growth promoting fungi and arbuscular mycorrhizal fungus *Glomus mosseae* and induction of systemic resistance to anthracnose disease in cucumber. *Plant and Soil*. 286, 209-217.

35. Chandanie, W. A., Kubota, M and Hyakumachi, M., 2009. Interactions between the arbuscular mycorrhizal fungus *Glomus mosseae* and plant growth-promoting fungi and their significance for enhancing plant growth and suppressing damping-off of cucumber (*Cucumis sativus* L.). *Applied Soil Ecology*. 41, 336-341.

36. Clark, R. B and Zeto, S. K., 2000. Mineral acquisition by arbuscular mycorrhizal plants. *Journal of Plant Nutrition*. 23, 867-902.

37. Cordier, C., Gianinazzi, S and Gianinazzi-Pearson, V., 1996. Colonization patterns of root tissues by *Phytophthora nicotiana* var. parasitica related to reduced disease in mycorrhizal tomato. *Plant and Soil*. 185, 223-232.

38. Cordier, C., Pozo, M. J., Barea, J. M., Gianinazzi, S and Gianinazzi-Pearson, V., 1998. Cell defense responses associated with localized and systemic

resistance to *Phytophthora parasitica* induced in tomato by an arbuscular mycorrhizal fungus. *Molecular Plant-Microbe Interactions*. 11, 1017-1028.

39. Conrath, U., Beckers, G. J. M., Flors, V., Garcia-Agustin, P., Jakab, G., Mauch, F., Newman, M. A., Pieterse, C. M. J., Poinssot, B., Pozo, M. J., Pugin, A., Schaffrath, U., Ton, J., Wendehenne, D., Zimmerli, L and Mauch-Mani, B., 2006. Priming: Getting ready for battle. *Molecular Plant-Microbe Interactions*. 19, 1062-1071.

40. Davis, R.M and Menge, J.A., 1980. Influence of *Glomus fasiculatum* and soil phosphorus on *Phytophthora* root rot of citrus. *Phytopathology*. 70, 447-452.

41. Dehne, H. W., Schönbeck, F and Baltruschat, H., 1978. Untersuchungenzumeinfluss der endotrophen Mycorrhiza auf Pflanzenkrankheiten: 3. Chitinase-aktivitat und ornithinzyklus (The influence of endo trophic mycorrhiza on plant diseases: 3 chitinase-activity and ornithinecycle). *Journal for Plant Diseases and Plant Protection*. 85, 666-678.

42. Dehne, H. W and Schönbeck, F., 1979. Untersuchungenzumeinfluss der endotrophenMycorrhiza auf Pflanzenkrankheiten: II. Phenols to ffwechsel und lignifizierung. *Journal ofPhytopathology*. 95, 210-216.

43. Dehne, H.W., 1982. Interaction between vesicular-arbuscular mycorrhizal fungi and plant pathogens. *Phytopathology*. 72, 1115-1119.

44. Declerck, S., Risede, J. M.,Rufyikiri, G and Delvaux, B., 2002. Effects of arbuscular mycorrhizal fungi on the severity of root rot of bananas caused by *Cylindrocladium spathiphylli*. *Plant Pathology*. 51, 109-115.

45. De La Noval, B., Pérez, E., Martínez, B., León, O., Martínezgallardo, N and Délano-Frier, J., 2007. Exogenous systemin has a contrasting effect on disease resistance in mycorrhizal tomato (*Solanum lycopersicum*) plants infected with necrotrophic or hemibiotrophic pathogens. *Mycorrhiza*. 17, 449-460.

46. De Román, M., Fernández, I., Wyatt, T., Sahrawy, M., Heil, M and Pozo, M. J., 2011. Elicitation of foliar resistance mechanisms transiently impairs root association with arbuscular mycorrhizal fungi. *Journal of Ecology*. 99, 36-45.

47. Elsen, A., Gervacio, D., Swennen, R and De Waele, D., 2008. AMF-induced biocontrol against plant parasitic nematodes in *Musa* sp.: a systemic effect. *Mycorrhiza*. 18, 251-256.

48. Fester, T and Hause, G., 2005. Accumulation of reactive oxygen species in arbuscular mycorrhizal roots. *Mycorrhiza*. 15, 373-379.

49. Fester, T. and Sawers, R., 2011. Progress and challenges in agricultural applications of arbuscular mycorrhizal fungi. *Critical Reviews in Plant Sciences*. 30, 459-470.

50. Fitter, A. H and Garbaye, J., 1994. Interactions between mycorrhizal fungi and other soil organisms. *Plant and Soil*.159, 123-132.

51. Filion, M., St-Arnaud, M and Fortin, J.A., 1999. Direct interaction between the arbuscular mycorrhizal fungus *Glomus intraradices* and different rhizosphere microorganisms. *New Phytologist*. 141, 525-533.

52. Filion, M., St-Arnaud, M. and Jabaji-Hare, S.H., 2003.Quantification of *Fusarium solani* f. sp *phaseoli* in mycorrhizal bean plants and surrounding mycorrhizosphere soil using real-time polymerase chain reaction and direct isolations on selective media. *Phytopathology*. 93, 229-235.

53. Finlay, R.D., 2008. Ecological aspects of mycorrhizal symbiosis: with special emphasis on the functional diversity of interactions involving the extraradical mycelium. *Journal of Experimental Botany*. 59, 1115–1126.

54. Forge, T., Muehlchen, A., Hackenberg, C., Neilsen, G and Vrain, T., 2001. Effects of preplant inoculation of apple (*Malus domestica* Borkh.) with arbuscular mycorrhizal fungi on population growth of the root-lesion nematode, *Pratylenchus penetrans*. *Plant and Soil*. 236, 185-196.

55. Frey-Klett, P., Garbaye, J and Tarkka, M., 2007. The mycorrhiza helper bacteria revisited. *New Phytologist*. 176, 22-36.

56. Fritz, M., Jakobsen, I., Lyngkjær, M. F., Thordal-Christensen, H and Pons-Kühnemann, J., 2006. Arbuscular mycorrhiza reduces susceptibility of tomato to *Alternaria solani*. *Mycorrhiza*.16, 413-419.

57. Franco-Correa, M., Quintana, A., Duque, C., Suarez, C., Rodriguez, M.X and Barea, J.M., 2010. Evaluation of actinomycete strains for key traits related with plant growth promotion and mycorrhiza helping activities. *Applied Soil Ecology*. 45, 209-217.

58. García-Garrido, J. M and Ocampo, J. A., 2002. Regulation of the plant defense response in arbuscular mycorrhizal symbiosis. *Journal of Experimental Botany*. 53, 1377-1386.

59. Gernns, H., Von Alten, H and Poehling, H. M., 2001. Arbuscular mycorrhiza increased the activity of a biotrophic leaf pathogen – Is a compensation possible, *Mycorrhiza*. 11, 237-243.

60. Gianinazzi-Pearson, V., Gollotte, A., Lherminier, J., Tisserant, B., Franken, P., Dumas- Gaudot, E., Lemoine, M. C., van Tuinen, D and Gianinazzi, S., 1995. Cellular and molecular approaches in the characterization of symbiotic events in functional arbuscular associations. *Canadian Journal of Botany*. 73, S526-S532.

61. Gianinazzi-Pearson, V., Dumas-Gaudot, E., Gollotte, A., Tahiri-Al-aoui, A and Gianinazzi, S., 1996. Cellular and molecular defense-related root responses to invasion by arbuscular mycorrhizal fungi. *New Phytologist*. 133, 45-57.

62. Goodman, R. N., Kiraly, Z and Zaitlin, M., 1967.The biochemistry and physiology of infections. In: Plant Disease. Van Nostrand, Princeton, NJ.

63. Gosling, P., Hodge, A., Goodlass, G and Bending, G.D., 2006.Arbuscular mycorrhizal fungi and organic farming. *Agriculture, Ecosystems and Environment*. 113, 17-35.

64. Graham, J. H., Leonard, R. T and Menge, J. A., 1981. Membrane-mediated decreases in root exudation responsible for phosphorus inhibition of vesicular-arbuscular mycorrhiza formation. *Plant Physiology.* 68, 548-552.

65. Graham, J. H and Menge, J. A., 1982. Influence of vesicular-arbuscular mycorrhizal fungi and soil phosporus on take-all disease of wheat. *Phytopathology.*72, 95-96.

66. Graham, J.H., 1982. Effect of citrus root exudates on germination of chlamydospores of vesicular-arbuscular mycorrhizal fungus *Glomus epigaeum. Mycologia.* 74, 831-835.

67. Graham, J. H and Egel, D. S., 1988. *Phytophthora* root rot development on mycorrhizal and phosphorous fertilized non-mycorrhizal sweet orange seedlings. *Plant Disease.*72, 611-614.

68. Gryndler, M., 2000. Interactions of arbuscular mycorrhizal fungi with other soil organisms. In: Kapulnik, Y., and Douds, D. D. (Eds.), Arbuscular Mycorrhizas: *Molecular Biology and Physiology.* Kluwer Press. pp. 239-262.

69. Graham, J. H., 2001. What do root pathogens see in mycorrhizas. *New Phytologist.* 149, 357-359.

70. Güimil, S., Chang, H.-S., Zhu, T., Sesma, A., Osbourn, A., Roux, C., Ioannidis, V., Oakeley, E. J., Docquier, M., Descombes, P., Briggs, S. P and Paszkowski, U., 2005. Comparative transcriptomics of rice reveals an ancient pattern of response to microbial colonization. *Proceedings of the National Academy of Sciences of the Unites States of America.* 102, 8066-8070.

71. Hayman, D. S., 1978. Endomycorrhizas. In: *Interactions Between Non-pathogenic Soil Microorganisms and Plants,* eds. Dommergues, Y. R., and Kurupa, S. V., Elsevier, Amsterdam.

72. Harrison, M. J., 1999. Molecular and cellular aspects of the arbuscular mycorrhizal symbiosis. *Annual Review of Plant Physiology and Plant Molecular Biology.* 50, 361-89.

73. Harrier, L. A and Watson, C. A., 2004. The potential role of arbuscular mycorrhizal (AM) fungi in the bioprotection of plants against soil-borne pathogens in organic and/or other sustainable farming systems. *Pest Management Science.* 60, 149-157.

74. Hause, B and Fester, T., 2005. Molecular and cell biology of arbuscular mycorrhizal symbiosis. *Planta,* 221, 184-196.

75. Hause, B., Mrosk, C., Isayenkov, S and Strack, D., 2007. Jasmonates in arbuscular mycorrhizal interactions. *Phytochemistry.* 68, 101-110.

76. Hao, Z., Fayolle, L., Van Tuinen, D., Chatagnier, O., Li, X., Gianinazzi, S and Gianinazzi-Pearson, V., 2012. Local and systemic mycorrhiza-induced protection against the ectoparasitic nematode *Xiphinema* index involves priming of defense gene responses in grapevine. *Journal of Experimental Botany.*doi:10.1093/jxb/ers046.

77. Heil, M and Ton, J., 2008. Long-distance signaling in plant defense. *Trends in Plant Science.* 13, 264-272.

78. Hernández-Montiel, L.G., Rueda-Puente, E.O., Cardoba-Matson, M.V., Holguín-Peña, J.R and Zulueta-Rodríguez, R., 2013. Mutualistic interaction of rhizobacteria with arbuscular mycorrhizal fungi and its antagonistic effect on Fusarium oxysporum in *Carica papaya* seedlings. *Crop Protection.* 47, 61-66.

79. Hiltner, L., 1904. UberneuereErlahrungen und Probleme auf demGebiet der bodenbakteriologie und unterbesondererBruksichtingung der Crundungung und Brache (On recent insights and problems in the area of soil bacteriology under special consideration of the use of green manure and fallowing). *Arb. Dtsch. Landwirt. Ges.* 98, 59-78.

80. Ingham, R.E., 1988. Interactions between nematodes and vesicular arbuscular mycorrhizae. *Agriculture, Ecosystems and Environment.* 24, 169-182.

81. Ijdo, M.,Cranenbrouck, S and Declerck, S., 2011. Methods for large-scale production of AM fungi: past, present, and future, *Mycorrhiza.* 21, 1-16.

82. Jaiti, F., Meddich, A and El Hadrami, I., 2007. Effectiveness of arbuscular mycorrhizal fungi in the protection of date palm (*Phoenix dactylifera* L.) against bayoud disease. *Physiological and Molecular Plant Pathology.* 71, 166-173.

83. Johnson, D., Krsek, M., Weillington, E.M., Stott, A.W., Cole, L., Bardgett, R.D., Read, D.J and Leake, J.R., 2005. Soil invertebrates disrupt carbon flow through fungal networks. *Science.* 309, 1047.

84. Jones, J. D. G and Dangl, J. L., 2006. The plant immune system. *Nature.* 444, 323-329.

85. Jung, S. C., Martinez-Medina, A., Lopez-Raez, J. A and Pozo, M. J., 2012. Mycorrhiza-induced resistance and priming of plant defenses.*Journal of Chemical Ecology.* 38, 651-664.

86. Kaplan, D. T., Keen, N. T and Thompson, I. J., 1980. Association of glyceollin with incompatible response of soybean roots to *Meloidogyne incognita. Physiological Plant Pathology.* 16, 309-318.

87. Kamińska, M., Klamkowski, K., Berniak, H and Sowik, I., 2010. Response of mycorrhizal periwinkle plants to aster yellows phytoplasma infection. *Mycorrhiza.* 20, 161-166.

88. Khaosaad, T., García-Garrido, J. M., Steinkellner, S and Vierheilig, H., 2007. Take-all disease is systemically reduced in roots of mycorrhizal barley plants. *Soil Biology and Biochemistry.* 39, 727-34.

89. Kiers E. T., Duhamel, M.,Yugandgar, Y., Mensah, J. A., Franken, O.,Verbruggen, E.,Felbaum, C. R.,Kowalchuk,G. A., Hart,M. M.,Bago, A., Palmer,T. M., West,S. A.,Vandenkoornhuyse, P., Jansa,J and Bücking, H., 2011. Reciprocal rewards stabilize cooperation in the mycorrhizal symbiosis. *Scienc..*333, 880-882.

90. Kloppholz, S., Kuhn, H and Requena, N., 2011. A secreted fungal effector of *Glomus intraradices* promotes symbiotic biotrophy. *Current Biology*. 21, 1204-1209.

91. Kohler, J., Caravaca, F., Carrasco, L and Roldán, A., 2007. Interactions between a plant growth-promoting rhizobacterium, an AM fungus and a phosphate-solubilising fungus in the rhizosphere of *Lactuca sativa*. *Applied Soil Ecology*. 35, 480-487.

92. Krishna, K. R and Bagyaraj, D. J., 1986, Phenoilcs of mycorrhizal and uninfected groundnut var. MGS-7. *Current Research*.15, 51-52.

93. Krishna, H., Singh, S.K., Sharma, R.R., Khawale, R.N., Grover, M and Patel, V.B., 2005. Biochemical changes in micropropagated grape (*Vitis vinifera* L.) plantlets due to arbuscular mycorrhizal fungi (AMF) inoculation during ex vitro acclimatization. *Scientia Horticulturae*. 106, 554-567.

94. Krishna, H., Das, B., Attri, B.L., Grover, M and Ahmed, N., 2010. Suppression of *Botryosphaeria* canker of apple by arbuscular mycorrhizal fungi. *Crop Protection*. 29, 1049-1054.

95. Lambais, M. R and Mehdy, M. C., 1993. Suppression of endochitinase-1,3-endoglucanase, and chalcone isomerase expression in bean VAM roots under different soil phosphate conditions. *Molecular Plant-Microbe Interaction*. 1, 75-83.

96. Larimer, A., Bever, J and Clay, K., 2010. The interactive effects of plant microbial symbionts: a review and meta-analysis. *Symbiosis*. 51, 139-148.

97. Lee, C. S., Lee, Y. J and Jeun, Y. C., 2005. Observations of infection structures on the leaves of cucumber plants pre-treated with arbuscular mycorrhiza *Glomus intraradices* after challenge inoculation with *Colletotrichum orbiculare*. *Journal of Plant Pathology*. 21, 237-243.

98. Linderman, R. G., 1985, Microbial interaction in the mycorrhizosphere. In: *Proc. 6th N. Am Conf. on Mycorrhizae*, ed. Molina, R., pp. 117–120.

99. Linderman, R.G., 1988. Mycorrhizal interactions with the rhizosphere microflora: The mycorrhizosphere effect. *Phytopathology*. 78, 366-371.

100. Linderman, R. G., 1992. VA mycorrhizae and soil microbial interactions. In: Bethelenfalvay, G. J., and Linderman, R. G. (Eds), *Mycorrhizae in Sustainable Agriculture*. ASA Special Publication No.54. Madison, WI, pp. 45–70.

101. Linderman, R.G., 1994. Role of VAM Fungi in Biocontrol. In: Pfleger, F.L., and Linderman, R.G. (Eds.), *Mycorrhizae and Plant Health. The American Phytopathological Society*. St. Paul, MN, USA, ISBN: 0-89054-158-2, pp. 1-27.

102. Liu, J., Blaylock, L. A., Endre, G., Cho, J., Town, C. D., Vandenbosch, K. A and Harrison, M. J., 2003. Transcript profiling coupled with spatial expression analyses reveals genes involved in distinct developmental stages of an arbuscular mycorrhizal symbiosis. *The Plant Cell*. 15, 2106-2123.

103. Li, H.-Y., Yang, G.-D., Shu, H.-R., Yang, Y.-T., Ye, B.-X., Nishida, I and Zheng, C., 2006. Colonization by the arbuscular mycorrhizal fungus *Glomus versiforme* induces a defense response against the root-knot nematode *Meloidogyne incognita* in the grapevine (*Vitis amurensis* Rupr.), which includes transcriptional activation of the class III chitinase gene VCH3. *Plant and Cell Physiology*.47, 154-163.

104. Liu, J., Maldonado-Mendoza, I., Lopez-Meyer, M., Cheung, F., Town, C. D and Harrison, M. J., 2007. Arbuscular mycorrhizal symbiosis is accompanied by local and systemic alterations in gene expression and an increase in disease resistance in the shoots. *The Plant Journal*. 50, 529-544.

105. Lioussanne, L., Jolicoeur, M and St-Arnaud, M., 2008. Mycorrhizal colonization with *Glomus intraradices* and development stage of transformed tomato roots significantly modify the chemotactic response of zoospores of the pathogen *Phytophthora nicotianae*. *Soil Biology and Biochemistry*. 40, 2217-2224.

106. Lin, S. C., Wang, C.Y and Su, C.C., 2012. Using arbuscular mycorrhizal fungus and other microorganisms for control of *Fusarium* wilt of banana. *Journal of Taiwan Agricultural Research*. 61, 241-249.

107. López-Ráez, J. A., Flors, V., García, J. M.and Pozo, M. J., 2010. AM symbiosis alters phenolic acid content in tomato roots. *Plant Signaling and Behavior*. 5, 1138-1140.

108. López-Ráez, J. A., Pozo, M. J and García-Garrido, J. M., 2011. Strigolactones: a cry for help in the rhizosphere. *Botany*.89, 513-522.

109. Lumini, E., Bianciotto, V., Jargeat, P., Novero, M., Salvioli, A., Faccio, A., Becard, G and Bonfante, P., 2007. Presymbiotic growth and sporal morphology are affected in the arbuscular mycorrhizal fungus *Gigaspora margarita* cured of its endobacteria. *Cellular Microbiology*. 9, 1716-1729.

110. Ludwig-Müller, J. 2010. Hormonal responses in host plants triggered by arbuscular mycorrhizal fungi., in H. Koltai and Y. Kapulnik (eds.), *Arbuscular mycorrhizas: Physiology and Function*. Springer Netherlands, Dordrecht. 169–190

111. Marschner, P., Crowley, D. E and Lieberei, R., 2001. Arbuscular mycorrhizal infection changes the bacterial 16s rDNA community composition in the rhizosphere of maize. *Mycorrhiza*. 11, 297-302.

112. Martínez-Medina, A., Pascual, J.A., Pérez-Alfocea, F.,Albacete, A and Roldán, A., 2010. *Trichoderma harzianum* and *Glomus intraradices* modify the hormone disruption induced by *Fusarium oxysporum* infection in melon plants. *Phytopathology*.100, 682-688.

113. Meyer, J. R and Linderman, R. G., 1986. Selective influences on populations of rhizosphere or rhizoplane bacteria and actinomycetes by mycorrhizas formed by *Glomus fasciculatum*. *Soil Biology and Biochemistry*. 18, 191-196.

114. Mehta, P and Bharat, N.K., 2013. Effect of indigenous arbuscular - mycorrhiza (*Glomus* spp) on apple (*Malus domestica*) seedlings grown in replant disease soil. *Indian Journal of Agricultural Sciences.* 83, 1173-1178.

115. Millner, P.D and Wright, S.F., 2002. Tools for support of ecological research on arbuscular mycorrhizal fungi (Review article). *Symbiosis.* 33, 101-123.

116. Miransari, M., 2010. Contribution of arbuscular mycorrhizal symbiosis to plant growth under different types of soil stress. *Plant Biology.* 12, 563-569.

117. Mosse, B., 1962. The establishment of vesicular-arbuscular mycorrhiza under aseptic conditions. *Journal of General Microbiology.* 27, 509-520.

118. Morandi, D., Baily, J. A., and Gianinazzi-Pearson, V., 1984.Isoflavonoid accumulation in soybean roots infected with vesicular-arbuscular mycorrhizal fungi. *Physiological Plant Pathology.* 24, 357-364.

119. Morandi, D., 1987. VA mycorrhizae, nematodes, phosphorus and phytoalexins on soybean. In: *Mycorrhizae in the Next Decade, Practical Application and Research Priorities, eds.,*

120. Moraes, R.M., Zita, D.A., Bedir, E., Dayan, F.E., Lata, H., Khan, I., and Pereira, A.M.S., 2004. Arbuscular mycorrhiza improves acclimatization and increases lignin content of micropropagated mayapple (*Podophyllum peltatum* L.). *Plant Science.* 166, 23-29.

121. Moller, K., Kristensen, K., Yohalem, D and Larsen, J., 2009. Biological management of gray mold in pot roses by coinoculation of the biocontrol agent *Ulocladiu matrum* and the mycorrhizal fungus *Glomus mosseae. Biological Control.* 49,120-125.

122. Mukerji, K G and Ciancio, A., 2007. "Mycorrizae in the integrated pest and disease management". In *General concepts in integrated pest and disease management,* Edited by: Ciancio, A and Mukerji, K G. Heidelberg: Springer. 245–266.

123. Naumann, M., Schuessler, A and Bonfante, P., 2010. The obligate endobacteria of arbuscular mycorrhizal fungi are ancient heritable components related to the Mollicutes. *Indian Society for Microbial Ecology.* 4, 862-871.

124. Niranjan, R., Mohan, V and Rao, V. M., 2007. Effect of indole acetic acid on the synergistic interactions of *Bradyrhizobium* and *Glomus fasciculatum* on growth, nodulation, and nitrogen fixation of *Dalbergia sissoo* Roxb.*Arid Land Research and Management.* 21, 329-342.

125. Norman, J.R and Hooker, J.E., 2000. Sporulation of *Phytophthora fragaria* shows greater stimulation by exudates of non-mycorrhizal than by mycorrhizal strawberry roots. *Mycological Research.* 104, 1069-1073.

126. Olsson, P.A., Thingstrup, I., Jakobsen, I and Baath, E., 1999. Estimation of the biomass of arbuscular mycorrhizal fungi in linseed field. *Soil Biology and Biochemistry.* 31, 1879-1887.

127. Olawuyi, O.J., Odebode, A.C., Oyewole, I.O., Akanmu, A.O and Afolabi, O., 2013. Effect of arbuscular mycorrhizal fungi on Pythium aphanidermatum causing foot rot disease on pawpaw (*Carica papaya* L.) seedlings. *Archives of Phytopathology and Plant Protection.* 47, 185-193.

128. Pacovsky, R. S., Bethelenfalvay, G. J and Paul, E. A., 1986. Comparisons between P-fertilized and mycorrhizal plants. *Crop Science.* 16, 151-156.

129. Paszkowski, U., 2006. Mutualism and parasitism: the yin and yang of plant symbioses. *Current Opinion in Plant Biology.* 9, 364-370.

130. Pinochet, J., Camprubí, A and Calvet, C., 1993. Effects of the root lesion nematode *Pratylenchus vulnus* and the mycorrhizal fungus *Glomus mosseae* on the growth of EMLA-26 apple rootstocks. *Mycorrhiza.* 4, 79-83.

131. Pinochet, J., Calvet, C., Camprubí, A and Fernandez, C., 1995. Growth and nutritional response of Nemared peach rootstock infected with *Pratylenchus vulnus* and the mycorrhizal fungus *Glomus mosseae*. *Fundamental and Applied Nematology.* 18, 205-210.

132. Pinochet, J., Calvet, C., Camprubí, A and Fernandez, C., 1996. Interactions between migratory endoparasitic nematodes and arbuscular mycorrhizal fungi in perennial crops: *A review. Plant and Soil.* 185, 183-190.

133. Pivato, B., Offre, P., Marchelli, S., Barbonaglia, B., Mougel, C., Lemanceau P., 2009. Bacterial effects on arbuscular mycorrhizal fungi and mycorrhiza development as influenced by the bacteria, fungi, and host plant. *Mycorrhiza.* 19, 81-90.

134. Pozo, M. J., Azcón-Aguilar, C., Dumas-Gaudot, E and Barea, J. M., 1998. Chitosanase and chitinase activities in tomato roots during interactions with arbuscular mycorrhizal fungi or *Phytophthora parasitica*. *Journal of Experimental Botany.* 49 ,729-1739.

135. Pozo, M. J., Cordier, C., Dumas-Gaudot, E., Gianinazzi, S., Barea, J. M and Azcón-Aguilar, C., 2002. Localized versus systemic effect of arbuscular mycorrhizal fungi on defence responses to *Phytophthora infection* in tomato plants. *Journal of Experimental Botany.*53, 525-534.

136. Pozo, M. J and Azcón-Aguilar, C., 2007. Unraveling mycorrhiza induced resistance. *Current Opinion in Plant Biology.*10, 393-398.

137. Pozo, M. J., Jung, S. C., López-Ráez, J. A and Azcón-Aguilar, C., 2010. Impact of arbuscular mycorrhizal symbiosis on plant response to biotic stress: The role of plant defence mechanisms, pp. 193–207, in H. Koltai and Y. Kapulnik (eds.), Arbuscular mycorrhizas: Physiology and Function. Springer *Netherlands*, Dordrecht.

138. Poovarasan, S., Sukhada, M., Paneerselvam, P., Saritha, B. and Ajay, K.M., 2013a. Mycorrhizae colonizing actinomycetes promote plant growth and control bacterial blight disease of pomegranate (*Punica granatum* L. cv Bhagwa). *Crop Protection.* 53, 175-181.

139. Poovarasan, S., T. R Usharani, Paneerselvam, P. Saritha B, Sukhada Mohandas, 2013b.Evaluation of actinomycetes isolated from VAM fungi for the biological control of *Fusarium* wilt of banana. In Third International Science Congress (ISC-2013-IAFS-90) at Karunya University, Coimbatore, Tamil Nadu, India,pp-36.

140. Prasad, S. S. K., 1971. Effects of amino acids and plant growth substances on tomato and its root knot nematode *Meloidogyne incognita* Chitwood. M.Sc. (Agric.) thesis, University of Agricultural Sciences, Bangalore, India.

141. Raj, H and Sharma, S. D., 2009. Integration of soil solarization and chemical sterilization with beneficial microorganisms for the control of white root rot and growth of nursery apple. *Scientia Horticulturae*. 119, 126-131.

142. Reddy, P. P., 1974.Studies on the action of amino acids on the root-knot nematode *Meloidogyne incognita*. Ph.D. thesis, University of Agricultural Sciences Banglore, India.

143. Reid, C. P. P., 1984, Mycorrhizae: a root-soil interface in plant nutrition. In: *Microbial-Plant Interactions*, ASA Special Publication, Vol. 47, ed., R. L. Todd, and J. E. Giddens, pp. 29–50.

144. Rillig, M.C., Ramsey, P.W., Morris, S and Paul, E., 2003. Glomalin, an arbuscular-mycorrhizal fungal soil protein, responds to land-use change. *Plant and Soil*. 253, 293-299.

145. Rillig, M.C and Mummey, D.L., 2006. Tansley review – mycorrhizas and soil structure. *New Phytologist*. 171, 41-53.

146. Ridgway, H.J., Kandula, J and Stewart, A., 2008. Arbuscular mycorrhiza improve apple rootstock growth in soil conducive to specific apple replant disease. *New Zealand Plant Protection*. 61, 48-53.

147. Rosendahl, S., 1985. Interactions between the vesicular-arbuscular mycorrhizal fungus *Glomus intraradices* and *Aphanomyces euteiches* root rot of peas. *Journal of Phytopathology*. 114, 31-40.

148. Saldajeno, M. G. B and Hyakumachi, M., 2011. The plant growth promoting fungus *Fusarium equiseti* and the arbuscular mycorrhizal fungus *Glomus mosseae* stimulate plant growth and reduce severity of anthracnose and damping-off diseases in cucumber (*Cucumis sativus*) seedlings. *Annals of Applied Biology*. 159, 28-40.

149. Salvioli, A and Bonfante, P., 2013. Systems biology and "omics" tools: cooperation for next-generation mycorrhizal studies. *Plant Science*. 203-204, 107- 114.

150. Schonbeck, F., 1979. Endomycorrhiza in relation to plant disease. In: Schipper, B., and Gams, W.(Eds), *Soil Borne Plant Pathogens*. Academic. New York, pp. 271–280.

151. Schellenbaum, L., Berta, G., Ravolanirina, F., Tisserant, B., Gianinazzi, S and Fitter A. H., 1991. Infuence of endomycorrhizal infection on root morphology in micropropagated woody plant species (*Vitis vinifera* L.). *Annals of Botany*. 68, 135–141.

152. Secilia, J. and Bagyaraj, D.J., 1987. Bacteria and actinomycetes associated with pot cultures of vesicular-arbuscular mycorrhizas. *Canadian Journal of Microbiology*. 33, 1069-1073.

153. Sharma, M.P., Gaur, A and Mukerji, K.G., 2007. Arbuscular Mycorrhiza Mediated Plant Pathogen Interactions and the Mechanisms Involved. In: Sharma, M.P., Gaur, A., and Mukerji, K.G. (Eds.), *Biological Control of Plant Diseases*. Haworth Press. Binghamton, USA, pp. 47-63.

154. Sikora, R. A., 1978. Einfluss der endotrophenmykorrhiza (*Glomus mosseae*) auf daswirt-parasit-verhaltnis*Von Meloidogyne incognita* in tomaten. *Zeitschrift fur Pflanzenkrankheiten und Pflanzenschutz.*85, 197-202.

155. Singh, Y. P., Singh, R. S and Sitaramaiah, K., 1990. Mechanisms of resistance of mycorrhizal tomato against root-knot nematodes. In: *Current Trends in Mycorrhizal Research*, eds., Jalali, B. L., and Chand, H., *Proc. Nat. Conf. Mycorrh.*, H.A.U., Hisar, India, pp. 96-97.

156. Singh, R., Adholega, A and Mukerji, K.G., 2000. Mycorrhiza in Control of Soil Borne Pathogens. In: Mukerji, K.G., Chamola, B.P., and Singh, J. (Eds.), *Mycorrhizal Biology*. Kluwer Academic/Plenum Publishers. New York, USA, pp. 173-196.

157. Smith, S. E and Giananizzi-Pearson, V., 1988. Physiological interactions between symbionts in vesicular-arbuscular mycorrhizal plants. *Annual Review of Plant Physiology and Plant Molecular Biology*. 39, 221-244.

158. Smith, S.E and Read, D.J., 1997. *Mycorrhizal Symbiosis*. 2nd Edn. Academic Press. London, UK, ISBN: 0-12-652840-3, pp, 605.

159. Smith, S.E. and Read, D.J., 2008. Mineral Nutrition, Toxic Element Accumulation and Water Relations of Arbuscular Mycorrhizal Plants. In: *Mycorrhizal Symbiosis*. 3rd Edn. Academic Press. London, ISBN-10: 0123705266, pp. 145-148.

160. Smith, S.E., Facelli, E., Pope S., Smith, F.A., 2010. Plant performance in stressful environments: interpreting new and established knowledge of the roles of arbuscular mycorrhizas. *Plant and Soil*. 326, 3-20.

161. Smith S.E., Jakobsen I., Gronlund M., Smith F.A., 2011. Roles of arbuscular mycorrhizas in plant phosphorus nutrition: Interactions between pathways of phosphorus uptake in arbuscular mycorrhizal roots have important implications for understanding and manipulating plant phosphorus acquisition. *Plant Physiology*. 156, 1050-1057.

162. Smith, S.E and Smith, F.A., 2011. Roles of arbuscular mycorrhizas in plant nutrition and growth: new paradigms from cellular to ecosystem scales. *Annual Review of Plant Physiology and Plant Molecular Biology*. 62, 227-250

163. Soto, M. J., Domínguez-Ferreras, A., Perez-Mendoza, D., Sanjuán, J and Olivares, J., 2009. Mutualism versus pathogenesis: the give-and-take in plant–bacteria interactions. *Cellular Microbiology*. 11, 381-388.

164. Song, Y.Y.,Zeng, R.S., Xu, J. F., Li, J., Shen, X and Yihdego, W. G., 2010. Interplant communication of tomato plants through underground common mycorrhizal networks. *PLoS ONE* 5: doi:10.1371/journal.pone.0013324. s0013003.

165. Song, F., Song, G., Dong, A and Kong, X., 2011. Regulatory mechanisms of host plant defense responses to arbuscular mycorrhiza. *Acta Ecologica Sinica*. 31, 322-327.

166. Srivastava, A.K., Ahmed, R., Kumar, S and Sukhada, M., 2001. Role of VA-mycorrhiza in the management of wilt disease of guava in the alfisols of Chotanagpur. *Indian Phytopathology*. 54, 78-81.

167. Strack, D., Fester, T., Hause, B., Schliemann, W and Walter, M.H., 2003. Arbuscular mycorrhiza: Biological, chemical and molecular aspects. *Journal of Chemical Ecology*. 9, 1955-1979.

168. Stockinger, H., Krüger, M and Schüssler, A., 2010. DNA barcoding of arbuscular mycorrhizal fungi.*New Phytologist*. 187, 461-474.

169. Sukhada Mohandas (2005). Final report of the National Agricultural Technology project (NATP) entitled : Studies on the interactive effect of VAM and other beneficialorganisms with the pathogens causing panama wilt of banana and root rot of papaya. Submitted to NATP,Coordinator, New Delhi.

170. Sukhada Mohandas, Manjula, R and Rawal, R.D., Lakshmikantha, H.C.,Chakraborty, S., and Ramchandra, Y.L., 2010. Evaluation of Arbuscular mycorrhiza and other biocontrol agents in managing *Fusarium oxysporum* f. sp. *cubense* infection in banana cv. Neypoovan. *Biocontrol Science and Technology*. 20, 165-181.

171. Sukhada Mohandas, Manjula, R and Rawal, R.D., 2011. Evaluation of arbuscular mycorrhiza and other biocontrol agents against *Phytophthora parasitica* var. nicotianae infecting papaya (*Carica papaya* cv. Surya) and enumeration of pathogen population using immunotechniques. *Biological Control*. 58, 22-29.

172. Sukhada Mohandas, Poovarasan, S., Paneerselvam, P., Saritha, B., Upreti, K.K., Kamal, R and Sita, T, 2013. Guava (*Psidium guajava* L.) rhizosphere *Glomus mosseae* spores harbor actinomycetes with growth promoting and antifungal attributes. *Scientia Horticulturae*. 150, 371-376.

173. Suresh, C. K., 1980, Interaction between vesicular arbuscular mycorrhizae and root-knot nematodes in tomato. *M.Sc. (Agric.) thesis*, University of Agricultural Sciences, Banglore, India.

174. Suresh, C. K and Bagyaraj, D. J., 1984. Interaction between vesicular-arbuscular mycorrhizae and a root-knot nematode and its effect on growth and chemical composition on tomato. *Nematologia Mediterranea*. 12, 31-39.

175. Sylvia, D. M and Sinclair, W. A., 1983. Phenolic compounds of resistance to fungal pathogens induced in primary roots of Douglas-fir seedlings by the ectomycorrhizal fungus *Laccarislaccata*. *Phytopathology*.73, 390-397.

176. Thomma, B. P. H. J., Nürnberger, T and Joosten, M. H. A. J., 2011. Of PAMPs and effectors: The blurred PTI-ETI dichotomy. *The Plant Cell*. 23, 4-15.

177. Tinker, P.B., Durall, D.M and Jones, M.D., 1994. Carbon use efficiency in mycorrhizas: theory and sample calculations. *New Phytologist*. 128, 115-122.

178. Toljander, J.F., Paul, L., Lindahl, B.D., Elfstrand, M and Finlay, R.D., 2007. Influence of AM fungal exudates on bacterial community structure. *FEMS Microbiology Ecology* .61, 295-304.

179. Toussaint, J. P., 2007. Investigating physiological changes in the aerial parts of AM plants: What do we know and where should we be heading. *Mycorrhiza*.17, 349-353.

180. Tylka, G. L., Hussey, R. S and Roncadori, R. W., 1991. Interactions of vesicular-arbuscular mycorrhizal fungi, phosphorus and *Heterodera glycien*on soybean. *Journal of Nematology*. 23, 122-123.

181. Umesh, K. C., Krishnappa, K and Bagyaraj, D. J., 1988. Interaction of burrowing nematode, *Radopholus similis* (Cobb, 1983) Thorne, 1949, and VA mycorrhiza, *Glomus fasciculatum* (Thaxt.) Gerd. and Trappe, in banana (*Musa acuminate* Colla.). *Indian Journal of Nematology*. 18, 6-11.

182. Van Wees, S. C. M., Van Der Ent, S and Pieterse, C. M. J., 2008. Plant immune responses triggered by beneficial microbes. *Current Opinion in Plant Biology*. 11, 443-448.

183. Verhagen, B. W. M., Glazebrook, J., Zhu, T., Chang, H. S., Van Loon, L. C and Pieterse, C. M. J., 2004. The transcriptome of rhizobacteria-induced systemic resistance in *Arabidopsis*. *Molecular Plant-Microbe Interaction*. 17, 895-908.

184. Vigo, C., Norman J. R and Hooker, J. E., 2000. Biocontrol of pathogen *Phytophthora parasitica* by arbuscular mycorrhizal fungi is a consequence of effects on infection loci. *Plant Pathology*. 49, 509-514.

185. Volpin, H., Elkind, Y., Okon, Y and Kalpulnik, Y., 1994. A vesicular arbuscular mycorrhizal fungus (*Glomus intraradices*) induces defence response in alfalfa roots. *Plant Physiology*. 104, 683-689.

186. Vos, C., Claerhout, S., Mk AndAwire, R., Panis, B., De Waele, D and Elsen, A., 2011. Arbuscular mycorrhizal fungi reduce root-knot nematode penetration through altered root exudation of their host. *Plant and Soil*. 354, 335-345.

187. Walters, D and Heil, M., 2007. Costs and trade-offs associated with induced resistance. *Physiological and Molecular Plant Pathology*. 71, 3-17.

188. Whipps, J. M., 2004. Prospects and limitations for mycorrhizas in biocontrol of root pathogens. *Canadian Journal of Botany*. 82 ,1198-1227.

189. Wick, R. L and Moore, L. D., 1984. Histology of mycorrhizal and non-mycorrhizal *Ilex crenata*'Helleri' challenged by *Thielaviopsis basicola*. *Canadian Journal of Botany*. 6,146-150.

190. Wright, D. P., Read, D. J and Scholes, J. D., 1998. Mycorrhizal sink strength influences whole plant carbon balance of *Trifoliumrepens* L. *Plant Cell Environment*. 21, 881-891.

191. Yao, M. K., Désilets, H., Charles, M. T., Boulanger, R and Tweddell, R. J., 2003. Effect of mycorrhization on the accumulation of rishitin and solavetivone in potato plantlets challenged with *Rhizoctonia solani*. Mycorrhiza.13, 333-336.

192. Yan Li, H., Yang, G.D., Shu, H.R.,Yang, Y.T., Ye, B.X., Nishida, I., and Zheng, C.C., 2006. Colonization by the Arbuscular Mycorrhizal Fungus *Glomus versiforme* induces a Defense Response against the Root-knot Nematode Meloidogyne incognita in the Grapevine (*Vitis amurensis Rupr.*), Which Includes Transcriptional Activation of the Class III Chitinase Gene VCH3. *Plant Cell Physiol*. 47(1), 154–163.

193. Yousefi, Z., 2011. Effects of Arbuscular Mycorrhizal Fungi on Mildew Infection Percentage in Apple Seedling. *Journal of basics and applied scientific research*. 1, 2458-2461.

194. Zhao, M., Li, M and Liu, R.J., 2010. Effects of arbuscular mycorrhizae on microbial population and enzyme activity in replant soil used for watermelon production. *International Journal of Engineering, Science and Technology*. 2, 17-22.

195. Zhu, H. H and Yao, Q., 2004. Localized and systemic increase of phenols in tomato roots induced by *Glomus versiforme* inhibits *Ralstonia solanacearum*. *Journal of Phytopathology*. 152, 537-542.

Chapter 10

Phytoremediation and Heavy Metal (lOID) Toxicity Management

Sebastián Meier[1]*, Fernando Borie[1], Naser Khan[2,3], Gustavo Curaqueo[1], Jorge Medina[1], Pablo Cornejo[1],Nanthi Bolan[2],

[1]*Scientific and Technological Nucleus of Bioresources, Departamento de Ciencias Químicas, Universidad de La Frontera, P.O. Box 54-D, Temuco, Chile*

[2]*Centre for Environmental Risk Assessment and Remediation (CERAR), Building X, University Blvd., University of South Australia, Mawson Lakes SA 5095, Australia*

[3]*School of Natural and Built Environments, University of South Australia, Building P, Materials Lane, Mawson Lakes, SA 5095, Australia*

*Address for the correspondence: sebastianmeier@gmail.com

ABSTRACT

High concentrations of metals in soils cause negative effects on microorganisms and plants. However, some microorganisms can adapt to metal stress, and some are able to assist plants in metal polluted soils. Among them arbuscular mycorrhizal fungi (AMF) are the only ones, which provide a direct link between soil and roots. They are widely recognized as plant growth promoter that assists by increasing acquisition of several nutrients and also encouraging plant growth in soils contaminated by heavy metals (oids) through a series of mechanisms. This chapter reports the role of AMF in phytoremediation of soils contaminated by metals (oids).

Keywords: *Heavy metal; arbuscular mycorrhizal fungi; metal tolerance echanisms; phytostabilization; phytoextraction.*

1.0 Introduction

Heavy metals have specific density equal to or greater than 5 g mL^{-1}, or atomic number greater than 20 excluding the elements in groups I and II of the Periodic Table (Adriano, 2001). Toxicity of an element is denoted by the term "Potentially Toxic Elements" (PTE) (Gadd, 1993). Some heavy metals (loids) do not have biological functions in plants, such as Cd, Hg and Pb (Adriano, 2001), which are also toxic when present even in low concentrations. Some heavy metals are micronutrients for plants such as Co, Cu, Mn and Zn, but behave toxic when present at high concentrations. Therefore, the accumulation and the migration of these contaminants through dust or leachates into non-contaminated areas are examples of events that contribute towards contamination of our ecosystem (Meier *et al.*, 2012b).

In the over last two decades, the global discharge of Cd was 22 Gg, Cu ~939 Gg, Pb ~783 Gg, and Zn ~1350 Gg (Singh *et al.*, 2003). McGrath *et al.* (2001) reported that 1.4 million sites were affected by heavy metals in Western Europe. This Figureure does not include Central and Eastern European countries (Gade, 2000). In USA, there are about 0.6 million brown fields, which were contaminated with heavy metals (McKeehan, 2000). In China, one-sixth of total arable land was polluted by heavy metals (Liu, 2006). Small industries in Bangladesh, India, and Pakistan, have been pumping untreated effluents, containing heavy metals, in the drains and rivers, resulting in severe water and soil pollution (McKeehan, 2000).

Metals(loids) enter into ecosystems through two major pathways. The first one is natural (lithogenic), in which metals are derived from primary minerals through geological weathering; which contributes only a small fraction of the total input (Wuana and Okieimen, 2011). The second one is anthropogenic, which is the major contributor of metals (loids) in the ecosystems. The anthropogenic activities related are industrial processes, mining, and agriculture practices. The first two of these produce metal enriched solid and effluent wastes. Some agricultural practices like 'fertilizing' with untreated biosolids may increase metals in soils (Wuana and Okieimen, 2011). The discharges of the metal-enriched wastes into the ecosystems pose risks and hazards. Firstly, the hazards may be caused through direct contact with contaminated soils, drinking of contaminated ground water and consuming toxic food (Wuana and Okieimen, 2011). In such cases, the heavy metals can be accumulated in the tissues of living organisms (bioaccumulation). Their concentrations increase as they pass from lower trophic levels to higher trophic levels (a phenomenon known as biomagnification). Secondly, irreversible soil degradation may be caused by the presence of heavy metals (oids). In such case, its physical, chemical and biological properties are degraded, thus limiting vegetation and agriculture use, and affecting the ecosystem functions and sustainability (Bolan *et al.*, 2003; Navarro *et al.*, 2008).

The metals (loids) in the soils undergo both chemical and biological transformations including retention, redox and methylation reactions. They are retained in the soil by sorption, precipitation and complexation, or removed by plant uptake and/or leaching (Adriano, 2004). Although most metals (oids) are not subject to volatilization losses, As, Se and Hg tend to form gaseous compounds through redox and methylation reactions (Frankenberger and Karlson, 1995). When metal solution concentration is low and sorption surface large, sorption/desorption process will govern the metal concentration in soil solution (Tiller, 1989; Bolan *et al.*, 1999). The fate of metals in the soil depends on both soil properties and environmental factors.

2.0 Phytoremediation of metal polluted soils

According to most legislative schemes, a soil may require remediation if the concentration of one or more metal(loid)s exceeds the specified threshold level in the soil profile (**Table 10.1**). For the above, in order to maintain quality of soils and keep them free from contamination continued efforts have been made to develop technologies that are easy to use, sustainable and economically feasible. The emerging physicochemical approaches include soil treatments such as washing (or in combination with electrokinetic techniques), vitrification, thermal treatment, and excavation and confinement of the soil in special dumps or landfill sites (Mulligan *et al.*, 2001). However, these approaches are intrusive in nature causing soils degradation; and and not economically viable particularly for large areas (Khan, 2005).

Since last decade, plants have been emerging as a cheaper technology for phytoremediation (Pilon-Smiths, 2005). Phytoremediation refers to a suite of technologies that use plants (and their associated microorganisms) to remove, transfer, stabilize, decrease, and/or decompose pollutants in the environment (McGrath *et al.*, 2001). Heavy metals cannot be removed by biological meansfrom soils, but bringing change in their oxidation state is possible (Reimann *et al.*, 2001). However, they can be immobilized or sequestered by biota eg. microorganisms and higher plants,which involves various metabolic mechanisms (Alkorta *et al.*, 2004). Phytoremediation has been widely used by governmental agencies and industries because of its low implementation cost and the appreciation by those who prefer 'green technologies'to chemicals or heavy machinery (Lewandowski *et al.*, 2006).

Table 10.1: Soil concentration ranges and regulatory guidelines for some heavy metals

Metal	Soil concentration range (mg kg^{-1})[1,2,3]	Threshold limit value (mg kg^{-1})[4]
Pb	1.0 – 6900	100-300[5]
Cr	0.05 – 345	50[6]-250[7]
As	0.1-50	5.6
Zn	150-5000	200-500[7]
Cu	30-850	60[6]-210[9]
Hg	0.01-1800	0.5
Cd	0.10-345	1-5[10]

1. Ryley *et al.*, 1992,
2. NJDEP, 1996,
3. Wuana and Okieimen, 2011
4. Hungarian Govermental regulation number. 2000.10/2000,
5. European Community Directive, 1986,
6. Australian Soil Resource Information System, 2009,
7. Ewers, 1991,
8. USEPA, 2007,
9. BOE, Royal Decree 1310/19,
10. Ministere de l'Environmement du Quebec.

Depending on the nature of the contaminant, soil conditions and required cleaning level, plants and their rhizosphere organisms can be used for phytoremediation. In the case of metal polluted soils, the principal ways of phytoremediation are:

   a. **Phytostabilization**: It is based on the reduction of pollutant mobility and bioavailability through its chemical immobilization and consequent prevention of its migration to groundwater or their entry into the food chain (Meier *et al.*, 2012b). Nevertheless, phytostabilization is not a permanent solution because the metals remain in the soil, only their movement is limited.

   b. **Phytoextraction**: It is also known as phytoaccumulation, phytoabsortion or phytosequestration. It is the most common and promising technique to remove heavy metals from soils. It is based on the extraction of metals by plants followed by translocation to another site or system. After harvesting, the plant material is commonly incinerated; and the ashes are treated as hazardous residues (Paz-Ferreiro, *et al.*, 2014). The recovery of plant-extracted metals from the final ashes is referred to as 'phytomining' (Sheoran *et al.*, 2009).

d. **Phytofiltration:** It is based on the use of plant roots (also known as rhizofiltration) or seedlings (blastofiltration) to extract pollutants, mainly metals, from water and aqueous waste streams (Prasad and Freitas, 2003). Plant roots or seedlings grown in aerated water extract, precipitate and concentrate toxic metals from polluted effluents (Elless *et al.*, 2005). Phytofiltration prevents movement of pollutants to underground waters (Paz-Ferreiro *et al.*, 2014).

The plants used in phytoremediation of metal contaminated soils must be genetically capable of growing in soils contaminated with metals (Marchiol *et al.*, 2004). The roots should have a deep, well-branched system in order to take up metals with higher efficiency. After harvest, dying shoots and roots of annual crops should easily be decomposed in the soil so that the synthesis of complexing agents (humic and fulvic acids) is minimized (Ernst, 2005). However, only a limited number of plants, known as 'metallophytes', fulfill this requirement (Baker, 1987). Metallophytes that are able to accumulate high amount of metals in their shoots are called 'hyperaccumulators' (McGrath and Zhao, 2003). Commonly, a plant is defined as a hyperaccumulator if that can store a metal at a level 100-fold greater (without yield reduction) than those of common plants (Chaney *et al.*, 1997).

Nevertheless, the phytoremediation techniques have some disadvantages, eg., it is a slow process, which could take several years or decades to reduce metal concentrations in soil to acceptable levels (McGrath and Zhao, 2003). This slowness could be due to the limit of growth orbiomass production of hyperaccumulator plants (Peuke and Rennenberg, 2005). The lack of knowledge about the interactions between soils, plants and microbial communities limiting the metal availability, further contributes to this issue (Bolan *et al.*, 2008).

Recently, a number of soil microbe-assisted strategies were suggested for enhancing the efficiency of phytoremediation processes (See review of Meier *et al.*, 2012b). In this sense, several studies showed that bacteria, fungi, ectomycorrhizal and ericoid AMF play an important role in the phytoremediation of metal-contaminated sites (Agerer, 2001; Gadd, 1993; Martino *et al.*, 2003). However, the most prominent symbiotic fungus for potential use in phytoremediation is the AMF due to its ubiquity in soil environments, and also because these fungi can employ several strategies that allow the plant to tolerate high metal concentration in the soil (Janousková *et al.*, 2005).

Although there have been a number of reviews that describe the contributions of the AMF to the phytoremediation of metal (loid) contaminated soils (Göhre and Paszkowski, 2006; Hildebrandt *et al.*, 2007) there has been no comprehensive compendium linking the AMF metal tolerance mechanisms to its environmental significance in detail. This chapter describes the AMF in metal-polluted soils and evaluates their role in the promotion of phytoremediation processes.

3.0 AMF in metal-polluted soils.

AMF belong to the phylum *Glomeromycota* (Schüßler *et al.*, 2001, Redecker *et al.*, 2013), and are a group of obligate biotroph commonly associated with the roots of most terrestrial plants forming the so-called AMF symbiosis, which involve 80-85% of vascular plants in almost all ecosystems (Barea *et al.*, 1987). The AMF expand the interface between plants and the soil environment and contribute to plant uptake of macronutrients P and N (as well micronutrients such as Cu and Zn). In turn, fungi obtain photosynthates from the plant for their metabolic functions (Hildebrandt *et al.*, 2007).

The ability of AMF to confer resistance to plants against metals has been reported in several studies (Barea *et al.*, 2002; Curaqueo *et al.*, 2014; Hildebrant *et al.*, 2007; Meier *et al.*, 2012b). Therefore, the manipulation and use of the AMF as a tool for polluted soils must be considered when phytoremediation programs are designed. However, in order to analyze the specific role of AMF in the host's exposure to metal stress and in the progression of the host's stress response depends on a variety of factors, including the plant species, the diversity of species, and the metal and its availability (Meier *et al.*, 2012b).

3.1 Presence and diversity of AMF in metal-polluted soils

The presence of AMF in metal polluted soils must be considered in terms of its ecological diversity (i.e. qualitative aspect), and functional compatibility with the endemic metallophytes and hyper accumulator plants in the ecosystem (evaluated quantitatively through the density of fungal infection). The presence of AMF in metal-polluted soils and their ability to form an effective mycorrhiza symbiosis have been extensively investigated (da Silva *et al.*, 2003;del Val *et al.*, 1999; Leyval and Weissenhorn, 1996), and found that the diversity of AMF ecotypes in metal polluted soils were low (Pawlowska *et al.*, 1996), because high metal concentrations reduce both density and diversity of fungal populations (del Val *et al.*, 1999). The AMF genera reported in the above noted literature were mainly *Glomus* and *Gigaspora* species (da Silva *et al.*, 2003; 2006). However, AMF taxonomic diversity in metal-polluted soils also includes other genera such as *Acaulospora* (González-Chávez *et al.*, 2002), *Entrophospora, Paraglomus,* and *Scutellospora* (da Silva *et al.*, 2003; 2006) **(Table 10.2).**

3.2 Tolerance mechanisms developed by AMF to alleviate metal stress

Most of the mechanisms that plants and AMF have developed to alleviate metal stress are quite similar, due to the strict biotrophy of AMF product of a co-evolution of both organisms (Meier *et al.*, 2012b). The mechanisms used by AMF are summarized in the **Figure 10.1** that depicts: immobilization of metal by chelating substances secreted to soil (mechanism 1, Ernst *et al.*, 1992); metals binding to biopolymers in the cell wall, such as chitin and glomalin (mechanism 2, González-Chávez *et al.*, 2004); immobilization of metal in the plasmatic membrane once it crosses the cell wall (mechanism 3, Ernst *et al.*, 1992); membrane transporter that mobilizes metals from the soil to the cytosol (mechanism 4); intracellular chelation through metallothioneins (MT, González-Guerrero *et al.*, 2006); exudation of low molecular weight organic acids (LMWOA) (mechanism 5, Meier *et al.*, 2012a); transference of metals from the cytosol by membrane transporters (mechanism 6); and confinement of metals into the vacuoles (mechanism 7, González-Guerrero *et al.*, 2008; Hall, 2002).

Table 10.2: AMF species and associated host plants reported growing in metal polluted soils.

AMF	Soil contaminant	Host plant	References
Acaulospora delicata	As	Holcus lanatus	González-Chávez et al., 2002
Acaulospora laevis	Cd, Cu	Zea mays	Liao et al., 2003
Acaulospora scrobiculata	Cu, Zn	"Tropical grassland"	da Silva et al., 2003
Acaulospora spinosa	Zn, Cu, Cd, Pb	Brachiaria sp.	da Silva et al., 2006
Acaulospora undulata	As	Holcus lanatus	Gonzáles-Chávez et al., 2002
Entrophospora infrequense	As, Cu, Zn	Holcus lanatus, "Tropical grassland"	da Silva et al., 2003; González-Chávez et al., 2002
Gigaspora gigantea	Zn, Cu, Cd, Pb	Brachiaria sp.	da Silva et al., 2006
Gigaspora rosea	Pb, Zn, Cd, Cu, As	Fragaria vesca, Holcus lanatus	González-Chávez et al., 2002; Turnau et al., 2001
Glomus microaggregatum	Cu, Zn	"Tropical grassland"	Da Silva et al., 2003
Glomus aggregatum	Calamine (Cd, Pb, Zn)	Festuca ovina, Leontodon hispidus	Pawlowska et al., 1996
Glomus albidum	Cu, Zn	"Tropical grassland"	da Silva et al., 2003
Glomus caledonium	Cd, Cu	Zea mays, Sorghum vulgare	González-Chávez et al., 2002; Liao et al., 2003
Glomus claroideum	Zn, Cd, Pb, As, Cu	Plantago lanceolata, Sorghum vulgare, Sorghum bicolor	del Val et al., 1999; González-Chávez et al., 2002; Orlowska et al., 2005
Glomus constrictum	Cd. Cr, Cu, Pb, Zn, As	Zea mays, Holcus lanatus	González- Chávez et al., 2002;
Glomus diaphanum	Cu, Zn	"Tropical grassland"	da Silva et al., 2003

Contd...

Table 10.2: Contd...

AMF	Soil contaminant	Host plant	References
Glomus etunicatum	Zn, Cd, Pb, Cu	*Plantago lanceolata,* "Tropical grassland"	da Silva *et al.,* 2003; Orlowska *et al.,* 2005.
Glomus fasciculatum	As	*Holcus lanatus*	González-Chávez *et al.,* 2002
Glomus geosporum	Cd, Cu	*Aster tripolium*	Carvalho *et al.,* 2006
Glomus intraradices	Cr, Pb, Cd, Cu, As	*Helianthus annuus, Zea mays, Agrostis capillaris, Pteris vittata.*	Sudová and Vosátka, 2007 Sudová *et al.,* 2008
Glomus manihotis	Cd, Cu	*Zea mays*	Liao *et al.,* 2003
Glomus mosseae	Pb, Zn Cd, Cu, Zn, Pb, Cd, Cr, Ni, Hg	*Tetraclinis articulata, Trifolium subterraneum, Sorghum vulgare, Viola calaminaria, Sorghum bicolor*	Curaqueo *et al.,* 2014; del Val *et al.,* 1999; González-Chávez *et al.,* 2002; Joner and Leyval, 1997; Tonin *et al.,* 2001.
Glomus occultum	Pb, Zn, Cd, Cu, As	*Fragaria vesca*	Turnau, 2001
Glomus sinuosum	Cu, Zn	"Tropical grassland"	da Silva *et al.,* 2003
Glomus sinuosum	Cu, Zn	"Tropical grassland"	da Silva *et al.,* 2003
Glomus tortuosum	Cu, Zn	"Tropical grassland"	da Silva *et al.,* 2003
Paraglomus occultum	Cu, Zn	"Tropical grassland"	da Silva *et al.,* 2003
Scutellospora gilmorei	Cu, Zn	"Tropical grassland"	da Silva *et al.,* 2003
Scutellospora gregaria	Zn, Cu, Cd, Pb	*Brachiaria sp.*	da Silva *et al.,* 2006
Scutellospora heterogama	Zn, Cu, Cd, Pb	"Tropical grassland"	da Silva *et al.,* 2003

A mechanism against metal stress, present exclusively in AMF, involves transporting metals by means of the fungal hyphae (mechanism 8, González-Chávez *et al.*, 2002). Additionally, membrane transporters in AMF arbuscules may carry metals to the interfacial matrix (the contact zone between the plasma membrane of the fungus and the plant cell) followed by their subsequent incorporation inside the plant (mechanism 9, Meier *et al.*, 2012b). This may explain how some plants can accumulate metals in their shoots (Ebbs and Kochian, 1998) and also accumulation of metals in AMF resistant structures (spores) (Ferrol *et al.*, 2009 -mechanism 10-). The storage of metals in spores has been described only in monoxenic culture (Ferrol *et al.*, 2009), however, recently Cornejo *et al.*, (2013) demonstrated the accumulation of Cu in mycorrhizal spores in soils contaminated with Cu (**Fig.10.2**). Thus, AMF symbiosis can play a role in protecting plants against toxicity in metal-polluted ecosystems. This may be considered as a biotechnological tool for phytoremediation programs.

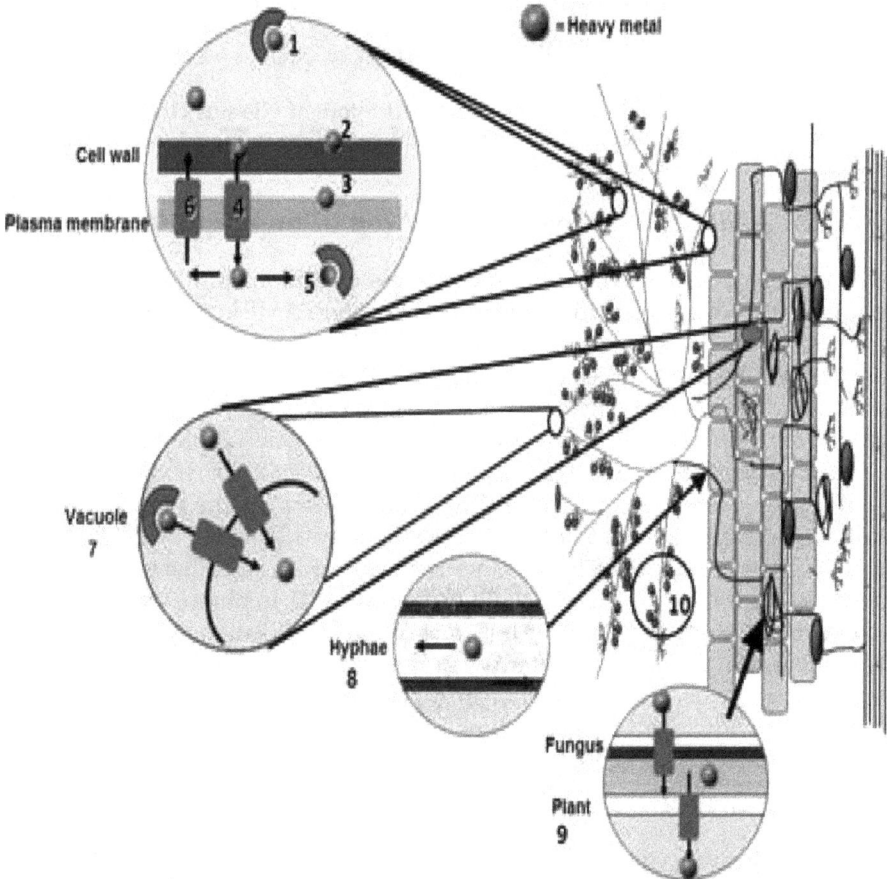

Fig. 10.1: Possible mechanisms by which AMF enhance resist/tolerate metal toxicity of plants. Adapted from Meier *et al.*, 2012b.

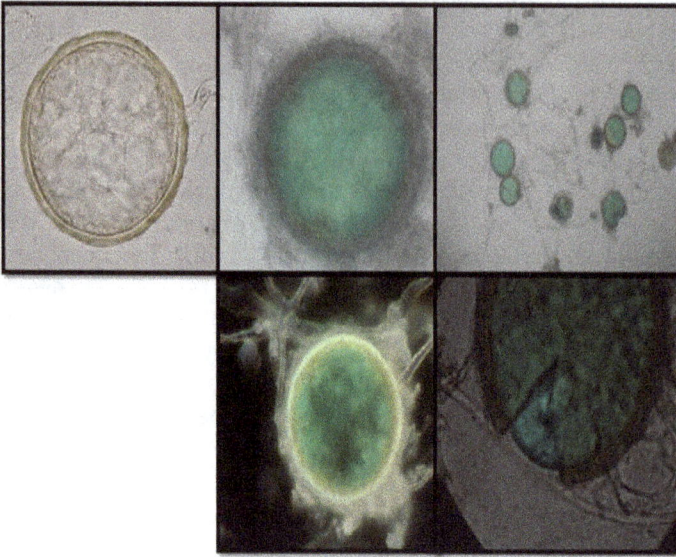

Plate 10.1 Copper accumulation in AMF spores. A) Spore of Glomus claroideum in Cu unpolluted soil. B) Spores of *G. claroideum* **in field condition in high Cu polluted soils (450 µg Cu kg⁻¹, Cornejo** *et al.,***2014).**

4.0 Role of AMF in the phytoremediation of metal-polluted soils

The benefits of AMF symbiosis under stress conditions produced by metals have already been mentioned in this chapter. The role of AMF in phytoremediation, however, remains unclear because some studies have findings that are contradictory. In, some studies it was shown that AMF employ mechanisms that allow accumulation of metal in plant roots (Giasson *et al.*, 2005) but its translocation to shoot was prevented thus promoting a specific type of phytostabilization (Audet and Charest, 2006; Citterio *et al.*, 2005; Giasson *et al.*, 2005; Janousková *et al.*, 2006). In contrast, some studies have shown that AMF promote phytoextraction, causing an increase in metal translocation to the shoots (Davies *et al.*, 2001; Khan *et al.*, 2000; Trotta *et al.*, 2006). Some other studies (Audet and Charest 2006) have proposed another mechanism to explain the role of AMF in phytoextraction when the metal concentration in soil is low and phytostabilization when the metal concentration is high. In this chapter we suggest that AMF role promoting either phytostabilization or phytoextraction in metal contaminated soils depends on multi factorial components: such as type of soil, metal, host, AMF ecotypes and diversity among others **(Fig.10 3)**.

4.1 AMF and phytostabilization

Phytostabilization can reduce the spread of metal (loid)s by inhibiting their dispersal through wind with dust, transport with surface water run-off and leaching to aquiferous. AMF participate in the immobilization of metals beyond the immediate vicinity of plant roots, creating an extensive zone of influence called the 'mycorrhizosphere' (Giasson *et al.*, 2005). In this zone, AMF can immobilize metals through mechanisms similar to those used by plants (**Fig.10.2, see mechanisms 1 to 7**). Among those the most outstanding ones are:

A) Extracellular mechanism

A.1) Metal sequestration by AMF hyphae: Its well known that the extra-radical mycelium (ERM) of AMF can contribute to the remediation of metal-contaminated soils. Joner *et al.* (2000) reported that the outer surface of ERM had a larger capacity for sorbing metals. Similarly, González-Chávez *et al.* (2002; 2009) found that the ERM could sorb and accumulate high levels of Cu (3–14 mg Cu g^{-1} dry hyphae). This high metal sorption capacity might be related to the presence of free amino acids in the fungal cell wall, and negatively charged hydroxyl, carboxyl, and other functional groups capable of binding cations, such as Cu^{2+} and Zn^{+2} among others (Zhou, 1999).

A.2) Metal sequestration through glomalin: The glomalin (*Glomalin-related soil protein*, GRSP) is a consortium of protein produced by the hyphae of AMF (Gadkar and Rillig, 2006). The GRSP is only operationally defined by its extraction method, and has been only poorly characterized (Nichols, 2003). Recent studies conclude that glomalin is a mixture of compounds containing glomalin-related stable proteins, lipids and humic materials (Gillespie *et al.*, 2011).

GRSP has a range of diverse soil functions. The most known function is assisting particle aggregation (Curaqueo *et al* 2011; Rillig and Mummey, 2006). GRSP can also sequester (bind) toxic elements through its functional groups (Cornejo *et al.*, 2008; González-Chávez *et al.*, 2004; González-Chávez *et al.*, 2009). González-Chávez *et al.* (2004) showed that GRSP could sequester up to 4300 mg of Cu, 1120 mg of Pb and 80 mg of Cd per kg GRSP from soils polluted with Cd, Cu, and Pb. Similarly, Vodnik *et al.* (2008), working with Pb-polluted soils, found that GRSP bound Pb ranged from 690 to 23400 mg kg^{-1} soil, representing 0.8–15.5% of the total Pb. Cornejo *et al.* (2008) observed a high correlation between the GRSP production and the content of Cu and Zn in some Cu-polluted soils (r = 0.89 and 0.76 for Cu and Zn, respectively), and the metal linked to GRSP corresponding to 1.44–27.5% of the total Cu content and 5.8% of the total Zn content in the soils. This indicates that GRPS plays an effective role in sequestering metals thus mitigating metal stress of plants.

b) Molecular metal homeostasis mechanisms

Metal tolerance in AMF occurs through several homeostatic mechanisms. It can immobilize metals on their surface through GRSP (as noted above), and various other mechanisms like chelation at intracellular level (González-Guerrero *et al.*, 2005; 2009). The cell wall constitutes the first protective barrier against the uptake of metals. However, in highly polluted soils a fraction of the metal ions may cross this barrier, and reach the fungal cytoplasm. Chelating agents in the cytoplasm can deactivate metals thus minimize cell damage. Metallothioneins (MTs) are the most effective metal chelators in AMF (Vasak and Hasler, 2000).

MTs are polypeptides comprised 70-75 amino-acids with a high content of cysteine, capable of forming stable complexes with cations through its sulphydryl groups (Kojima, 1991). MTs are involved in maintaining cellular homeostasis against high concentrations of metals (Cobbett and Goldsbrough, 2002). However, the production of MTs may be induced by metals as well as by hormones (Haq *et al.*, 2003). In case of AMF various studies were made on the interaction between the MTs and metals and their contribution in the phytostabilization processes. This has been confirmed through experiments in several organisms. For example, the insertion of the gene *CUP1* (from yeast) in cauliflower resulted in an increase of the Cd tolerance and up to 16-fold accumulation (Hasegawa *et al.*, 1997). Transgenic tobacco plants (inoculated with AMF) with the same gene were exposed to four different Cd doses. These plants showed a lower Cd concentration in their shoots and leaves compared to non-transgenic and uninoculated tobacco of the same variety (Janousková *et al.*, 2005).

So far, four AMF genes, involved in maintaining cellular homeostasis against metals, have been characterized: *i)GrosMT1* in *Gigaspora rosea* (Stommel *et al.*, 2001), *ii) GinZnT1*(Zn transporter) in *Glomus intraradices* involved in vacuolar Zn compartmentalization (González-Guerrero *et al.*, 2005); *iii)GmarMT1* in *Gigaspora margarita* (BEG 34), which codes for MTs that regulate the fungal redox potential and protect it against the oxidative stress produced by some metals, such as Cd or Zn (González-Guerrero *et al.*, 2006; 2007); and *iv)GintABC1*, which codes for a polypeptide of 434 amino acids that participates actively in Cu and Zn detoxification (González-Guerrero *et al.*, 2006).

4.2 AMF and phytoextraction

Phytoextraction is a recent, cost-effective and attractive green technology for the remediation of soils affected by mining processes (Shah and Nongkynrih, 2007). The process uses hyper accumulator plants, which are able to take up, tolerate, and develop phytomass even in soils with high metal concentrations. Hyper accumulator plants can accumulate metals in their shoots (Ginocchio and Baker, 2004). Therefore, it is possible to extract metals simply by harvesting the shoots. This provides a considerable advantage over phytostabilization (where the metals remain in soil in a harmless or non-bioavailable form). However, phytoextraction is a slow process as numerous cycles of 'plantation-harvest' will be required over few to many years to decrease metal concentrations to an acceptable level. The slowness of the process is mainly due to the interaction of the two components: soil and plant: (Barceló and Poschenrieder, 2003).

i) **Soil:** Complex chemical interactions between plant, metals and soil make the absorption of metal(loid)s by plants difficult, and reduce the effectiveness of the process (Adriano, 2004).In some cases, phytoextraction can be improved by increasing metal availability with application of synthetic chelating substances, such as diethylene triamine pentaacetic acid (DTPA), ethylene diamine tetraacetate (EDTA), and ethylene diamine dissuccinate (EDDS), (Liu *et al.*, 2008; Sudová *et al.*, 2007). These substances can accelerate the absorption of metals in shoots and roots, while the presence of AMF promote plant growth (Adel and Hashem, 2009). However, these chelating substances are usually non-biodegradable and non-selective in their action, thus their use would also increase the risk of leaching of metals, thus cause contamination of subterranean waters (Barona *et al.*, 2001).

ii) **Plant:** Hyper accumulators should produce a large biomass because the amount of metals extracted is proportional to the rate of plant growth (Shah and Nongkynrih, 2007). Unfortunately, there is no plant hyper accumulator identified in the wild that can produce large biomass as well and also has both of these properties (Li *et al.*, 2003).

Another aspect to consider is that most hyper accumulator plants, including families like *Brassicaceae, Plumbaginaceae, Juncaceae, Caryophyllaceae, Juncaginaceae, Amaranthaceae,* and some members of the family *Fabaceae* do not develop arbuscular mycorrhizae. Exceptions were Ni hyperaccumulators belonging to the *Asteraceae* family (Turnau and Mesjasz-Przybylowics, 2003) and some Pteridophytes (Gaur and Adholeya, 2004). Among the above-mentioned plants, *Pteris vittata* L. is a well known as hyper accumulator, which is fast-growing perennial plant that can produce a large phytomass and, an extensive root system. In the absence of AMF it can uptake as high as 60.4 mg As kg^{-1}biomass; this, however, may increase up to 88.1 mg As kg^{-1} biomass in the presence of AMF (Leung *et al.*, 2006). These characteristics make it potentially useful for phytoextraction programs (Ma *et al.*, 2001). However, the information available on AMF-As interactions in the context of *P. vittata* is still limited. More studies are required in this area, particularly on the role of AMF in increasing the efficiency of phytoremediation of As-contaminated soils (Liu *et al.*, 2009).

Despite the many studies carried out over last two decades, our knowledge of the role of AMF in phytoextraction processes remains incomplete. Many studies have shown that AMF can enhance metal transfer from soil to plant (translocation factor). But, Weissenhorn and Leyval (1995) indicated that high metal accumulation (as a result of AMF colonization) might inhibit plant growth.

The mycotrophic status of hyper accumulator plants belonging to the *Brassicaceae* family, particularly the *Thlaspica erulescens*, have been receiving world wide attention in relation to phytoremediation (Fischerová *et al.*, 2006). DeMars and Boerner (1996) reported that18.9% of the 946 species of *Thlaspica erulescens*form AM associations. Similarly, Vögel-Mikus *et al.* (2006) and Pongrac *et al.* (2009) found that *T. praecox* too form AMF association although the colonization rate was low and variable. They concluded that the AMF association in *T. praecox* plays an important role in enhancing phytoextraction by increasing the translocation of Cd, Pb and Zn.

There should be increased efforts to identify more metallophytes that are hyper accumulaor, especially the native ones. (Ginocchio and Baker, 2004). The death of plants after hyper accumulation of metals deserves further study. There are three hypotheses (opinions) about the existence or domination of phytoextraction and/or phytostabilization under the influence of AMF in metal-polluted soils. One group of scientists favours the view of phytoextraction; the other group supports phytostabilization; and a third group favours existence of both processes. According to the last hypothesis, AMF promote 'phytoextraction' when the concentration of a metal in a soil is low, and 'phytostabilization'when the concentration is high.

5.0 Use of cropping plants for the restoration of metal contaminated sites.

Nevertheless all said before in this chapter,the phytoremediation is only part of the solution, and there is a great need for comprehensive for agronomic approaches to deal with contaminated soils. As a remediation process, phytoremediation can only be considered as an element to be included in sustainable production systems adapted to contaminated environments (Tang *et al.,* 2012).Considering that even contaminated soils are a potential resource for plant production, it is necessary to propose sets of agricultural scenarios and recommendations to permit plant growth and avoid dissemination of toxic elements and develop suitable strategies to restrict the metal accumulation in such crops and its further transfer to the food chain (Wong, 2003). The main principle is to use phytoexclusion strategy, which is involved in the selection and breeding of low-accumulating cultivars, reduction of bioavailable metals in the soil, and restriction of their potential uptake and translocation by plants (Tang *et al.,* 2012).

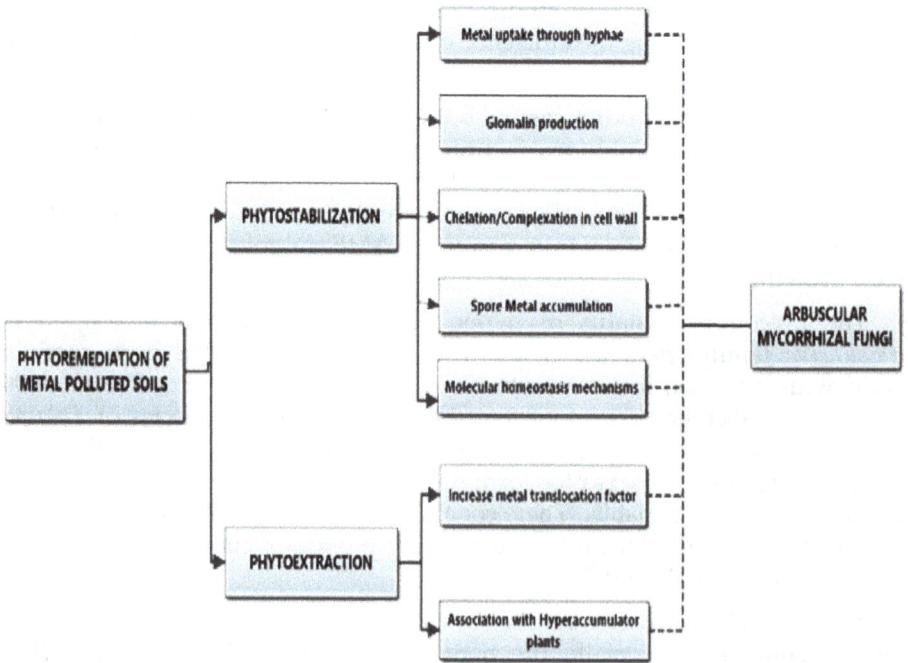

Fig. 10. 2: Contribution of AMF to the phytoremediation of metal polluted soils

On the other hand, although the existence of several reports informing the accumulation capacity of cereals and vegetables growing in metal contaminated soils, there is scarce information in the area of fruit crops. Rufyikiri, *et al.,* (2000) showed that mycorrhiza inoculated banana plants had greater resistance to aluminium toxicity in banana crop than in non-mycorrhizal plants. Micropropagated banana plants, were inoculated with AMF *Glomus intraradices*, grown for 40 days in pots filled with sand, and continuously irrigated with a nutrient solution containing up to 180 muM of aluminium. The water and nutrient uptake were measured once a week for 24 h, and root AMF colonization, biomass production, and mineral content of roots and shoots were measured during harvest. The AMF colonized roots were large, and not significantly influenced by aluminium treatment. The effects of aluminium on both mycorrhizal and non-mycorrhizal plants showed decrease in biomass production, water and nutrient uptake, and magnesium content of roots and shoots; greater aluminium content in roots than in shoots; and increase in potassium and phosphorus content, particularly in roots. The significant positive effect of arbuscular mycorrhizal fungi on plant growth was observed with aluminium treatment, and was most pronounced at the highest concentration. The benefits, compared with nonmycorrhizal plants, included increase in shoot dry weight, uptake of water and of most nutrients, and in calcium, magnesium and phosphorus content, particularly in roots, decrease in aluminium content in root and shoot; and delay in the appearance of aluminium-induced leaf symptoms. These results indicate that AMF could be effective in alleviating aluminium toxicity to banana plants. Yang and Goulart, (2000) showed that the mycorrhizal infection reduced short-term aluminum uptake and increased root cation exchange capacity of high bush blueberry plants. There are reports related to the exposition fruits to metals through Cu-based pesticides (Khan *et al.,* 2010), which has resulted in elevated accumulation of Cu in soils of vineyard and apple orchard and other fruit crops (Chopin *et al.,* 2008). Graham,*et al.,*(1986) studied citrus toxicity in fungicidal copper soil. The carrizo citrange (*Ponicircus trifolata x Citrus sinensis)* seedlings were potted in sand soil with pH 6.8 amended with the eight rates of Cu (0-300 μg-g^{-1} of soil) as a basic copper sulphate. Double acid extractable copper concentration in soil ranged from 3-248 μg-g^{-1} . Seedlings inoculated with AMF species *Glomus intraradices* reduced the Cu concentration than the un-inoculated seedlings. The minimum Cu concentration 9-34 μg-g^{-1} of soil was observed compared to un-inoculated seedlings. The Cu induced plants showed reduction of P uptake in mycorrhizal plants was more closely related to the inhibition of hyphal developments outside the root than to development of vesicle and arbuscules inside roots. Therefore, we believe that special attention should be given to restore wildlife communities after the remediation of metal contaminated soils using metallophytes or hyper accumulator plants. Diversified crops and fruit trees should be planted, and agriculture should be integrated with forestry and animal husbandry, appropriate to local conditions (Pulford and Watson, 2003). Assessment and monitoring should be made in order to ensure toxic substances are not transferred and accumulated through food chains, if the sites are used for agriculture and animal husbandry purposes.

6.0 Conclusions and future strategies

The past studies improved our understanding of mycorrhizal biology and also the metal tolerance of plants and fungi.There is no doubt that AMF play important role in promoting the survival and also the growth of some plants in metal-polluted soils. However, contradictory views about the existence or domination of phytoextraction and/or phytostabilization suggest that it is still premature to decide about using AMF in the field. This suggests that, more studies are necessary to remove the contradictory views.

The future researches should focus on the process optimization, determination of physiological mechanisms involved in metal sorption, translocation, and metabolism in the plants; and describing the genetic control mechanisms. Understanding of physiological mechanisms involved in metal tolerance by AMF, including intracellular metal chelation and the role played by MTs, is also necessary. A database of AMF, containing their diversity in metal-polluted soils and functional compatibility with metallophytes and hyper accumulator plants will be helpful for decision making.

7.0 Acknowledgement

Sebastian Meier thanks Fondecyt Project number 313039 for supporting this research.

References

1. Adriano, D.C., 2001. Trace Elements in Terrestrial Environments: Biogeochemistry, Bioavailability, and Risk of Metals, Springer-*Verlag*, New York.

2. Adriano, D., Wenzel, W., Vangronsveld, J., Bolan, N., 2004. Role of assisted natural remediation in environmental cleanup. *Geoderma*. 122, 121-142.

3. Adel, R. and Hashem, M., 2009. Effect of microbial inoculation and EDTA on the uptake and translocation of heavy metal by corn and sunflower. *Chemosphere*. 76, 893-899.

4. Agerer, R., 2001. Exploration types of ectomycorrhizae. A proposal to classify ectomycorrhizal mycelial systems according to their patterns of differentiation and putative ecological importance. *Mycorrhiza*. 11, 107-114.

5. Alkorta, I., Hernández, A., Becerril, J.M., Albizu, I. and Garbisu, C., 2004. Recent findings on the phytoremediation of soils contaminated with environmentally toxic heavy metals and metalloids such as zinc, cadmium, lead, and arsenic.*Rev. Environ. Sci. Bio/Technol*. 3, 71–90.

6. Audet, P. and Charest, C., 2006. Effects of AM colonization on 'wild tobacco' plants grown in zinc-contaminated soil. *Mycorrhiza*. 16, 277-283.

7. Australian Soil Resource Information System (ASRIS), 2009. http://www. asris.csiro.au/index other.html

8. Baker, A.J.M., 1987. Metal tolerance. *New Phytol.* 106, 93-111.

9. Barceló, J. and Poschenrieder, Ch., 2003. Phytoremediation: principles and perspectives. *Contrib. Sci.* 2, 333-344.

10. Barea,J.M., Azcón-Aguilar, C and Azcón, R., 1987. VA mycorrhiza improve both symbiotic N_2-fixation and N uptake from soil as assessed with a 15N technique under field conditions. *New Phytol.*106, 717-721.

11. Barona, A., Aranguiz, I., Elias, A., 2001. Metal associations in soils before and after EDTA extractive decontamination: implications for the effectiveness of further clean-up procedures. *Environ. Pollut.* 113, 79-85.

12. Barea, J.M.,Azcón, R. and Azcón-Aguilar, C., 2002. Mycorrhizosphere interactions to improve plant fitness and soil quality.*Anton. Leeuw. Int. J G.* 81, 343-351.

13. BOE, 1990. BOE Royal Decree 1310/1990 of 29 October, which regulatesthe use of sewage sludge in agriculture. BOE No. 262 de 1 de noviembre de 1990, Madrid, Spain, 32339–32340.

14. Bolan, N., Adriano, D., Duraisamy, P. and Mani, A., 2003. Immobilization and phytoavailability of cadmium in variable charge soil. Effect of phosphate addition. *Plant Soil.* 250, 83-94.

15. Bolan, N., Ko, B., Anderson, C., Vogeler, I., Mahimairaja, S. and Naidu, R., 2008. Manipulating bioavailability to manage remediation of metal-contaminated soils. *Develop Soil Sci.* 32, 657-678.

16. Bolan, N., Khan, M., Tillman, R., Naidu, R. and Syers, J., 1999. The effects of anion sorption on sorption and leaching of cadmium. *Aust. J. Soil Res.* 37, 455-460.

17. Carvalho, L., Caçador, I. and Martinis-Loução, M., 2006. Arbuscular mycorrhizal fungi enhance root cadmium and copper accumulation in the roots of the salt marsh plant *Aster tripolium* L. *Plant Soil.* 285, 161-169.

18. Chaney, R.L., Malik, M., Li, Y.M., Brown, S.L., Brewer, E.P., Angle, J.S and Baker, A.J.M., 1997. Phytoremediation of soil metals. *Curr. Opin. Biotech.* 8, 279-284.

19. Chopin, E. I. B., Marin, B., Mkoungafoko, R., Rigaux, A., Hopgood, M. J., Delannoy, E., Cancés, B. and Laurain, M., 2008. Factors affecting distribution and mobility of trace elements (Cu, Pb, Zn) in a perennial grapevine (*Vitis vinifera* L.) in the Champagne regionof France. *Environ. Pollut.* 156, 1092-1098.

20. Citterio, S., Prato, N., Fumagalli, P., Massa, N., Santagostino, A., Sgorbati, S. and Berta, G., 2005. The arbuscular mycorrhizal fungus *Glomus mosseae* induces growth and metal accumulation changes in *Cannabis sativa* L. *Chemosphere.* 59, 21-29.

21. Cobbett,C. andGoldsbrough,P.,2002.Phytochelatinsandmetallothioneins: roles in heavy metal detoxification and homeostasis. *Annu. Rev. Plant. Biol.*53, 159-182.

22. Cornejo, P., Meier, S., Borie, G., Rillig, M. and Borie, F., 2008. Glomalin-related soil protein in a Mediterranean ecosystem affected by a copper smelter and its contribution to Cu and Zn sequestration. *Sci. Total Environ.*406, 154-160.

23. Cornejo, P., Pérez-Tienda, J., Meier, S., Valderas, A., Borie, F., Azcón-Aguilar, C. and Ferrol, N. 2013. Copper compartmentalization in spores as a survival strategy of arbuscular mycorrhizal fungi in Cu-polluted environments. *Soil Biol. Biochem.* 57,925-928.

24. Curaqueo, G., Barea, J.M., Acevedo, E., Rubio, R., Cornejo, P., Borie, F., and 2011.Effects of different tillage system on arbuscular mycorrhizal fungal propagules and physical properties in a Mediterranean agroecosystem in central Chile. *Soil Tillage Res.* 113, 11-18.

25. Curaqueo, G., Schoebitz, M., Borie, F., Caravaca, F., Roldán, A., 2014. Inoculation with arbuscular mycorrhizal fungi and addition of composted olive-mill waste enhance plant establishment and soil properties in the regeneration of a heavy metal-polluted environment. *Environ Sci Pollut R.* 21,7403-7412.

26. Davies, F.T., Puryear, J.D., Newton, R.J., Egilla, J.N., Saraiva Grossi, J.A., 2001. Mycorrhizal fungi enhance accumulation and tolerance of chromium in sunflower (*Helianthus annuus*). *J. Plant. Physiol.* 158, 777-786.

27. da Silva, S., Trufem, S., Saggin, O. and Maia, L., 2003. Arbuscular mycorrhizal fungi in a semiarid copper mining area in Brazil. *Mycorrhiza.* 15, 47-53.

28. da Silva, S., Siqueira, J. and Fonsêca, C., 2006. Fungos micorrízicos no crescimento e na extração de metais pesados pela braquiária em solo contaminado. *Pesqui. Agropecu. Bras.* 41, 1749-1757.

29. DeMars, B., Boerner, R. 1996. Vesicular arbuscular mycorrhizal development in the *Brassicaceae* in relation to plant life span. *Flora.* 191, 179-189.

30. del Val, C., Barea, J.M. and Azcón-Aguilar, C., 1999. Diversity of arbuscular mycorrhizal fungus population in heavy-metal contaminated soil. *Appl. Environ. Microbiol.* 65, 718-723.

31. Ebbs, S. and Kochian, L., 1998. Phytoextraction of zinc by oat (*Avena sativa*), barley (*Hordeum vulgare*) and Indian mustard (*Brassica juncea*). *Environ. Sci. Technol.* 32, 802-806.

32. Elless, P., Poynton, Y., Williams, A., Doyle, P., Lopez, C. and Sokkary, D., 2005. Pilot-scale demonstration of phytofiltration for drinking arsenic in New Mexico drinking water. *Water Res.* 39, 3863-3872.

33. Ernst, W., Verkleij, J and Schat, H., 1992. Metal tolerance in plants. *Acta Bot. Neerl.* 41, 229-248.

34. Ernst, W., 2005. Phytoextraction of mine wastes, options and impossibilities. *Chemie der Erde.* 65, 29-42.

35. European Community Directive.1986. "Protection of the environment, and in particular of the soil, when sewage sludge is used in agriculture," *Journal European Commission.* 181, 6–12.

36. Ewers, W., 1991.Standards, guidelines and legislative regulations concerning metals and their compounds, in Metals and Their Compounds in the Environment: Occurrence, Analysis and Biological Relevance, E. Merian, Ed., VCH, Weinheim, Germany.

37. Ferrol, N., González Guerrero, M., Valderas, A., Benabdellah, K., Azcón-Aguilar, C., 2009. Survival strategies of arbuscular mycorrhizal fungi in Cu-polluted environments. *Phytochem. Rev.* 8, 551-559.

38. Ferreiro, P., Lu, H, Fu, S., Gascó, G. and 2014.Use of phytoremediation and biochar to remediate heavy metal polluted soils: a review. *Solid Earth.* 5, 65–75.

39. Fischerová, Z., Tlustos, P., Száková, J. and Sichorová, K., 2006. A comparison of phytoremediation capability of selected plant species for given trace elements. *Environ. Pollut.* 144, 93-100.

40. Frankenberger, W.T. and Karlson., 1995. Volatilization of selenium from adewatered seleniferous sediment: a field study. *J. Ind. Microbiol.* 14, 226-232.

41. Gadd, G.M., 1993. Interaction of fungi with toxic metals. *New Phytol.* 124, 25-60.

42. Gade, L.H., 2000. Highly polar metal—Metal bonds in "early-late" heterodimetallic complexes. *Angewandte Chemie-International Edition.*39, 2658-2678.

43. Gaur, A., Adholeya, A., 2004. Prospects of arbuscular mycorrhizal fungi in phytoremediation of heavy metal contaminated soils. *Current Sci.* 86, 528-534.

44. Gaur, A., Adholeya, A., 2004. Prospects of arbuscular mycorrhizal fungi in phytoremediation of heavy metal contaminated soils. *Current Sci.* 86, 528-534.

45. Gadkar, V., Rillig, M., 2006. The arbuscular mycorrhizal fungal protein glomalin is a putative 16 homolog of heat shock protein 60. *FEMS Microbiol. Lett.* 263, 93-101.

46. Ginocchio, R. and Baker, A., 2004. Metallophytes in Latin America: a remarkable biological and genetic resource scarcely known and studied in the region. *Rev. Chil. Hist. Nat.* 77, 185-194.

47. Giasson, P., Jaouich, A., Gagné, S. and Moutoglis, P., 2005. Arbuscular Mycorrhizal Fungi Involvement in Zinc and Cadmium Speciation Change and Phytoaccumulation. *Remediation.* 15, 75-81.

48. Gillespie, A., Farrel, R., Walley., F., Ross, A., Leinweber, P., Eckhardt, K.U. Regier, T. and Blyth, R., 2011. Glomalin-related soil protein contains non-

mycorrhizal-related heat-stable proteins, lipids and humic materials. *Soil. Biol. Biochem.*43, 766-777.

49. González-Chávez, C., D'Haen, J., Vangronsveld, J. and Dodd, J., 2002. Copper sorption and accumulation by the extraradical mycelium of different *Glomus* spp. (arbuscular mycorrhizal fungi) isolated from the same polluted soil. *Plant Soil.* 240, 287-297.

50. González-Chávez, M., Carrillo-González, R., Wrigth, S. and Nichols, K., 2004. The role of glomalin, a protein produced by arbuscular mycorrhizal fungi, in sequestering potentially toxic elements. *Environ. Pollut.* 130, 317-323.

51. González-Guerrero, M., Azcón-Aguilar, C., Mooney, M., Valderas, A., MacDiarmid, CW., Eide, DJ. and Ferrol, N., 2005. Characterization of a *Glomus intraradices* gene encoding a putative Zn transporter of the cation diffusion facilitator family. *Fungal Genet. Biol.* 42, 130-140.

52. Göhre, V. and Paszkowski, U., 2006. Contribution of the arbuscular mycorrhizal symbiosis to heavy metal phytoremediation. *Planta.*223, 1115-1122.

53. González-Guerrero, M., Azcón-Aguilar, C. and Ferrol, N., 2006. GintABC1 and GintMT1 are involved in Cu and Cd homeostasis in *Glomus intraradices*. 5th International Conference on Mycorrhiza. "*Mycorrhiza for science and society*". Granada, Spain, 27, July.

54. González-Guerrero, M., Cano, C., Azcón-Aguilar, C. and Ferrol, N., 2007. *GintMT1* encodes a functional metallothionein in *Glomus intraradices* that responds to oxidative stress. *Mycorrhiza.* 17, 327-335.

55. González-Guerrero, M., Melville, L., Ferrol, N., Lott, J., Azcón-Aguilar, C., and Peterson, R., 2008. Ultrastructural localization of heavy metals in the extraradical mycelium and spores of the arbuscular mycorrhizal fungus *Glomus intraradices*. *Can. J. Microbiol.* 54, 103-110.

56. González-Chávez, M.C., Carrillo-González, R. and Gutíerrez-Castorena, M.C., 2009. Natural attenuation in a slag heap contaminated with cadmium: The role of plants and arbuscular mycorrhizal fungi. *J. Hazard. Mater.* 161, 1288-1298.

57. González-Guerrero, M., Benabdellah, K., Ferrol, N. and Aguilar, C., 2009. Mechanisms underlying heavy metal tolerance in arbuscular mycorrhizas. In: C, Azcón-Aguilar., JM, Barea., S, Gianinazzi., V, Gianinazzi-Pearson. (Eds.), Mycorrhizas: functional processes and ecological impact. Springer. Berlin, Germany. 107-122.

58. Graham,J.H.,Timmer,L.W.and Fardelmann,D.,1986.Toxicity of fungicidal copper in soil to citrus seedlings and vesiclular- arbuscular mycorrhizal fungi, *Phytopatholog.*, 76,66-70.

59. Hasegawa, I., Terada, E., Sunairi, M., Wakita, H., Shinmachi, F., Noguchi, A., Nakajima, M. and Yazaki, J., 1997. Genetic improvement of heavy metal

tolerance in plants by transfer of the yeast metallothionein gene (*CUP1*). *Plant Soil.* 196, 277-281.

60. Haq, F., Mahoney, M. and Koropatnick, J., 2003. Signaling events for metallothionein induction. *Mutat. Res.* 533, 211-226.

61. Hirrel, M., Mehravaran, H. and Gerdemann, J., 1978. Vesicular arbuscular mycorrhizae in the Chenopodiaceae and Cruciferae: do they occur.*Can. J. Bot.* 56, 2813-2817.

62. Hildebrandt, U., Regvar, M. and Bothe, H., 2007. Arbuscular mycorrhiza and heavy metal tolerance. *Phytochemistry.* 68, 139-146.

63. Hungarian Govermental regulation, 2000. Number 10/2000.

64. Janousková, M., Pavlíková, D., Macek, T. and Vosátka, M., 2005. Arbuscular mycorrhiza decreases cadmium phytoextraction by transgenic tobacco with inserted metallothionein. *Plant Soil.* 272, 29-40.

65. Janousková, M., Pavlíková, D. and Vosátka, M., 2006. Potential contribution of arbuscular mycorrhiza to cadmium immobilization in soil. *Chemosphere.* 65, 1959-1965.

66. Joner, E.J. and Leyval, C ., 1997. Uptake of 109Cd by roots and hyphae of *Glomus mosseae/Trifolium subterraneum* mycorrhiza from soil amended with high and low concentrations of cadmium. *New Phytol.,* 35, 353–360.

67. Joner, E., Briones, R. and Leyval, C., 2000. Metal-binding capacity of arbuscular mycorrhizal mycelium. *Plant Soil.*226, 227-234.

68. Khan, A.G., 2005. Role of soil microbes in the rhizospheres of plants growing on trace metal contaminated soils in phytoremediation. *J. Trace Elem. Med. Biol.* 18, 355-364.

69. Khan, A.G., Kuek, C., Chaudhry, T.M., Khoo, C.S. and Hayes, W.J., 2000. Role of aplants, mycorrhizae and phytochelators in heavy metal contaminated land remediation. *Chemosphere.* 41, 197-207.

70. Khan, S., Rehman, S., Khan, A., Khan, M. and Shah, M., 2010. Soil and vegetables enrichment with heavy metals from geological sources in Gilgit, northern Pakistan. *Ecotox. Environ. Safe.* 73, 1820-1827.

71. Kojima, Y., 1991. Definitions and nomenclature of metallothioneins. *Method. Enzymol.*205, 8-10.

72. Leyval, C. and Weissenhorn, I., 1996. Tolerance to metals of arbuscular mycorrhizal fungi from heavy metal polluted soils. A summary of results. In C, Azcón-Aguilar., and J. M., Barea. (Eds.). Mycorrhizae in integrated systems: from genes to plant development. European Commission, Brussels, Belgium.452-454.

73. Lewandowski, I., Schmidt, U., Londo, M. and Faaij, A., 2006.The economic value of the phytoremediation function assessed by the example of cadmium remediation by willow (*Salix* ssp).*Agr. Syst.* 89, 68-89.

74. Leung, H., Ye, Z. and Wong, M., 2006. Interactions of mycorrhizal fungi with *Pterisvittata* (As hyperaccumulator) in As-contaminated soils. *Environ. Pollut*, 139, 1-8.

75. Li, Y., Chaney, R., Brewer, E., Roseberg, R., Angle, J., Baker, A., Reeves, R. and Nelkin, J., 2003. Development of a technology for commercial phytoextraction of nickel: economic and technical considerations. *Plant Soil* .249, 107-115.

76. Liao, J., Lin, X., Cao, Z., Shi, Y. and Wong, M., 2003. Interactions between arbuscular mycorrhizae and heavy metals under sand culture experiment. *Chemosphere*. 50, 847-853.

77. Liu, Y., 2006. Shrinking Arable Lands Jeopardizing China's Food Security. http://www.worldwatch.org/node/3912

78. Liu, D., Islama, E., Li, T., Yang, X., Jin, X . and Mahmood, Q., 2008. Comparison of synthetic chelators and low molecular weight organic acids in enhancing phytoextraction of heavy metals by two ecotypes of Sedum alfredii Hance. J. *Hazard. Mater*. 153, 114-122.

79. Liu, Y., Christie, P., Zhang, J. and Li, X., 2009. Growth and arsenic uptake by Chinese brake fern inoculated with an arbuscular mycorrhizal fungus. *Environ. Exp. Bot*. 66, 435-441.

80. Ma, L., Komar, K., Tu, C., Zhang, W., Cai, Y. and Kennelley, E., 2001. A fern that hyperaccumulates arsenic. *Nature*. 409, 579-579.

81. McGrath,S.P.,Lombi,E.,Zhao,F.J.andDunham,S.J.,2001.Phytoremediation of heavy metal contaminated soils: Natural Hyperaccumulation versus Chemically Enhanced Phytoextraction. *J. Environ. Qual*. 30, 1919-1926.

82. McGrath, S.P., Zhao, F.J. and Lombi, E., 2001. Plant and rhizosphere process involved in phytoremediation of metal-contaminated soils. *Plant Soil*. 231,207-214.

83. McKeehan, P., 2000. Brownfields: The Financial, Legislative and Social Aspects of the Redevelopment of Contaminated Commercial and Industrial Properties.http://md3.csa.com/discoveryguide/brown/overview.php

84. McGrath, S.P. and Zhao, F.J., 2003. Phytoextraction of metals and metalloids from contaminated soils. *Curr. Opin. Biotech*. 14, 277-282.

85. Martino, E., Perotto, S., Parsons, R. and Gadd, R., 2003. Solubilization of insoluble inorganic zinc compounds by ericoid mycorrhizal fungi derived from heavy metal polluted sites. *Soil Biol. Biochem*. 35, 133-141.

86. Marchiol, L., Assolari, S., Sacco, P. and Zerbi, G., 2004. Phytoextraction of heavy metals by canola (*Brassica napus*) and radish (*Raphanus sativus*) grown on multicontaminated soil. *Environ. Pollut*. 132, 21-27.

87. Meier, S., Alvear, A., Borie.F., Aguilera, P., Ginocchio, R. and Cornejo, P., 2012a. Influence of copper on root exudate patterns in some metallophytes and agricultural plants. *Ecotoxicol. Environ. Saf*.75, 8-15.

88. Meier, S., Bolan, N., Borie, F. and Cornejo, P., 2012b.Phytoremediation of metal polluted soils by arbuscular mycorrhizal fungi. *Crit. Rev. Env. Sci. Technol.* 42, 741-775.

89. Ministere de l'Environmement du Quebec., 2001. Politique de Protection des Soil et de Rehabilitation des Terrains Contamines. Publications of the MEQ, Collection Terrains Contamines, Quebec, Canada.

90. Mulligan, C., Young, R. and Gibbs, B., 2001. Remediation technologies for metal-contaminated soils and ground-water: an evaluation. *Eng. Geol.* 60, 193-207.

91. Navarro, M., Pérez-Sirvent, C., Martínez-Sánchez, M., Vidal, J., Tovar, P and Bech, J., 2008. Abandoned mine sites as a source of contamination by heavy metals: A case study in a semi-arid zone. *J. Geochem. Explor.* 96, 183-193.

92. Nichols, K., 2003. Characterization of glomalin a glycoprotein produced by arbuscular mycorrhizal fungi. Ph D Dissertation. University of Maryland, College Park, Maryland, 285.

93. NJDEP, 1996. Soil Cleanup Criteria, New Jersey Department of Environmental Protection, Proposed Cleanup Standars for Contaminated Sites. NJAC 7:26D.

94. Orlowska, E., Ryszka, P., Jurkiewicz, A. and Turnau, K., 2005. Effectiveness of arbuscular mycorrhizal fungal (AMF) strains in colonization of plants involved in phytostabilisation of zinc wastes. *Geoderma.* 129, 92-98.

95. Pawlowska, T.E., Blaszkowski, J. and Ruhling, A., 1996. The mycorrhizal status of plants colonizing a calamine spoil mound in southern Poland. *Mycorrhiza.* 6, 499-505.

96. Paz-Ferreiro, J., Lu, H., Fu, S., Méndez, A. and Gascó, G., 2014. Use of phytoremediation and biochar to remediate heavy metal polluted soils: *a review. Solid Earth.* 5, 65-75.

97. Peuke, A. and Rennenberg, H., 2005. Phytoremediation "viewpoint". *EMBO J.* 6, 497-501.

98. Pilon-Smits, E., 2005. Phytoremediation. *Annu. Rev. Plant. Biol.* 56, 15-39.

99. Pongrac, P., Sonjak, S., Vogel-Mikuš, K., Kump, P., Nečemer, M. and Regvar, M., 2009. Roots of metal hyper accumulating population of *Thlaspi praecox* (Brassicaceae) harbour arbuscular mycorrhizal and other fungi under experimental conditions.*Int. J. Phytoremediat.***11**, 347-359.

100. Prasad, M. and Freitas, H., 2003. Metal hyperaccumulation in plants Biodiversity prospecting for phytoremediation technology. *Electron. J. Biotechn.* 6, 275-321.

101. Pulford, I.D. and Watson, C., 2003. Phytoremediation of heavy metal-contaminated land by trees –*a review. Environ. Inter.* 29, 529-540.

102. Reimann, C., Koller, F., Kashulina, H., Niskavaara, G. and Englmaier,P., 2001. Influence of extreme pollution on the inorganic chemical composition of some plants. Environ. Pollut. 115, 239-252.

103. Redecker, D., Schüßler, A., Stürmer, S., Morton, J. and Walker, C., 2013. An evidence-based consensus for the classification of arbuscular mycorrhizal fungi (*Glomeromycota*).An evidence-based consensus for the classification of arbuscular mycorrhizal fungi (*Glomeromycota*). *Mycorrhiza.* 23, 515-531.

104. Rillig, M. and Mummey, D., 2006. Mycorrhizas and soil structure. *New Phytol.* 171, 41-53.

105. Rufyikiri, G., Declerck, S., Dufey, J.E. and Delvaux, B., 2000. Arbuscular mycorrhizal fungi might alleviate aluminium toxicity in banana plants. *New Phytologist.* 148(2),343-352.

106 Ryley, R.G., Zachara, J.M. and Wobber, F.J., 1992. Chemical contaminants on DOE lands and selection of contaminated mixtures for sustrate science research. US-DOE, Energy Resource Subsurface Science Program, Washington, DC, USA.

107. Schüßler, A., Gehrig, H., Schwarzott, D. and Walker, C., 2001. Analysis of partial Glomales SSU rRNA gene sequences: implications for primer design and phylogeny. *Mycol. Res.* 105, 5-15.

108. Shah, K. and Nongkynrih, J., 2007. Metal hyperaccumulation and bioremediation. *Biol. Plantarum.* 51, 618-634.

109. Sheoran, V., Sheoran, A. and Poonia, P., 2009. Phytomining: *A review.* *Miner. Eng.* 22, 1007-1019.

110. Singh, O., Labana, V., Pandey, S., Budhiraja, G. and Jain, R., 2003. Phytoremediation: an overview of metallic ion decontamination from soil. *Appl. Microbiol. Biot.*, 61, 405-412.

111. Smith, S.E. and Read, D.J., 1997. Mycorrhizal Symbiosis. San Diego, *Academic Press.*

112. Stommel, M., Mann, P. and Franken, P., 2001. EST-library construction using spore RNA of the arbuscular mycorrhizal fungus *Gigaspora rosea*. *Mycorrhiza.* 10,281-285.

113. Sudová, R., Pavlíková, D., Macek, T. and Vosátka, M., 2007. The effect of EDDS chelate and inoculation with the arbuscular mycorrhizal fungus *Glomus intraradices* on the efficacy of lead phytoextraction by two tobacco clones. *Appl. Soil. Ecol.* 35, 163-173.

114. Sudová, R. and Vosátka, M., 2007. Differences in the effects of three arbuscular mycorrhizal fungal strains on P and Pb accumulation by maize plants. *Plant Soil.* 296, 77-83.

115. Sudová, R., Doubková, P. and Vosátka, M., 2008. Mycorrhizal association of *Agrostis capillaris* and *Glomus intraradices* under heavy metal stress:

Combination of plant clones and fungal isolates from contaminated and uncontaminated substrates. *Appl. Soil. Ecol.* 40, 19-29.

116. Tang, Y.., Deng, T., Wu, Q., Wang, S., Qiu, R., Wei, Z., Guo, X., Wu, Q., Lei, M. and Chen, T., Echevarria, G., Sterckeman, T., Simonnot, M., Morel, L., 2012. Designing cropping systems for metal-contaminated sites: *a review.* *Pedosphere.* 22,470-488.

117. Tiller, K.G., 1989. Heavy metals in soils and their environmental significance pollution. *Adv Soil Sci.* 9, 113-141.

118. Tonin, C., Vandenkoornhuyse, P., Joner, E., Straczek, J. and Leyval, C., 2001. Assessment of arbuscular mycorrhizal fungal diversity in the rhizosphere of *Viola calaminaria* and effect of these fungi on heavy metal uptake by clover. *Mycorrhiza.* 10, 161-168.

119. Trotta, A., Falaschi, P., Cornara, L., Minganti, V., Fusconi, A., Drava, G. and Berta, G., 2006. Arbuscular mycorrhizae increase the arsenic translocation factor in the As hyperaccumulating fern *Pteris vittata* L. *Chemosphere.* 65, 74-81.

120. Turnau, K., Ryszka, P., Gianinazzi-Pearson, V. and van Tuinen, D., 2001. Identification of arbuscular mycorrhizal fungi in soils and roots of plants colonizing zinc wastes in southern Poland. *Mycorrhiza.* 10, 169-174.

121. Turnau, K. and Mesjasz-Przybylowics, J., 2003. Arbuscular mycorrhiza occurrence in *Berkheya coddii* another Ni-hyperaccumulating members of Asteraceae from ultramafic soils in South Africa. *Mycorrhiza.* 13, 185-190.

122. USEPA. PART 503, 2007.Standards for the use or disposal of sweage sludge.Electronic Code of Federal Regulations.http://www.epa.gov/epacfr40/chapt-I.info/chi-toc.htm.

123. Vasak, M. and Hasler, D., 2000. Metallothioneins: new functional and structural insights. *Curr. Opin. Chem. Biol.* 4, 177-183.

124. Vögel-Mikus, K., Pongrac, P., Kump, P., Necemer, M. and Regvar, M., 2006. Colonization of a Zn, Cd and Pb hyperaccumulator *Thlaspi praecox* Wulfen with indigenous arbuscular mycorrhizal fungal mixture induces changes in heavy metal and nutrient uptake. *Environ. Pollut.* 139, 362-371.

125. Vodnik, D., Grčman, H., Maček, I., van Elteren, J.T. and Kovačevič, M., 2008. The contribution of glomalin related soil protein to Pb and Zn sequestration in polluted soil. *Sci. Total Environ.* 392, 130-136.

126. Wauna, R. A. and Okieimen, F. E., 2011. Heavy Metals in Contaminated Soils: A Review of Sources, Chemistry, Risks and Best Available Strategies for Remediation.*ISRN Ecology.* , 1-20.

127. Weissenhorn, I. and Leyval, C., 1995. Root colonization of maize by a Cd-sensitive and a Cd-tolerant *Glomus mosseae* and cadmium uptake in sand culture. *Plant Soil.*175, 233-238.

128. Wong. M.H., 2003. Ecological restoration of mine degraded soils, with emphasis on metal contaminated soils. *Chemosphere.* 50, 775-780.

129. Yang ,W.Q. and Goulart, B.L., 2000. Mycorrhizal infection reduces short-term aluminum uptake and increases root cation exchange capacity of highbush blueberry plants. *Hortscience.* 35(6),1083-1086.

130. Zhou, J.L., 1999. Zn biosorption by *Rhizopus arrhizus* and other fungi. *Appl. Microbiol. Biotechnol.* 51, 686-693.

Part V

Interaction of Arbuscular Mycorrhiza with other beneficial Microbes

Chapter 11

Arbuscular Mycorrhizal Spore Associated Microbes and their Interaction

P. Panneerselvam*[1], B. Saritha [1] and P. Ravindrababu[2]

[1]Division of Soil Science and *Agricultural Chemistry,*
Soil Microbiology Lab,
ICAR -Indian Institute of Horticultural Research,
Bengaluru – 560 089, India

[2]Sreenidhi Institute of Science and *Technology (SNIST),Ghatkesar, Hyderabad –*
501301, India.

*Address for the Correspondence: panse@iihr.ernet.in

ABSTRACT

Arbuscular mycorrhizal fungal (AMF) spores provide a long-term reservoir of beneficial bacteria, which help the AMF root colonization, spore germination and extraradical hyphal growth by producing stimulatory compounds or by influencing nutrients acquisition. These bacteria are generally called as mycorrhiza associated bacteria and are mostly fungi-specific but can also be host specific. The AMF spore associated bacteria are known to secrete metabolites, change soil pH, produce cell wall degrading enzymes and growth promoting substances for stimulation of mycorrhizal spore germination and colonization of host plants. In addition to supporting mycorrhizal colonization, these bacteria act as plant growth promoters, nutrient solubilizers and effective biocontrol agents in some of the fruits seedlings. Many findings strongly suggest the combined application of AMF with their associated bacteria for better performance of plant growth rather than single application of AMF or the bacteria. In this chapter, the different genera of AMF spore associated bacteria, their role on plant growth promotion, nutrient solubilization, antagonistic potential, enhancement of mycorrhizal colonization and yield improvement in some selected fruit crops are discussed in detail.

Keywords: *AMF, Mycorrhiza associated bacteria, colonization, antagonistic potential, nutrient solubilisation*

1.0 Introduction

AMF play a key role in liaising between the plants and the surrounding soil, thereby helping in mobilizing mineral nutrients from soil to plant. During the AMF colonization of crop plants, the fungal spores and hyphae provide sites for free living bacteria to thrive on their surfaces. Bacteria involved in mycorrhizal establishment and/or its functioning were defined as Mycorrhiza associated bacteria (MAB) by Garbaye (1994) and are currently the most investigated group among mycorrhizal helper bacteria interacting with mycorrhizas (Frey-Klett *et al.*, 2007). These associated bacteria play a key role in promoting mycorrhizal colonization of host plants by the secretion of metabolites that help in the proliferation of the fungal hyphae and promote the colonization of host plant roots. Though AMF encompass 18 genera, the MAB associations have been reported only in limited genera, more commonly in *Glomus* sp. These MAB produce many known and unknown metabolites, which will allow easy proliferation of the fungal hyphae and colonization of the host-plant roots (Schrey *et al.*, 2005). In the case of AMF, many examples of MAB have been described in the literature since the first mention by Mosse (1962) in the genus *Glomus* (**Table 11.1**). The MAB strains that have been identified to date belong to many bacterial groups and genera, such as gram-negative Proteobacteria (*Agrobacterium, Azospirillum, Azotobacter, Burkholderia, Bradyrhizobium, Enterobacter, Pseudomonas, Klebsiella* and *Rhizobium*), gram-positive Firmicutes (*Bacillus, Brevibacillus* and *Paenibacillus*) and gram-positive Actinomycetes (*Rhodococcus, Streptomyces* and *Arthrobacter*). Many plant models have been used to study the MAB effect, including herbaceous and woody plant species, mainly from temperate ecosystems. Only a few studies have focused on tropical plant species.

The production of volatile compounds by AMF associated bacteria positively influence the germination of AMF spores. Other mechanisms by which the associated bacteria influence AMF colonization of roots include a change in the pH, production of cell wall degrading enzymes and growth promoting substances by AMF spore associated bacteria. Recent findings (Panneerselvam *et al.*, 2012a and 2013, Sukhada *et al.*, 2013; Poovarasan *et al.*, 2013, Saritha *et al.*, 2014a) clearly documented the beneficial effects of AMF associated bacteria in terms of growth promotion, nutrient mobilization, disease suppression, enhancement of mycorrhizal colonization and improvement of yield in perennial tropical fruit crops like sapota, guava and banana. They have also reported that at the time of

AMF application, inclusion of their associated bacteria is very essential for getting more benefits than AMF alone. In view of this, the beneficial effect of MAB on fruit crops production are discussed in detail in this chapter.

2.0 AMF spore associated bacteria in fruit crops

Mycorrhizal associated bacteria were isolated from *Glomus* spores colonizing the root system of guava and sapota orchards of Karnataka, India (Paneerselvam *et al.*, 2012a ; 2013). The mycorrhizal spores found to be more at a depth of 15-20 cm of the rhizosphere) (Saritha *et al.*, 2014b) (In press). The major group of bacterial isolates from these spores were *P. putida*, *P. aeruginosa*, *Brevibacillus* sp. and *Bacillus subtilis*. Sukhada *et al.*, (2013) reported *Streptomyces fradiae*, *Streptomyces avermitilis*, *Streptomyces cinnamonensis* and *Leifsonia poae* from *Glomus* species colonizing roots of guava plants.

Some mycorrhizal fungi harbour bacterium-like organisms (BLOs) in their cytoplasm (MacDonald, 1982) but these BLOs cannot be identified as they cannot be grown on cell-free media. Bianciotto *et al.* (1996) used a combined morphological and molecular approach to demonstrate that the cytoplasm of the AMF *Gigaspora margarita* contains bacterial endosymbionts. Analysis of the small-subunit rRNA gene sequence of the BLOs in *G. margarita* spores led to the conclusion that these endosymbionts were closely related to the genus *Burkholderia*. In order to determine whether intracellular bacteria occur sporadically in individual AMF isolates or as a common feature in the family Gigasporaceae, Bianciotto *et al.* (2000) investigated two geographically separated isolates of the species *G. margarita* and five other species in the genera *Gigaspora* and *Scutellospora*. The results showed that all investigated species except *G. rosea* contained endosymbiotic bacteria closely related to the genus *Burkholderia*. Further studies by Minerdi *et al.* (2001) demonstrated that *Burkholderia* sp. in *G. margarita* contained genes involved in N fixation.

3.0 Bacterial interaction with mycorrhiza

Bharadwaj *et al.* (2012) observed that the exudates of *G. irregulare* contain carbohydrates, amino acids and unidentified compounds, which acted as stimulants for the growth of AMF associated bacteria. The association of bacteria with AMF spores depended on AMF exudates rather than on root exudates of host plants (Roesti *et al.*, 2005). These studies suggested that association of bacteria on the surface of AMF spores could vary from genus to genus. Some mechanisms are being proposed by Frey-Klett *et al.* (2007) to understand the success of mycorrhiza associated bacteria which involved the ability to produce growth factors that stimulate fungal spore germination, mycelial growth, increased root branching and root colonization, and by reducing the soil-mediated stress through detoxification of antagonistic substances and inhibition of competitors and antagonists. These bacteria produced some specific enzymes, which softened the root tissue for easy colonization and root-fungus recognition.

Paenibacillus validus associated with *G. intraradices* was found to support the growth and sporulation of the *Glomus* sps . (de Boer *et al.*, 2005) and this association was highly efficient in sustaining fungal growth and germination of new spores by release of sugars and unidentified compounds in the rhizosphere of plant system. This finding indicated that AMF can grow independently of the host plant in the presence of their closely associated bacteria (Hildebrandt *et al.*, 2006). Besides change in soil pH, production of cell wall degrading enzymes and growth-promoting substances these mycorriza associated bacteria are also associated with different mechanisms for stimulation of mycorrhizal colonization in host plants (Nazir *et al.*, 2010; Bharadwaj *et al.*, 2012). During the time of AMF colonization in crop plants, the fungal spores and hyphae provide sites for some bacteria to live (Schuessler *et al.*, 1994). The beneficial effects of AMF in the rhizosphere are the result of synergistic interactions among the rhizosphere microbes and which in turn enhance the plant growth. Thus, the relationship between AMF and their associated bacteria may be of great importance for sustainable agriculture (Barea *et al.*, 1997). The interaction of AMF with bacterial association may give positive or negative effects (Sylvia *et al.*, 1998) depending on the associated bacteria. In general, not all mycorrhiza associated bacteria are useful for establishment of AMF colonization in crop plants. These associated bacteria play very important role in promoting mycorrhizal colonization in host plants by the secretion of metabolites helpful in the easy proliferation of the fungal hyphae and colonization of host plant roots (Schrey *et al.*, 2005). The production of volatile compounds by AMF associated bacteria positively influences the germination of spores by supplying nitrogen through nitrogen fixation and other mineral nutrient uptake.

The studies suggest that release of active diffusible molecules and physical contact between bacteria and mycorrhizal fungi are important for the establishment of their interactions as identified in rhizobia and legume plants (Oldroyd and Downie, 2008). Here, the partners release active diffusible molecules that are reciprocally perceived (Bouwmeester *et al.*, 2007), also activating Ca-mediated responses (Bonfante *et al.*, 2009; Kosuta *et al.*, 2008; Navazio *et al.*, 2007), while a physical contact between the fungus and the plant is required to elicit many of the plant responses that precede fungal colonization (Genre *et al.*, 2008; Genre *et al.*, 2005). Recent data also suggest that production of volatile organic compounds (VOCs) from all the members of the underground consortium may be important for inter- and intra-organismic communication (Tarkka and Piechulla, 2007). Bacterial VOCs affect soil fungi, including the mycorrhizal ones (Tarkka and Piechulla, 2007). These bioactive molecules are important determinants for symbiosis establishment. Thus, integration of descriptive data with genomics, proteomics, and metabolomics will offer glimpse of the mechanisms taking place between plants and fungi, there is a need to identify the molecules that conFigureure the basis of interaction between MABs and their fungal and plant hosts.

Table. 11.1: List of Some Mycorrhiza Associated Bacteria (MAB)

AMF	Species of AMF	Bacteria associated with AMF	Reference
Glomus Associated Bacteria	*Glomus mosseae*	*Paenibacillus* sp. *Pseudomonas* sp. *Bradyrhizobium japonicum Pseudomonas fluorescens Brevibacillus* sp.	Budi *et al.* (1999) Barea *et al.* (1998) Xie (1995) Gamalero *et al.* (2004) Vivas *et al.* (2003)
	Glomus mosseae	*Pseudomonas putida Brevibacillus parabrevis Providencia rettgeri Pseudomonas aeruginosa Brevibacillus* sp. *Bacillus subtilis*	Panneerselvam *et al.* (2012a) Panneerselvam *et al.* (2013)
	Glomus mosseae	*Streptomyces fradiae Streptomyces avermitilis Streptomyces cinnamonensis Leifsonia poae Streptomyces canus*	Sukhada *et al.* (2013)
	Glomus fasciculatum	*Azotobacter chroococcum*	Bagyaraj and Menge (1978)
	Glomus intraradices	*Bacillus subtilis Pseudomonas monteilii Rhizobium Agrobacterium rhizogenes, Pseudomonas fluorescens Rhizobium leguminosarum Streptomyces coelicolor Pseudomonas monteilii*	Toro *et al.* (1997) Duponnois and Plenchette (2003) Requena *et al.* (1997) Fester *et al.* (1999) Abdel-Fattah and Mohamedin (2000)
	Glomus constrictum Glomus geosporum	*Cellvibrio, Chondromyces, Flexibacter, Lysobacter, Pseudomonas*	Roesti *et al.* (2005)
	Glomus clarum	*Azotobacter diazotrophicus*	Paula *et al.* (1992)
	Glomus deserticola	*Alcaligenes denitrificans*	Will and Sylvia (1990)
From mixed spores of *Glomus* sp.	*Glomus mosseae Glomus intraradices*	*Paenibacillus brasiliensis*	Artursson (2005)
From mixed spore of *Glomus* sp.	*Glomus fasciculatum Glomus mosseae*	*Rhizobium meliloti*	Azcón *et al.* (1991)

Table Contd....

Table 11.1: Contd....

AMF	Species of AMF	Bacteria associated with AMF	Reference
	Glomus fasciculatum Glomus mosseae Glomus caledonium	Bacillus coagulans	Mamatha et al. (2002)
Gigaspora Associated Bacteria	Gigaspora margarita	P. polymyxa and J. lividum Paenibacillus sp. Brevibacillus sp. Bacillus sp. Candidatus Glomeribacter gigasporarum Azospirillum brasilense	Cruz et al. (2008) Long et al (2009) Long et al (2009) Rao et al. (1985)
Gigaspora and Scutellospora Associated Bacteria	Gigaspora margarita Scutellospora persica Scutellospora castanea	Candidatus Glomeribacter gigasporarum	Bianciotto et al. (2003)
Endogone Associated Bacteria	Endogone sp.	Pseudomonas sp.	Mosse (1962)
Mixed spores of AMF	AMF mixed species	Pseudomonas putida	Meyer and Linderman (1986
Mixed spores of AMF	AMF mixed species	Pseudomonas sp.	Babana and Antoun (2005)
Mixed spores of AMF	AMF mixed species	Bacillus mycoides	von Alten et al. (1993)

Filippi *et al.* (1998) used transmission electron microscopy to show that bacteria were associated with the surfaces of peridium-covered spore clusters (sporocarps), surfaces of spores, hyphae of *Glomus mosseae*. Bacterial cells were found embedded in spore walls, possibly by tunnels produced in the walls by the bacteria. The study determined that large numbers of bacteria, actinomycetes, and non-mycorrhizal fungi, some of which displayed chitinolytic abilities, occurred on the sporocarp surfaces and in the sporocarp homogenate. Cruz and Ishii (2012) isolated three closely associated bacteria, *Bacillus* sp., *B. thuringiensis* and *Paenibacillus* sp. from *Gigaspora margarita* spores. Roesti *et al.* (2005) isolated the bacteria Cellvibrio, *Chondromyces, Flexibacter* and *Pseudomonas* from the surface of *G. constrictum* and *G. geosporum* spores. They suggested that the most of the organisms which were found to be associated with the above mentioned *Glomus* spores were having the ability to degrade the chitin cell wall of spores. Sukhada *et al.* (2013) reported that the chitin degrading activity was very intensive with the *Glomus* associated actinobacteria *Streptomyces avermitilis* which showed three times more activity in comparision with the other 4 isolates of *streptomyces* sp.

4.0 MAB and their role in plant growth promotion and nutrient cycling

4.1 Effect of MAB on AMF Spore germination and root colonization

As early as 1962, Mosse (Mosse, 1962) showed that some MAB and their culture filtrates were able to stimulate arbuscular mycorrhizal fungal spore germination of *G. mosseae*. Direct contact between the spores and bacteria was necessary for the induction of spore germination in *Glomus clarum* (Xavier and Germida, 2003), indicating a ligand–receptor interaction between the two microbes. These spore germination stimulatory bacteria were accompanied by other bacterial isolates producing antagonistic volatiles, suggesting the presence of a complex bacterial consortium on the *G. clarum* spore surface that regulates germination. In contrast, volatile compounds produced by different species of *Streptomyces* were proved to promote the germination of *G. mosseae* spores (Tylka *et al.*, 1991). Working with the *Paenibacillus validus–Glomus intraradices* interaction, Hildebrandt *et al.* (2002) showed that the obligately symbiotic *G. intraradices* could grow and sporulate in fungus bacterium cocultures. A specific carbon source, raffinose, was detected in bacterial cultures and mycelial growth was supported by this sugar (Hildebrandt *et al.*, 2006).

Plants are known to produce some chemical compounds which act as signals for attracting microbial colonization in rhizosphere and these chemical compounds include flavonoids as reported by Lagrange *et al.* (2001) and Akiyama *et al.* (2002). Some reports say that the bacteria associated with mycorrhiza have the capability to induce the plant to release these flavonoids facilitating mycorrhizal colonization by indirect means. For example, MAB strain of *Bradyrhizobium japonicum* produces Nodulation factors which found to stimulate the flavonoid production thereby improving the mycorrhiza colonization in host plants (Xie, 1995). It was demonstrated that mycorrhiza colonization was increased even by the artificial application of nod factors and falvonoids. This suggests that flavonoids may play a major role in stimulating the mycorrhiza colonization by inducing nod factors. Similar to nod factors, some MAB effectors such as cell-wall digesting enzymes (Mosse, 1962) are found to facilitate root colonization by increasing the rate of penetration and spreading of the fungus within the root cortex.

Xavier and Germida (2003) observed that the bacteria from AMF spore cell walls were able to promote *G. clarum* germination and this association was found to improve mycelial extension and mycorrhizal colonization (Garbaye and Bowen, 1989; Gryndler and Vosatka, 1996; Schrey *et al.*, 2005). Significant increase in AMF mycelial development was reported in host plants when the AMF was co-inoculated with *P. putida* or *P. validus* (Gryndler and Vosatka, 1996). Assessing mycorrhization improvement in sapota seedlings, the co-inoculation significantly increased percentage of root colonization by 89% and spore number compared to other isolates. Studies involving inoculation of *P. putida* with *G. mosseae* to sapota and guava seedlings in the nursery by Panneerselvam *et al.* (2012a and 2013) revealed that co-inoculation of AMF and its associated bacteria in sapota seedlings (cv Cricket ball) significantly increased AMF spore number by 30.0% and root colonization by 13.4% when compared to AMF application alone. Similarly guava seedlings treated with AMF and its associated bacteria improved the AMF spore number by 27.5% and root colonization by 24.1% when compared to AMF application alone(**Plate 11.2**).

Table. 11.2: Effect of MAB on AMF spore proliferation and Root colonization in Sapota and Guava(180DAP) (Panneerselvam *et al.*, 2012a ; 2013)

Treatments	Sapota seedlings		Guava seedlings	
	AMF spore load (number g soil-1)	Mycorrhizal colonization(%)	AMF spore load (number g soil-1)	Mycorrhizal colonization(%)
AMF alone	10.5	67.5	14.7	65.4
MAB alone	7.4	29.7	11.2	47.9
AMF +MAB	15.0	78.0	20.3	86.2
Control	5.00	21.5	6.1	33.7
SEd	0.12	0.75	0.17	0.68
CD(p=0.05)	0.27	1.63	0.37	1.47

Many researchers also reported that the application of MAB positively improved lateral root development (Garbaye, 1994; Vivas *et al.*, 2003a; Schrey *et al.*, 2005), which might be due to the production of auxins or auxin-related substances by the bacteria. In the host plants, the formation of novel root tips may lead to the establishment of more mycorrhizae. For example, *Paenibacillus* sp. EJP73 and *Burkholderia* sp. EJP67, two strains isolated from mycorrhizae, were found to promote dichotomous root branching in Scots Pine (*Pinus sylvestris*) seedlings (Aspray *et al.*, 2006). Similarly, co-inoculation of *P. fluorescens* 92 or *P. fluorescens* P190r with *G. mosseae* BEG12 led to strongly increased plant growth (Gamalero *et al.*, 2004).

One of the symbiotic effects of AMF and soil bacteria on plant performance is by inducing systemic resistance which is also influenced by plant hormones (Van Wees *et al.*, 2008). AMF can also alter the amount of plant hormones, jasmonic acid (JA) and abscisic acid (ABA) (Ludwig-Muller, 2000; Hause *et al.*, 2007; Grunwald *et al.*, 2009). Both hormones are necessary for the establishment of AMF symbiosis (Isayenkov *et al.*, 2005). In addition, the newly classified plant hormones, strigolactones, can also cause mycorrhization in plants, which is also influenced by mitochondrial activities (Akiyama *et al.*, 2005; Akiyama and Hayashi, 2006; Besserer *et al.*, 2009). Under the nutrient deficiency stress, plants may increase the level of strigolactones, resulting in the reduction of shoot branching and enhancement of mycorrhization. The elucidation of interactions between auxin, cytokinins and strigolactones may provide more details regarding root branching (Shimizu-Sato *et al.*, 2009) as well as the establishment of mycorrhizal establishment.

Fungus–bacterium cocultures are easily produced and used in screening their symbiotic effects in field inoculations. A significant correlation has been shown to exist between increases in mycelial biomass and promotion of mycorrhiza establishment (Garbaye and Bowen, 1989; Gryndler and Vósatka, 1996; Founoune *et al.*, 2002). The arbuscular mycorrhizal colonization of roots with *Glomus fistulosum* and the growth rate of the hyphae in the soil substrate were significantly higher when the fungus was coinoculated with *Pseudomonas putida* or with the low-molecular-weight fraction of the bacterial culture supernatant (Vósatka and Gryndler, 1999), indicating that the effective substances were in this fraction.

4.2 Effect of MAB on Plant Growth Promotion

The enhancement of plant growth in mycorrhiza fungus-treated plants seems to be indirect, mainly through larger root systems, which enhance higher water and nutrient uptake. Many scientific observations indicate that early mycorrhizal inoculation and colonization of perennial crop seedlings reduces transplanting shock, thus improving establishment rates in the field. Recent findings indicated that application of *G. mosseae* along with *P. putida* produced significantly increased sapota seedling height (16.0cm), total biomass (1.91g/plant), total leaf area (189 sq cm.) and mycorrhizal colonization (89.7%) compared with other isolates (Panneerselvam *et al.*, 2013). The pooled nursery data of inoculation effect of AMF and its associated bacteria in sapota seedlings showed that co-inoculation of Mycorrhiza associated bacteria(MAB) with AMF significantly increased the total biomass by 11.4% when compared to AMF inoculation alone. The effect of co-inoculation of MAB isolate *Pseudomonas putida* along with *Glomus mosseae* on shoot and root growth in sapota seedlings (180DAP) is shown in **Plate 11.1**

A   B

Plate 11.1: Increased shoot growth (A) and root branching (B) in MAB (*P. putida*) and AMF treated sapota seedlings compared to control plant on the respective left . (Saritha, Ph.D work under publication).

Streptomyces avermitilis   *Streptomyces cinnamonensis*   *Streptomyces canus*

Plate 11.2: Growth promotion by selected Actinomycetes strains isolated from *Glomus mosseae* on 8 month old guava *cv* Arka Kiran seedlings under nursery conditions. Controls on the left. (Source: Poovarasan Ph.D work underpublication).

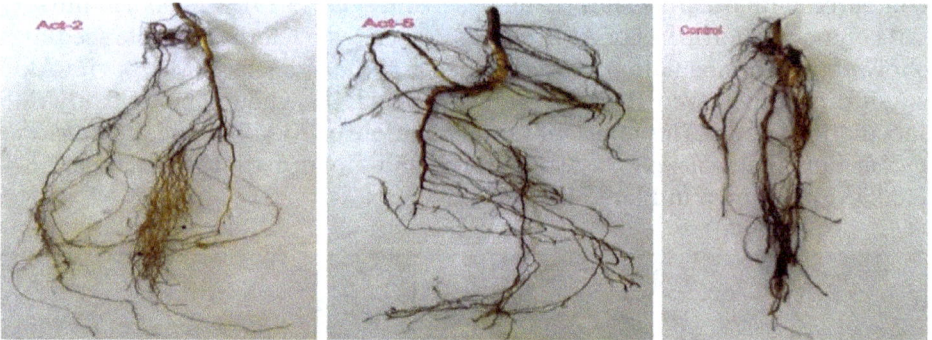

Plate 11.3: Root growth promotion by *Glomus mosseae* associated actinomycetes (Act 2 – *Streptomyces avermitilis* and Act 5 – *Streptomyces canus*) in guava cv Arka Kiran seedlings under glass house conditions(Poovarasan, *et al.,* 2013).

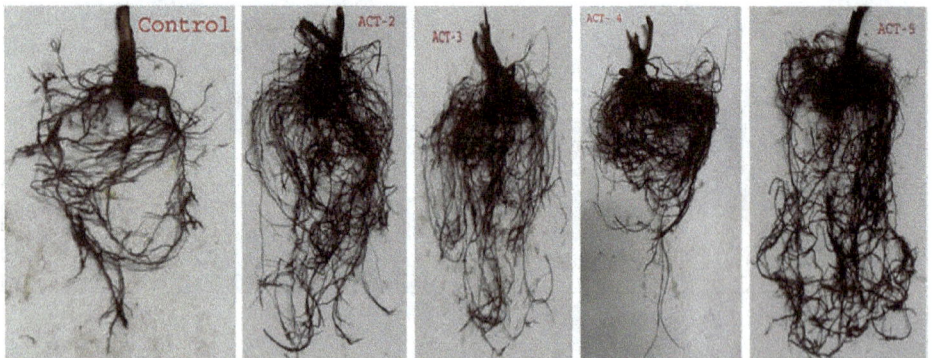

Plate 11. 4: Root growth promotion in pomegranate cv Bhagva by mycorrhizae associated actinomycetes (Act-2-*S. avermitilis*, Act-3 *S. cinnamonensis*, Act-4 L. *poea* and Act-5- S. *canus*) under glass house condition. (Poovarasan, *et al.,* 2013).

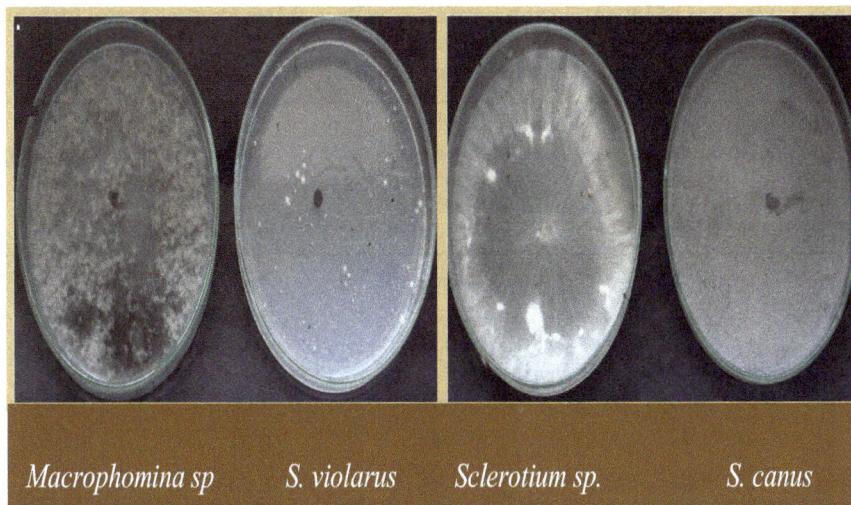

| Macrophomina sp | S. violarus | Sclerotium sp. | S. canus |

Plate 11.5: Antagonistic Potential of *Streptomyces* sp. isolated from *Glomus mosseae* against *Macrophomina* sp. and *Sclerotium* sp. *Source: Saritha et al., 2014 (Under Publication)*

Similar studies with guava seedlings showed that there was an increase of 14.2% in total biomass due to co-inoculation of AMF.with their associated bacteria over AMF application alone **(Table 11.3)**. Panneerselvam *et al.*, 2012a reported that in guava seedlings along with increased growth promoting activity, application of *G. mosseae* and its MAB isolate *Pseudomonas putida* significantly increased the soil phosphatase activities by 7.5%, leaf chlorophyll by 11.3% and total phenol by 12.1% over individual inoculation of *G. mosseae* (Table 11.4) **(Plate 11.3)**. The same was observed with the inoculation of *Glomus* associated *Streptomyces* sp. to the pomegranate seedlings wherein there was significant increase in biomass by 68-277% (Poovarasan *et al.*, 2013) **(Plate 11.4).**

Table. 11. 3: Effect of inoculation of AMF and its associated bacteria on growth promotion in Sapota and Guava (180DAP) (Panneerselvam *et al.*, 2012a ; 2013)

Treatments	Sapota seedlings			Guava seedlings		
	Shoot dry weight(g plant⁻¹)	Root dry weight (g plant⁻¹)	Total biomass (g plant⁻¹)	Shoot dry weight(g plant⁻¹)	Root dry weight(g plant⁻¹)	Total biomass (g plant⁻¹)
AMF alone	1.2	0.37	1.47	10.8	4.2	15.1
MAB alone	1.1	0.31	1.42	10.6	4	14.6
AMF +MAB	1.3	0.36	1.66	12.5	5	17.6
Control	1	0.3	1.3	8.6	3.6	12.3
SEd	0.01	0	0.01	0.07	0.03	0.1
CD(p=0.05)	0.02	0	0.02	0.15	0.06	0.22

Table 11.4: Effect of MAB on phosphatase activity, leaf chlorophyll and total phenol in Guava seedling (300DAP).
(Panneerselvam *et al.*, 2012a)

Treatments	Acid phosphatase activity (µg PNP g^{-1} soil h^{-1})	Alkaline phosphatase activity (µg PNP g^{-1} soil h^{-1})	Total Chlorophyll (mg g^{-1} fresh leaf)	Total phenol (µg g^{-1} fresh leaf)
AMF alone	69.0	64.3	1.33	187.2
MAB alone	54.3	58.5	1.20	161.4
AMF +MAB	75.3	69.0	1.50	213.2
Control	44.8	43.4	1.04	159.5
SEd	0.49	0.42	0.01	1.19
CD(p=0.05)	1.07	0.93	0.02	2.59

These investigations clearly proved that inclusion of AMF associated bacteria is highly beneficial when applying AMF for growth promotion of sapota and guava seedlings (Panneerselvam *et al.*, 2012 a; 2013). In continuation of this experiment, a study was conducted for the past three years on field application of AMF and its associated bacteria in sapota crop at IIHR, Bengaluru. Results indicated that the mixed inoculum of AMF and its associated bacteria increased fruit yield by 22.5% over AMF application alone and 56.0% over un-inoculated control (**Figure 11.1**).

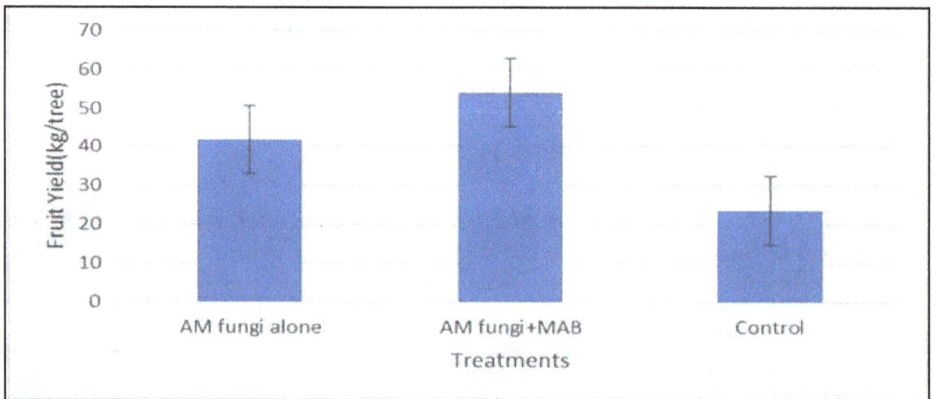

Fig. 11.1. Effect of AM fungi and its associated bacteria in sapota fruit bearing

Source: Saritha *et al.*, 2014 (Unpublished results)

The growth promotion observed due to the inoculation of *G. mosseae* with their associated bacteria might be associated with higher production of growth hormones. Panneerselvam *et al.* (2012a and 2013) reported that the MAB isolates from *Glomus* sp. of sapota and guava cropping system could produce growth hormones (IAA and GA3) in the range of 135.0 to 605.3 ng ml-1 and 222.5 to 799.0 ng ml^{-1} respectively (**Figure 11.2**).

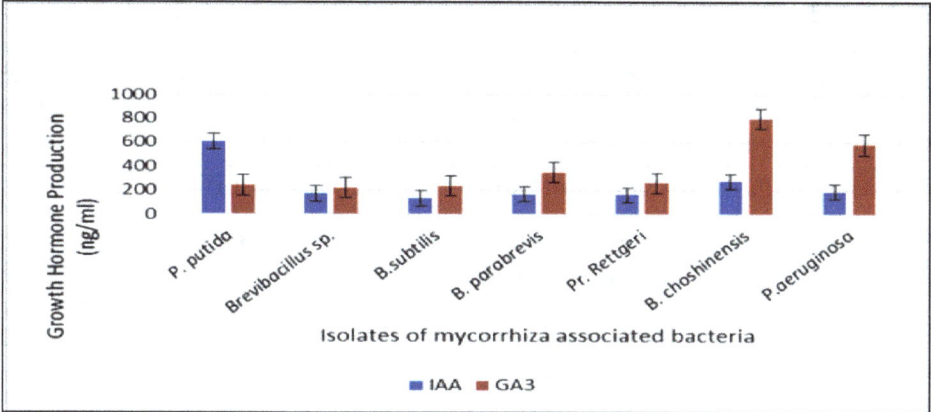

Fig. 11.2. Growth Hormone Production of mycorrhiza associated bacteria(MAB)

Source: Panneerselvam *et al.,* 2012a: 2013

The actinobacterial isolates from *Glomus* sp. were also able to produce IAA (5.3- 10.1µg ml⁻¹) and GA3 (6.3-12.01µg ml⁻¹) as reported by Sukhada *et al.* (2013) (**Figure 11.3**).

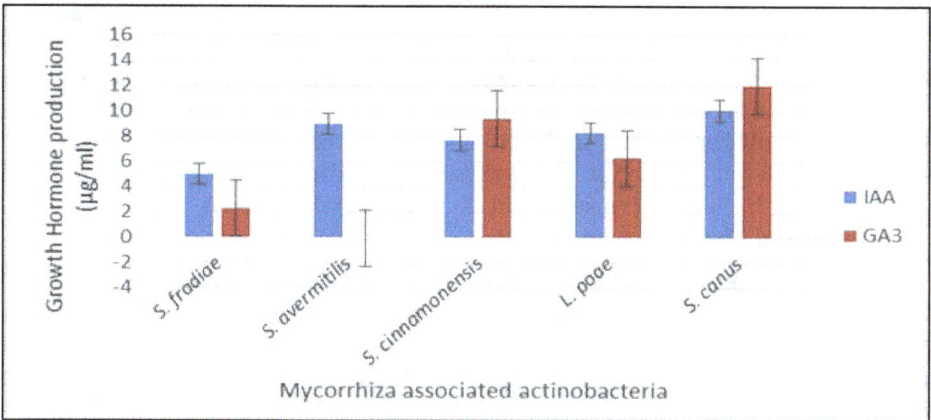

Fig. 11.3. Growth hormone production by mycorrhiza associated actinobacteria

Source: Sukhada *et al.,* 2013

Many *Pseudomonas* sp. are also considered to be plant growth-promoting rhizobacteria, which can stimulate plant growth (Kloepper *et al.* 1980; Panneerselvam *et al.* 2006). The combined inoculation of *Pseudomonas* and AMF has been shown to increase leaf area, root length and dry weight of plants in many crops (Peng *et al.*, 1993; Panneerselvam *et al.*, 2008) and had positive effects on the growth of seedlings. Similarly, dual inoculation of *G. fasciculatum* and *Azotobacter* sp.

increased total root length of air-layered litchi shoots by 81.39% over uninoculated control (Sharma *et al.*, 2009). Many researchers have shown that application of AMF increases plant height, canopy volume, mean leaf area, number of new shoots and dry biomass in various plants (Smith and Smith, 2011; Abdel-Fattah and Arsar-Wasea, 2012; Shamshiri *et al.*, 2012).

4.3 Effect of MAB on Nutrient Solubilization and Mobilization

AMF have been shown to improve productivity in low fertility soils (Jeffries, 1987) and are particularly important for increasing the uptake of slow diffusing ions (Jacobsen *et al.*, 1992), immobile nutrients such as P, Zn and Cu (Lambert *et al.*, 1979; George *et al.*, 1994, 1996; Ortas *et al.*, 1996; Liu *et al.*, 2002). The increased efficiency of mycorrhizal roots versus non-mycorrhizal roots is caused by the active uptake and transport of different nutrients by mycorrhizae.

There are some bacteria (P solubilizing bacteria) in the soil that are able to enhance P uptake by AMF species and plant through enhancing the solubility of soil P, present in organic and inorganic forms (Zabihi *et al.*, 2010; Salimpour *et al.*, 2010). Organic and inorganic P are made available by phosphatase and organic acid producing bacteria respectively. This significantly increases P uptake by AMF hyphae (Smith and Read, 2008). The soil inorganic form of P, which is not available to plants, is strongly bound in the insoluble structures of P and is also attached to the clay surface layers (Zabihi *et al.*, 2010; Salimpour *et al.*, 2010), wherein P solubilizing bacteria helps to get unvailable P to available one. The synergistic effects of AMF and P solubilizing soil bacteria has been indicated (Kim *et al.*, 1998). Under limited P availability, the interaction effects between P solubilizing bacteria and AM fungi result in the enhanced plant colonization by AMF and the increased bacterial population in the rhizosphere. The coinoculation of AMF and P solubilizing bacteria increased plant N and P uptake, relative to control plants (Artursson *et al.*, 2006).

The combined inoculation of AMF and P solubilizing bacteria was the most efficient treatment, significantly enhancing P absorption by plant. P solubilizing bacteria increased the amount of available P from the rock phosphate, absorbed by AMF hyphae (Barea *et al.*, 2005; Zabihi *et al.*, 2010; Salimpour *et al.*, 2010). This can be of particular significance for the proper development of P fertilization. The ability of MAB to solubilise zinc has been studied by number of workers. Panneerselvam *et al.* (2012a) reported that all the MAB isolates of guava rhizosphere could solubilize phosphorous and zinc. *Pseudomonas putida* among them, recorded significantly higher solubilisation of zinc phosphate (163.2 ppm) compared to Zinc oxide (117.1ppm). Phosphate solubilization activity was found to be higher with *Bacillus subtilis* (21.2mg from 100 mg of TCP in broth) when compared to *P.putida* **(Table 11.5)**.

Panneerselvam *et al.* (2013) reported that *Pseudomonas putida* isolated from *G. mosseae* spores from sapota rhizosphere recorded significantly higher zinc phosphate solubilization (206.0 µg ml^{-1}) and Zinc oxide (232.4 µg ml^{-1}). Zinc phosphate solubilization by *P. fluorescens* was investigated by Di Simine *et al.* (1998). They found that gluconic acid and 2-keto gluconic acids produced in the culture broth helped in the solubilization of the zinc salts. Since all the cultures showed a shift in pH towards the acidic range, which suggested that organic acid might be involved in the solubilization of zinc compounds. Though the soil is rich in total zinc, it is not available to the plants. Under this situation, application of *G. mosseae* along with *P. putida* could be useful to alleviate zinc deficiency in fruit crops cultivation

Table. 11. 5: Studies on ability of MAB on Phosphorous and Zinc Solubilization in guava rhizosphere (Panneerselvam *et al.*, 2012a)

Treatments	P solubilization from 100 mg of TCP in broth (mg)	Zinc Oxide (ppm)	Zinc Carbonate (ppm)	Zinc Phosphate (ppm)
P. putida	18.6	117.1	11.0	163.6
B. subtilis	21.2	85.7	13.2	143.2
Brevibacillus sp.	12.6	22.4	10.6	126.6
P. aeruginosa	14.7	23.1	9.2	82.1
SEd	0.11	1.09	0.06	1.01
CD(p=0.05)	0.23	2.38	0.14	2.19

6.4 Effect of MAB on Soil aggregation and Glomalin production

Soil structure plays an important role which supports plant and animal life, maintains the quality of environment and water, nutrient and gas fluxes. Soil structure is often expressed as the degree of stability of aggregates being a critical factor which moderates physical, chemical, and biological processes leading to the soil dynamics (Bronick and Lal, 2005). Soil aggregates results from a combination of primary mineral particles with organic and inorganic materials. The principle behind the formation of soil aggregation is complex. The formation of soil aggregation is governed by many factors, of which soil microbes play a major role which was proved by many researchers. Soil aggregation is a complex process which is mainly dependent upon secretion of microbial polysaccharides or gummy substances that hold soil particles together. Among the many soil microbes, AMF are considered to be primary soil aggregators and its influence on soil structure is given in **Fig. 11.4**.

Glomalin , a glycoprotein produced by AMF hyphae have a cementing capacity to maintain soil particles together. They play a key role in the soil aggregation, affecting the carbon (C) dynamics in agro ecosystems (Singh, 2012). In general AMF takes carbon from the plant to grow and then produce glomalin, which is firmly incorporated into the hyphae and spore walls in large amounts (Driver *et al.*, 2005) and is positively correlated with soil aggregate stability (Wright and Upadhyaha, 1998). Soil quality is also a result of combined action of extracelluar hyphae and glomalin (Rilling *et al.*, 2002; Bedini *et al.*, 2008). Glomalin levels in the soil ranged from 1.6 to 2.3 mg / g soil (Bedini *et al.*, 2010). Total Glomalin content in mycorrhiza inoculated rhizosphere soil is estimated by extraction procedures by Wright and Jawson (2001). In guava seedlings, application of *G. mosseae* along with *P. putida* increased total glomalin content by 40% compared with the uninoculated control, whereas *G. mosseae* alone increased it by 22% (Panneerselvam *et al.*, 2012a). The increase in glomalin with *G. mosseae* with their associated bacteria *P. putida* inoculation resulted in better soil aggregation, which has a positive influence on guava seedling growth. This finding clearly indicated that mycorrhiza associated bacteria also assisted in glomalin production by AMF

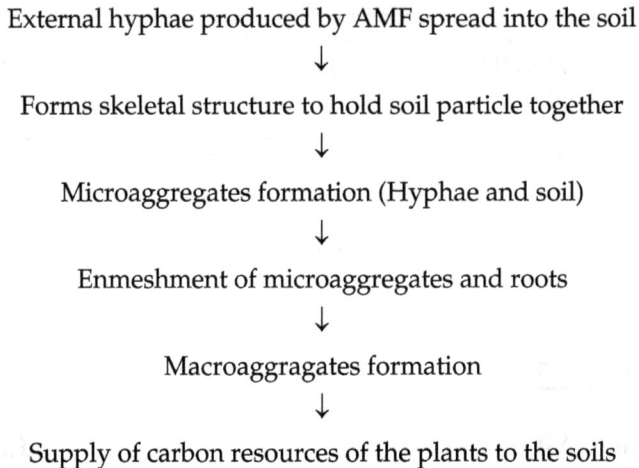

External hyphae produced by AMF spread into the soil

↓

Forms skeletal structure to hold soil particle together

↓

Microaggregates formation (Hyphae and soil)

↓

Enmeshment of microaggregates and roots

↓

Macroaggragates formation

↓

Supply of carbon resources of the plants to the soils

Fig.11.1 Influence of soil structure by AMF (Miller and Jastrow 1990).

5.0 MAB and their role in biotic stress management

5.1 Impact of MAB inoculation on plant disease suppression

In general, the bacteria associated with mycorrhiza possess antagonistic activity against many soilborne pathogens (Xavier and Germida, 2003, Sukhada *et al.* 2013). The *Paenibacillus* sp. B2, isolated from the mycorrhizosphere of *G. mosseae* showed antagonistic activity against soilborne pathogens and this bacteria also reported to produce antibiotic compounds *viz.*, polymyxin B1 and two other polymyxin-like compounds (Budi *et al.*, 2000; Selim *et al.*, 2005). *Bacillus simplex, B. niacini, B. drententis, Paenibacillus* sp. and *Methylobacterium* sp. isolated from surface disinfected spores of *G. mosseae* showed antagonism against various soilborne pathogens (*P. nicotianae* particularly, but also *F. solani* and three stains of *F. oxysporum*) (Lioussanne, 2007). Application of bacterial antagonistic *E. agglomerans* and *B. subtilis* along with arbuscular mycorrhizal fungus *Glomus intraradices* significantly reduced the infection of apple plants with *P. cactorum* and increased fruit yield and tree trunk growth (Utkhede and Smith, 2000).

Similarly, recent work on suppression of white root rot disease in Japanese apricot (*Prunus mume*) seedlings and also in grape and Figure seedlings was investigated by Cruz *et al.* (2014). The synergistic effect of AMF *Gigaspora margarita* and its MAB *Paenibacillus rhizospherae* indicated that there is significant decrease in *Rosellinia necatrix* infection rates and also increased AMF colonization in the presence of the MAB *Paenibacillus rhizospherae* before (74.7%) and after (57.2%) the disease infection. This indicates that the AMF and bacteria could be used as biocontrol agents against white root rot disease.

Many scientific reports documented that actinobacteria, particularly *Streptomyces* sp, have shown to be excellent bio-control agents against soil borne fungal pathogens. Poovarasan *et al.* (2013) reported that *Glomus* associated *Streptomyces* sp. had strong antagonsitic activity against *Xanthomonas axonopodis pv punicae* (Xap). Among different actinobacteria, *S. fradiae* and *S. cinnamonensis* showed significant growth inhibition (86.6%) of *X. axonopodis pv punicae* followed by *S. avermitilis* (83%) in the total plant bio-assay under poly-house condition. Large scale utilization of these actinomycetes on-farm would ensure control of bacterial blight of pomegranate in effective manner.

Similarly, the different *Streptomyces* sp. isolated from the surface sterilized *Glomus mosseae* had strong antagonistic activity against many soil borne pathogens viz., *Fusarium sp.*, *Alternaria* sp., *Macrophomina* sp. and *Sclerotium* sp. (Sukhada *et al.*, 2013, Panneerselvam *et al.*, 2012b, Panneerselvam *et al.*, 2014) (**Plate 11.5**).

In-vitro and *In-vivo* studies were conducted to study the effect of suppression on Banana wilt causing *Fusarium oxysporum* f. sp. *Cubense* by the three *Streptomyces* sp. isolated from *Glomus* spores (Panneerselvam *et al.*, 2014). Of the three isolates, *S. canus* (91.7%) showed strong antagonistic activity followed by *S. avermitilis* and *S. cinnamonensis* (84.7%) under *in-vitro* condition. *In-vivo* studies on three month old Banana suckers (cv.Elakkibale (syn Neypoovan)) revealed that wilt incidence was greatly reduced in the *Streptomyces* sp. inoculated suckers when compared to uninoculated control.

The actinobacterial potential of antagonism was further confirmed by molecular studies in which the genes responsible for its antagonistic traits viz., Chitinase, Phenazine and Pyrrolnitrin were detected from the *Streptomyces* genomic DNA. This experiment indicated that these *Streptomyces* sp. can be effective in controlling *F. oxysporum* f. sp. *cubense* in banana cultivation. Also it was found that these *Streptomyces* sp. efficiently suppressed the growth of *Alternaria solani* by 74-94% under *in-vitro* (Saritha *et al.*, 2014a). The above findings clearly revealed that the mycorrhizal associated actinobacteria can be explored and used as efficient biocontrol agents in plant disease suppression.

5.2 Impact of MAB inoculation on plant parasitic Nematodes

Nematode is one of the major pests occurring in patches in horticultural orchard. In areas infested with nematodes the establishment of plants may be difficult due to high mortality and poor development of the transplanted seedlings (Campos *et al.*, 1990). Studies in many crops have demonstrated that AMF inoculation may limit the plant damage from nematodes. The AMF association in crop plants has been reported to improve the tolerance of plants to nematodes in many ways (Pinochet *et al.*, 1995). The enhancement of plant growth response in mycorrhiza treated plants seems to be indirect, mainly through larger root system and thus, an increased capacity for water and nutrient uptake rather than a direct inhibitory effect over the nematode. In plants, phenol metabolism and hypersensitivity are resistant mechanisms against the invading pathogenic organisms.

There are several reports on the effect of microbes on nematode population. Banana plants inoculated with AMF (*Glomus mosseae*) suppressed nematode (*Pratylenchus coffeae*) population buildup and reduced the damage in roots, caused by the nematodes. Some phenols are known to form complex amino acids and chlorogenic acid complexes (Clarke *et al.*, 1959) which are highly toxic to the parasites. The phenols in mycorrhizal roots are associated in the reduced reproduction of nematode (Singh *et al.*, 1990). Brandao *et al.* (2004) reported that early mycorrhizal inoculation of sour sop (*Annona muricata*) seedlings in the nursery could protect the young trees and reduce the severity of nematode (*Pratylenchus coffeae*) infection in the field. *Pseudomonas fluorescens* is identified as one of the mycorrhiza helper bacteria for arbuscular mycorrhiza (Frey-Klett *et al.*, 2007). *Pseudomonas* can produce hydrogen cyanide and this repels bacterivorous nematodes. Also 2,4-DAPG, an antibiotic compound produced by *P. fluorescens*, acts as nematicide (Neidig *et al.*, 2011). There are many examples where *P. fluorescens* combined with mutualistic fungi was more successful than single inoculations of either bacteria or fungi (Tayal *et al.*, 2011; Walker *et al.*, 2012).

6.0 Conclusion and future strategies

AMF play an important role in the terrestrial ecosystems by forming obligate symbiotic associations with most of the plant root systems. The beneficial effect of AMF spore associated bacteria *viz.*, nutrient solubilization, phytohormone production, growth promotion, disease management, yield improvement and enhancemnet of mycorrhizal colonization in some selected fruit crops discussed in this chapter suggest that mycorrhization of fruit crops with their suitable associated bacteria is one of the essential intervention for healthy seedlings production as well as improvement of yield under field condition. The mycorrhiza associated bacteria particularly actinobacteria having strong antagonistic potential against various soil borne fungal pathogens revealed that these bacteria can be used as effective biocontrol agents to tackle many root wilt diseases in fruit crops production. There are number of research findings that proved the beneficial effects of AMF in fruit crops production. Recent classification on AMF indicates that out of eighteen genera only limited genera have been exploited so far for fruit crop production. Similarly, the research on spore associated bacteria was also confined to some genera of AMF viz., *Glomus*, *Gigaspora* and *Scutellospora* spp etc. The available information on AMF spore associated bacteria revealed that the association of bacteria on spore surface may vary from genus to genus; therefore identification of efficient specific bacteria to different genus is essential. Many studies have documented the AMF spore associated bacteria, but their practical field application has not been explored much. Hence, researchers should focus their attention towards understanding the interaction effect between AMF and their associated bacteria on enhancement of mycorrhizal colonization, spore proliferation, nutrient mobilization, abiotic and biotic stress management in fruit crops production. The recent understanding of AMF spore associated bacteria and their beneficial role with AMF has given new dimension to mycorrhizal research, however more research is required in following aspects

❖ The beneficial bacterial diversity on different AMF spores should be identified, characterized and conserved for future use

❖ Understanding the interactions between AMF spores associated bacteria and AMF are required to identify the efficient compatible beneficial bacteria

❖ More research is required to understand the host preference of AMF and identify suitable spore associated bacteria for different fruit crops

❖ In general, combined application of AMF and MAB is much better for improvement of plant growth than individual isolate. Hence, more in-depth research is needed to identify the suitable AMF and bacterial consortium for different fruit crops

❖ The molecular understanding or chemical dialogue between AMF spore associated bacteria and AMF to be studied in detail

❖ The mycorrhiza associated bacteria particularly actinobacteria has strong antagonistic activity against many phytopathogens, hence research should focus in this aspect to identify efficient biocontrol agent in fruit crops production

❖ Location/crop specific AMF package and efficient delivery mechanism to be standardized for different fruit crops/ rootstocks.

References

1. Abdel-Fattah, G.M. and Mohamedin, A.H., 2000. Interactions between a vesicular-arbuscular mycorrhizal fungus (*Glomus intraradices*) and *Streptomyces coelicolor* and their effects on sorghum plants grown in soil amended with chitin of brawn scales. *Biol. Fert. Soils.* 32, 401–409.

2. Abdel-Fattah, G.M. and Arsar-Wasea, A.A., 2012. Arbuscular mycorrhizal fungal application to improve growth and tolerance of wheat (*Triticum aestivum L*), plants grown in saline soil. *Acta. Physiol. Plant.* 34, 267–277.

3. Akiyama, K., Matsuoka, H. and Hayashi, H., 2002. Isolation and identification of a phosphate deficiency-induced C-glycosyl-flavonoid that stimulates arbuscular mycorrhiza formation in melon roots. *Mol. Plant. Micro. Inter.* 15, 334–340.

4. Akiyama, K., Matsuzaki, K. and Hayashi, H., 2005. Plant sesquiterpenes induce hyphal branching in arbuscular mycorrhizal fungi. *Nature.* 435, 824–827.

5. Akiyama, K. and Hayashi, H., 2006. Strigolactones: chemical signals for fungal symbionts and parasitic weeds in plant roots. *Ann. Bot.* 97, 925–931.

6. Artursson, V., 2005. Bacterial-fungal interactions highlighted using microbiomics: potential application for plant growth enhancement. PhD Thesis, Swedish University of Agricultural Sciences.

7. Artursson, V., Finlay, R.D. and Jansson, J.K., 2006. Interactions between arbuscular mycorrhizal fungi and bacteria and their potential for stimulating plant growth. *Environ. Microbiol.* 8(1), 1–10.

8. Aspray, T.J., Eirian, J.E., Whipps, J.M. and Bending, G.D., 2006. Importance of mycorrhization helper bacteria cell density and metabolite localization for the Pinus sylvestris-Lactarius rufus symbiosis. *FEMS Microbiol. Ecol.* 56, 25–33.

9. Azcón, R., Rubio, R. and Barea, J.M., 1991. Selective interactions between different species of mycorrhizal fungi and *Rhizobium meliloti* strains, and their effects on growth, N$_2$-fixation (15N) and nutrition of *Medicago sativa* L. *New Phytol.* 117, 399-404.

10. Bagyaraj, D.J. and Menge, J.A., 1978. Interaction between a VA mycorrhiza and *Azotobacter* and their effects on the rhizosphere microflora and plant growth. *New Phytol.* 80, 567–573.

11. Barea, J.M., Azcon-Aguilar, C. and Azcon, R., 1997. Interactions between mycorrhizal fungi and rhizosphere microorganisms within the context of sustainable soil–plant systems. In: Gange, A.C., Brown, V.K.(Eds.). Multitrophic interactions in terrestrial systems. Oxford (UK): *Blackwell Science.* pp. 65–77.

12. Barea, J.M., Andrade, G., Bianciotto, V., Dowling, D., Lohrke, S., Bomfante, P., O'Gara, F. and Azcón-Aguilar, C., 1998. Impact on arbuscular mycorrhizal function formation of *Pseudomonas* strains used as inoculants for the biocontrol of soil-borne plant pathogens. *Appl. Environ. Microbiol.* 64, 2304–2307.

13. Babana, A.H. and Antoun, H., 2005. Biological system for improving the availability of Tilemsi phosphate rock for wheat (*Triticum aestivum* L.) cultivated in Mali. *Nutr. Cycl. Agroecosys.* 72,147–157.

14. Barea, J.M., Pozo, M.J., Azcon, R. and Azcon-Aguilar, C., 2005. Microbial cooperation in the rhizosphere. *J. Exp. Bot.* 56, 1761–1778.

15. Bedini, S., Cristani, C., Avio, L., Sbrana, C., Turrini, A. and Giovannetti, M.,2008. Influence of organic farming on arbuscular mycorrhizal fungal populations in a Mediterranean agro-ecosystem. Proceedings of 16th IFOAM Organic World Congress, June 16-20, Modena, Italy.

16 Besserer, A., Becard, G., Roux, C. and Sejalon-Delmas, N., 2009. Role of mitochondria in the response of arbuscular mycorrhizal fungi to strigolactones. Pl. Signal Behave. 4, 75-77.

17. Bedini, S., Turrini, A., Rigo, C., Argese, E. and Giovannetti, M., 2010. Molecular characterization and glomalin production of arbuscular mycorrhizal fungi colonizing a heavy metal polluted ash disposal island, downtown Venice. *Soil Biol. Biochem.* 42, 758–765.

18. Bharadwaj, D.P., Lundquist, P. and Alstro¨m, S., 2008. Arbuscular mycorrhizal fungal spore-associated bacteria affect mycorrhizal

colonization, plant growth and potato pathogens. *Soil. Biol. Biochem.* 40, 2494–250.

19. Bharadwaj, D.P., Alstro"m, S. and Lundquist, P., 2012. Interactions among Glomus irregulare, arbuscular mycorrhizal spore-associated bacteria, and plant pathogens under *in vitro* conditions. *Mycorr.* 22, 437–447.

20. Bianciotto, V., Bandi, C., Minerdi, D., Sironi, M., Tichy, H.V. and Bonfante, P., 1996. An obligately endosymbiotic mycorrhizal fungus itself harbors obligately intracellular bacteria. *Appl. Environ. Microbiol.* 62, 3005–3010.

21. Bianciotto, V., Lumini, E., Lanfranco, L., Minerdi, D., Bonfante, P. and Perotto, S., 2000. Detection and identification of bacterial endosymbionts in arbuscular mycorrhizal fungi belonging to the family Gigasporaceae. *Appl. Environ. Microbiol.* 66, 4503–4509.

22. Bianciotto, V., Lumini, E., Bonfante, P. and Vandamme, P., 2003. 'Candidatus *Glomeribacter gigasporarum*' gen. nov., sp. nov., an endosymbiont of arbuscular mycorrhizal fungi. *Int. J. Syst. Evol. Microbiol.* 53, 121–124.

23. Bouwmeester, H.J., Roux, C., Lopez-Raez, J.A. and B'ecard, G., 2007. Rhizosphere communication of plants, parasitic plants and AM fungi. *Tr. Pl. Sci.* 12, 224–230.

24. Bonfante, P., Balestrini, R., Genre, A. and Lanfranco, L., 2009. Establishment and functioning of arbuscular mycorrhizas. In: Deising, H. (Ed.). The Mycota, V: Plant Relationships. Berlin: Springer-*Verlag*. *2nd ed.* pp. 257–272.

25. Brandao, J.A.C., Cavalcante, U.M.T., Pederosa, E.M.R. and Maia, L.C., 2004. Interaction between arbuscular mycorrhizal fungi and *Pratylenchus coffeae* on nursery soursop (*Annona muricata*) seedlings. *Nematol. Bras.* 28(1), 27-33.

26. Bronick, C.J. and Lal, R., 2005. Soil structure and management : *a review. Geoderma.* 124, 3-22.

27. Budi, S.W., Van Tuinen, D., Martinotti, G. and Gianinazzi, S., 1999. Isolation from *Sorghum bicolor* mycorrhizosphere of a bacterium compatible with arbuscular mycorriza development and antagonistic towards soilborne fungal pathogens. *Appl. Environ. Microbiol.* 65, 5148–5150.

28. Budi, S.W.,van Tuinen,D., Arnould, C., Dumas-Gaudot,E., Gianinazzi-Pearson, V. and Gianinazzi, S., 2000. Hydrolytic enzyme activity of *Paenibacillus* sp. strain B2 and effects of the antagonistic bacterium on cell integrity of two soil-borne pathogenic fungi. *Appli. Soil. Ecol.* 15, 191–199.

29. Campos, P.V., Srivapalan, P. and Gnanapragasam, N.C., 1990. Nematode parasites of coffee, cocoa and tea. In: Luc, M., Sikora, R.A., Bridge, J. (Eds.). Plant Parasitic Nematodes in Subtropical and Tropical Agriculture. CAB International, Wallingford, pp. 387-430.

30. Clarke, R.A., Kuchenze, R.T. and Quackenbush, F.W., 1959. The nature

and fungitoxicity of an aminoacid addition product of chlorogenic acid. *Phytopathol.* 49, 594-598.

31. Cruz, A. F., Horii, S., Ochiai, S., Yasuda, A. and Ishii, T., 2008. Isolation and analysis of bacteria associated with spores of *Gigaspora margarita.J. Appl. Microbiol.*104, 1711–1717.

32. Cruz, A.F. and Ishii, T., 2012. Arbuscular mycorrhizal fungal spores host bacteria that affect nutrient biodynamic and biocontrol of soil borne plant pathogens. *Biol. Open.* 1, 52–57.

33. Cruz, A.F., Soares, W.R.O. and Blum, L.E.B., 2014. Impact of the Arbuscular Mycorrhizal Fungi and Bacteria on Biocontrol of White Root Rot in Fruit Seedlings. J. *Pl. Physiol. Pathol.* 2, 1.

34. De Boer, W., Folman, L.B., Summerbell, R.C. and Boddy, L., 2005. Living in a fungal world: Impact of fungi on soil bacterial niche development. *FEMS Microbiol. Rev.* 29, 795–811.

35. Di Simine, C.D., Sayer, J.A. and Gadd, G.M., 1998. Solubilization of zinc phosphate by a strain of *Pseudomonas fluorescens* isolated from forest soil. *Biol. Fert. Soils.* 28, 87–94.

36. Driver, J.D., Holben, W.E. and Rillig, M.C., 2005. Characterization of glomalin as a hyphal wall component of arbuscular mycorrhizal fungi. *Soil. Biol. Biochem.* 37, 101–106.

37. Duponnois, R. and Plenchette, C., 2003. A mycorrhiza helper bacterium enhances ectomycorrhizal and endomycorrhizal symbiosis of Australian *Acacia* species. *Mycorr.* 13, 85–91.

38. Fester, T., Maier, W. and Strack, D., 1999. Accumulation of secondary compounds in barley and wheat roots in response to inoculation with an arbuscular mycorrhizal fungus and co-inoculation with rhizosphere bacteria. *Mycorr.* 8, 241–246.

39. Filippi, C., Bagnoli, G., Citernesi, A.S. and Giovannetti, M., 1998. Ultrastructural spatial distribution of bacteria associated with sporocarps of *Glomus mosseae. Symb.* 24, 1–12.

40. Founoune, H., Duponnois, R. and Bâ, A.M., 2002a. Ectomycorrhization of Acacia mangium, Willd. and Acacia holosericea, A. Cunn. Ex G. Don in Senegal. Impact on plant growth, populations of indigenous symbiotic microorganisms and plant parasitic nematodes. *J. Arid. Environ.* 50, 325–332.

41. Frey-Klett, P., Garbaye, J. and Tarkka, M., 2007. The Mycorrhiza associated bacteria revisited. *New Phytol.* 176: 22–36.

42. Garbaye, J. and Bowen, G.D., 1989. Stimulation of mycorrhizal infection of *Pinus radiata* by some microorganisms associated with the mantle of ectomycorrhizas. *New Phytol.* 112, 383–388.

43. Garbaye, J., 1994. Helper bacteria: a new dimension to the mycorrhizal symbiosis. *New Phytol.* 128, 197–210.

44. Gamalero, E., Trotta, A., Massa, N., Copetta, A., Martinotti, M.G. and Berta, G., 2004. Impact of two fluorescent pseudomonads and an arbuscular mycorrhizal fungus on tomato plant growth, root architecture and P acquisition. *Mycorr.* 14, 185–192.

45. George, E., Romheld, V. and Marschner, H., 1994. Contribution of mycorrhizal fungi to micronutrient uptake by plants. In: Biochemistry of Metal Micronutrients in the Rhizosphere. Monthey, JA., Crowley, D.E., Luster, D.G. (Eds.). *CRC Press, Boca Raton FL*, pp. 93-109.

46. George, E., Gorgus, E., Schmeisser, A. and Marschner, H., 1996. A method to measure nutrient uptake from soil by mycorrhizal hyphae. *In: Mycorrhizas in Integrated System from Genes to Plant Development.* Azcon-Aguilar., Barea, M. (eds). Luxembourg. European Community.

47. Genre, A., Chabaud, M., Timmers, T., Bonfante, P. and Barker, D.G., 2005. Arbuscular mycorrhizal fungi elicit a novel intracellular apparatus in Medicago truncatula root epidermal cells before infection. *Pl. Cell.* 17, 3489–3499.

48. Genre, A., Chabaud, M., Faccio, A., Barker, D.G. and Bonfante, P., 2008. Prepenetration apparatus assembly precedes and predicts the colonization patterns of arbuscular mycorrhizal fungi within the root cortex of both *Medicago truncatula* and *Daucus carota. Pl. Cell.* 20, 1407–1420.

49. Gryndler, M. and Vosatka, M., 1996. The response of *Glomus fistulosum*-maize mycorrhiza to treatments with culture fractions from *Pseudomonas putida. Mycorr.* 6, 207–211.

50. Grunwald, U., Guo, W.B., Fischer, K., Isayenkov, S., Ludwig-Müller, J., Hause, B., Yan, X.L., Küster, H. and Franken, P., 2009., Overlapping expression patterns and differential transcript levels of phosphate transporter genes in arbuscular mycorrhizal, P-i-fertilised and phytohormone-treated *Medicago truncatula* roots. *Planta.* 229, 1023–1034.

51. Hause, B., Mrosk, C., Isayenkov, S. and Strack, D., 2007. Jasmonates in arbuscular mycorrhizal interactions. *Phyto-chem.* 68, 101-110.

52. Hildebrandt, U., Janetta, K. and Bothe, H., 2002. Towards growth of arbuscular mycorrhizal fungi independent of a plant host. *Appl. Environ. Microbiol.* 68, 1919–1924.

53. Hildebrandt, U., Ouziad, F., Marner, F.J. and Bothe, H., 2006. The bacterium *Paenibacillus validus* stimulates growth of the arbuscular mycorrhizal fungus *Glomus intraradices* up to the formation of fertile spores. *FEMS Microbiol. Lett.* 254, 258–267.

54. Isayenkov, S., Mrosk, C., Stenzel, I., Strack, D. and Hause, B., 2005. Suppression of allene oxide cyclase in hairy roots of *Medicago truncatula* reduces jasmonate levels and the degree of mycorrhization with *Glomus intraradices. Pl. Physiol.* 139, 1401-1410.

55. Jacobsen, I., Abbott, L.K. and Robson, A., 1992. External hyphae of vesicular arbuscular mycorrhizal fungi associated with *Trofoluim subterraneum* L. I. spread of hyphae and phosphorus inflow into roots. *New phytol.* 120, 371-380.

56. Jeffries, P., 1987. Use of mycorrhiza in agriculture. *Crit. Rev. Biotechnol.* 5, 319-357.

57. Kim, K.Y., Jordan, D. and McDonald, G.A., 1998. Effect of phosphate - solubilizing bacteria and vesicular arbuscular mycorrhizae on tomato growth and soil microbial activity. *Biol. Fert. Soils.* 26, 79-87.

58. Kloepper, J.W., Leong, J. and Schroth, M.N., 1980. *Pseudomonas siderophores*: A mechanism explaining disease suppressive soils. *Curr. Microbiol.* 4, 317–320.

59. Kosuta, S., Hazledine, S., Sun, J., Miwa, H. and Morris, R.J., 2008. Differential and chaotic calcium signatures in the symbiosis signaling pathway of legumes. *Proc. Natl. Acad. Sci. USA.* 105, 9823–9828.

60. Kuster,H. and Franken, P., 2009. Overlapping expression patterns and differential transcript levels of phosphate transporter genes in arbuscular mycorrhizal, Pi-fertilised and phytohormone-treated *Medicago truncatula* roots. *Planta.* 229, 1023–1034.

61. Lambert, D.H., Baker, D.E. and Cole, H.Jr., 1979. The role of mycorrhizae in the interactions of phosphorus with zinc, copper and other elements. *Soil Sci. Soc. Am. J.* 43, 976-980.

62. Lagrange, H., Jay-Allgmand, C. and Lapeyrie, F., 2001. Rutin, the phenolglycoside from eucalyptus root exudates, stimulates *Pisolithus* hyphal growth at picomolar concentration. *New Phytol.* 149, 349–355.

63. Liu, A., Hamel, C., Elmi, A., Costa, C., Ma, B. and Smith, D.L., 2002. Concentrations of K, Ca and Mg in maize colonised by arbuscular mycorrhizal fungi under field conditions. *Can. J. Soil Sci.* 82(3), 271-278.

64. Lioussanne, L., 2007. Rôles des modifications de la microflore bactérienne et de l'exudation racinaire de la tomate par la symbiose mycorhizienne dans le biocontrôle sur le *Phytophthora nicotianae*. Doctoral thesis. University of Montreal, Montreal. [In French].

65. Long, L., Yao, Q., Ai, Y. and Zhu, H., 2009. Analysis of bacterial colonization associated with *Gigaspora margarita* spores by green fluorescence protein (GFP) marked technology. Weishengwu Xuebao. 49(5), 617-623.

66. Ludwig-Muller, J., 2000. Hormonal balance in plants during colonization by mycorrhizal fungi. In: *Arbuscular mycorrhizas: physiology* and *function.* Douds, D.D., Kapulnik, Y. (Eds.).Dordrecht, The Netherlands: Kluwer Academic Publishers. pp. 263–285.

67. MacDonald, R.M., Chandler, M.R. and Mosse, B., 1982. The occurrence of bacterium-like organelles in vesicular-arbuscular mycorrhizal fungi. *New Phytol.* 90, 659–663.

68. Mamatha, G., Bagyaraj, D.J. and Jaganath, S.,2002. Inoculation of field-established mulberry and papaya with arbuscular mycorrhizal fungi and a mycorrhiza helper bacterium. *Mycorr*.12, 313–316.

69. Meyer, J.R. and Linderman, R.G.,1986. Response of subterranean clover to dual-inoculation with vesicular-arbuscular mycorrhizal fungi and a plant growth-promoting bacterium, *Pseudomonas putida. Soil Biol. Biochem*.18, 185–190.

70. Miller, R.M. and Jastrow, J.D., 1990. Hierarchy of root and mycorrhizal fungal interactions with soil aggregation. *Soil. Biol. Biochem*. 22, 579–584.

71. Minerdi, D., Fani, R., Gallo, R., Boarino, A. and Bonfante, P., 2001. Nitrogen fixation genes in an endosymbiotic *Burkholderia* strain. *Appl. Environ. Microbiol*. 67, 725–732.

72. Mosse, B., 1962. The establishment of vesicular-arbuscular mycorrhiza under aseptic conditions. *J. Gen. Microbiol*. 21, 509-520.

73. Navazio, L., Moscatiello, R., Genre, A., Novero, M. and Baldan, B., 2007. A diffusible signal from arbuscular mycorrhizal fungi elicits a transient cytosolic calcium elevation in host plant cells. *Pl. Physiol*.144, 673–681.

74. Nazir, R., Warmink, J.A., Boersma, H. and van Elsas, J.D.,2010. Mechanisms that promote bacterial fitness in fungal-affected soil microhabitats. *FEMS Microbiol. Ecol*. 71, 169–185.

75. Neidig, N., Paul, R. J., Scheu, S. and Jousset, A., 2011. Secondary metabolites of *Pseudomonas fluorescens* CHA0 drive complex non-trophic interactions with bacterivorous nematodes. *Microb. Ecol*. 61, 853–859.

76. Oldroyd, G.E. and Downie, J.A., 2008. Coordinating nodule morphogenesis with rhizobial infection in legumes. *Annu. Rev. Plant Biol*. 59, 519–546.

77. Ortas, I., Harries, P.J. and Rowell, D.L., 1996. Enhanced uptake of phosphorus by mycorrhizal sorghum plants as influenced by forms of nitrogen. *Pl. Soil*. 184, 255-264.

78. Paula, M.A., Urquiaga, S. and Sijqueira, J.O. 1992. Synergistic effects of vesicular-arbuscular mycorrhizal fungi and diazotrophicus bacteria on nutrition and growth of sweet potato (*Ipomoea batatas*). *Biology* and *Fertility of Soils* .14, 61–66.

79. Panneerselvam, P., Thangaraju, M. and Jayarama., 2006. Induction of defense mechanisms in *Coffea arabica* L. by native rhizobacterial isolates against coffee leaf rust (*Hemileia vastatrix*) disease. *Trop. Agric. Res*. 18, 173–181.

80. Panneerselvam, P., Thangaraju, M., Senthilkumar, M. and Jayarama., 2008. Microbial consortium and its effect on controlling coffee root-lesion nematode (*Pratylenchus coffeae*). *J. Biol. Cont*. 22, 425–432.

81. Panneerselvam, P., Sukhada M., Saritha, B., Upreti, K.K., Poovarasan, S., Ajay, M. and Sulladmath, V.V., 2012a. *Glomus mosseae* associated bacteria

and their influence on stimulation of mycorrhizal colonization, sporulation and growth promotion in guava (*Psidium gujava* L.) seedlings. *Biol. Agri. Hort.* 28(4), 267-279.

82. Panneerselvam, P., Saritha,B., Poovarasan, S., Sukhada Mohandas. and Upreti, K. K., 2012b. Antagonistic Potential of mycorrhizal associated *actinomycetes* (MAA) against *Fusarium* and *Alternaria*. In: *National symposium on Blending Conventional* and *Modern Plant Pathology for sustainable Agriculture held during 4-6 December*, 2012 at Indian Institute of Horticultural Research, Bengaluru, pp. 137.

83. Panneerselvam, P., Saritha, B., Sukhada Mohandas, Upreti, K. K., Poovarasan, S., Sulladmath, V.V. and Venugopalan, V., 2013. Effect of mycorrhiza associated bacteria on enhancing colonization and sporulation of *Glomus mosseae* and growth promotion in sapota (*Manilkara achras* (Mill.) Forsberg) seedlings. *Biol. Agric. Hort.* 29(2), 118-131.

84. Panneerselvam, P., Prajwala, M., Sneha, P., Sindhu, B.L., Saritha, B., Poovarasan, S. and Ganeshamurthy, A.N., 2014. *Actinobacteria-* A potential Biocontrol Agent against banana wilt causing *Fusarium oxysporum f. sp.* Cubense. In: *National conference on Productivity* and *Sustainability Role of Agriculturally Important Microorganisms* April 10-12, 2014, GKVK Campus, Bangalore, pp. 105-106.

85. Peng, S., Eissenstat, D.M., Graham, J.H., Williams, K. and Hodge, N.C., 1993. Growth depression in mycorrhizal citrus at high phosphorus supply. *Pl. Physiol.* 101, 1063–1071.

86. Pinochet, J., Calvet, C., Camprubi, A. and Fernandez, C., 1995. Growth and nutritional response of Nemared peach rootstock infected with *Pratylenchus vulnusm* and the mycorrhizal fungus *Glomus mosseae. Fundam. app. NemalO.* 18(3), 205-210.

87. Poovarasan, S., Sukhada M., Panneerselvam, P., Saritha, B. and Ajay, K.M., 2013. Mycorrhizae colonizing actinomycetes promote plant growth and control bacterial blight disease of pomegranate (*Punica granatum* L. cv Bhagwa). *Crop Prot.* 53, 175-181.

88. Rao, N.S.S., Tilak, K.V.B.R. and Singh, C.S. 1985. Effect of combined inoculation of *Azospirillum brasilense* and vesicular-arbuscular mycorrhiza on pearl millet (*Pennisetum americanum*). *Pl. Soil.* 84, 283–286.

89. Requena, N., Jimenez, I., Toro, M. and Barea, J.M. 1997. Interactions between plant-growth-promoting rhizobacteria (PGPR), arbuscular mycorrhizal fungi and *Rhizobium* sp. in the rhizosphere of *Anthyllis cytisoides*, a model legume for revegetation in mediterranean semi-arid ecosystems. *New Phytol.* 136, 667–677.

90. Rilling, M.C., Wright, S.F. and Eviner, V.T., 2002. The role of arbuscular mycorrhizal fungi and glomalin in soil aggregation: comparing effects of five plant species. *Pl. Soil.* 238, 325-333.

91. Roesti, D., Ineichen, K., Braissant, O., Redecker, D., Wiemken, A. and Arango, M., 2005. Bacteria associated with spores of the arbuscular mycorrhizal fungi *Glomus geosporum* and *Glomus constrictum*. *Appl. Environ. Microbiol*. 71, 6673–6679.

92. Salimpour, S., Khavazi, K., Nadian, H., Besharati, H. and Miransari, M., 2010. Enhancing phosphorous availability to canola (*Brassica napus* L.) using P solubilizing and sulfur oxidizing bacteria. *Aust. J. Crop. Sci.* (in press).

93. Saritha, B., Panneerselvam, P. and Ravindrababu, P., 2014. Studies on native Arbuscular Mycorrhizal (AM) Fungi and its helper bacteria for their potential for nutrient mobilization and plant growth promotion in sapota (*Manilkara achras* (Mill.) Forsberg*)*. (PhD Work unpublished)

94. Saritha, B., Panneerselvam, P., Sindhu, B.L., Sneha, P., Prajwala, M., Poovarasan, S. and Ganeshamurthy, A.N., 2014a. Bio-control potential of *Streptomyces* sp. Against plant pathogenic *Alternaria solani*. In: *National conference on Productivity* and *Sustainability Role of Agriculturally Important Microorganisms April 10-12, 2014,* GKVK Campus, Bengaluru, pp. 92.

95. Saritha, B., Panneerselvam, P., Sukhada, M., Sulladmath, V.V., Ravindrababu, P. and 2014b. Studies on host preference of *Glomus* spp and their synergistic effect on sapota (*Manilkara achras* (Mill*)* Forsberg) seedlings growth. *Pl. Arch.* 14(2), 701-706.

96. Schuessler, A., Mollenhauer, D., Schnepf, E. and Kluge, M., 1994. *Geosiphon pyriforme*, and endosymbiotic association of fungus and cyanobacteria: The spore structure resembles that of arbuscular mycorrhizal (AM) fungi. *Bot. Acta.* 107, 36–45.

97. Schrey, S.D., Schellhammer, M., Ecke, M., Hampp, R. and Tarkka, M.T., 2005. Mycorrhiza helper bacterium *Streptomyces* AcH 505 induces differential gene expression in the ectomycorrhizal fungus *Amanita muscaria*. *New Phytol.* 168, 205–216.

98. Selim, S., Negrel, J., Govaerts, C., Gianinazzi, S. and van Tuinen, D. 2005. Isolation and partial characterization of antagonistic peptides produced by *Paenibacillus* sp. strain B2 isolated from the *Sorghum* mycorrhizosphere. *Appl. Environ. Microbiol.* 71, 6501–6507.

99. Sharma, S.D., Kumar, P., Raj, H. and Bhardwaj, S.C., 2009. Isolation of arbuscular mycorrhizal fungi and *Azotobacter chroococcum* from local litchi orchards and evaluation of their activity in the air layers system. *Sci. Hort.* 123, 117–123.

100. Shimizu-Sato, S., Tanaka, M. and Mori, H., 2009. Auxin-cytokinin interactions in the control of shoot branching. *Pl. Mol. Biol.* 69, 429–435.

101. Shamshiri, M.H. and Usha, K., Singh, B., 2012. Growth and nutrient uptake responses of kinnow to vesicular arbuscular mycorrhizae. *ISRN Agron.* doi: 10.5402/2012/535846.

102. Singh, Y.P., Singh, R.S. and Sitaramaia, K., 1990. Mechanism of resistance of mycorrhizal tomato against root knot nematode. In: *Trends in mycorrhizal Research (Eds.)*. Jalali, B.L., Chand, H. Haryana Agricultural University, Hissar. pp. 96-97.

103. Singh, P.K., 2012. Role of glomalin related soil protein produced by arbuscular mycorrhizal fungi: a review. *Agri. Sci. Res. J.* 2, 119–125.

104. Smith, S.E. and Read, D.J., 2008. Mycorrhizal symbiosis (3rd Ed) *Academic Press*, London.

105. Smith, S.E. and Smith, F.A., 2011. Roles of arbuscular mycorrhizas in plant nutrition and growth: New paradigms from cellular to ecosystem scales. *Annu. Rev. Pl. Biol.* 62, 227–250.

106. Sukhada, M., Poovarasan, S., Panneerselvam, P., Saritha, B., Upreti, K.K., Ranveer, Kamal. and Sita, T., 2013. Guava (*Psidium guajava* L.) rhizosphere *Glomus mosseae* spores harbor actinomycetes with growth promoting and antifungal attributes. *Sci. Hort.* 150, 371–376.

107. Sylvia, D.M., Fuhrmann, J.J., Hartel, P.T. and Zuberer, E., 1998. Principles and applications of soil microbiology. *Prentice Hall*, London. pp. 550.

108. Tarkka, M.T. and Piechulla, B., 2007. Aromatic weapons: Truffles attack plants by the production of volatiles. *New Phytol.* 175, 381–383.

109. Tayal, P., Kapoor, R. and Bhatnagar, A.K., 2011. Functional synergism among *Glomus fasciculatum*, *Trichoderma viride* and *Pseudomonas fluorescens* on *Fusarium* wilt in tomato. *J. Plant Pathol.* 93, 745–750.

110. Toro, M., Azcón, R. and Barea, J.M., 1997. Improvement of arbuscular mycorrhiza development by inoculation of soil with phosphate-solubilizing rhizobacteria to improve rock phosphate bioavailability (P32) and nutrient cycling. *Appl. Environ. Microbiol.* 63, 4408–4412.

111. Tylka, G.L., Hussey, R.S. and Roncadori, R.W., 1991. Interactions of vesicular-arbuscular mycorrhizal fungi, phosphorus, and Heterodera glycines on soybean. *J. Nematol.* 23, 122 –133.

112. Utkhede, R.S. and Smith, E.M., 2000. Impact of chemical, biological and cultural treatments on the growth and yield of apple in replant-disease soil. *Aus. Pl. Pathol.* 29, 129-136.

113. Van Wees, S.C., Van der Ent, S. and Pieterse, C.M., 2008. Plant immune responses triggered by beneficial microbes. *Curr. Opin. Pl. Biol.* 11, 443– 448.

114. Vivas, A., Marulanda, A., Ruiz-Lozano, J.M., Barea, J.M. and Azcón, R., 2003. Influence of a *Bacillus* sp. on physiological activities of two arbuscular mycorrhizal fungi and on plant responses to PEG-induced drought stress. *Mycorr.* 13, 249–256.

115. Vivas, A., Voros, I., Biro, B., Campos, E., Barea, J.M. and Azcon, R., 2003b. Symbiotic efficiency of autochthonous arbuscular mycorrhizal fungus

(*G. mosseae*) and *Brevibacillus* sp. isolated from cadmium polluted soil under increasing cadmium levels. *Environ. Pollut.* 126, 179–189.

116. Von Alten, H., Lindemann, A. and Schönbeck, F., 1993. Stimulation of vesicular-arbuscular mycorrhiza by fungicides or rhizosphere bacteria. *Mycorr.* 2, 167–173.

117. Vosatka, M. and Gryndlar, M., 1999. Treatment with culture fractions from *Pseudomonas putida* modifies the development of *Glomus fistulosum* mycorrhiza and the response of potato and maize plants to inoculation. *Appl. Soil. Ecol.* 11, 245-251.

118. Walker, V., Couillerot, O., Von Felten, A., Bellvert, F., Jansa, J. and Maurhofer, M., 2012. Variation of secondary metabolite levels in maize seedling roots induced by inoculation with *Azospirillum*, *Pseudomonas* and *Glomus* consortium under field conditions. *Pl. Soil.* 356, 151–163.

119. Will, M.E. and Sylvia, D.M., 1990. Interaction of rhizosphere bacteria, fertilizer, and vesicular-arbuscular mycorrrhizal fungi sea oats. *Appl. Environ. Microbiol.* 56, 2073–2079.

120. Wright, S.F. and Upadhyaya, A., 1998. A survey of soils for aggregate stability and glomalin, a glycoprotein produced by hyphae of arbuscular mycorrhizal fungi. *Pl. Soil.* 198, 97–107.

121. Wright, S.F. and Jawson, L., 2001. A pressure cooker method to extract glomalin from soils. *Soil Sci. Soc. Am. J.* 65 (6), 1734-1735.

122. Xavier, L.J.C. and Germida, J.J., 2003. Bacteria associated with *Glomus clarum* spores influence mycorrhizal activity. *Soil Biol. Biochem.* 35, 471–478.

123. Xie, Z-P., 1995. Rhizobial nodulation factors stimulate mycorrhizal colonization of nodulating and non nodulating soybeans, *Pl. Physiol.* 108, 1519-1525.

124. Zabihi, H.R., Savaghebi, G.R., Khavazi, K., Ganjali, A. and Miransari, M., 2010. *Pseudomonas* bacteria and phosphorus fertilization, affecting wheat (*Triticum aestivum* L.) yield and P uptake under greenhouse and field conditions. *Acta. Physiol. Pl. in press.*

Chapter 12

Beneficial Soil Microbes and their interaction with Arbuscular Mycorrhizal Fungi in Nutrient Uptake

Poovarasan S[1], Sukhada Mohandas*[1] and Sita T[2]

[1] *ICAR–Indian Institute of Horticultural Research, Hessaraghatta, Bengaluru 560089,India*

[2] *St' Martin's College of Engineering, Dollapally Hyderabad, India*

**Address for correspondence: sukhada.mohandas@gmail.com*

ABSTRACT

Fruits are the natural source of nutrients and vitamins in human diet. Practices followed in fruit crop cultivation play a major role in retaining or improving the natural character of the fruits. Since increasing use of chemical fertilizers affects fruit quality, arbuscular mycorrhizal fungi (AMF) and other beneficial microbes are being used in recent times for improving the fruit quality and also to maintain a sustainable production system. Combined inoculation of these bio-inoculants results in significant improvement in the plant height, fruit size, leaf/fruit nutrient content, number of fruits per plant and also fruit quality. Absorption of insoluble nutrients by the plants may be achieved through the interactional effect of AMF with nutrient solublizing bacteria. The combined inoculation facilitates colonization of both strains in rhizosphere by secreting growth factors. Nutrients which get solubilized by chemicals released by the bacteria become more available for absorption by AMF hyphae. The increase in population of the beneficial microbes provides more benefit to the plants which may not be achieved through single inoculation. The interactional effect may also help to remediate the soil contaminated by the application of chemical fertilizer. The present review discusses the studies utilizing AMF with other microbes in fruit crop cultivation for sustainable fruit crop development.

Keywords: *Soil Microbes, Rhizosphere Trichoderma Psuedomonas interaction Fruit cultivation*

1.0 Introduction

Presently, area under fruit crops in India is about 6.7 million hectare with a production of 76.4 million tons, which contributes 30% of share in total production and it is expected to reach 115 million tons by 2017. Synthetic fertilizers play a major role in agricultural practices when the soils lack natural nutrients and other growth promoting substances. Most of the chemical fertilizers are the combination of heavy metals and radionuclide. Once it starts accumulating in soil the environment of the soil will not be favourable in the long run for the cultivation of plants. There are research reports stating that fruits cultivated in soil with low fertility would produce malnutrition in humans. Intake of fruits from less fertile soils produced 23 to 70% population with protein energy malnutrition and 17 to 54% peoples with chronic energy deficiencies . Excessive use of chemical fertilizers will be toxic for the human population at large (Sylviane *et al.*, 2002) as it includes the contamination of ground water (Shrestha and Ladha 1998) soil (Topbas *et al.*, 1998; Malakoff, 1998) and air (Shaviv, 2000) and effects soil productivity (Crawford *et al.*, 2006). Therefore the nutrient management through the biological inputs is the most appropriate approach for managing the nutrient content in soil.

In commercially important fruit crops, plant growth is a key factor to be considered for bringing about yield improvement. Plant growth is highly dependent on the soil available nutrients and other important biotic factors which play a major role in maintaining the soil fertility. The population of beneficial microbes like AMF and other plant growth promoting bacteria in soil indirectly influence the growth of plants. AMF are important for their establishment of mycelial network between the plants and the soil (Dames and Ridsdale, 2012) which promote plant growth and help plants resist diseases. (Smith and Read, 1997; Hayat *et al.*, 2010). The plant growth promoting bacteria and fungi also help in AMF spore germination and hyphal growth (Xavier and Germida, 2003). These beneficial interactions operate not only in nutrient exchange but play major role to remediate the contaminated soil (Vivas *et al.*, 2003, 2006a, 2006b; Zhuang 2007). Agricultural productivity is highly dependent on the soil fertility and its nutritional value. Soil with good health promotes the plant growth and increases the yield. As fruit crops are commercial cash crops, AMF and other beneficial organisms can be used effectively to increase the yield and income of growers. There are many reports highlighting the importance of beneficial soil micro-organisms in bringing about improvement in plant growth by recycling the existing soil nutrients.(Ahmad *et al.*, 2008; Beneduzi *et al.*, 2008; Hameeda *et al.*, 2008). Among the microorganisms which interact synergistically in promoting growth and nutrition there are both fungi, such as *Trichoderma* spp. and *Gliocadium* spp., and plant growth promoting rhizobactera (PGPR), like *Pseudomonas* spp. and *Bacillus* spp. (Kloepper *et al.*, 1991; Linderman, 1994). In the present study we review the interactional effect of AMF with these beneficial organisms in nutrient management of fruit crops.

2.0 Mechanism of AMF and microbes interaction

The interactions with AMF and other beneficial microbes occur in the zone of soil surrounding the root and fungal hyphae commonly called as mycorrhizosphere. Two main groups of bacteria interact with AMF in the mycorrhizosphere, saprophytes and symbionts both groups consist of detrimental, neutral and beneficial bacteria. Mutualism is an association between two or more species where both species derive benefit. Sometimes, it is an obligatory lifelong interaction involving close physical and biochemical contact, such as those between plants and mycorrhizal fungi.

Microorganisms play an important role in agricultural systems. Plant growth benefits may be attributed mainly to three mechanisms as follows (i) AMF and other microbes acting as biofertilizers assist plant nutrient uptake (ii) acting as phytostimulators promote plant growth usually by producing plant hormones (iii) acting as biological control agents (such as *Trichoderma, Pseudomonas*, and *Bacillus*) protect plants against phytopathogenic organisms and improve plant growth.

Interaction of AMF with *Trichoderma harzianum* is known to increase the levels of several hormones like zeatin, IAA, ACC and salicyclic acid, jasmonic acid and abscisic acid in the host plant (Martinez- Medina *et al.*, 2011) which would improve the plant growth and help in imparting resistance against disease in host plant. *Azospirillum brasilense* also improves plant growth development by producing of auxins, cytokinins, and gibberellins which change the physiology, architecture and root surface area through production of more root hairs, leading to an increase in mineral uptake. (Tabelsi and Mhamdi, 2013). The PGPR containing ACC deaminase are present in various soils offer promise as a bacterial inoculum for improvement of plant growth, particularly under unfavourable environmental conditions such as flooding, heavy metals, phytopathogens, drought and high salt. PGPR containing ACC deaminase can hydrolyze ACC, the immediate precursor of ethylene, to F-ketobutarate and ammonia, and in this way promote plant growth. Inoculation of crops with ACC deaminase-containing PGPR may assist plant growth by alleviating deleterious effects of salt stress.

The use of phosphate solubilising bacteria as inoculants increases the P uptake by plants. Among the heterogeneous and naturally abundant microbes inhabiting the rhizosphere, the phosphate solubilising microorganisms (PSM) including fullfill P demands of plants and also facilitate plant growth by other mechanisms. The most efficient PSM belong to genera word of *Bacillus* and *Pseudomonas* amongst bacteria. Iron is an essential growth element for all living organisms. The scarcity of bioavailable iron in soil habitats and on plant surfaces foments a furious competition. Under iron-limiting conditions PGPB produce low-molecular-weight compounds (siderophores) to competitively acquire ferric ion. Siderophores (Greek: "iron carrier") are small, high-affinity iron chelating compounds secreted by microorganisms such as bacteria, fungi and grasses. Microbes release siderophores to scavenge iron from these mineral phases by formation of soluble Fe^{3+} complexes that can be taken up by active transport mechanism.

Number of surface components have been demonstrated which play a major role in physical interaction between bacterial and fungal hyphae. Artursson and Janson (2003) isolated *Bacillus cereus* strain from Swedish soil containing abundant arbuscular mycorrhizal fungi. Most of the bacterial cells were isolated from the fungal hyphae in significantly higher number indicating that the colonization of bacteria on fungal hyphae is dependent on the structural features of the bacteria.

To know the factors responsible for the bacterial attachment on the fungal hyphae, five bacterial strains tagged with green fluorescent protein were used to colonize on the vital and non-vital hyphae of the AMF *Glomus claroideum* (Toljander *et al.*, 2005). The study revealed that the electrostatic attraction between the fungal hyphae of bacterial cell wall is playing key role. Bianciotto *et al.*, (1996) demonstrated two step mechanism for the confirmation of AMF and PGPR interaction, First step mechanism confirmed the electrostatic attraction and the second step mechanism confirmed the strong bonding due to the formation of extra cellular polysaccharides or cellulose fibrils by bacterial cells. This was confirmed by Bianciotto *et al.* (2001) experiment, a bacterial mutant inhibited in the cellulose fibrils production showed less ability to attach with AMF hyphae surfaces compared with non-mutated cells.

Several microorganisms reported to be good root colonizers, for example, some *Pseudomonas* sp are also capable of adhering to AMF hyphal surface, suggesting that the mechanism involved could be fairly similar, close to cell to cell contact (Sen *et al.*1996). One possible way that attachment could benefit both partners would be through facilitation of certain metabolic interaction such as nutrient and carbon exchange and this would rely on close cell contact between the bacterial and fungal components.

There is a distance-related effect of the root on the bacterial community. Combined bio-inoculation of diacetyl-phloroglucinol producing PGPR strains and AMF can synergistically improve the nutritional quality of plant without negatively affecting mycorrhizal growth. Bacteria together with AMF may create more indirect synergism for plant growth including nutrient acquisition and enhancement of root branching (Trabeslsi and Mhamdi, 2013). AMF can support modification in microbial community structure within mycorrhizosphere. The identification of key microbial populations that are correlated with improved biomass production will help to understand the role of the microbial community in supporting plant growth, suppressing plant pathogen invasion, and other AMF functional abilities. AMF may inhibit pathogen proliferation through the formation of a bacterial community that limits the pathogen invasion.

3.0 Nutrient management in fruit crops through AMF and plant growth promoting microorganisms.

AMF and plant growth promoting bacteria play major role in nutrient and soil health management in fruit crop cultivation (Panneerselvam *et al.*, 2013). The inoculation effects of indigenous AMF and other beneficial organisms have been studied on many fruit crops (Declerck *et al.*, 2002; Jaizme- Vega *et al.*, 2003, 2004, 2005; Esitken *et al.*, 2010; Sharma *et al.*, 2011). But the combined effect of these inoculants on fruits crops are less studied. The synergistic benefits of AMF and PGPR combined inoculation have been demonstrated in the recent findings (Adesemoye *et al.* 2009; Zarei *et al.* 2006; Kim *et al.* 2010; Tajini *et al.* 2011).

3.1 Apple (*Malus domestica* Borkh)

Bharat (2013) studied the co-inoculation effect of AMF along with bio-control fungi (*Trichoderma*) during the apple re-plantation in soil treated with or without soil fumigant formaldehyde. The plants inoculated with AMF plus *Trichoderma* showed enhanced shoot, root growth, number of leaves per branch, total leaf area and plant total dry biomass etc. The dual inoculation also improved the nutrient P and K content in inoculated plant leaves. Raj and Sharma (2009) have also observed reduction in root rot incidence and enhanced growth in apple seedlings after inoculation with AMF. The growth promotion and disease suppression was found more when AMF were inoculated along with growth promoting bacterium *Azotobacter* spp. The reduction in disease incidence was attributed to the increased growth and disease resistance in mycorrhizal inoculated seedlings than that in non-mycorrhizal ones.

3.2 Avocado (*Persea americana* Mill)

Osoria vega *et al.* (2012) reported the effect of phosphate solublizing microorganism (PSM) along with AMF *G.fasiculatum* in avocado planted soil. The combined inoculation increased soil soluble P level (0.2mg L^{-1}) than the plants inoculated only with AMF.

3.3 Banana (*Musa* spp.).

Rodriguez-Romero *et al.*(2005) discussed the combinational effect of *G.maniotis* and *Bacillus* sp on banana cv. Grande Naine under nursery condition. The plants inoculated with both showed significant plant growth with increased shoot height and leaf area. The combined inoculated soil showed increased N, P and K content than the plant and soil inoculated with individual strains. *G.mosseae or G.fasciculatum* along with phosphate solubilising bacteria (PSB) (*Bacillus polymyxa* and *Pseudomonas striata*) and *Azospirillum brasilense* individually and in different combinations improved plant and soil nutrient levels and as a result plant growth and yield in banana under field conditions (Sukhada,1996). The available phosphorus level in the rhizosphere soil increased in all PSB treatments and available nitrogen increased in all GF + *Azospirillum* treatments. Leaf P and leaf Zn also increased significantly in *Azospirillum* treatment and in combined inoculation of AMF and*Azospirilum* and PSB. Bunch content was the best in GM+PSB treatment. *G.mosseae* +PSB treatments which received no basal dose of P and GF+ *Azospirillum* treatment which received half the level of nitrogen was on par with plants which received 300 g of P (super phosphate) per plant and full dose of nitrogen. A saving in the cost of chemical fertilizers application was noticed with combined inoculations. Sukhada *et al*. (2010) studied the interactional effect of AMF *G.mosseae, P.flourescense* and *T.harzianum* on banana growth improvement and on *Fusarium* infection. The plants pre-inoculated with *G.mosseae* + *T.harzianum* showed 60-70% improvement in plant height and girth respectively and corresponding increase in bunch weight over plants not pre colonized with BCAs but challenged with FOC, which finally succumbed to the disease. ELISA studies for evaluating *Fusarium* population revealed that the *Fusarium* population was reduced to 0.58 OD in 7 months in *G. mosseae* and *T . harzianum* treatment compared to a level of 1.9 OD in *Fusarium* alone treated plants. Beneficial effect of BCAs may be due to the overall protection provided by them by causing physical modifications in the cell wall, growth promotion and through induction of disease resistance. Adriano-Anaya *et al*. (2011) studied interactional effect of free N_2 fixing bacteria with AMF for the improvement of colonization in banana cv. "Great dwarf" rhizosphere. The results showed that the addition of nitrogen fixing bacteria to the mycorrhizosphere improved plant growth parameters and resistance against nematode infections. Aggangan *et al*. (2013) evaluated the interactional effect of AMF with nitrogen fixing bacteria on banana cv. Lakatan for the growth improvement and resistant against the root infecting nematode *Radopholus similis* and *Meloidogyne incognita*. The individual AMF inoculated plants showed greater plant growth than the control plants. Likewise the AMF and N fixing bacterial inoculated plants showed increased plant height, total leaf area. The nematode infection was gradually decreased in the combined inoculated plants.

3.4 Citrus (*Citrus* spp.).

Trichoderma aureoviride Rifai inoculated with the arbuscular mycorrhizal fungus *Glomus intraradices* enhanced growth of *Citrus reshni* more than the *G. intraradices* used alone (Camprubi *et al*., 1995). The inoculation effect of *Gigaspora* with *Azotobacter chrococcum* on citrus seedlings on growth improvement and wilt resistance after soil solarization was reported by Harender Raj and Sharma (2010). Compared with control plants from un-solarized soil, the plants inoculated with *Gigaspora* and *A.chrococcum* from the solarized soil showed 71.4% increase in normal shoot height, 74.3% increase in root height. Sharma *et al*.(2011) studied the effect of two AMF sp (*G.fasliculatum* and *G. macrocarpum*) with two *Azotobactor chrococcum strains* on the growth improvement of citrus seedlings on solarized and natural soil with mulching practices. The study showed the plants inoculated with *G.fasiculatum* and *A.chrococcum* strain had improved the plant growth with increased shoot, root growth and leaf nutrient (N, P, K and Zn) content in solarized soil with black plastic mulching sheet. Sitole, (2013) reported the effect of mycorrhizal inoculum with bacterial sp on citrus seedlings on growth improvement and resistance against *phytopthora* infection. The plants inoculated with mycorrhizae and *Bacillus cereus* significantly increased the seedling shoot, root growth and total plant biomass.

3.5 Guava *(Psidium guajava L.)*

Ibrahim *et al.* (2010) studied the effect of co-inoculation of AMF (*Glomus mosseae,. G.fasciculatum* and *G. aggregatum*) with *Bacillus megaterium* on balady guava tree. The dual inoculation considerably increased the shoot height and the total leaf area and the mineral (N, P, K, Ca, Mg) content of the leaves and fruit vitamin C. The fruit size, weight and fruit yield per plant also increased significantly with dual inoculation. Panneerselvam *et al.*(2013) studied the growth promotional effect *Glomus mosseae* and plant growth promoting bacteria which are isolated from the *G.mosseae* spores on guava seedlings. Out of four bacterial isolates *Pesudomonas putida* was further studied. The seedlings inoculated with *G.mosseae P. putida* showed increased plant height and total biomass. The interaction of *G.mosseae* + *P. putida* had significantly increased the contents of leaf chlorophyll, phenol, and total glomalin, acid and alkaline phosphatase activities.

Different bio-inoculants enhanced the growth of guava seedlings compared to uninoculated controls. FYM + AMF and FYM+ PGPR bacteria treated seeds showed maximum germination followed by the maximum plant height and yield (Pathak *et al.* (2013).

3.6 Grape *(Viṭis vinifera L.)*

Waschkies *et al.*(1994) studied the interactional effect of *G.mosseae* with *Pseudomonas fluorescens* on replant disease of grape vine cv. 5C. The inoculation of *G.mosseae* and *P.flourescense* improved the plant shoot, root growth and leaf nutrient content. The plant survival rate in non-replant soil with AMF and bacterial inoculation was comparatively more than the plant growth in replant soil without inoculation. Rizk-Alla and Tolba, 2010 had studied the interactional effect of AMF (*Glomus sp, Gigaspora, Acaulospora*) with few soil conditioners (Humic acid and Nile fertile) for the growth improvement of Monukka grape vine in the calcareious soil. The interactional effect of AMF with soil conditioner and acid producing bacteria showed the enhanced growth response in total number of leaf, leaf area, vine growth, NPK% of the leaf and shoot diameter. This treatment also improved the long fibrous root growth in the inoculated plants. Freitas *et al.* (2011) compared the organic cultivation and conventional practices to enhance the AMF and other microbial activity in soil for the improvement of seedless grape growth. These organic practices highly improved the plant growth, number of AMF colonization, microbial carbon level, total CO_2 emission and FDA activity of the inoculated soil.

3.7 Litchi *(Litchi chinensis Sonn.)*

Interactional effect of AMF and *Azospirillum brasilense* on the development of litchi marcots under nursery condition by using soil+ vermi compost as substrate was reported by Sagar and Roy (2013) . Out of 3 sp of AMF, *Sclerocystis pakistanica* with *A.brasilense* had improved plant height, shoot length with diameter and number of leaves per plant in the soil plus vermicompost substrate. Similar effect of dual inoculation of AMF sp with *A.chrococcum* for the root development in air-layered litchi shoots was observed by Sharma *et al.* (2009). Out of four AMF sp *G.fasiculatum* with *A.chrococcum* showed 81.3% increase in root growth in inoculated air-layered litchi shoots.

3.8 Mango (*Mangifera indica* L.)

Sharma *et al.*(2011) studied the effect of AMF *Glomus fasiculatum* and *Azotobacter chrococcum* on mango under nursery condition with different agricultural practices. The seedlings inoculated with *G. fasiculatum* and *A. chroococcum* showed enhanced seedling's height, diameter, leaf area and total root length. The solarization and black polyethene mulched practices improved the soil and leaf N, P and K content in the AMF and *Azotobacter* inoculated mangostones and saplings. Duttu and Kundu, (2012) studied the effect of mixed inoculum containing AMF, *Azotobacter*, *Azospirillum*, PGPR, PSB in different combinations on mango seedlings. Combined application of all above mentioned bio-inoculants recorded maximum N, P and K uptake in leaf. The inoculated soil showed increased organic carbon and N, P and K content.

3.9 Pineapple (*Ananas comosus* L. Merr.)

Naher *et al.*(2013) reviewed the individual and dual inoculation effect of AMF *G.mosseae* and PGPR bacteria for the growth improvement micropropagated pineapple seedlings. Compared with individual inoculated treatments, the plant inoculated with both AMF fungi and PGPR bacteria had shown increased plant total dry mass and phosphorus content in leaf.

3.10 Peach (*Prunus persica* (L.) Stokes

The signal transduction between the AMF and PGPR bacteria induced the secretion of important PR proteins to facilitate plant growth and enhance the plant defense mechanisms in peach (Alizadeh *et al.*, 2013). The inoculation of both *G.fasiculatum* and *Azotobacter* increased plant dry matter and N and K content in the leaf of peach seedlings than the control and single culture inoculated plants (Godara *et al.*, 1998).

3.11 Passion fruit (*Passiflora edulis* Sims. f. *flavicarpa* Deg.)

Dual inoculation effect of *Glomus caledonium* or *Gigaspora margarita* with red green algae extracts for the growth development of passion fruit cuttings were reported by Kuwada *et al.*(2006). The cuttings inoculated with 25 % methyl extract of red algae, *Gelidium amansii* and *G.caledonium* significantly improved the plant height and shoot and root biomass.

3.12 Plum (*Prunus cerasifera* Ehrh.)

Xueming *et al.*(2014) studied the combinational effect of AMF and phosphate solublizing fungi (PSF) on salt-tolerance of beach plum plants under salt stress condition. Efficient salt tolerance was observed in plants inoculated with both AMF and PSF in 1% NaCl concentration. Compared with single inoculations dual inoculation showed significant variation in ion concentration (Na^+/K^+), plant growth parameters, available phosphate level, phosphatase enzyme concentration and improved root colonization of AMF inoculated plants.

3.13 Pomegranate (*Punica granatum* L.)

The combined effect of mycorrhizal inoculation on pomegranate cultivar "Ganesh' under pot culture and field conditions was evaluated by Sukhada *et al* ,(2000) and found that combined inoculation of AMF, Azospirillum and phosphate soluble microorganisms benefitted plant growth and nutrient uptake better than their individual treatments. Poovarasan *et al*.(2013) reported the effect of *Streptomyces* sp. isolated from the mycorrhizal spores *G.mosseae* on pomegranate seedlings under glasshouse condition. Out of five isolates *S.canus* inoculated seedlings showed enhanced growth of plant shoot, root, and plant total dry mass than the control and other treatments.

3.14 Papaya (*Carica papaya* L.)

Mamatha *et al.* (2002) studied the effect of AMF and PGPR bacteria on papaya seedlings. The plants inoculated with both AMF+PGPR showed statistically more fruit yield (with 50% recommended P) than the control plants with 100% of recommended P. Combined effect of *Glomus claroideum*+ *Azospirillum brasilense* on papaya cv. Red Marodol in low P soil was reported by Alarcon *et al.* (2002). The seedling inoculated with *G.claroideum* and *A.brasilense* showed increased plant height, total dry mass and leaf area and root acid phosphatase activity than the un-inoculated plants. The plants inoculated with *G.claroideum* + *A.brasilense* showed increased number of bacterial population than the plants inoculated with *A. brasilense* only.

A   B   C

Plate 12.1. Papaya cv Surya grown in field inoculated with *Glomus mosseae* (B) and a combination of and *Glomus mosseae* and *Trichoderma harzianum* (C) The improvement in growth of combined inoculation over Control (A) and *G.mosseae* alone can be seen. (Source: Sukhada Mohandas NATP project report ,2005).

Jaizme-Vega *et al.* (2005) studied the enhancement of AMF effect by using PGPR bacteria in the growth improvement of papaya and against the nematode infection. Application of AMF with PGPR improved the shoot height and root length and uptake of N, P and K in inoculated seedlings than the plants inoculated either AMF or PGPR alone. Moreover the plants inoculated with *G.mosseae* + PGPR bacteria gradually reduced the further nematode reproduction. Nutrient uptake and plant growth improved significantly in papaya cv Solo inoculated with AMF and *Trichoderma* (Sukhada *et al.*, 2011)**(Plate 12.1).**

3.15 Sapota *(Manilkara achras* (Mill) Forsberg)

Panneerselvam *et al.* (2013) studied the co-inoculation effect of *G.mosseae* with *P.putida* on sapota seedlings. Compared with uninoculated plants the seedlings inoculated with *G.mosseae* and *P.putida* showed increased plant height, leaf area and total plant dry mass after 180 days after sowing. The leaf chlorophyll, phenol, and total glomalin, acid and alkaline phosphatase activities increased in dual inoculations.

3.16 Strawberry (*Fragaria×ananassa* Duchesne), **Blueberry** *(Vaccinium corymbosum* L.)

The dual inoculation effect of AMF with other beneficial soil bacteria on growth improvement of strawberry was reported by Gryndler *et al.* (2002). The results of the study showed that the plants inoculated with *G.etunicatum* or *G.fasciculatum* with bacterial isolates M30 showed significant increase in shoot, root dry mass and enhanced nutrient (P, K) content in shoot biomass than the other bacterial isolates. Malusa *et al.* (2007) evaluated the interactional effect of AMF with *Trichoderma* and other PGPR bacteria against mixture of fertilizers applied through foliar fertilization for growth improvement in three strawberry varieties. Compared with fertilizer mixture, the plants inoculated with AMF and *Trichoderma* and other PGPR bacteria showed significantly increased plant shoot height, shoot girth total number of leaf with increased leaf area. Moreover the mineral (N, P, K, Ca and Mg) content of inoculated plants were significantly increased. Camprubi *et al.* (2007) studied the effectiveness of dual inoculation of AMF with *Trichoderma aureoviride* and PGPR bacteria (*Bacillus* sp.) on improvement of strawberry fruit yield in disinfested soil by solarization and metham-sodium treatment under polyhouse condition. The plants inoculated with *Bacillus* plus *G.intraradicus* showed increased fruit yield (28% and 14%) than the control and plants which received single inoculum. Chauhan *et al.*, (2010) tested the effect of AMF on strawberry plants, co-inoculated with *Trichoderma viridae*. The plants colonized by *A.laevis* + *T.viridea* showed maximum increase in plant height , fresh shoot weight , dry shoot weight , fresh root weight , total chlorophyll and phosphorus content in roots as compared to control. Triple inoculation recorded maximum leaf area over control. Lingua *et al.* (2013) reported that the AMF inoculation with *Pseudomonas* sp. significantly increased the anthocyanin concentration in the strawberry fruit along with plant growth under the low fertilizer condition.

Lingua *et al.*(2013) studied the interactional effect of AMF with two *Pseudomonas* sp. to increase the anthocyanin content in strawberry fruit with less fertilization. Out of two strains of *Pseudomonas* sp., *Pseudomonas fluorescence* with AMF significantly increased the anthocyanin and other metabolite content in strawberry fruits with 70% of natural fertilizers. This combination also increased the nitrogen uptake level in inoculated leaves. Bona *et al.*(2014) reported the interactional effect of AMF and plant growth promoting rhizobacteria on strawberry plants. The study showed that the plants inoculated with AMF (*Rhizophagus intraradices, Glomus aggregatum, Glomus viscosum, Claroideoglomus etunicatum* and *Claroideoglomus claroideum*) mix and *Pseudomonas* sp improved the plant growth by means of increased shoot, root growth, flower size, fruit size and number of fruits per plant over the control plants. The increased sugar uptake, ascorbic acid and folic acid content were observed in the AMF+PGPB inoculated plants than the un-inoculated control. Arriagada *et al.*(2012) studied the effect of AMF (*Gigaspora rosea; Glomus claroideum; G. deserticola; G. viscosum; G. intraradices* and *G. constrictum*) interaction of with saprobe fungi (*Coriolopsis rigida, Phanerochaete chrysosporium, Trametes versicolor; Trichoderma harzianum* and *Penicillium chrysogenum*) for the growth improvement on blueberry and enhancement of soil enzyme activities. The plants inoculated with *G. viscosum* + *P. chrysosporium,* and *G. intraradices* + *C. rigida* significantly increased the plant shoot and root height respectively. The enhanced dehydrogenase enzyme activity was observed with the application of *G. claroideum* + *T. versicolor.*

4.0 Conclusion and future strategies

Combined inoculation of AMF and other beneficial microbes are more beneficial in improving plant growth and development in fruit crops than individual treatments. The use of effective AMF and other plant growth promoting microorganism in a combined form will enhance the plant growth and also stop or reduce the chemical input into soil. The interactional effects of AMF with other beneficial microbes influence the plant to sustain in adverse conditions also. The combined inoculation with tested microbial combination will be effective in getting maximum benefit in field applications.

The exact proportion of both the organisms which would give maximum benefit to the plant needs to be worked out. The mode of action of both the microbes needs be studied.The accurate management of these interactions, by tailoring appropriate mycorrhizosphere systems relevant to plant growth and health in an integrated approach, is important in sustainable horticulture.

References

1. Adesemoye, A.O., Torbert, H.A and Kloepper, J.W., 2009. Plant growth promoting rhizobacteria allow reduced application rates of chemical fertilizers. *Microb Ecol.* 58,921–929.

2. Adriano-Anaya, M.L., Gutiérrez-Miceli, F.A., Dendooven, L and Salvador-Figureueroa, M.,2011. Biofertilization of banana (*Musa* spp.)

with free-living N2 fixing bacteria and their effect on mycorrhization and the nematode *Radopholus similis*. *Journal of Agricultural Biotechnology and Sustainable Development*. 3(1),1-6.

3. Aggangan, N.S., Tamayao, P.J.S., Aguilar, E.A., Anarna, J.A and Dizon, L.V., 2013. Arbuscular Mycorrhizal Fungi and Nitrogen Fixing Bacteria as Growth Promoters and as Biological Control Agents against Nematodes in Tissue-Cultured Banana var. Lakatan, *Philippine Journal of Science*, 142 (2), 153-165.

4. Ahmad, F., Ahmad, I. and Khan, M.S.,2008. Screening of free-living rhizospheric bacteria for their multiple plant growth promoting activities. *Microbiological Research*. 163,173-181.

5. Alarcón, A., Davies Jr, F.T., Egilla, J.N., Fox, T.C., Estrada-Luna, A.and Ferrera-Cerrato, R.,2002. Short term effects of *Glomus claroideum* and *Azospirillum brasilense* on growth and root acid phosphatase activity of *Carica papaya* L. under phosphorus stress. *Rev. Latinoam. Microbiol.* 44(1),31-37.

6. Alizadeh, H., Behboudi, K., Ahmadzadeh, M., Javan-Nikkhah, M., Zamioudis, C. Pieterse, C.M.J.,2013. Induced systemic resistance in cucumber and *Arabidopsis thaliana* by the combination of *Trichoderma harzianum* Tr6 and *Pseudomonas* sp. Ps14. *Biol. Control*. 65, 14–23.

7. Arriagada, C., Manquel, D., Cornejo, P., Soto, J., Sampedro, I.and Ocampo, J.,2012. Effects of the co-inoculation with saprobe and mycorrhizal fungi on *Vaccinium corymbosum* growth and some soil enzymatic activities. *Journal of Soil Science and Plant Nutrition*.12 (2), 283-294.

8. Artursson, V and Jansson, J.K. (2003). Use of bromodeoxyuridine immunocapture to identify active bacteria associated with arbuscular mycorrhizal hyphae. *Appl Environ Microbiol*. 69, 6208–6215.

9. Beneduzi, A., Peres, D., Vargas, L.K., Bodanese-Zanettini, M.H.and Passaglia, L.M.P.,2008. Evaluation of genetic diversity and plant growth promoting activities of nitrogen-fixing bacilli isolated from rice fields in South Brazil. *Applied Soil Ecology*. 39, 311 – 320.

10. Bharat, N.K., 2013. Effect of indigenous AMF and BCAs on health of apple seedlings grown in replant disease soil. *Indian Phytopath*. 66 (4), 381-386.

11. Bianciotto, V., Minerdi, D., Perotto, S. and Bonfante, P.1996a. Cellular interactions between arbuscular mycorrhizal fungi and rhizosphere bacteria. *Protoplasma*. 193,123–131.

12. Bianciotto, V., Andreotti, S., Balestrini, R., Bonfante, P. and Perotto, S. 2001.Extracellular polysaccharides are involved in the attachment of *Azospirillum brasilense* and *Rhizobium leguminosarum* to arbuscular mycorrhizal structures.*Eur J Histochem*. 45, 39–49.

13. Bona, E., Lingua, G., Manassero, P., Cantamessa, S., Marsano, F., Todeschini, V., Copetta, A., D'Agostino, G., Massa, N., Avidano, L. and Gamalero,E.,Berta, G.,2014. AM fungi and PGP pseudomonads increase flowering, fruit production, and vitamin content in strawberry grown at low nitrogen and phosphorus levels. *Mycorrhiza.* 2014 Aug 30.

14. Camprubi, A., Calvet, C. and Estaún V.,1995. Growth enhancement of *Citrus reshni* after inoculation with *Glomus intraradices* and *Trichoderma aureoviride* and associated effects on microbial populations and enzyme activity in potting mixes. *Plant and Soil.*173, 233–238.

15. Camprubi, A., Estaun, V., El- Bakali, M.A., Garcia-Figureueres, F.and Calvet, C.,2007. Alternative strawberry production using solarization, metham sodium and beneficial soil microbes as plant protection methods. *Agron. Sustain. Dev.* 27, 179-184.

16. Chauhan, S., Kumar, A., Mangla, M and Aggarwal, A.,2010. Response of Strawberry plant (*Fragaria ananassa* Duch) to inoculation with arbuscular mycorrhizal fungi and *Trichoderma viride*. *Journal of Applied and Natural Science.* 2 (2), 213-218.

17. Crawford, E.W., Jayne, T.S. and Kelly, V.A., 2006. Alternative approaches for promoting fertilizer use in Africa. *Agriculture and Rural Development Discussion* Paper 22. Washington, DC, World Bank.

18. Dames, J. F. and Ridsdale, C.J.,2012. What we know about arbuscular mycorhizal fungi and associated soil bacteria. *Afric J Biotechn.* 11(73), 13753-13760.

19. Declerck, S., Risede, J.M. and Delvaux, B., 2002. Greenhouse response of micropropagated bananas inoculated with *in-vitro* monoxenically produced arbuscular mycorrhizal fungi, *Sci. Hort.* 93, 301–309.

20. Dutta, P. and Kundu, S.,2012. Effect of biofertilizers on nutrient status and fruit quality of Himsagar mango grown in new alluvial zones of West Bengal. *Journal of Crop and Weed.* 8(1), 72-74.

21. Esitken, A., Yildiz, H.E., Ercisli, S., Donmez, M.F., Turan, M.and Gunes, A.,2010. Effects of plant growth promoting bacteria (PGPB) on yield, growth and nutrient contents of organically grown strawberry. *Scientia Horticulturae.* 124,62–66.

22. Freitas, N.O., Yano-Melo, A.M., Barbosa, da Silva, F.S., Franklin, de Melo, N.and Maia, L.C., 2011. Soil biochemistry and microbial activity in vineyards under conventional and organic management at Northeast Brazil. *Sci. Agric.* (Piracicaba, Braz.).68(2),223-229.

23. Godara, R.K., Awasthi, R.P.and Kaith, N.S.,1998.Interaction effect of VA mycorrhizae and *Azotobacter* inoculation on growth and macronutrients of peach seedlings. *Hary. J. Hort. Sci.* 27,235-240.

24. Gryndler, M ., Vosátka, M., Hršelová, H., Catská, V., Chvátalová., I and Jansa, J., 2002. Effect of dual inoculation with arbuscular mycorrhizal

fungi and bacteria on growth and mineral nutrition of strawberry, *Journal of plant nutrition*. 25,6, 1341-1358.

25. Hameedaa, B., Harinib, G., Rupelab, O.P., Wanib, S.P., Reddy, G.,2008. Growth promotion of maize by phosphate solubilizing bacteria isolated from composts and macrofauna. *Microbiological Research*. 163, 234-242.

26. Harender raj and Sharma, S.D., 2010. Combination of soil solarization, vesicular-arbuscular mycorrhiza and *Azotobacter chrococcum* for the management of seedling wilt of citrus. *Indian Phytopath*. 63 (3) , 282-285.

27. Hayat, R., Ali, S., Amara, U., Khalid, R. and Ahmed, I.,2010. Soil beneficial bacteria and their role in plant growth promotion: a review. *Annals of Microbiology*. 60, 579-598.

28. Ibrahim, H. I .M., Zaglol, M.M.A. and Hammad, A.M.M., 2010. Response of balady guava trees cultivated in sandy calcareous soil to bio fertilization with phosphate dissolving bacteria and / or VAM Fungi. *Journal of American Science*. 6(9),399-404.

29. Jaizme-Vega, M.C., Rodríguez-Romero, A.S., Marín Hermoso, S. and Declerck, S.,2003. Growth of micropropagated bananas colonized by root-organ culture produced arbuscular mycorrhizal fungi entrapped in Ca-alginate beads, *Plant Soil* .254, 329–335.

30. Jaizme-Vega M.C., Rodríguez-Romero, A.S.and Piñero Guerra, M.S.,2004. Potential use of rhizobacteria from the Bacillus genus to stimulate the plant growth of micropropagated banana, *Fruits*. 59, 83–90.

31. Jaizme-Vega, M.C., Rodríguez-Romero, A.S.and Barroso Núñez, L.A.,2005. Effect of the combined inoculation of arbuscular mycorrhizal fungi and plant- growth promoting rhizobacteria on papaya (*Carica papaya* L.) infected with the root-knot nematode *Meloidogyne incognita*. *Fruits*. 61 (3),1-2. 190-192.

32. Kim, K., W. Yim., P. Trivedi., M. Madhaiyan., H.P. Deka Boruah., Md. Rashedul Islam., G. Lee. and T.M. Sa., 2010. Synergistic effects of inoculating arbuscular mycorrhizal fungi and *Methylobacterium oryzae* strains on growth and nutrient uptake of red pepper (*Capsicum annuum* L.). *Plant Soil*. 327,429-440.

33. Kloepper J.W., Zablowicz, R.M., Tipping, B. and Lifshitz, R., 1991. Plant growth mediated by bacterial rhizosphere colonizers. In: D.L.Keister and B.Gregan (Eds.) the rhizosphere and plant growth. *BARC Symp*. 14, 315-326.

34. Kuwada, K., Wamocho, L.S., Utamura, M., Matsushita, I. and Ishii, T.,2006. Effect of Red and Green Algal Extracts on Hyphal Growth of Arbuscular Mycorrhizal Fungi, and on Mycorrhizal Development and Growth of Papaya and Passion fruit. *Agronomy journal*.98, 1340-1344.

35. Linderman, R.G., 1994. Role of VAM fungi in biocontrol. In: F.L. Pfleger and R.G. Linderman (Editors), *Mycorrhizae and Plant Health*. APS Press, St. Paul, pp. l-26.

36. Lingua, G., Bona, E., Manassero, P., Marsano, F., Odeschini, V., Cantamessa, S., Copetta, A., Agostino, G.D., Gamalero, E. and Berta, G.,2013. Arbuscular Mycorrhizal Fungi and Plant Growth-Promoting *Pseudomonads* Increases Anthocyanin Concentration in Strawberry Fruits (*Fragaria x ananassa* var. Selva) in Conditions of Reduced Fertilization. *Int. J. Mol. Sci.* 2013, 14, 16207-16225.

37. Malakoff, D., 1998. Coastal ecology: death by suffocation in the Gulf of Mexico. *Sci.* 281.

38. Malusa, E., Sas-Paszt, L., Popinska, W. and Zurawicz, E., 2007. The Effect of a Substrate Containing Arbuscular Mycorrhizal Fungi and Rhizosphere Microorganisms (*Trichoderma, Bacillus, Pseudomonas* and *Streptomyces*) and Foliar Fertilization on Growth Response and Rhizosphere pH of Three Strawberry Cultivars. *International Journal of Fruit Science.* 6,4, 25-41.

39. Mamatha, G., Bagyaraj, D.J. and Jaganath, S.,2002. Inoculation of field established mulberry and papaya with VAM fungi and mycorrhiza helper bacterium. *Mycorrhiza.* 12, 313-316.

40. Martínez-Medina, A., Roldán, A., Albacete, A. and Pascual ,J.A., 2011.The interaction with arbuscular mycorrhizal fungi or *Trichoderma harzianum* alters the shoot hormonal profile in melon plants. *Phytochemistry.* 72(2-3),223-9.

41. Naher, U.A., Othman, R.and Panhwar, Q.A., 2013. Beneficial effects of mycorrhizal association for crop production in the tropics - a review. *Int J Agr Biol.* 15,1021-1028.

42. Osorio Vega, N.W., Serna Gómez, S.L. and Montoya Restrepo, B.E.,2012. Use Of Soil Microorganisms As A Biotechnological Strategy To Enhance Avocado (*Persea Americana*)-Plant Phosphate Uptake And Growth. *Rev. Fac.Nal.Agr.Medellín.* 65(2), 6645-6657.

43. Panneerselvam, P., Saritha, B., Sukhada Mohandas., Upreti, K.K., Poovarasan, S., Sulladmath, V.V. and Venugopalan, R.,2013. Effect of mycorrhiza associated bacteria on enhancing colonization and sporulation of *Glomus mosseae* and growth promotion in sapota (*Manilkara Achras* (Mill.) Forsberg) seedlings. *Biological Agriculture and Horticulture.* 29(2), 118–131.

44. Pathak, D. V., Surender Singh. and R. S. Saini.,2013. Impact of bio-inoculants on seed germination and plant growth of guava (*Psidium guajava*L). *International Journal of Horticulture and Floriculture.* 1 (3), 021-022.

45. Poovarasan, S., Sukhada Mohandas., Panneerselvam,P., Saritha,B. andAjay,K.M.,2013. Mycorrhizae colonizing actinomycetes promote plant growth and control bacterial blight disease of pomegranate (*Punica granatum* L. cv Bhagwa). *Crop Protection.* 175–181.

46. Raj, H. and Sharma, S.D., 2009. Investigation on soil solarization and chemical sterilization with beneficial microorganisms for control of white root rot and growth of nursery apple. *Scientia Hortic.*119 , 126-131.

47. Rizk-Alla, M.S. and Tolba, H.I.,2010. The Role of some Natural Soil Conditioner and AM Fungi on Growth, Root Density and Distribution, Yield and Quality of Black Monukka Grapevines Grown on Calcareous Soil. *Journal of American Science.* 6(12),253-263.

48. Rodriguez Romero, A.S., Guerra, M.S. and Jaizme Vega, M.D.,2005. Effect of arbuscular mycorrhizal fungi and rhizobacteria on banana growth and nutrition. *Agron Sust Devel.* 25, 395–399.

49. Sagar, P.and Roy, A.K.,2013. Efficiency of Arbuscular Mycorrhizal Fungi for Litchi [*Litchi Chinensis* (Gaertn.) Sonn] Marcots inoculation in nursery. *J. Mycopathol, Res* .51(1),141-144.

50. Sen, R., Nurmiaho-Lassila, K., Haahtela, K., and Korhonen, K. 1996. Specificity and mode of primary attachment of *Pseudomonas fluorescens* strains to the cell walls of ectomycorrhizal fungi. In *Mycorrhizas in Integrated Systems:From Genes to Plant Development.* Azcón-Aguilar, C., and Barea, J.M. (eds). Brussels, Belgium: ECSC-EC-EAEC Press.

51. Shrestha, R.K and Ladha, J.K., 1998. Nitrate in groundwater and integration of nitrogen-catch crop in rice-sweet pepper cropping system. *Soil Sci Soc Am J* .62,1610–1619.

52. Shaviv, A., 2000. Advances In Controlled Release Fertilizers, *Advances in Agronomy.* 71,1-49.

53. Sharma, S.D., Kumar, P., Gautam, H.R.and Bhardwaj, S.K.,2009. Isolation of arbuscular mycorrhizal fungi and *Azotobacter chroococcum* form local litchi orchards and evaluate their activity in air-layers system. *Scientia Hortic.* 123, 117–123.

54. Saharan,B.S., and Nehra,V., 2011. Plant Growth Promoting Rhizobacteria: A *Critical ReviewLife Sciences and Medicine Research*, Volume 2011 p1-30.

55. Sharma, S.D., Kumar, P., Bhardwaj, S.K.and Chandel, A., 2011. Symbiotic effectiveness of arbuscular mycorrhizal technology and Azotobacterization in citrus nursery production under soil disinfestation and moisture conservation practices. *Scientia Horticulturae.*132,27–36.

56. Sitole, P., 2013. Investigating the role of mycorrhizal fungi and associated bacteria in promoting growth of citrus seedlings. *Thesis.*

57. Smith, S.E. and D.J. Read., 1997. Mycorrhizal symbiosis. *Academic Press,* San Diego, CA.

58. Sukhada, M.,1996. Utilisation of biofertilisers for banana cultivation. In Banana, improvement, production and utilization. *Proceedings of Nat.Conf. on Banana, Trichi 24-26 Sept*, . Eds. H.P Singh and K.L.Chadha 1996.

59. Sukhada, M., Kumar, B.P.and Manamohan, M.,2000. Response of pomegranate cv Ganesh to inoculation with beneficial microbes. In: *National Seminar on Hitech Horticulture*, Bengaluru June 26-28.

60. Sukhada, M., Manjula, R., Rawal, R.D., Lakshmikantha, H.G., Chakarborty, S and Ramachandra, Y.L., 2010. Evaluation of arbuscular mycorrhizal and other biocontrol agents in managing *Fusarium oxysporum* f. sp. cubense infection of banana cv. Neypoovan. *Biocontrol Science and Technology*. 20(2), 165-181.

61. Sukhada M., Manjula R.and Rawal R. D.2011. Evaluation of arbuscular mycorrhiza and other biocontrol agents against *Phytophthora parasitica* var. *nicotianae* infecting papaya (*Carica papaya* cv. Surya) and enumeration of pathogen population using immunotechniques. *Biol. Control* 58 22–29.

62. Sylviane, N., Vaucheret Mike, S.V., Juliette, W. and Steve, T., 2002. *DEATH IN MALL DOSES Cambodia's pesticide problems and solutions, Pesticide use in Cambodia*.

63. Tajini, F., M. Trabelsi. and J.J. Drevon., 2011. Co-inoculation with Glomus intraradices and Rhizobium tropici CIAT899 increases P use efficiency for N2 fixation in the common bean (*Phaseolus vulgaris* L.) under P deficiency in hydroaeroponic culture. *Symbiosis*. 53,123-129.

64. Topbaş, M.T., Brohi, R.A. and Karaman, M.R., 1998. Environmental Pollution, *Environment Ministry*.

65. Toljander, J.F., Artursson, V., Paul, L.R., Jansson, J.K. and Finlay, R.D. (2005) Attachment of different soil bacteria vitality to arbuscular mycorrhizal fungal extraradical hyphae is determined by hyphal and fungal species. *FEMS Letters* (in press).

66. Trabelsi, D. and Mhamdi R., 2013. Microbial Inoculants and Their Impact on Soil Microbial Communities: A *Review BioMed Research International* .,Page.11.

67. Vivas, A., Azcon, R., Biro, B., Barea, J.M. and J.M. Ruiz Lozano, J.M., 2003. Influence of bacterial strains isolated from, lead-polluted soil and their interactions with Arbuscular mycorrhizae on the growth of Trifolium pratense L. under lead toxicity. *Can. J. Microbiol*. 49,577-588.

68. Vivas, A., B. Biro, T. Nemeth., J.M. Barea. and R. Azcona., 2006a. Nickel-tolerant Brevibacillus brevis and arbuscular mycorrhizal fungus can reduce metal acquisition and nickel toxicity effects in plant growing in nickel supplemented soil. *Soil Biol. Biochem*. 38,2694-2704.

69. Vivas, A., B. Biro, J.M. Ruiz-Lozano, J.M. Barea, and R. Azcon. 2006b. Two bacterial strains isolated from a Zn-polluted soil enhance plant growth and mycorrhizal efficiency under Zn-toxicity. *Chemosphere*. 62,1523-1533.

70. Xavier, L.J.C. and J.J. Germida., 2003. Bacteria associated with Glomus clarum spores influence mycorrhizal activity. *Soil Biol. Biochem*. 35,471-478.

71. Xueming, Z., Zhenping, H., Yu, Z., Huanshi, Z.and Pei, Q.,2014. Arbuscular mycorrhizal fungi (AMF) and Phosphate-solublizing fungus (PSF) on tolerance of beach plum (*Prunus maritima*) under salt stress. *AJCS*. 8(6), 945-950.

72. Zarei, M., N. Saleh-Rastin., H. Ali Alikhani. and N. Aliasgharzadeh., 2006. Responses of lentil to co-inoculation with phosphate solubilizing rhizobial strains and arbuscular mycorrhizal fungi. *J. Plant Nutr.* 29,1509-1522.

73. Zhuang, X. L., Chen, J., Shim, H.and Bai, Z., 2007. New advances in plant growth-promoting rhizobacteria for bioremediation. *Environ. Int.* 33, 406–413.

Part VI

AMF inoculum Production and Mycorrhization of Crop Plants

Chapter 13

Arbuscular Mycorrhizal Inoculum Production and application

K.Kumutha, M.Srinivasan, R.Vinuradha ,L.Srimathi Priya and D.Subhashini[1]

Department of Agricultural Microbiology, Tamil Nadu Agricultural University, Coimbatore-03, Tamil Nadu (India)

1Central Tobacco Research Institute, Rajahmundry (ICAR) –533105 (A.P.),India

*Address for the Correspondence: kkumuthatnau@gmail.com

ABSTRACT

Production of Arbuscular Mycorrhizal Fungi (AMF) inocula is the most challenging task as they are obligate symbionts and depend on a living host for their multiplication. Production systems have evolved considerably during the recent decade from inocula produced in raised bed nurseries, earthen pots and other containers to on farm production. On-farm inoculua production is very beneficial for fruit crop cultivation. Many host species have been tested and found to be good for inoulum multiplication. Different substrates have been used in soil and substrate-based production techniques as well as substrate-free culture techniques (hydroponics and aeroponics) and *in-vitro* cultivation methods have all been attempted for the large-scale production of AMF. This chapter describes the principal *in-vivo* and *in-vitro* production methods that have been developed so far and presents the parameters that are critical for optimal production,

Keywords: *Propagules, hydroponics, aeroponics, root organ culture, trap culture, carrier material, on-farm production*

1.0 Introduction

Today, mycorrhiza is the most wide spread symbiotic association existing in the ecosystems throughout the world. Members of several fungal taxa are involved in mycorrhizal associations and possess a few common characteristic features such as: ubiquitous distribution, strong biotrophic dependence on host plants and rarely free living saprophytes. To date, seven types of mycorrhizae are recognized *viz.*, Ectomycorrhiza, Ectendomycorrhiza, Endomycorrhiza, Arbutoidmycorrhiza,Monotropoidmycorrhiza, Ericoidmycorrhiza, Orchid mycorrhiza (Smith and Read, 1997). Frank was the first person to coin the term "Mykorrhizen" in the year 1885 for fungal association in the trees of pines which means "fungal-root" in Greek. Paleobotany dates these associations to the Devonian era approximately 400 million years back. Allen (1991) defined a mycorrhiza as "a mutualistic symbiosis between plant and fungus localized in a root or root-like structure in which energy moves primarily from plant to fungus and inorganic resources move from fungus to plant".

2.0 Inoculum production systems with commercial application

Production systems of AMF have evolved considerably during recent years, from relatively simple technologies to more complex ones, for example, *in-vitro* methods (Jarstfer and Sylvia, 1994). At present, inoculum is produced for commercial purposes in the following ways:

(i)   Nursery plots with soil (Sieverding, 1991), in which inoculated plants are cultured in open field or nursery beds. Advantages: simple, adapted for local use and low cost; Disadvantages: limited in application, easily contaminated and not well adapted for the development of an industrial activity.

(ii) Containers (pots), Cement tanks or Bricklined tank (**Plate 13.2)** with different substrates (Feldmann and Idczak, 1994; Feldmann and Grotkass, 2002). Advantages: low technology input, undesirable contaminations fairly easily eliminated and reasonable costs; Disadvantages: not pure and limited in its industrial development.

(iii) Aeroponic systems (Jarstfer and Sylvia, 1994), where pre-inoculated plant roots are continuously misted with nutrient solution sprayed within cultivation boxes. Advantages:easier control of contaminants, carrier free inoculum and adapted for microplants; Disadvantages: relatively complicated technological setup.

(iv) *In-vitro,* on roots transformed with *Agrobacterium rhizogenes* (Declerck *et al.,* 1996). Advantages: pure cultures and permits industrial development; Disadvantages: high technological investment, high costs, not all AMF successfully culturable in this system and suitability of inoculum produced *in-vitro,* in particular its competitive ability towards other microbes in field soil has yet to be tested.

3.0 On-farm production of AMF

"There are two principal ways of ensuring that the benefits in terms of crop production are obtained from mycorrhizal associations: (1) by inoculating with selected, efficient mycorrhizal fungi and (2) by promoting the activity of effective indigenous mycorrhizal fungi by proper cultural practices". Cultural practices that increase the activity of indigenous AMF are: reduced tillage, crop rotations, cover crops, and phosphorus management. An AMF available as inoculum may be more effective on a given crop relative to the effectiveness of the indigenous AMF community. One of the main agronomic situations in which inoculation is desirable, however, is to take advantage of the benefits of out planting a pre-colonized seedling (Rice *et al.*, 2002). Even though inocula of AMF are commercially available, production of AMF inoculum on the farm is an attractive alternative. Purchasing the large amounts of inoculum necessary for large scale agriculture may be cost prohibitive. Producing the inoculum on-site saves processing and shipping costs included in the price of commercial inocula. These factors are the primary reason why most on-farm methods have been utilized in developing nations. Another benefit of on-farm production of inoculum is that locally adapted isolates, which may be more effective than introduced ones in certain situations (Sreenivassa, 1992), can be produced when the farmer's indigenous AMF communities are used as starter inocula.

Inocula of AMF are available commercially in a variety of forms ranging from high concentrations of AMF propagules in carrier materials to potting media containing inoculum at low concentrations. Effective AMF inoculum may also be grown on the farm using a variety of methods but, recently a method suitable for temperate climates was developed (Douds *et al.*, 2006). This method entails mixing compost with vermiculite to decrease the effective concentration of available nutrients in the compost, notably of P (Douds *et al.*, 2008a). Bahiagrass (*Paspalum notatum Flugge*) seedlings, colonized by specific AMF, are then transplanted into bags containing the compost and vermiculite mixture. The AMF proliferate in the media as the bahiagrass grows throughout the summer and the bags are weeded and watered as needed. The bahiagrass is then frost killed. The AMF over winter in the media and the inoculum are ready for use in the following spring. Inoculum produced in this fashion has increased the yield of green pepper (Douds *et al.*, 2003) and strawberries (Douds *et al.*, 2008b).

Sukhada *et al.* (2004) suggested on farm production of mycorrhizal inoculum on finger millet (*Eleucine coracana* L.). Large scale inoculum production on farm is carried out by clearing a known farm area of weeds followed by digging and loosening the soil. The soil with 50% moisture content is solarized t for three weeks by covering it with transparent polyethylene sheet 150 mm thick. Starter culture of AMF of *Glomus mosseae* or *G. fasciculatum* containing 30-50 spores g^{-1} soil is applied by hand in close rows having 15 cm distance between them.. Ten kgs of starter inoculum is wa used for 25 m^{-2} area. Seeds of finger millet as host plant are sown on them and covered. The land is irrigated regularly with a rose can. The seeds germinate within a fortnight and AMF colonize the root system. After 12 weeks the roots are checked for root colonization by AMF. The inoculum is harvested by cutting the shoots and digging the roots up to 9 cm below ground. The soil clumps are removed and roots are cut into small pieces and mixed in the soil. The inoculum containing 20-30 spores per gram soil is used for field inoculations **(Plate.13.3).**

Large scale production may also be achieved in single pots of various materials (e.g., earthenware or plastic) and sizes or scaled up to medium size bags and containers and to large raised or grounded beds or cement tanks . The production process is often conducted under controlled or semi-controlled conditions in greenhouses or performed in growth chambers for the easy handling and control of parameters such as humidity and temperature.

4.0 Trap culture of AMF

Spores of AMF isolated directly from the field may not be viable, loose or changed in appearance of their structural characteristics in response to root pigments, soil temperature, moisture and microbial activity. It is e important to identify and obtain abundant healthy spores for the establishment of monospecific cultures. In tropical environments with high microbial activity and organic matter content the spores undergo either structural changes or degradation, which makes identification of these spores very difficult or rather impossible. Different types of trap cultures are used for establishment of naturally occurring spores *viz.*, soil trap culture, root trap culture and plant trap cultures.

Soil trap culture is widely used in assessing the diversity of AMF flora in natural soils (Muthukumar and Udaiyan, 2002). The soil collected from an area is mixed with disinfected substrate usually sand or soil or expanded montmorillonite clay and placed in disinfected containers and planted with host plant. The plants are maintained for a period of three months for the establishment of mycorrhizae and maximum fungal sporulation, in two to four cycles. In root trap culture, the roots of the host plant collected from their natural habitat are cut into 2-3 cm long bits and layered 5 cm below the surface of the disinfected substrate. Though it is identical to soil trap culture, there are more chances of obtaining single-species culture if root fragments are sufficiently small.

The failure of this method has poor chances of initiating new mycorrhiza and not useful for the members of Gigasporaceae. Plant trap cultures are also done by means of taking small plants or seedlings of shrubs and tree from their natural environment and planting in a suitable disinfected substrate after washing the soil, spores and extramatrical fungal hyphae adhered to the roots. This method has the unique advantage of confirmation of AMF symbionts to their associated host but possibility of contamination of adherent propagules.

An important consideration in trap culture is that continued maintenance of cultures beyond the cessation of root growth introduces contaminating organisms. Trap cultures are generally stored for 30 days prior to spore extraction for monospecific culture initiations.

5.0 Production parameters

Mass production in sand or soil or other substrate based production systems is most often initiated with a single identified species or a consortium of selected identified AMF species while on-farm production is also sometimes started with species that are indigenous to the site of application and not always identified to the species level. The starter inoculum to initiate production usually consists of isolated spores or a mixture of spores and mycorrhizal root pieces. To obtain a mixed inoculum, roots may be dried and chopped into fine pieces while spores are most often obtained by wet sieving and decanting. The soil containing AMF hyphae may also be used in mixed inocula. International culture collections such as INVAM (International Culture Collection of Vesicular Arbuscular Mycorrhizal Fungi) and BEG (International Bank for the Glomeromycota) can in most cases guarantee the delivery of well identified monospecies and offer a clear traceability of the organism via a repository identification code.

5.1. Host plants

Although AMF are non-host specific, recent evidences (Walker *et al.*, 1998; Kramadibrata *et al.*, 2000; Herrera-Peraza *et al.*, 2001) point out the possible existence of host inclination in some AMF. A single host species that could maintain compatible associations universally under all environmental conditions is ideal for the obligate nature of AMF association, which varies with respect to edaphic and environmental factors. A suitable host plant should be highly mycotrophic with the ability to sustain a wide range of AMF genotypes, high photosynthetic efficiency, increased root production, adapted to different soil types, tolerant to drought conditions and resistance to diseases and pests.

Many plant species such as corn (*Zea mays*), sorghum (*Sorghum vulgare*), peanut (*Arachis hypogia*), strawberry (*Fragaria* spp.), onion (*Allium* spp.), millet (*Pennisetum americanum*), lettuce (*Lactuca sativa*), *Coleus* sp., *Stylosanthes* spp., bahia grass (*Paspalum notatum*), Kudzu (*Purpurea phaseoloides*), big blue stem (*Andropogon geradii*), red clover (*Trifolium pratense*), fescue (*Festuca arundinacea*), *Plantago lanceolata*, Rhodes grass (*Chloris gayana*), etc., have been experimented as host plant for the symbiotic association of AMF. But two or more hosts can be used to enhance the possibility of invigorating the sporulation of the full complement of native AMF in trap cultures.

Host dependent sporulation of AMF species is an important determinant for inoculum production. For example, high intraradical colonization levels are important for the production of mixed spore root inoculum, while this might not always be needed for the achievement of a spore inoculum. Although correlation between intraradical colonization and extraradical sporulation has sometimes been reported (Douds, 1994), this relation was not found in all cases and is dependent on the plant or fungus association and particular culture conditions (Hart and

Reader, 2002).

5.2 Substrates and amendments

Various substrates either pure or mixed have been used to propagate large scale production of AMF. The most frequent and universal substrate used for culture of AMF is soil with low P content often sandy. For instance, relatively inert substrates (e.g, vermiculite and perlite) have been used to dilute nutrient rich soil and compost. Conversely, compost or other organic substrates such as peat can be added to nutrient deficient soils. Many different organic amendments have been reported to influence AMF root colonization. For example, chitin and humic substances enhanced colonization levels, whereas cellulose reduced colonization by the AMF. Inert substrates have also been used as carrier medium to support roots and fungal growth under conditions where plant feeding was mainly provided by a nutrient solution.

Soil less media has inimitable benefits like unvarying composition, low mass and improved aeration than soil media. High levels of P in the rhizosphere in soil-less media should be avoided, as soil-less media lack P buffering capacity. Enrichment of nutrients, mixing of slow release of fertilizers or the use of sparingly-available form of P provides excellent cultures of AMF in soil-less media.

To improve plant growth response, sand alone is not a appropriate substrate because it lacks buffering capacity. So to meet out the optimal nutritional levels for plant growth, soluble nutrients are continually added. The nutrient levels best suited for plant growth may be excess for optimal growth of the fungi due the lack of nutrient fixation by the medium. This issue can be solved in sand-based semi-hydroponic system by using an automated watering and fertilization regime which produces maximum AMF propagules in shortest possible time. The greatest advantage of this system is the absence of soil or other particulates that clutter spore extracts in extraction methods and increased probability of lower background microbial activity on root, hyphae and spore surfaces. It is easier to retrieve spores and hyphae for a variety of assays.

Expanded clay is also recommended for mass production of AMF. A mixture of expanded clay aggregates (Lesica®), sand and pieces of mineral wool (Grodania®) or mixture of attapulgite clay (Terra green®) and vermiculite has been used for hyphal production of several *Glomus* species (Joner *et al.*, 2000).

5.3 Nutrition

Manipulation of nutrient regimes has been demonstrated to impact AMF propagule production. The nutrient content of the substrate as well as the addition of macronutrients and micronutrients may influence the AMF directly but also indirectly by the plant responses to nutrient availability. Although it remains largely unclear as to which extent the used plant or AMF associations differ in their nutrient requirements, optimal nutrient regimes should support initial colonization, promote adequate plant (root) growth, and optimize the AMF propagule production. In contrast, other studies reported that AMF 1 strains of *G. intraradices* were tolerant to high P levels. Furthermore, the timing of nutrient addition might influence colonization levels and propagule production, as nutrient requirements to culture AMF might differ throughout time. Whereas high P availability often suppresses colonization, the addition of P in later stage might enhance AMF growth and sporulation

6.0 Formulation of the AMF inocula

Basically, the formulation procedure consists of placing fungal propagules (root fragments colonized with AMF, fragments of fungal mycelium, and spores) in a given carrier (perlite, peat, inorganic clay, zeolite, vermiculite, sand, etc.) for a given application. The final configuration of the formulation will result from a more or less technologically complex procedure, determined by the microbe involved, the way of producing inoculum, and the target inoculum application (bare-root plants, containerized plants, cuttings, seeds, potting mixes, soils, etc.). Fungal propagules must be formulated in such a way that they can be stored and distributed under a wide range of temperatures without losing viability. Formulation should be simple and economical and the formulated inocula should be easy to transport and apply. Some companies producing AMF inocula have adopted the approach of one type of formulation (i.e., single fungal species) for all markets, while others produce a range of products for their target buyers. The list of production units and companies producing AMF in India and a few in China are given in **Table13.1 and 13.2.**

7.0 Quality control of AMF inoculum

Most of the scientists working in the areas of mycorrhizal fungi mention that, *"it is increasingly likely that greater regulations and control over the production and selling of AMF inoculant and will be introduced in the coming years"* (Feldmann and Schneider, 2009). This statement ensures that so far there has been no investigation made on quality control of AMF inoculant exclusively. In this context, the following criteria should be fulfilled by the companies: (*i*) plants to be inoculated must be able to form mycorrhizas; (*ii*) the AMF inoculum must be free of agents that could negatively affect normal plant growth and development; (*iii*) the shelf life of the inoculum should be sufficient to suit the end user markets.

7.1 Starter culture

The inoculum from which a crude inoculum is started can be a pure isolate obtained from another researcher, a culture collecting and curating organization such as INVAM, or a reliable commercial culture producing firm. Or, an isolate can be made from a specific soil by the person producing the inoculum. The procedure for obtaining an isolate from soil is described in Appendix 1.

The amount of starter inoculum to use will depend on its quality. The culture must be highly infective, contain at least four infective propagules per gram, and be free of pathogenic microorganisms. The aim is to inoculate the inoculum-production medium at a rate of 500 infective AMF propagules per kilogram of medium. Other qualities of a starter inoculum are discussed in the section on production of root inoculum. Photosynthate to support the formation and development of AMF on its roots without adverse effects on itself. Consequently, environmental variables such as light intensity, soil and air temperature, and soil water status should be favorable for normal plant function.

7.2 Pore density

It has been proved that even 1 spore g^{-1} of the starter inoculum to the seedling initially is enough to result in 90 per cent colonization in maize plant (Sylvia *et al.*, 1993). The spore extracted by wet sieving and decanting method tested for the portion of living spores when used improves the performance of the inoculum (Vosatka and Dodd, 2002). Infectivity of the inoculum depends mostly on the spore load. The number of infection units in an inoculum of AMF depends on the number of spores, colonized root fragments and mycelial fragments, which can actually lead to root colonization under favourable conditions. The relevant number of propagules can be determined with the MPN (most probable number) standardized estimations, IP inoculum potential assay and spore counts.

7.3 Carrier material

AMF can be bound to a wide range of carrier materials (Jarstfer and Sylvia, 1994). Information about the carrier material is important for the subsequent inoculation procedure. For instance, the inoculation of plants on "roof tops" requires a carrier, which effectively protects the AMF propagules during the "blowing-up" procedure of the substrate. In the experience of Gianinazzi and Miroslav Vosatka, (2004), turf substrates or expanded clays did not survive this rough procedure but other materials like lava were resistant enough and guaranteed effective AMF colonization after a short time. Similarly attapulgite clays, expanded clays or other inert carriers have been very successful carriers in Horticulture (companies Biorize, France; Mycotec, Germany) and Landscaping (Plantworks, UK).

8.0 *In-vitro* cultivation methods of AMF

Many cultivation s and inoculum production techniques of the plant beneficial AMF have been developed in the last decades. Mass production of contaminant-free AMF inoculum remained a bottleneck for application in agriculture. Production of monoxenic AMF culture under *in-vitro* conditions is one of the most promising ways for understanding of the AMF symbiosis and high number of spore propagules in a shortest time with contamination free inoculum.

It is really a challenging goal for the culture the AMF under axenic conditions. Mosse in 1962, for the first time, established the culture by inoculating the germinating resting spores of *Endogone* sp. She demonstrated the necessity of *Pseudomonas* sp. for germination, appresorium formation and even penetration of *Endogone* sp. into the root. Later Mosse and Hepper (1975) opened the path of *in-vitro* culture, when they used excised roots of tomato (*Lycopersicum esculentum* Mill.) and red clover (*T. pratense* L.) *in-vitro* and succeeded in establishing AMF symbiosis. Isolated root can be propagated continuously in different solid and liquid media with high reproducibility. Clonal roots of some 15 plants have been established and this list has enlarged during the last decades. Based on the above approach, the monoxenic cultivation system has become a valuable tool to produce contaminant-free AMF, allowing the realization of large-scale production under strictly controlled conditions. There are three prerequisites for obtaining such dual cultures. They are suitable host root, an efficient mycorrhizal fungus, and a suitable cultivation medium.

8.1 Root organ culture method

A. Establishment on Transformed Roots

Currently, the use of genetically transformed roots by Ri plasmid of *Agrobacterium rhizogenes* has given a new impetus to the culture production of these obligate symbionts. The *A. rhizogenes,* a soil dwelling, gram negative bacterium produces a neoplastic plant disease syndrome known as 'Hairy Root' which has ability to grow rapidly showing two-fold multiplication within 48 hours in *Atropa belladona* and *Nicotiana tabaccum* (Ryder *et al.*,1985). They are also highly stable and capable of secondary metabolite production. This characteristic is due to presence of integrated copies of transfer DNA (T-DNA) which occurs in Ri (root inducing) plasmid, which is a large plasmid (Chilton *et al.*,1982) The TR-DNA locus of T-DNA of Ri plasmid also carry a gene responsible for the production of certain amino-acid derivatives called opines like agropine, mannopine, cuccumopine and mikimopine, which is indicative of transformation. Similarly loci encoding for auxin and other loci responsible for the production of roots are still the areas of further investigations (Tepfer, 1989). The rapid and stable root growth due to modification by *A. rhizogenes* infection is very important and favourable for the mass production of arbuscular-mycorrhizal culture.

B. Production of Transformed Roots

Initiation of transformed root requires previously surface sterilized with classical disinfectants (sodium hypochlorite, ethanol), and then thoroughly washed in sterile distilled water. Hairy root induction occurs after 7 days at 28°C in the dark on MS medium. Hairy roots produced by *A. rhizogenes* infection can be sub-cultured as excised roots. The tips (5 cm) of emerged roots can be transferred to a rich medium such as modified White medium (Becard, and Fortin, 1988) or Strullu and Romand medium (Strullu and Romand,1987) with suitable antibiotics like cefotaxime to make the roots free from original bacterial inoculum and thereby establish the isolation of a single root piece clonal culture. An example of the production of such culture from carrot disc is described. Root segments of mature carrot plant, after surface sterilization with sodium hypochlorite (2% available chlorine) for 5 min., were inoculated with freshly grown bacterium (48 hrs. old in yeast extract mannitol broth) at distal surface, because this bacterium has a higher virulence on the distal surface than apical part due to the higher endogenous auxin level (Ryder *et al.*,1985) The treated stem segments were placed in Murashige and Skoog (MS) medium and incubated in dark for two days at 27°C. Thereafter, the carrot discs were transferred to MS medium amended with cefotaxime (250 mgL^{-1}) and again incubated under the same conditions. After 8-10 days, few transformed roots were found to be proliferated which were later excised and maintained as a clonal culture. Successful transformation was later confirmed either by the detection of *agropine synthase* in hairy root tissues (Srinivasan *et al.*, 2014).

C. Selection of Mycorrhizal Propagules

Two types of fungal inoculum can be used to initiate monoxenic cultures: either extraradical spores or propagules from the intraradical phase (i.e., mycorrhizal root fragments and isolated vesicles) of the fungus. selection of the type of mycorrhizal propagules is very important for infecting hairy roots. The inoculum should be healthy, viable and contamination-free. Different workers have used different types of inocula. Mosse and Hepper, (1975) used sporocarp of *Glomus mosseae* (Mosse, and Hepper (1975), *Glomus intraradices* (Puri and Adoleya, 2013) *Gigaspora margarita* (Diop *et al.*, 1992) *Glomus etunicatum* (Schreiner and Koide, 1993), *Glomus versiforme* (Declerck *et al.*, 1996) *Glomus fistulosum* (Gryndler *et al.*, 1998), *Glomus macrocarpum* (Chandra and Potty 2011), *Gigaspora decipiens* and *Glomus clarum* (Costa *et al.*, 2013). Strullu and Romand (1987) as well as Zhipeng and Shiuchien (1991) used infected root segments colonised with mycorrhiza.

The isolation of AMF spores generally involves wet sieving and decanting (Gerdemann, and Nicolson, 1963) followed by density gradient centrifugation in order to further purify the spores (Faurlan *et al.*, 1980). The latter technique removes dead spores and other soil debris, which is the main source of contamination.

D. Surface sterilization of spore propagules

Surface sterilization process are keys for the establishment of *in-vitro* based AMF inoculum production. Before being used as *in-vitro* inoculum, spores must be surface sterilized. This step is critical because success depends on the elimination of all contaminants. A thorough perusal of literature reveals that different workers have used different techniques of spore sterilization (Mosse 1962). Most of them have used chloramine-T (2%), gentamycin and streptomycin sulphate. Diop, 2003 modified the procedure of sterilization of Mertz *et al.* (1979) which was simple, easy and highly effective than the processes described by earlier workers. In the process, the spores were treated with 2% chloramine-T containing 2% tween-20 and 95 % ethanol by in an injection vial under light vacuum developed by taking out air from the solution by injection syringe. Light vacuum was able to remove liquefied gases and air droplets adhered to the spore surface. Thereafter, the spores were removed on sterile Whatmann No. 1 filter paper placed in a funnel fitted with rubber tube and stop cock. The spores were washed thoroughly with sterile distilled water to remove tween-20 and chloramine-T. The spores were then treated with sterile antibiotic solution (gentamycin 100 µg mL^{-1} + streptomycin sulphate 200 µg mL^{-1}) for 20 min. They were either utilized immediately or were stored at 4°C for 3 to 4 months in a centrifuge tube.

E. Dual culture

Solid nutrient media are good substrate for the establishment of dual culture of root and fungus. Since Mosse and Hepper, 1975 first established monoxenic mycorrhiza with *Glomus mosseae* at least 27 AMF species have been successfully cultured monoxenically in association with *Agrobacterium rhizogenes* T-DNA transformed excised host roots of many plant species (Fortin *et al.*, 2002). Becard and Fortin (1988), later, improved the methodology and adopted it for the study of initial events of AMF ontogenesis. They directly inoculated a single ungerminated spore of *Gigaspora margarita* on a single root system. Since the germ tube growth is negatively geotropic in nature, inserted a single spore into the medium at the bottom of Petri plate, so that germ tube could grow towards the surface of the medium and come in contact with properly placed roots (Watrud *et al.*,1978) Thereafter, the process of inoculation was simplified, understanding the negatively geotropic nature of the germ tube. A single spore was placed near the single transformed root in the same Petri plate and incubated the whole Petri plate vertically in such a way that germ tube grow upward and immediately come in contact with the root. The most widely used plant fungus combination is carrot (*Daucus carota*) and *G.intraradices* (Becard and Piche,1992) because of the vigor of this system and the ease with which it can be propagated. To maximize infection the medium should contain minimal phosphorus and low pH (5.5). Furthermore the use of gellan gum (Clerigal,sigma. USA) as a gelling agents provides a completely transparent medium which allow growth of the microscopic AMF hyphae. Infection can take 2-3 weeks to become established, after which the extrradical hyphae will begin to grow throughout the Petri dish, forming a dense hyphal mat throughout the medium over a period of 30 days. AMF co- culture grow well in the dark at 27°C in Petri dishes sealed with parafilm to conserve moisture and prevent contamination. The direct transfer of a portion of the *G.intraradices* infected carrot root to a new Petri dish containg fresh medium will propagate the co-culture (**Plate.13.1**).

F. Mass production in bioreactors

Airlift Bioreactor was designed by Jolicoeur *et al.*, 1999. The airlift bioreactors were made of 1.2^{-L} total volume autoclavable polycarbonate jars with a modified cover. A central glass draught tube (ID 4 2.54 cm) was installed 2 cm above the bottom and below the liquid level. Medium circulation and oxygenation were provided by a gas flow rate of 5 mL min^{-1} through a porous (2 mm) stainless steel sparger which generated fine bubbles at the bottom of the draught tube. This gas flow rate yielded an initial oxygen mass transfer coefficient (kLao) of 8h. A stainless steel screen mesh (mesh size of 20) was installed 3 cm above the bottom to cover the annular section in order to prevent root circulation with the fluid drag. Sterile air filters (bacterial air vent) and a liquid condenser ensured sterility and minimal liquid losses by evaporation throughout the cultures' duration. A dissolved oxygen probe was installed in the bioreactor. . Cultures in this bioreactor were performed using an initial volume of 500 mL of liquid M medium. Medium and bioreactors were steam sterilized separately for 35 min (121°C, 1 bar). All bioreactor cultures were fed with air enriched with 2% CO_2 . Cultures were grown at 23 ± 1°C under continuous light.

9.0 Selection of the culture system

Basically, two culture systems are used: the mono-compartmental system in square or round Petri plates, and the bi-compartmental system in round Petri

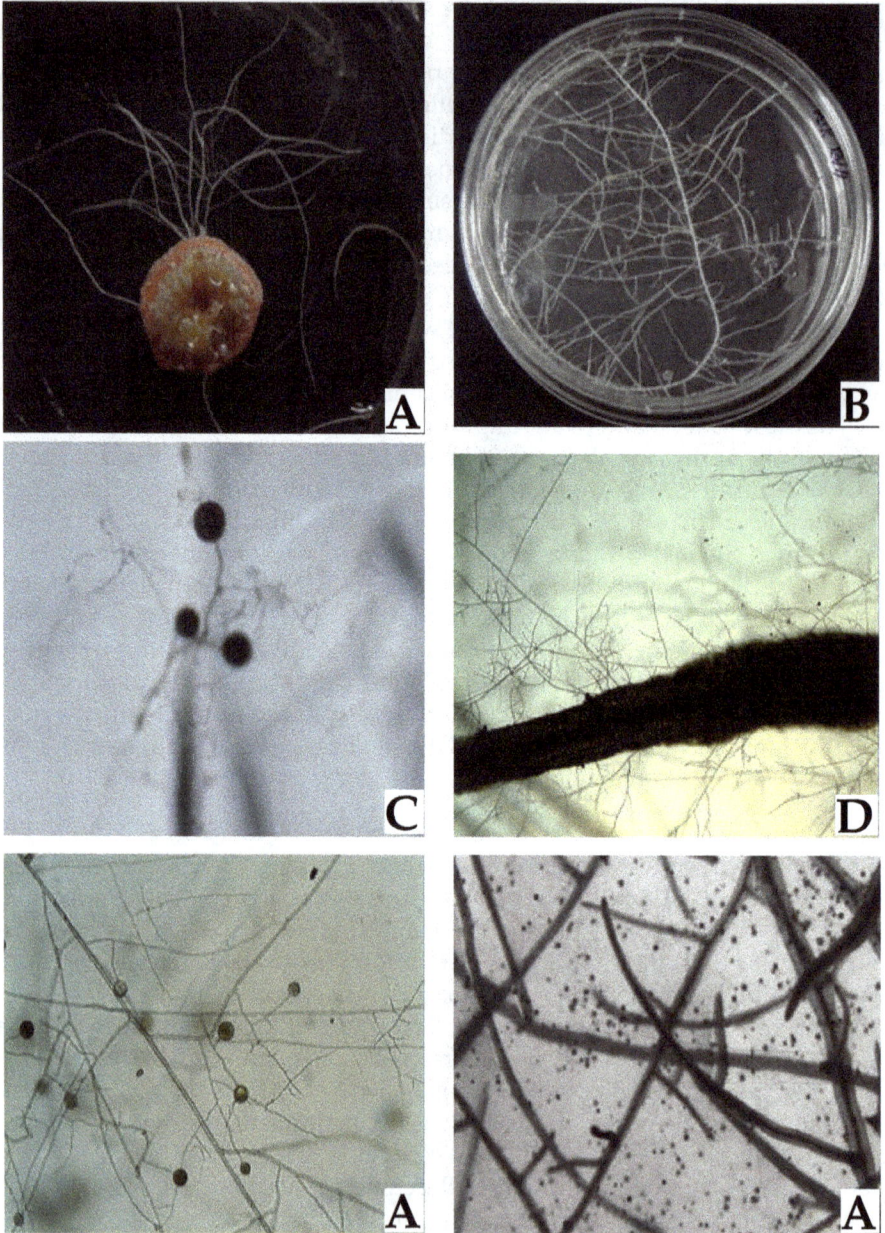

Fig. 13.1: Development of root organ cuture. A Hairy root induction from carrot after 1-2 week incubation on the MS medium, B. Mass multiplication of transformed hairy root on the MSR medium, *C. In vitro Glomus intraradice* spore germination and germ tube growth, D. Hyphal network growth E. Formation of new *Glomus intraradices* germ tubes, F. Mass production of *Glomus intraradices* spores

plates. The first system consists of a mono-compartmental Petri plate filled with a growth medium on which is placed a contaminant-free, actively growing excised root together with AMF propagules. This system was developed in the mid-1970s (Mosse and Hepper,1975) and since then has been applied with success to numerous *Glomus* species.

The second system (also named split system) consists of a bi-compartmental Petri plate, with a proximal compartment in which the mycorrhizal root develops and containing a synthetic growth medium and a distal compartment in which only the mycelium is allowed to grow on a similar synthetic medium, but lacking C source. Both compartments are physically separated by a plastic wall. Roots crossing the partition are trimmed at regular intervals. This system was developed by St-Arnaud *et al.* (1996). In the bi-compartmental system, the spore and mycelium density produced in the distal compartment is markedly higher in comparison to the proximal compartment, which is probably related to the absence of the root and the difference in availability of C (Fortin *et al.* 2002), making this system more productive than the mono-compartmental system. It is particularly adapted to the culture of *Glomus* species. Recently, Douds (2002) has increased the spore production of a *G. intraradices* strain by repeated harvesting and gel replacement of the distal compartment.

9.1 Cultivation parameters

a. Selection of cultivation media

The most important factor for successful AMF formation in root organ culture is the adjustment of appropriate culture medium for dual cultivation of both the components. Since the root, which needs rich nutrient medium for its growth and the AMF grows normally in relatively poor nutrient conditions, there must be such a balance in the nutrient medium that the growth of the host as well as fungal symbionts should progress during dual culture establishment. Two culture media are frequently used to culture AMF on ROC : the minimal medium (M-medium Bécard and Fortin,1988) and the modified Strullu Romand (MSR) medium (Strullu and Romand, 1986 modified by Declerck *et al.,* 1998). Both media result from the little modification of media usually used for *in-vitro* plant culture and are equally successful for a range of AMF. However Miller-Wideman and Watrud (1984) used 1/10 diluted MS medium for co-culturing *G. margarita* on tomato seedling explant. Although, they had success in establishing the symbiosis, they felt difficulty in root growth evaluation due to poor nutritional status of culture medium. It is more appropriate to use White's medium for root initiation, since this medium has been developed only for root organ culture. Moreover, presence of ammonium ions (in MS medium) was found to be detrimental to root growth because of steep fall in pH of the medium due to this additive. On the contrary, in White's medium, the additive nitrate counteracts the acidification in the culture medium following root growth that buffers the culture medium and retains pH at 6 for several months. Likewise, establishment of mycorrhizae greatly depends upon presence/absence and concentration of sodium sulphate, P, and sucrose in the culture medium. Mosse and Philips (1971) reported detrimental effect of sodium on AMF establishment a they observed internal development of *Endogone mosseae* in the roots of *Trifolium parviflorum* decrease when sodium was present in the medium.

Reducing the concentration of sucrose in the medium also benefits the mycorrhizal colony and arbuscules development, although the root diameter is reduced. The media used by Mosse (1988) although, rich in sucrose and P, provided nutritional conditions to stimulate "independent" growth of *G. intraradices* but simultaneously suppressed mycorrhizal infection of transformed carrot roots. In this direction the medium modified by Abdul-Khaliq and Bagyaraj (2000) with less in sucrose and P, with no or least ammonium salts, sometime also replacing ammonium salts with nitrate salts, was found suitable for hairy root culture preparation. It has also been convenient for dual culture of *G. intraradices* with transformed roots and/or normal roots of carrot (Chabot,1992) as well as for *G. margarita* with transformed roots of tomato.

b. Temperature and pH

Spore germination and the establishment of monoxenic cultures of AMF may be affected by several factors, such as the presence of root exudates or volatiles, flavonoids, substrate composition, humidity, light, CO_2, nutrient availability, presence of contaminants, temperature and pH (Maia *et al.*, 2010).

The pH of the substrate influences the germination of AMF, the development of hyphae, the percentage of root colonization, the formation of intraradical structures, and the density of spores (Panwar *et al.*, 2011). AMF species require different pH ranges for development, with the optimal conditions for germination and mycelial growth depending on the species, but the most favorable range for most species is between pH 6.0 and 7.0 (Postma *et al.*, 2007).

Temperature is another important factor that influences spore germination, establishment of root colonization (Wu and Zou, 2010) and spore density (Panwar *et al.*, 2011). The ideal temperature for germination of *A. laevis* is 20°C, while the optimum for the growth of hyphae ranged from 15 to 25°C (Tommerup, 1983). Temperatures between 18 and 25°C are ideal for germination of spores of *Glomus versiforme*, whereas temperatures above 35°C or below 15°C are detrimental (Siqueira *et al.*, 1985). Recently established monoxenic cultures of *Gigaspora decipiens* and *Glomus clarum* in transformed carrot hairy root. He also demonstrated that the sporulation was mainly influenced by temperature and pH of the medium. The highest spore production of *G. decipiens* occurred in a pH of 6.5, whereas for *G. clarum* at a pH of 4.0. (Costa *et al.* 2013).

c.Advantages, and disadvantages

The most obvious advantage shared by all *in-vitro* cultivation systems is the absence of undesirable microorganisms, which makes them more suitable for large-scale production of high-quality inoculum. While cross contaminations by other AMF are evidently excluded (if the starter inoculum is monospecific), the contamination by other microorganisms may occur either at the establishment of the cultivation process or at later stages of culture. Therefore, it may be useful to control the cultures visually, by standard plate-counting techniques and by molecular techniques. The cultures may be placed in a growth chamber requiring minimal space for incubation with no light required in the case of ROCs. The possibility to follow sporulation dynamics during cultivation also provides a means to control the level of spore production and to determine the optimal harvesting time. Factors that influence optimal production (*e.g.*, nutrient availability, presence of contaminants) can be more easily detected and controlled in (liquid) *in-vitro* cultures. Other advantages of the ROC systems are the low requirements in the follow-up of the cultures. Once successfully initiated, the cultures may be maintained for periods exceeding 6 to 12 months without intervention.

As a disadvantage, the diversity (in terms of genera) of AMF that have been grown *in-vitro* is lower than under pot cultivation systems. Another disadvantage of *in-vitro* production is the costs associated with the production systems, requiring skilled technicians and laboratory equipments such as sterile work flows, controlled incubators for ROC, and growth chambers for plant systems.

9.2 Innovations and Future Developments in AMF inoculum production

One innovative technique is the ready and easy to use inoculum in which fungal propagules are extracted from growing media, concentrated and mixed with carriers such as peat, sand, vermiculite or expanded clay. Products are available in powdered form containing a specified number of active fungal propagules per volume of inoculum. Liquid inoculum dedicated to horticultural use and isolated spores are also available. Aeroponic inoculum production at large scale has been investigated by Souza *et al.* (1996) but has not reached commercialization. Bioreactor assays with liquid AMF root organ culture propagation (Jolicoeur *et al.*, 1999) may eventually become suitable for commercialization for research needs. However, as the fungi are produced in association with *Agrobacterium* transformed roots, it is unlikely that its used can be allowed for field inoculation.

Knowing the performance variability between AMF strains, the improvement of commercial inoculum quality will almost certainly come from the selection of higher performance mycorrhizal fungal strains better adapted to the plant host or crop to be colonized and to specific environmental growing conditions (Adholeya, 2003).

9.3 Seed treatment and entrapment of *in-vitro* produced AMF propagules

The widely used method for obtaining AMF inoculum has been the use of root segments of plant hosting AMF and /or the infected soil containing AMF propagules (Gaur and Adholeya, 2000). However, these inocula are voluminous, heavy and often contaminated by other organisms (Vallino *et al.*, 2009). Successful entrapment of *in-situ* and *in-vitro* produced AMF propagules including spores, intraradical hyphae and vesicles have been achieved in seeds (Strullu and Plenchette, 1991; Declerck *et al.*, 1996), resulting in successful germination and colonization of AMF in the plantlets (Jaizme - Vega *et al.*, 2003; Plenchette and Strullu, 2003). Often, natural polysaccharide gels like alginates are used as inoculant carriers for AMF (Kapoor *et al.*, 2008). Entrapment of AMF directly on the seeds would save space and require much less quantity of inoculum as the antagonist is established at the infection sites of the pathogens (Declerck *et al.*, 1996; Plenchette and Strullu, 2003). Seed treatment is best accomplished using either powdered or liquid mycorrhizal inoculants applied so that the inoculum adheres directly to the surface of the seed and the most important factor for reintegrating mycorrhizae into the crop land environment is to place mycorrhizal propagules near seed or near the root systems of target plants. When one of these colonizing units touches or comes into very close proximity of living root tissue in this case the sprouted seed, they are activated by minute amounts of specialized root exudates and begin the mycorrhizal colonization process.

9.4 Constraints and Regulations in AMF inoculum production

The following constraints and regulations were notified by Feldmann and his colleagues at present so that it had to be taken into great care at the time of high quality AMF inoculum production and utilization in large scales in order to benefit both the producers and end users.

9.5 Cost of inoculum versus fertilizers

The obligate biotrophic nature of AMF, unlike other fungi implies the establishment of a plant propagation system, either under greenhouse conditions or *in-vitro* laboratory propagation. These techniques result in high inoculum production costs, which still remains a serious problem since they are not competitive with production costs of phosphorus fertilizer. Even if farmers understand the significance of sustainable agricultural systems, the reduction of phosphorus inputs by using AMF inocula alone cannot be justified except, perhaps, in the case of high value crops. This could be the case of organic crop farmers, who can sell their products at premium price.

9.6 Sanitary control

Another serious problem in commercializing inoculum comes from the need to control the biological composition of the product, especially from invading phytopathogenic microorganisms. At present, the inoculum produced using the pot culture variants, either in greenhouses, growth chambers, or fields, is never completely free from external microorganisms. This is a problem even though the producers attempt to control pathogens with various agrochemicals. Farmers are usually aware of the risk of pathogens, so they avoid using inoculum containing host root residues. In most commercial inoculum, colonized root segments are chopped into 0.08 to 0.12 inch long pieces so segments that remain are difficult to detect. When colonized roots are directly incorporated to carriers, their surface sterilization with a light solution of disinfecting product can be done without affecting the effectiveness of an inoculum. When roots and rhizosphere material are used for inoculum preparation, handling with clean apparatus is advised.

9.7 Efficiency of inoculum

In the field application of any microbial inoculum, it is essential to verify that the inoculated microorganisms possess the characteristics and the potential described by the inoculum manufacturers. With AMF inoculum, such evaluations can be done using several approaches such as morphological identification of AMF spores to confirm the fungus identity, and by estimating the mycorrhizal root colonization level of test plants (Dalpe, 2004). Tentative molecular techniques have been developed for the detection of AMF inoculum strains and discrimination from indigenous AMF strains naturally occurring in soils. These techniques are not yet totally reliable due to the large genetic heterogeneity in AMF and, as such, these techniques are not routinely used for the detection of AMF. Similar situations are observed with the discrimination among strains when using internal transcribed spacer (ITS) sequences of ribosomal DNA genes (rDNA). Reliable molecular techniques to trace the inoculated strains using rep-PCR and specific primers developments are under study. Such a technological breakthrough would greatly facilitate both fundamental and applied research on mycorrhizae as well as improve quality control of commercial inoculum.

9.8 Official registration for commercial products

The commercialization of mycorrhizal inoculum is subjected to regional or national registration at agriculture departments and usually falls under the country's Fertilizer Act. In Canada, mycorrhizal inoculum are considered to be supplements: products, "other than fertilizers, manufactured, sold or represented for use in the improvement of the physical condition of the soil or to aid plant growth or crop yields." In the USA, registration of an AMF inoculum may fall either in the fertilizer or the pesticide sectors, depending on the vocation of the proposed mycorrhizal product. Application for registration is required for such products and extensive information is attached to the registration request: (a) a list of ingredients and possible contaminants in the proposed inoculum; (b) the minimum concentration of each ingredient including the active mycorrhizal fungi and the purpose of each of them; (c) official material safety data sheets; (d) the product label, showing the name and address of producer, the number of viable fungal propagules or the symbiotic efficiency expressed as percentage of colonization expected by the inoculum, recommended plant host, soil conditions for effectiveness, recommended application rate, storage conditions, and expiration date; (e) manufacturing process and (f) the testing protocol. Such quality control is important to exclude poor quality microbial inocula from the market (Dalpe, 2004).

One inoculant form, which is produced in containers using light expanded clay aggregates (LECA) as substrate (Dehne and Backhaus, 19861, appears to be of interest. This porous material can be easily separated from the host roots and, contains infective mycelium and spores (Feldmann and Idczak, 1992). The aggregates can be surface-sterilized and applied to field grown crops in small quantities (SO-150 kg ha⁻¹) (Baltruschat, 1987).

10.0 Patents

Over 40 patents involving AMF have been deposited in the last two decades. Many of them concerned the beneficial properties of AMF. Some AMF inoculum production systems have been proposed, and several patented (Gianinazzi *et al.*, 1990b; Sieverding and Barea, 1991; Mulongoy *et al.*, 1992; Jarstfer and Sylvia, 1992; Sylvia and Jarstfer, 1994; Thompson, 1994; Lovato *et al.*, 1995). Other patents focused on inoculum preparation (Sylvia and Jarstfer, 1992), on formulation and applications (Cano and Bago, 2007; Fernandez *et al.*, 2006), and on cultivation methods either in substrate-free (Mosse and Thompson, 1981) or under *in-vitro* culture conditions (Declerck *et al.*, 2009, WO/2009/ 090220; Fortin *et al.*, 1996; Mugnier *et al.*, 1986; Wang, 2003).

The most recent patents on production methods involved *in-vitro* cultures. Two patents are based on ROC (Fortin *et al.*, 1996, US Pat. No. 5554530; Wang, 2003, US Pat. No. 6759232), while another patent is based on autotrophic *in-vitro* mycorrhizal plants (Declerck *et al.*, 2009, WO/2009/ 090220). Fortin *et al.*, (1996, US Pat. No. 5554530) cultivated the AMF in a compartmented bioreactor. Wang (2003, US Pat. No. 6759232) utilized a container in which the AMF was temporarily exposed to a nutrient solution. In the third patent on *in-vitro* cultivation, deposited recently by Declerck *et al.* (2009, WO/2009/ 090220), pre-mycorrhized *in-vitro* produced plants (Voets *et al.*, 2009) grew in a slightly inclined growth tube in which the nutrient solution was continuously flowing on the mycorrhized root system.

11.0 Conclusion and Future strategies

Production of AMF inoculum in large scale have been taken up by many companies world wide. For nursery application and green house plant production these inocula will be highly useful. For largescale field applications on-farm inoculum production is more suitable. There are chances of the inoculum getting contaminated in open field, but all precautions should be taken to prevent such contaminations by not allowing trespassing by workers and cordoning off the production area. Organic crop farmers who have started using AMF inoculum in larger-scale production are potentially a new clientele base. However, at this point, they are having difficulties with inoculum application. There is a need to educate farmers on on-farm inoculaproduction.

For large scale application of inoculum, future research focusing on achieving good contact between seed and inoculum is needed. Regardless of the method of inoculum application, new users should establish a portion of their crop without inoculum in order to assess the benefits obtained in the crop established with inoculum. The commercial inocula production needs to be intensified and made more authentic. A more in-depth evaluation of the available commercial

Table. 13.1: Existing production unit of AMF in India 2011-12.
Source: Regional Centre of Organic Farming, Nagpur

S. No.	Name of the units with complete Address , Tel no	AMF Production in metric tons/ L
A.	**Goa**	
1.	Cosme Biotech Pvt Ltd.Panaji, Goa. Ph. No 08323265255, 09372153684	592.14
B.	**Andhra Pradesh**	
1.	Sri Biotech Laboratory India Pvt Ltd Medak, Hyderabad. 040-23701153,0963347551	5.0
2.	RadderBiotech,Vijaywada. 0866-2412987	10.50
3.	Sri Arbindo Inst Of Rural Dev , KVK, Nalgonda . 09441703198	-
4.	Rover Biotech, Vijayawada M.No. 0866-2582359, 09440882288	50 L (liquid)
5.	IPM Biocontrol Labs (P) Ltd, Secunderabad	-
6.	K.N. Biosciences (India) Pvt Ltd 09866019248	38.45
7.	Pralhad Biofertilizer, Karimnagar. 08121707889	50 L
8.	KCP Sugar & Industries Corporation Vuyur, Krishna 08676-232001/02	17.0
9.	Pratishtha Industries ltd, Nalgonda. 040-27974989	10.0
10.	Lotus Biotech,West Godavari. 09949894621	1.0
11.	Pragathi Biofertilisers, Nellore	1.0
12.	Biotech Agriscience, Kukatpally, Hyderabad. 040-23357381	10.0
13.	Kisan Bioformulations, Vijayawada 0866-2543004, 09490333003	50.0
C.	**Gujarat**	
1.	Krishak Bharti Co. Ltd, Surat	12.05

Contd...

Table 13.1: contd...

S. No.	Name of the units with complete Address , Tel no	AMF Production in metric tons/ L
2.	Director of Research PG Studies, Navsari	409.00 L
3.	Pratishtha Industries ltd, Nalgonda. 040-27974989	10.0
4.	Lotus Biotech,West Godavari. 09949894621	1.0
D.	**Maharashtra**	
1.	Environmental Protection Res. Found., Sangli. 02332301857	40.00
2.	Choudhari Agrotech, Nagpur. 07123095145	100.00
3.	Microplex India , Wardha 07152253146	3.035
4.	Vasant Dada, Sugar Institute , Manjari, Pune. 020-26902100	33114 L
5.	Krishak Bharti Cooperative Ltd, Lanja, Rantagiri. 02612862766	1.2
6.	Trinity Agrotech & Phytochem. Pvt. Ltd,	1470 L
7.	Shakti Biotech, Aurangabad	1040 L
8.	Krishi Vigyan Kendra, Babhaleshwar, Ahmadnagar. 02422252414	5.41
9.	Manshya Enviro Biotech Pvt Ltd, Pune. 02026930094	3.60
10.	Shankar Maharshi Mohite Patil, Solapur, 9860775376	2.50 /18.9
11.	College of Agriculture, Pune	6.36
12.	Sahkar maharshi Bhausaheb Thorat Sah. Sakhar Karkhana, Ahmadnagar. 02422225370	10.21
13.	Biocare, Nagpur 0712-2224344	5.0
14.	KVK, Amaravati, 9922410177	50

Contd...

Table 13.1: contd...

S. No.	Name of the units with complete Address , Tel no	AMF Production in metric tons/ L
15.	Vidarbha Biotech lab Yawatmal. 07232242333	54.0
16.	Vasant Biotech, Pusad	20.0
17.	DEENI Chemicals Chandrapur. 07172287711	25.5
18.	INORA pune020-22951753	25.5
19.	Nikubioresearch, Pune	20.0
20.	Ramkrusha Bajaj College of Agri. Wardha	3.5
21.	Mahalaxmi Biotech, Jalgaon	5.0
22.	Anandniketan College Of Agri, Warora	5.5
23.	NCS Crop Sci Pvt, Nagpur 0712-2520204	5.0

Table 13.2 List of some Companies supplying AMF

S. No.	Company Name	AMF in metric tons/Litres	Cost/ Metric ton
1.	Indo Gulf company 51/57, Dontad Street, 1st Floor, Off No.11, Mumbai, Maharashtra, India	10 Kilograms (Min. Order) 1 Metric Ton per Week (Supply Ability)	Info Not available
2.	Ambika Biotech and Agro Services,7 Industrial Area, Jaggakheri,, Mandsaur, Madhya Pradesh, India.	10 Kilograms (Min. Order) 200 Metric Ton/Metric Tons per Year (Supply Ability)	Info Not available
3.	Xiamen Vastland Chemical Co., Ltd. Room 205, No. 999, 1001, Anling Road, Huli District, Xiamen, Fujian, China (Mainland)	1 Metric Ton (Min. Order) 1000 Metric Ton/Metric Tons per Month (Supply Ability)	US $200-400 / Metric Ton (FOB Price)
		1 Metric Ton (Min. Order) 2000 Metric Ton/Metric Tons per Month (Supply Ability)	US $50-250 / *Metric Ton (FOB Price)*
		500 Kilograms (Min. Order) 1000 Metric Ton/Metric Tons per Month (Supply Ability)	US $3000-7000 / *Metric Ton (FOB Price)*
4.	Groundwork BioAg, 4 Hashaked St., Raanana, Israel	900 Grams (Min. Order) 25 Metric Ton/Metric Tons per Year (Supply Ability)	US $49-149 / Pack (FOB Price)

Table contd...

* *Source : http://www.alibaba.com/showroom/mycorrhiza.html.*

Table 13.2: contd...

S. No.	Company Name	AMF in metric tons/Litres	Cost/ Metric ton
5.	Microblend Fertilizer, No 702, Lebuh Bandar, Bandar Putra Kulim, Kulim, Kedah, Malaysia	1 Twenty-Foot Container (Min. Order) 1 Twenty-Foot Container per Day (Supply Ability)	Info not available
6.	Neo Herbal Healthcare Pvt. Ltd. D6 Saisanskrit Apartments, Near Balol Nagar Circle, Ranip, Ahmedabad, Gujarat, India	1 Ton (Min. Order) 0.1 Metric Ton/Tons per Month (Supply Ability)	US $500-550 / *Kilogram (FOB Price)*
7.	Xinya Biotech Co.,Ltd, 3F.-1, No.562-1, Sec. 2, Wenxin Rd., Xitun Dist., Taichung City 40758, Taiwan (R.O.C.), Taiwan	50 Kilograms (*Min. Order*)	Info not available
8.	Dr. Rajan Laboratories,# 108 / 33, Valmiki Street, East Tambaram 600059,India Tamil Nadu,Chennai.	100 Kilograms (Min. Order) 10 Metric Ton/Metric Tons per Month (Supply Ability)	US $10 / *Kilogram (FOB Price)*
9.	2504 Tianjin Haohang International Trade Co., Ltd. , Sanlian Building, Shiyijing Road, Hedong District, Tianjin, China (Mainland)	1000 Liters (Min. Order) 300000000 Liter/Liters per Month (Supply Ability)	US $0.049-0.055 / *Liter (FOB Price)*
10.	Shenzhen JYO Technology Co., Ltd.No. 9, Lane 4, Tangkeng New Village, Shiyan Street, Baoan District, Shenzhen, Guangdong, China (Mainland)	1 Piece (Min. Order) 100000 Piece/Pieces per Month (Supply Ability)	US $1-2000 / *Piece (FOB Price)*
11.	Mikskaar AS, Katusepapi 411412 ,Estonia,Harjumaa,Tallinn.	Not available	Info not available
12.	Sundaram Overseas Operation, F-339, Raghuleela Mega Mall B/H Poisar Bus Depot, Boraspada Road, Kandivali [W], Mumbai - 400067 Maharastra, India Call: +91 22 426611, Mumbai, Maharashtra, India	500 Liters (*Min. Order*)	Info not available
13.	Zhangjiagang Kangyuan New Material Co., Ltd. Building E, Changxing Machinery, Economic Development Zone, Zhangjiagang, Suzhou, Jiangsu, China (Mainland)	1 Ton (*Min. Order*)	Info not available

Contd...

Table 13.2: contd...

S. No.	Company Name	AMF in metric tons/Litres	Cost/ Metric ton
14.	Tai'an Guangyuan International Trade Co., Ltd. Room 17098, Floor 17, No. 96, Great Wall Road, Taian, Shandong, China (Mainland	20 Metric Tons (Min. Order) 10000 Metric Ton/Metric Tons per Month (Supply Ability)	US $200-400 / *Metric Ton (FOB Price)*
15.	Ennov Infra Solutions Pvt. Ltd. 61-b, Udhyog Vihar, Greater Noida, Uttar Pradesh - 201306, India, Uttar Pradesh, India	12 Cartons (Min. Order) 34000 Carton/Cartons per Month (Supply Ability)	US $9-13 / *Carton (FOB Price)*
16.	Shenzhen Aosion Photoelectricity Co., Ltd. Room 2016, 2017 (Work Place), 20/F, Unit/Block A, Building 1, Charming Times Garden, Intersection of Xixiang Avenue and Xinhu Road, Xixiang Street, Baoan District, Shenzhen, Guangdong, China (Mainland	1000 Pieces (Min. Order) 10000 Piece/Pieces per Week (Supply Ability)	US $2-6 / *Piece (FOB Price)*

Plate. 13.1 : Production of substrate based AMF inoculum in cement tanks

Plate 13.2 AMF colonized maize seedlings.

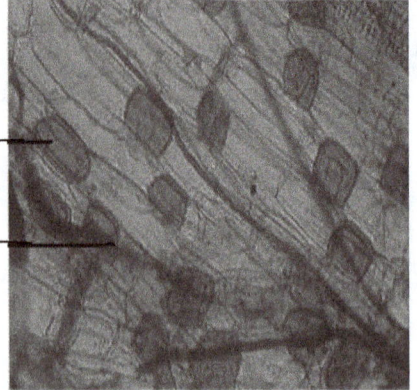

Plate13.2a Colonisation of maize roots by AMF . Fungal hypha and vesicle can be seen inside the root.

Plate 13.3 Chlamydospores of AMF in inoculum.

Plate 13.3a A single chlamydospore.

C

Plate.13.3. On-farm AMF inoculum production on finger millet *(Eleucine coracana* L.)*.*

A. Solarisation of the soil for three weeks in bright sunlight, sowing fingermillet seeds and raising closely planted crop.B. Harvesting the inoculum after 12 weeks. The shoots are cut at the soil surface and root is dug up to 9 cm below ground and root pieces along with the soil containing the propagules is the inoculum. (Sukhada,*et al.*,2004)

References

1. Abdul-Khaliq A. and Bagyaraj, D.J., 2000. Advances in Mass Production Technology of Arbuscular Mycorrhiza *Ind. J. Exp. Biol.* 38,1147-1151.

2. Adholeya, A., 2003. Commercial production of AMF through industrial mode and its large scale application. In: 5th International conference on mycorrhizal fungi. Montreal Quebec, Canada, 240.

3. Allen, M., 1991. The ecology of mycorrhizae. University Press, Cambridge, UK.180.

4. Baltnmchat, H., 1987. Field inoculation of maize with vesicular-arbuscular mycorrhizal fungi by using expanded clay as carrier material for mycorrhiza. J. *Plant Dis. Prot.*, 94: 419-430.

5. Becard, G.and Fortin, J.A., 1988. Early events of vesicular-arbuscular mycorrhizal formation on Ri-T-DNA transformed roots. *New Phytol.* 108, 211-218

6. Becard, G. and Y. Piche., 1992. Establishment of vesicular arbuscular mycorrhiza in root organ culture: review and proposed methodology. In : Techniques for the study of mycorrhiza.Edited by J. Norris, D. Read, and A. Varma. Academic Press,New York. 89-108.

7. Chilton, M. D., D. A. Tepfer, A. Petit, C. David, F. Casse-Delbart. and J .Tempe. 1982. *Agrobacterium rhizogenes* inserts T-DNA into the genomes of the host plant root cells. *Nature.* 295, 432-434.

8. Chabot, S., G. Becard and Y. Piche., 1992. Life cycle of *Glomus intraradices* in root organ culture. *Mycologia.* 84:315–321.

9. Chandran, P. R and V.P. Potty., 2011. Initiation of hairy roots from Canavalia sp. using *Agrobacterium rhizogenes* 15834 for the co-cultivation of Arbuscular mycorrhizal fungi, *Glomus microcarpum J. Agric. Technol.* 7(2), 235-245.

10. Costa, F. A., L. S. M. Haddad., M. C. M. Kasuya., W. C. Oton, M. D. Costa and A. C. Borges., 2013. *In-vitro* culture of *Gigaspora decipiens* and *Glomus clarum* in transformed roots of carrot: the influence of temperature and pH. *Acta Scientiarum.* 35, 315-323.

11. Dalpe, Y., 2004. Vesicular arbuscular mycorrhizae. Lewis Publications, CRC Press, 287-301.

12. Declerck, S., 1996. Entrapment of *in-vitro* produced spores of *Glomus versiforme* in alginate beads: *in-vitro* and *in- vivo* inoculum potentials. *J. Biotechnol.* 51-57.

13. Declerck, S., D.G. Strullu and C. Plenchette., 1996. *In-vitro* mass - production of the arbuscular mycorrhizal fungus, *Glomus versiforme*, associated with Ri T - DNA transformed carrot roots. *Mycological Research.* 100 , 1237 - 1242.

14. Declerck, S., D.G. Strullu and C. Plenchette., 1998. Monoxenic culture of the intraradical forms of *Glomus* sp. isolated from a tropicalecosystem: a proposed methodology for germplasm collection. *Mycologia,* 90, 579-585.

15. Diop, T.A., G. Becard and Y. Piché., 1992. Long-term *in-vitro* culture of an endomycorrhizal fungus, *Gigaspora margarita* on Ri T-DNA transformed roots of carrot. *Symbiosis,* 12, 249- 259.

16. Diop, T.A., 2003. *In-vitro* culture of Arbuscular mycorrhizal fungi: advances and future prospects. *Afr. J .Biotechnol.* 2, 692-697.

17. Douds, D.D. 1994. Relationship between hyphal and arbuscular colonization and sporulation in a mycorrhiza of *Paspalum notatum. New Phytol.* 126, 233–237.

18. Douds, D.D., 2002. Increased spore production by *Glomus intraradices* in the split-platemonoxenic culture system by repeated harvest, gel replacement and resupply of glucoseto the mycorrhiza. *Mycorrhiza.* 12, 163-167.

19. Douds, D.D. Jr., 2002. Increased spore production by *Glomus intraradices* in the split-plate monoxenic culture system by repeated harvest, gel replacement, and resupply of glucose to the mycorrhiza. *Mycorrhiza,* 12, 163-167.

20. Douds, D.D., and N.C. Johnson., 2003. On-farm production of mycorrhizal fungi. In: Contributions of arbuscular mycorrhizas to soil biological fertility. (L.K. Abbott and D. Murphy, Eds.). Kluwer Academic Press, Dordrecht, Netherlands. 23-45.

21. Douds, D.D., G. Nagahashi., P.E. Pfeffer., C. Reider and W.M. Kayser., 2006. On-farm production of AM fungus inoculum in mixtures of compost and vermiculite. *Biores. Tech.* 97, 809–818.

22. Douds, D.D., G. Nagahashi., C. Reider and P. Hepperly., 2008a. Choosing a mixture ratio for the on-farm production of AM fungus inoculum in mixtures of compost and vermiculite. *Compost Sci. Utiliz.* 16, 52–60.

23. Douds, D.D., G. Nagahashi., J.E. Shenk., and K. Demchack., 2008b. Inoculation of strawberries with AM fungi produced on-farm resulting increased yield. *Biol. Agric. Hortic.* 26, 209–219.

24. Feldmann, F. and Idczak, E., 1992. Inoculum production of vesicular-arbuscular mycorrhizal fungi for use in tropical nurseries. In: J.R. Norris, D.J. Read and A.K. Varma (Editors), Methods in Microbiology. Academic Press, London, pp. 339-357.

25. Feldmann, F and E. Idczak., 1994. Inoculum production of VA-mycorrhizal fungi. In: Techniques for mycorrhizal research. (J.R. Norris, D.J. Read and A.K. Varma, Eds.). Academic Press, San Diego, 799-817.

26. Feldmann, F and C. Schneider., 2009. Quality control of arbuscular mycorrhizal fungal inoculum. *Can. J. Bot.* 2, 1366-1380.

27. Fortin, J.A., G.Becard., S. Declerck., Y. Dalpe., M. St. Arnaud., A.P. Coughlan and Y. Piche., 2002. Arbuscular mycorrhiza on root-organ cultures. *Can. J. Bot.* 80, 1-20.

28. Furlan, V., 1980. Media for density gradient extraction of endomycorrhizal spores *Trans. Br. Mycol. Soc.* 75, 336-338.

29. Gaur, A and A. Adholeya., 2000. Response of three vegetable crops to VAM fungal inoculation in nutrient deficient soils amended with organic matter. *Symbiosis.* 29 , 19 - 31.

30. Gerdemann, J.W and T.H. Nicolson., 1963. Spores of mycorrhizal *Endogone* species extracted from soil by wet-sieving and decanting. *Trans. Br. Mycol. Soc.* 46, 235-244.

31. Gianinazzi, S., Trouvelot, A. and Gianinazzi-Pearson, V., 1990. Role and use of mycorrhizas in horticultural crop production. 27 August- 1 September, XXIII International Horticulture Congress, Florence, pp. 25-30.

32. Gianinazzi and Miroslav Vosatka., 2004. Inoculum of arbuscular mycorrhizal fungi for production systems: science meets business. *Can. J. Bot.* 2, 1264-1271.

33. Gryndler, M., H. Hrselova., I. Chvatalova and M. Vosátka., 1998. *In-vitro* proliferation of *Glomus fistulosum* intraradical hyphae from mycorrhizal root segments of maize. *Mycol. Res.* 102, 1067-1073.

34. Hart, M.M and R.J. Reader., 2002. Taxonomic basis for variation in the colonization strategy of arbuscular mycorrhizal fungi. *New Phytol.* 153, 335–344.

35. Herrera – Peraza, R., Cuenca, G and C. Walker., 2001. Scutellospora crenulata, a new species of Glomales from La Gran Sabana, Venezuela. *Can. J. Bot.* 79 , 674 – 678.

36. Jarstfer, A.G. and Sylvia, D.M., 1992. Inoculum production and inoculation strategies for vesicular-arbuscular mycorrhizal fungi. In: B. Meting (Editor), Soil Microbial Ecology; Applications in Agriculture and Environmental Management. Marcel Dekker, New York, pp. 349-377.

37. Jarstfer, A.G and D.M. Sylvia., 1994. Aeroponic culture of VAM fungi. In: Mycorrhiza: Structure, function, molecular biology and biotechnology. (A.K. Varma and B. Hock, Eds.). Springer, Berlin, 427-441.

38. Jaizme - Vega, M. C., A. S. Rodriguez – Romero., C.M. Hermoso and S. Declerck., 2003. Growth of micropropagated bananas colonised by root - organ culture produced arbuscular mycorrhizal fungi entrapped in Ca - alginate beads. *Plant and Soil.* 254 , 329 – 335.

39. Jolicoeur, M., R.D. Williams., C. Chavarie., J.A. Fortin and J. Archambault., 1999. Production of *Glomus intraradices* propagules, an arbuscular mycorrhizal fungus, in an airlift bioreactor. *Biotechnol. Bioeng.* 63, 224-232.

40. Joner, E.J., R. Briones and C. Leyval., 2000. Metal binding capacity of arbuscular mycorrhizal mycelium. *Plant Soil.* 226, 227 – 234.

41. Kapoor, R., D. Sharma and A.K. Bhatnagar., 2008. Arbuscular mycorrhizae in micropropagation systems and their potential applications. *Scientia Horticulturae*, 116 ,227 – 239.

42. Kramadibrata, K., C. Walker., D. Schwartzott and A. Schubler., 2000. A new species of *Scutellospora* with a coiled germination shield. *Ann. Bot.* 86, 21 – 27.

43. Lovato, P.E., Schilepp, H., Trouvelot, A. and Gianinazzi, S., 1995. Application of arbuscular mycorrhizal fungi (AMF) in orchard and ornamental plants. In: A. Varma and B. Hock (Editors), Mycorrhiza Structure, Function, Molecular Biology and Biotechnology. Springer, Heidelberg, pp. 521-559.

44. Maia, L. C., B.S. Silva., B.T. Goto., O.J. Siqueira., F.A. Souza., E. J. B. N. Cardoso and S. M. Tsai., 2010. Estrutura, ultraestrutura egerminação de glomerosporos. In: (Ed.) Micorrizas 30 anos de pesquisas no Brasil. Lavras: Editora UFLA. 75-116.

45. Mertz, S. M., J. J. Heithaus and R. L. Bush., 1979. Mass production of axenic spores of the endomycorrhizal fungus *Gigaspora margarita*. *Trans. Br. Mycol. Soc.* 72,167-169.

46. Miller-Wiideman, M.A. and Watrud, L.S. *Can. J. Bot.* 1984, 30:642-646.

47. Mosse, B., 1962. The establishment of vesicular-arbuscular mycorrhiza under aseptic conditions. *J. Gen. Microbiol.* 27, 509-520.

48. Mosse, B and Phillips, M., the influence of phosphate and other nutrients on the development of vesicular-arbuscular mycorrhizal in culture. *J. Gen. Microbiol.* 1971, 69,157-166.

49. Mosse, B and C.M. Hepper., 1975. Vesicular-arbuscular infections in root organ cultures. *Physiol. Plant Pathol.* 15, 215-233.

50. Mulongoy, K., Gianinazzi, S., Roger, P.A. and Dommergues, Y., 1992. Biofertilisers: agronomic and environmental impacts and economics. In: E. Da Silva et al. (Editors), Microbial Technology: Economic Social Aspects. Cambridge University Press, Cambridge, pp. 59-69.

51. Muthukumar, T and K. Udaiyan., 2002. Arbuscular Mycorrhizal Fungi composition in semi arid soils of western ghats, Southern India. *Curr. Sci.*82 , 624 – 628.

52. Panwar, V., M.K. Meghvansi and S. Siddiqui., 2011. Short-term temporal variation in sporulation dynamics of arbuscular mycorrhizal (AM) fungi and physico-chemical edaphic properties of wheat rhizosphere. *J. Biol. Sci.* 18(3), 247-254.

53. Plenchette, C and D.G. Strullu., 2003. Long - term viability and infectivity of intraradical forms of *Glomus intraradices* vesicles encapsulated in alginate beads. *Mycological Research.* 107 , 614 - 616.

54. Postma, J., P.A. Olsson and U. Falkengrengrerup., 2007. Colonisation by arbuscular mycorrhizal and fine endophytic fungi in four woodland grasses - variation in relation to pH. *Soil Biol. Biochem.* 40(9) , 2260-2265.

55. Puri, A and A. Adholeya., 2013. A new system using *Solanum tuberosum* for the co-cultivation of *Glomus intraradices* and its potential for mass producing spores of arbuscular mycorrhizal fungi. *Symbiosis.* 59, 87-97.

56. Rice, R.W., L.E. Datnoff., R.N. Raid and C.A. Sanchez., 2002. Influence of vesicular arbuscular mycorrhizae on celery transplant growth and phosphorus use efficiency. *J. Plant Nutr.* 25, 1839–1853.

57. Ryder, M.H., M.E. Tate and A Kerr.,1985. Virulence properties of strains of *Agrobacterium* on the apical and basal surface of carrot root discs. *Pl. Physiol.* 77,215-221.

58. Schreiner, R.P and R.T. Koide., 1993. Stimulation of vesicular arbuscular 75. mycorrhizal fungi by mycotrophic and non mycotrophic plant root systems. *Appl. Environ. Microbiol.* 59, 2750–2752.

59. Sieverding, E., 1991. Vesicular-arbuscular mycorrhiza Management: Technical Cooperation, Federal Republic of Germany.

60. Sieverding, E. and Barea, J.M., 1991. Perspectivas de la inocuIacion de sistemas de produccion vegetal conhongos formadores de micorrizas VA. In: J. Olivares and J.M. Barea (Editors), Fijaci6n y Movilizaci6n Biol6gica de Nutrientes. Coleccion Nuevas Tendencias. Vol. II, C.S.I.C., Madrid, pp. 221-245.

61. Sylvia, D.M. and Jarstfer, G.J.,1994. Production of inoculum and inoculation with arbuscular mycorrhizal fungi. In: A.D. Robson, L.K. Abbott and N. Malajczuk (Editors), Management of Mycorrhizas in Agriculture, Horticulture and Forestry. Kluwer, Dordrecht, pp. 23 1-238.

62. Siqueira, J.O., D.M. Sylvia., J. Gibson and D.H. Hubbell., 1985. Spores, germination and germ tubes of vesicular-arbuscular mycorrhizal fungi. *Can. J. Microbiol.* 31, 965-972.

63. Smith, S.E and D.J. Read., 1997. In : Mycorrhizal symbiosis. Academic Press, San Diego, California, 78-86.

64. Souza, E.S., H.A. Burity., A.C. Espirito Santo and M.L.R. Silva., 1996. Alternative for arbuscular mycorrhizal fungi inoculum. *Plant Soil.*22, 231-251.

65. Sreenivassa, M.N., 1992. Selection of an efficient vesicular arbuscular mycorrhizal fungus for Chilli (*Capsicum annuum*). *Sci. Hortic.*50, 53–58.

66. Srinivasan, M., Kumar, K., Kumutha, K and Marimuthu, P., 2014. Influence of acetosyringone concentration on induction of carrot hairy root by *Agrobacterium rhizogenes. Afr. J. Microbiol.* Res. 8 (26), 2486-2491.

67. Strullu, D.G and C. Romand.,1986. Méthodes d'obtention d'endomycorhizes à vésicules et arbuscules en conditions axéniques. *C R Acad. Sci. Paris.* 303 , 245-250.

68. Strullu, D.G and C.Romand., 1987. Culture axénique de vésicules isolées à partir d'endomycorhizeset réassociation *in-vitro* à des racines de tomate. *C. R. Acad Sci Paris*. 305, 15-18.

69. Strullu, D.G and C. Plenchette., 1991. The entrapment of *Glomus* sp. in alginate beads and their use as root inoculum. *Mycological Research*. 95 ,1194 - 1196.

70. St-Arnaud, M., C. Hamel., B. Vimard., M. Caron and J.A.Fortin., 1996. Enhanced hyphal growth anspore production of the arbuscular mycorrhizal fungus *Glomus intraradices* in an in vitro system in the absence of host roots. *Mycol. Res*. 100 , 328 – 332.

71. Sukhada Mohandas, Chandre Gowda, M.J and Manamohan2004. Popularization of Arbuscular Mycorrhizal (AM) Inoculum Production and Application On farm, *Acta Hort*. 638, 279-283.

72. Sylvia, D.M., L.C. Hammond., J.M. Bennett., J.H. Hass and S.B.Linda., 1993. Field response of maize to VAM fungus of varied spore density and water management. *Agron. J*. 85, 193-198.

73. Tepfer, D. A and J. Tempe., 1989. Production d'agropine par des racines formées sous l'action d'*Agrobacterium rhizogenes*, souche A4. *C. R. Acad. Sci*. 292, 153-156.

74. Tommerup 1883 Temperature relations of spore germination and hyphal growth of vesicular mycorrhizal fungi in soil. *Transactions of the British Mycological Society*. 81(2), 381-387.

75. Thompson, J.P., 1994. What is the potential for management of mycorrhizas in agriculture?. In: A.D. Robson, L.K. Abbott and N. Malajczuk (Editors), Management of Mycorrhizas in Agriculture, Horticulture and Forestry. Kluwer, *Dordrecht, pp*. 191-200.

76. Vallino, M., D. Greppi., M. Novero., P. Bonfante and E. Lupotto., 2009. Rice root colonisation by mycorrhizal and endophytic fungi in aerobic soil. *Annals of Applied Biology* .154 ,195 -204.

77. Vosatka, M and J.C. Dodd., 2002. Ecological considerations for successful application of arbuscular mycorrhizal fungi inoculum. In : Mycorrhizal technology in agriculture. (S. Gianinazzi, H. Schuepp, J.M. Barea and K. Haselwandter, Eds.). Birkhauser, Basel, 235–247.

78. Watrud, L.S., J.J. Heithaus and E. Jaworski., 1978. Geotropism in the endomycorrhizal fungus *Gigaspora margarita*. *Mycologia*. 70, 449-452.

79. Wu, Q.S and N.Z. Zou., 2010. Beneficial roles of arbuscular mycorrhizas in citrus seedlings at temperature stress. *Scientia Horticulturae*. 125 (3), 289-293.

80. Zhipeng, Z. and Shiuchien, K. *Acta Microbiologica* 1991, 31:32-35.

Chapter 14

Arbuscular Mycorrhizal Fungi in Micropropagation

Vijayalakshmi and Sukhada Mohandas*

*Division of Biotechnology, ICAR–Indian Institute of Horticultural Research,
Hessaraghatta, Bengaluru 560089,India*

**Address for the correspondence: sukhada.mohandas@gmail.com*

ABSTRACT

The inoculation of arbuscular mycorrhizal fungi (AMF) to the roots of
micropropagated plantlets called mycorrhization, plays a beneficial
role in plant establishment by improving pre and post-transplanting
performance.There are two types of mycorrhization techniques, *in-vitro*
and *ex-vitro*. *In-vitro* mycorrhization the inoculation of AMF to the roots
of micropropagated plantlets is done on agar medium. Contamination
of the inoculum, behavior of the host *in-vitro,* and obligate nature of the
endophyte make it difficult for the establishment of mycorrhizal host
symbiosis *in-vitro.* But *Ex-vitro* mycorrhization or *in-vivo* inoculation is
relatively easier, feasible and easily implementable compared to *in-vitro*
mycorrhization. In *ex-vitro* mycorrhization, the AMF spores are inoculated
to the micropropagated plantlets after transplanting into pots or nursery.
The utilization of AMF in *in-vitro* and *ex-vitro* plays important role in
the development of a superior root system, increased photosynthetic
efficiency, and increased water conducting capacity, enhanced nutrient
uptake, averting attack by harmful soil borne pathogens and in alleviating
environmental stresses and acclimatization in horticultural crops. Different
techniques such as, soil based, soil less inoculum and surface sterilized
AMF propagules (entrapment in polymer gel, alginate hydrogel, aeroponic,

hydroponic, root organ culture) are used for mycorrhization of seedlings. The usage of this technique successfully in a number of horticultural crops during pre and post acclimatization of tissue cultured plants is discussed.

Keywords: *AMF, Micropropagation, root organ culture, Mycorrhization, In-vitro, Ex-vitro.*

1.0 Introduction

Fruit crops propagated *in-vitro* do not have microbes associated with them which are beneficial or harmful. The micropropagated plants face transplantation shock as they are not equipped enough to face the challenges of pathogens in the soil. Therefore, micropropagated plants are inoculated with beneficial microorganisms which play a major role in their establishment. These beneficial microorganisms are biocontrol agents or biofertilizers. Inoculation of AMF to the roots of micropropagated plantlets plays a beneficial role on their pre and post-transplanting performance such as development of a superior root system, increased photosynthetic efficiency, increased water conducting capacity, enhanced nutrient uptake, averting attack by harmful soil borne pathogens and in alleviating environmental stresses. This present chapter discusses the multiple roles played by AMF in the establishment of micropropagated plantlets of different fruit crops with different mycorrhization techniques and their limitations.

Rai (2000) and Kapoor *et al.* (2008) reviewed the advances in mycorrhization in micropropagation and their potential application. In their natural environments some microorganisms, particularly beneficial bacteria and fungi, could improve plant performance under stress environments, and consequently enhance yield (Brown, 1974; Lazarovits and Nowak, 1997; Creus *et al.*, 1998). Successful AMF inoculation at the beginning of acclimatization period (Branzanti *et al.*, 1992; Estrada-Luna and Davies, 2003) or even during *in-vitro* conditions (Mathur and Vyas, 1995) has been demonstrated. The benefits associated with AMF inoculation of micropropagated plantlets are plenty. They help in getting increased volume of roots. Biopriming of micropropagated plantlets with AMF helps in better development of root system and therefore better establishment of plants. Extended mycelia of AMF increase the absorption surface, photosynthetic efficiency and storage of photosynthates by improving P nutrition in plants (Marschner, 1995). Increased root hydraulic activity, osmotic adjustment, cell wall elasticity and stomatal conductance, plays an important role in the water economy of the plantlets. The mycorrhizal symbiosis improves the hydraulic conductivity of the root at low soil water potential which ultimately influences the water potential, transpiration rate and leaf resistance. Enhanced nutrient uptake such as P, Ca, Cu, Mn and Zn, AMF develop intensively inside roots and within the soil by forming an extensive extraradical mycelium which helps the plant in exploiting mineral nutrients and water from the soil. In plants, particularly those with restricted/ weak root system, hyphal connections act as a bridge between roots and nutrient sites in soil and facilitate efficient uptake of immobile nutrients by host plants (Azcon-Aguilar and Barea, 1996). AMF has increased access to insoluble forms of

P through the release of organic acids from fungi such as oxalic, formic, lactic, citric and malic acid (Jones, 1998). AMF can increase the capacity of the soil to supply limiting nutrients. Organic forms of P may be accessible to AMF through the release of extracellular acid phosphatases. AMF provides protection against harmful soil borne pathogens such as *Fusarium, Phytophtora, Aphanomyces*, Verticillium and nematodes by bringing about morphological and physiological changes in the host roots. Several genes and corresponding protein products involved in plant defense responses have been extensively studied in AMF symbiosis and have been shown to be spatially and temporally expressed (Harrier and Watson, 2004). These include callose deposition, phytoalexins, b-1–3 glucanases, chitinases and PR-pathogenesis related proteins (Pozo *et al.*, 1999; Ruiz-Lozano *et al.*, 1999; Slezack *et al.*, 1999; Salzer *et al.*, 2000; Ruiz-Lozano *et al.*, 2001; Guillon *et al.*, 2002; Cordier *et al.*, 1996). Protection against environmental stresses such as drought, toxic metals, saline soil, root pathogens, high soil temperature and adverse pH (Barea *et al.*, 1993).

2.0 Mycorrhization and its methods

In-vitro mycorrhization (Pons *et al.*, 1983) is the inoculation of AMF to the roots of micropropagated plantlets growing on agar medium. *In-vitro* mycorrhization is a tough process. Contamination of the inoculum, behavior of the host *in-vitro*, and obligate nature of the endophyte make it difficult for the establishment of mycorrhizal host symbiosis *in-vitro* (Schubert *et al.*, 1990; Rai, 2001). *Ex-vitro* inoculation or *in-vivo* mycorrhization is relatively easier, feasible and easily implementable method compared with *in-vitro* mycorrhization. In *ex-vitro* inoculation, the AMF spores are inoculated to the micropropagated plantlets after transplanting into pots (Puthur *et al.*, 1998; Subhan *et al.*, 1998; Rai, 2001: Muthur and Vysa.,2007) **Fig. 14.1.**

AMF symbioses have been established successfully on agar medium using seedlings (Hayman, 1983) and root organ culture in vegetable and few fruit crops like banana (Jaizme *et al.*,2003) and Strawberry (Elmeskaoui *et al.*, 1995). Vestberg and Estaun (1994) suggested that an inoculation protocol should be designed for each plant species, taking into account root growth and development rate, and number of transplants after the *in-vitro* stage. The inoculation technique differs depending on the substratum or the nature of the inoculum used (Trouvelot *et al.*, 1986). Selection of quantity (Morandi *et al.*, 1979; Daniels *et al.*, 1981; Ravolanirina *et al.*, 1989a; Guillemin *et al.*, 1992; Morte *et al.*, 1996) and quality of inoculum is an important point both for *in-vitro* and *in-vivo* mycorrhization (Vestberg and Uosukainen, 1996). The inoculum should be pure and be able to exhibit the desired biological effect.

Apart from specificity, the length of mycorrhization also depends on the type of AMF inoculum used, mycelia, or spores. It has been shown that the spore dormancy of AMF differs among species and genera (Juge *et al.*, 2002). So, it is advisable to use germinated AMF spores for effective mycorrhization.

The main problem to producing substantial qualities of AMF inocula is their obligate nature; this continues to be a major limitation. There are soil-free inocula

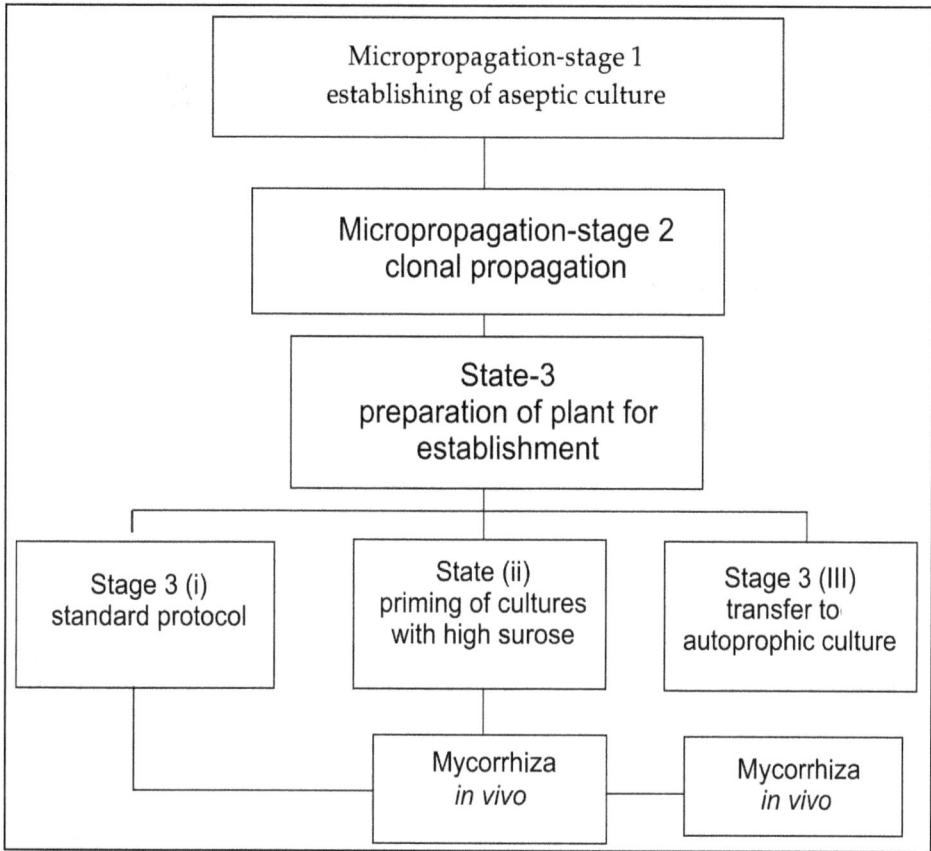

Fig. 14.1. Flow diagram showing alternative strategies for mycorrhization of micro-plants *in-vitro* and *in-vivo* (Nishi Mathur and Anil Vysa,2007).

like produced via aeroponic cultures, root organ culture, nutrient film technique (NFT) and polymer-based technique and soil based inocula multiplied on living host plant in pot culture and on-farm which is used in *ex-vitro* micropropagation.

3.0 *Ex-vitro* micropropagation techniques and AMF

In-vitro propagation techniques are increasingly applied for the growth and multiplication of cells, tissues and organs of plants on defined solid or liquid media under aseptic and controlled environment. The commercial technology is primarily based on micropropagation, in which rapid proliferation is achieved from tiny stem cuttings, auxiliary buds, and to a limited extent from somatic embryos, cell clumps in suspension cultures and bioreactors. The cultured cells and tissue can take several pathways. The pathways that lead to the production of true-to-type plants in large numbers are the preferred ones for commercial multiplication. Mass propagation of economically important tree species has been of a great interest (Lewu *et al.*, 2006 and Ozyigit *et al.*, 2007). Micropropagation of fruit crops is challenging as the tissues remain recalcitrant for months. Intensive research involving optimizations of growth factors have been carried out. Several authors have attempted to optimize the various steps involved in micropropagation of fruit crops including initiation and establishment of the culture *in-vitro*, use of supplements and growth regulators to promote development, and field transplantation. Desiccation and wilting are the main causes for the low survival during the transition from culture vessel to greenhouse or field conditions. Therefore acclimatization or hardening in the nursery is very important.

Although the most common means of producing AMF inocula employ matrixes like sand, soil, or a mixture of the two, inoculum can be produced in non-solid matrixes. Several techniques are used for obtaining AMF inocula under soil less cultures like the flowing solution culture technique, the flowing nutrient film technique, the stationary solution technique, and the aeroponic technique.

The most widely used method for AMF inoculum production is by using aeroponic cultures (Mohammad *et al.*, 2000). In this method, pure and viable spores of selected AMF are used to inoculate the cultured plants, which are later transferred into a controlled aeroponic chamber (Singh and Tilak, 2001) where humidity inside the chamber helps in rapid absorption of nutrients by the roots and improves aeration of both the plant and the fungus (Carruthers, 1992). The nutrient solution is provided in the form of a mist. Lack of physical substrate leads to extensive root growth, colonization and sporulation of the fungus and makes it an ideal system for obtaining sufficient amounts of clean AMF propagules (Abdul *et al.*, 2001). Jarstfer and Sylvia (1995) tested three types of aeroponic systems and chambers, they are an atomizing disc, pressurized spray through micro-irrigation nozzle, or ultrasonically generated fog of nutrient solution with a droplet size 3 to 10 μm in diameter, and from these they concluded that the pump and spray nozzle systems were the most versatile and reliable for aeroponic production of AMF.

Root organ culture was developed by White (1934) by obtaining continuous culture of non-transformed tomato (*Solanum lycopersicum* Mill.) roots in a liquid medium. After root induction, micropropagated plantlets were grown on cellulose plugs, in contact with the mycorrhizal root organ culture. They were then placed in growth chambers under an atmosphere enriched with 5000 ppm CO_2, and fed with a minimal medium. After 20 days of tripartite culture, all the plantlets were infected with mycorrhizal fungi.

Hairy roots formed by transforming tomato roots with *Agrobacterium rhizogenes* a gram negative soil inhabiting bacteria are used to multiply AMF successfully *in-vitro* (Abdul *et al.*, 2001). This technique could be used to multiply and maintain pure culture of mycorrhiza but is time consuming and input intensive for large scale application micropropagated plants. Similarly, nutrient film technique (NFT) is useful for producing limited quantities of clean root inoculum, but its usefulness in spore production is equivocal. In this technique roots of plants are bathed with a thin film of flowing nutrient solution or static aerated nutrient medium. The inoculum produced by this method is ideal for the production of easily harvestable solid mats of roots with more concentrated and less bulky form of inoculum than that produced by plants grown in soil-based or other solid media (Abdul *et al.*, 2001; Chellappan *et al.*, 2001).

Polymer based inocula are produced by entrapping AMF spores, vesicles or mycorrhizal roots in materials such as natural semi-synthetic and synthetic polysaccharide gels including kappa-carrageenan, agar and alginates. Calcium alginate is most widely used as carrier choice for encapsulation of AMF. In some cases, spores of AMF are introduced directly in synthetic seeds, which germinate under suitable conditions and colonizing the plantlet.

There are mycorrhizal helper bacteria (MHB) like strains of *Bacillus* spp. and *Pseudomonas fluorescens* which promote formation of AMF symbiosis in various crop plants by improving the susceptibility of roots to AMF during *in-vitro* hardening. These symbiotic associations have been successful in improving the growth and survival rates of micropropagated plantlets (Davey *et al.*, 1993; Hendrickson *et al.*, 1993; Webster *et al.*, 1995).

4.0 *Ex-vitro* Micropropagation techniques and AMF

Acclimatization of *in-vitro* propagated plants to the outside environmental conditions is done to ensure increased survival rate of plants. AMF have been successfully used to improve acclimatization, survival and growth of many micropropagated plant species including fruits such as banana (Yano, *et al.*, 1999); guava (Estrada, *et al.*, 2000); mayapple (Moraes, *et al.*, 2004) and grapes (Krishna, *et al.*, 2005; Krishna, *et al.*, 2006a and b) and strawberry (Borkowska, 2002) .

The soil based system has been adapted to the reproduction of different AMF strains for increasing propagule numbers. Soil based cultures are produced either in pots containing sterilized soil or on farm. On-farm soil is sterilized by solarization for nearly four weeks in bright sunlight. For propagation of AMF using the soil based system, starting fungal inocula usually composed of spores and colonized root segments are incorporated into the soil and seeds of good host plant are sown (Sukhada *et al.*, 2004). The fungi become established and spread within the substrate and colonize the root seedlings. Both colonized substrate roots then serve as mycorrhizal inocula. Bagyaraj (1992) found that a mixture of perlite and soilrite mix (1:1 v/v) was the optimal substrate and *Chloris gayana* (Rhodes grass) the optimal host for mass propagation of mycorrhizal inocula. Sukhada *et al.* (2004) found finger millet (*Eleucine coracana* L.) as the suitable host for AMF inocula production. In addition, pesticide captan and furadan added to the pot cultures at half the recommended level checks other microbial contaminants with no effect on the mycorrhizal fungi. This technique is very useful for the production of "clean" mycorrhizal inoculum (without other microbial contaminants) with high potentiality in a short span of time.

The trap plants commonly used for pot culture or on-farm culture of AMF are, *Sorghum halepense, Paspalum notatum, Panicum maximux,Cenchrus cilliaris,Zea mays, Trifolium subterroneum,Allium cepa* and *Chloris gayan, Eleusine coracana* (Chellappan *et al* ., 2001; Bhagyaraj 1992; Sukhada ,2004). The inoculum consists of spores , hyphal segments and infected root pieces and generally takes 3-4 months to produce on host plants. The inocula produced in pot culture or on farm has certain drawbacks that like bulky nature of inocula, transport problems, risk of contamination, presence of impurities and lack of genetic stability of inocula

(Abdula *et al.*, 2001). Large scale production of AMF inoculum requires control and optimization of both host growth and fungal development. By far, large scale production of soil based AMF inoculum in monitored clean environment seems to be the most feasible method for making inocula available for crop production. (**Fig. 14.2**). Molecular identification techniques are now available to identify the spores in case of contamination.

5.0 Mycorrhization of *in-vitro* and *ex-vitro* plants

Numbers of fruit crops have been subjected to *in-vitro* and *ex-vitro* mycorrhization techniques before cultivation in field. As the *in-vitro* mycorrhization is cumbersome and needs sophisticated culture conditions, *ex-vitro* mycorrhization which is relatively easy to follow has been widely used in hardening micropropagated plants.

5.1. Apple (*Malus* **spp.**), Peach (*Prunus persica* L. Stokes) **and Plum** (*Prunus cerasifera* **Ehrh.)**

Ex-vitro Application

Branzanti *et al.* (1992) inoculated micropropagated plants of 2 apple rootstocks (M9, M26) and one cultivar (Golden) with AMF during a very early weaning stage of acclimatization and supplied with nutrient solution at different P concentrations. Phosphate fertilization containing high level of P (40 ppm) had no effect on the growth response of mycorrhizal apple plants. At lower levels of 8 and 4 ppm P mycorrhizal plants maintained the same growth rate as with 40 ppm P. Phosphate fertilization had no influence on endomycorrhizal infection. No difference was observed in the mineral contents of mycorrhizal and nonmycorrhizal plants, or between plants receiving different levels of P. At the lower fertilization rates of P, endomycorrhizal infection not only improved growth but also homogeneity of Golden and M26 plants. Growth and leaf mineral content of two apple clones were increased substantially in *in-vitro* micropropagated plants hardened with AMF (Sivaprasad and Sulochana,2004).

Sbrana *et al.* (1994) inoculated apple, peach and plum rootstocks with AMF *Glomus* sp. strain A6 after transplanting from *in-vitro* to *in-vivo* culture. Plants were inoculated when the root length in apple rootstock M25, was 0.1–1.5 cm. Plants showed maximal growth increase and survival. Mycorrhizal infection of the Mr S. 2/5 rootstock induced earlier growth renewal after transplanting than controls. Mycorrhizal inoculation during transplantation from *in-vitro* to *in-vivo* culture could enhance both the growth and the survival of plants by providing them with nutrients.

Fortuna *et al.* (1996) studied the effects of phosphate fertilization and inoculation with the AMF *G. mosseae* (Nicol. and Gerd.) (Gerdmann and Trappe), *G. intraradices* Schenck and Smith or *G. viscosum* Nicolson on plantlets micropropagated from MM 106 apple and Mr.S. 2/5 plum rootstocks. Unfertilized and non-mycorrhizal plantlets showed no apical growth during the post *in-vitro* acclimatization phase, whereas P fertilization induced early resumption of shoot

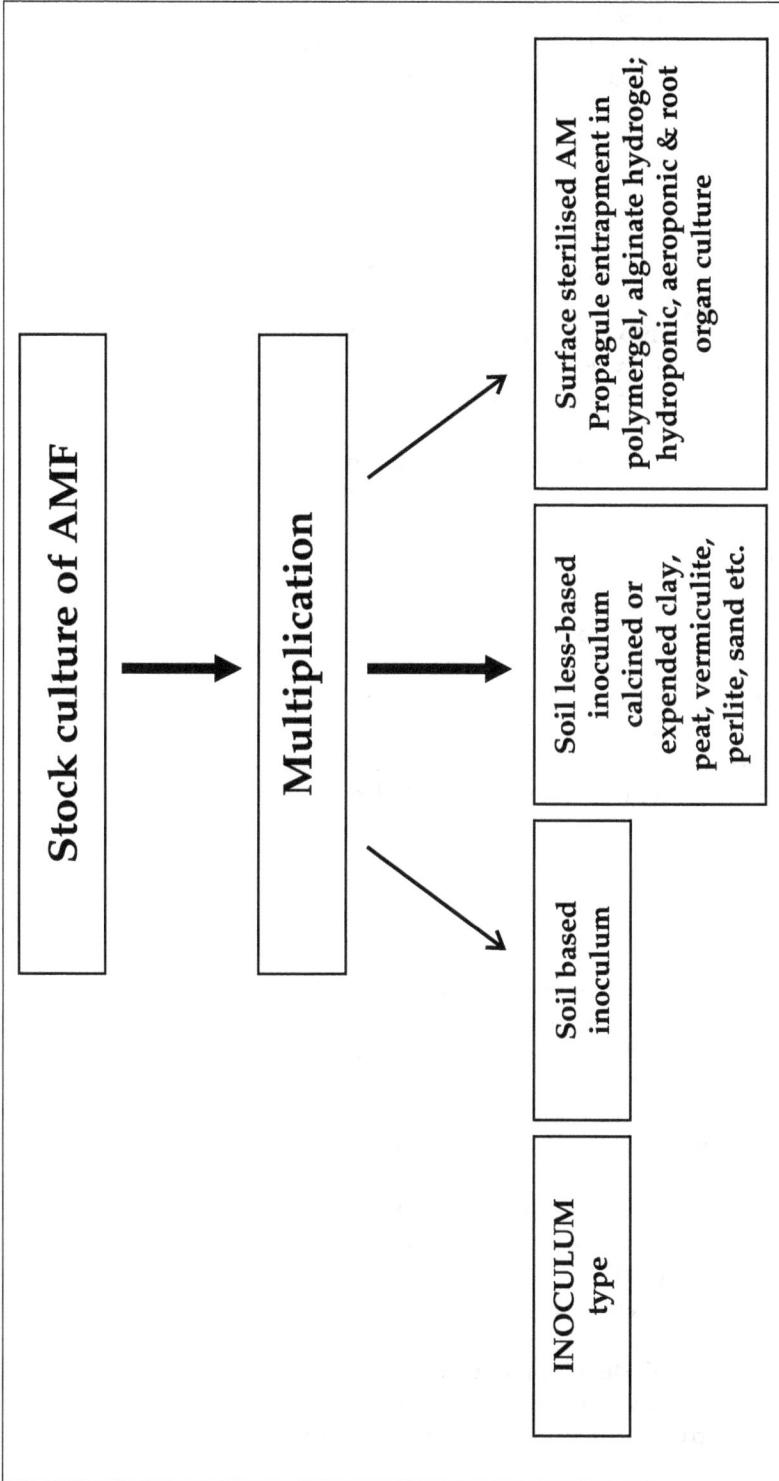

Fig. 14.2. Flow diagram shows AMF inoculum production techniques in *in-vitro* and *ex-vitro* mycorrhization (Azcon and Barea, 1997).

apical growth. Growth enhancement and percentage of actively growing apices of mycorrhizal-inoculated plantlets were comparable to those obtained in plantlets fertilized with P. Furthermore, tissue P concentrations of mycorrhizal plantlets were similar to those of plantlets fertilized with P.

5.2 Avocado (*Persea americana* Mill)

Ex-vitro Application

The acclimatization phase during micropropagation of avocado raises problems concerning the survival and development of plantlets *ex-vitro*. Comparison of potting mixes showed that survival of plantlets was highest in a soil - sand substratum and was increased by inoculation with *Glomus* sp in a peat - perlite mix. Mycorrhizal infection by *Glomus* sp during the acclimatization process also improved development of micropropagated avocado plants growing in these 2 substrata. Inoculation with other AMF showed that *G. deserticola*, and to a lesser extent *G. mosseae*, improved plant development in the soil - sand mix. Mycorrhiza formation, therefore, appears to play a key role in favouring *ex-vitro* development of micropropagated plants of avocado (Azcon -Aguilar *et al.*, 1992).

Micropropagated avocado plants generally exhibited low survival and very slow rate of growth during acclimatization. Inoculation of *G. fasciculatum* on micropropagated avocado plantlets improved the formation of well developed root system, shoot growth, shoot root ratio and NPK content in plant tissues which helped plants to tolerate environmental stress after transplanting. It showed that inoculation of AMF seems to be the key factor for subsequent growth and development of micropropagated plants of avocado (Sivaprasad and Sulochana ,2004).

5.3 Banana (*Musa* spp.)

In-vitro Applications

Jaizme *et al.* (2003) designed banana (cv. Grande Naine) plantlet production with AMF root organ culture using ca-alginate entrapped AMF strains *Glomus proliferum*, *Glomus versiforme* and *Glomus intraradices* strains. Ca-alginate entrapped ROC-produced AMF spores were able (1) to colonize the root system of a micropropagated banana cultivar under nursery conditions and form arbuscules and vesicles (2) to increase plant P nutrition and biomass, and (3) to proliferate in the commercial nursery substrate, therefore increasing the fungal inoculum biomass. The entrapment of ROC-propagated spores, adaptable to a wide range of Glomeromycetes, form alternative pathogen-free inoculum. Koffi *et al.* (2009) developed an *in-vitro* culture system associating autotrophic micropropagated banana plants with an AMF (*G. intraradices*). Intraradical root colonization, with production of arbuscules and vesicles, as well as extraradical development with production of new spores was observed. Extraradical hyphae extended from the mycorrhizal root system and developed in the growth medium, producing spores, fan-like and horsetail-like structures.

Ex-vitro Application

AMF has been reported to cause reduced transplant mortality (Biermann and Linderman 1983), reduced disease occurrence in host (Caron 1989), improved water relations (Gianinazzi *et al.*, 1990), increased drought tolerance (Davies *et al.*, 1992), increased nutrient uptake (Pearson and Jakobsen 1993), and overall growth (Wang *et al.*, 1993) of micropropagated banana plants. It has been reported that inoculation of AMF to the roots of micro propagated plantlets plays a beneficial role in their post-transplanting performance (Kapoor *et al.*, 2008).

The effect of AMF strains *Acaulospora scrobiculata, Glomus clarum* or *Glomus etunicatum* on micropropagated banana plantlets was evaluated during the acclimatization period (Yano, *et al.*, 1999). After cultivation in a greenhouse for 3 months, height, leaf area, fresh weight and dry matter of root and shoots, level of AMF colonization, nutrient level, photosynthesis and transpiration rate, water potential and stomatal conductance were measured. The number of AMF spores produced in each treatment was also determined. Plantlets inoculated with AMF had greater height, leaf area and fresh weight of shoots and roots, as well as higher rates of photosynthesis and transpiration than controls. Plants inoculated with *Glomus* were superior in most of the evaluated parameters.

Mathew *et al.* (2003) inoculated micropropagated plantlets of banana cultivars Dwarf Cavendish and Robusta during the secondary hardening stage with AMF. All the AMF increased the growth of both the cultivars, but the positive influence of the fungi was more evident for Dwarf Cavendish plantlets inoculated with *Glomus fasciculatum*. The mycorrhizal fungi heavily colonized the root system of both the cultivars. Both the cultivars inoculated with. *G. fasciculatum* exhibited a higher plantlet height (60.7% over control) and leaf area (2.2 times over control besides increasing the psuedostem girth (39.6%) over control and shoot biomass production. The symbiotic association also increased the shoot P concentration. Thus, mycorrhiza formation appears to be the key factor in improving the vigour and growth of micropropagated banana plantlets, which aids in the acclimatization process. Thaker and Jasrai (2002) regenerated Banana plantlets through shoot-tip culture were subjected to routine hardening procedure. Plantlets inoculated with AMF had greater height, leaf area and fresh weight of shoots and roots, as well as higher rates of photosynthesis and transpiration than controls. Plants inoculated with *Glomus* were superior in most of the evaluated parameters.

Utilization of AMF, *G. mosseae* and *G. fasiculatum* in the establishment of micropropagated banana (*Musa Paradisiacal* L.) cv Dwarf Cavendish in field was reported by Sowmya (2002) and Khade and Rodriguez.(2009) *G.mosseae* and *G.faciculactum* colonized well in the micropropagated banana plantlets. During the early stage of acclimatization mycorrhizal inoculation helped in producing the stronger root system before transfer to the field. It also helped in enhanced plant growth ,production of growth hormones, physiological status like increased level of chlorophyll pigments and photosynthesis rate, nutrients level Fe,Mn,Zn and in addition to P uptake, increasing the enzymatic activity like active phosphates . (Sowmya,2002) **Plate 14.1.**

5.4 Citrus (*Citrus* spp.)

Ex-vitro Application

Citrus is an important crop which is infested by many diseases such as yellow shoot, cancroids spot, and exocortis uncontrolled in certain regions (Wang, *et a.,l* 2005; Liu *et al.*, 2006). Micropropagation was imminent to get disease free citrus plantlets. But then micropropagated plants had difficulty in establishments. Therefore, Wu *et al.* (2011) tested if inoculations with two AMF were able to increase growth, photosynthesis, and nutrient uptake of micropropagated citrus plantlets during the acclimatization stage. The two mycorrhizal fungi successfully colonized the roots of citrus plantlets after an acclimatization period of 170 days, and the mycorrhizal developments were apt to low levels. Inoculation with G. *mosseae* improved acclimatized growth performance; increased photosynthetic rates, transpiration rates, and stomatal conductance; and stimulated the accumulations of nitrogen (N), phosphorus (P), calcium (Ca), copper (Cu), zinc (Zn), and manganese (Mn) in leaves and roots compared to the non-inoculated treatment. These observations suggested that only G.*mosseae* was the more efficient fungus, exhibited better adaptation to transplanted conditions, and therefore was appropriate to be introduced into the micropropagation protocol of citrus. AMF G. *mosseae* (BEG 116) and that of a Sicilian native mixture of *Glomus* species, have been successfully used to improve acclimatization, survival and growth of many micropropagated fruit citrus lemon (Quatrini *et-al.*, 2003).

5.5 Cherry

Ex-vitro Application

Lovato *et al.* (1994) inoculated micropropagated plants of common ash and wild cherry AMF during a 20-day weaning period, after which they were transferred to two different substrata supplemented with slow-release fertilizer. After a 13-week growth period, the stem height and diameter of the ash plants which had been inoculated with G.*intraradices* were three times greater than those of un-inoculated control plants. Increasing the peat content of the substratum improved growth of ash. Four weeks after being transferred to pots, shoots of wild cherry inoculated with G. *intraradices* or G. *deserticola* were taller and stems thicker than those of control plants, whereas those inoculated with *Gigaspora rosea* had shorter shoots and thinner stems than the controls. These beneficial effects of fungal inoculation on plant development disappeared after 13 weeks.

5.6 Custard apple (*Annona cherimola* L.)

Ex-vitro Application

The micropropagated custard apple is more dependent on mycorrhizal root colonization for optimum growth than plants derived from seeds. The effect of AMF on growth was observed when they were introduced after acclimatization period. The response of fruit crops to inoculation with G. *intraradices* showed improved survivability; quality plantlets and micropropagation also provided stress tolerance to plantlets. When they were transplanted to field, the mycorrhizal plants were consistently healthier than non-mycorrhizal plants (Sivaprasad and Sulochana,2004).

5.7 Guava *(Psidium guajava* L.)

Ex-vitro Application

The effect of AMF on growth, nutrient uptake and gas exchange of micropropagated guava plantlets was determined during acclimatization and plant establishment. Guava plantlets were asexually propagated through tissue culture and grown in a glasshouse for 18 weeks. Half of the plantlets were inoculated with a mixed endomycorrhiza isolate from Mexico, ZAC-19, containing *Glomus diaphanum, G. albidum* and *G. claroides.* All micropropagated guava plantlets survived transplant shock after 6 weeks; the mycorrhizal plantlets had greater shoot growth rates and leaf production than non-mycorrhizal plantlets with greater shoot length, leaf area, leaf stem, root dry mass, photosynthesic rate and stomatal conductance. Roots of inoculated guava planlets were heavily colonized with arbuscules, vesicles and endospores and the guava plantlets were highly mycotrophic with a mycorrhizal dependency index of 103% (Estrada *et al.,* 2000).

5.8 Grape *(Vitis vinifera* L.)

Ex-vitro application

Schubert *et al.*(1990) inoculated micropropagated grape *(Vitis bilandieeri* X *Vitis rupestris)* 1103 P with several AMF inocula with amended phosphorus. They observed increased plant growth in all unamended soil inoculated with AMF but positive effect of AMF inoculation decreased with increased P levels except two species *G. fasiculatum* and *G. monosporum* which showed higher growth response at increased levels (Lovato *et al.,*1992) inoculated growth medium containing micropropagated grapevine plantlets in a greenhouse with 2 commercial inoculants and observed that mycorrhizal grapevine plants showed a 3-fold increase in shoot growth as compared to control plants. All the root pieces were colonized up to 30%. Improved establishment of mycorrhizal tissue culture derived plantlets during acclimatization is commonly attributed to the enhanced vegetative growth due to better supply of nutrients like phosphorus and micronutrients, better cuticle development, high biomass accumulation and other physiological changes taking place at rapid rate.

Hare Krishna *et al.* (2005) studied the biochemical status of micropropagated grape plantlets in response to six single and a mixed strains of AMF during *ex-vitro* hardening under glass- house conditions. The histochemical studies revealed that the mycorrhizal inoculation resulted in accumulation of different biochemicals in the plant system such as chlorophyll, carotenoids, proline, phenol and enzymes like polyphenol oxidase (PPO) and nitrate reductase (NR). The mycorrhizal plantlets showed enhanced survival and improved tolerance against stresses experienced during weaning phase. The mycorrhizal plants also exhibited improved physiological and nutritional status and had higher relative water content (RWC) and photosynthetic rate. These plantlets also accumulated higher N, P, Mg and Fe concentrations, which may primarily be as a result of biochemical changes brought about by mycorrhizal association. Mycorrhizal plantlets also showed better hardening under glasshouse conditions.

Krishna *et al.* (2006) carried out a glasshouse study during 2002-2003 to evaluate the efficacy of different AMF strains for their utility as bio-hardening agent(s) for tissue culture derived grape plantlets. All the strains were found to significantly enhance the *ex-vitro* survival of micropropagated plantlets over control. Growth parameters like plant height, root length, shoot and root dry weights, leaf number, leaf area and physiological parameters such as relative water content and photosynthetic rate registered significantly higher values in inoculated plantlets than the control. Furthermore, the nutritional status of the mycorrhizal plantlets was also found to be improved. The different AMF species tried were *Acaulospora laevis, A. scrobiculata, Entrophospora colombiana, Gigaspora gigantea, Glomus manihotis, Scutellospora heterogama* and a mixed AMF inocula. *Acaulospora laevis* followed by mixed inocula were found to be majorly responsive as they affected most of the parameters and also resulted in maximum hardening success of tissue cultured plantlets.

5.9 Kiwi fruit (*Actinidia deliciosa* (A Chev) Liang et Ferguson)

Ex-vitro Application

Schubert *et al.* (1992) evaluated the effects of inoculation with a AMF (*Glomus* sp strain E3) and fertilization with a complex (NPK) fertilizer on growth of micropropagated kiwifruit. Plantlets were inoculated with AMF and fertilizer added when transplanting to a peat-based medium. Root fungal colonization was present in roots from all treatments 80 d after inoculation. Fertilization did not affect percent root colonization, which was increased by increasing the amount of inoculum. Plant growth was assessed in terms of leaf area and plant weight. Plant growth was low in the absence of fertilization. Increasing fertilization rates increased the growth of inoculated and non inoculated plants. AMF inoculation induced larger growth in plants receiving intermediate fertilizer rates, but reduced growth, in respect to non inoculated plants, at the highest fertilization rates.

5.10 Olive (*Olea europaea* L.).

Ex-vitro Application

Meddad Hamza (2010) tested two native Algerian AMF (*Glomus mosseae* and *Glomus intraradices*) for their effect on the growth of micropropagated olive tree. The effect of inoculation of plantlets with *G. mosseae* was also compared with chemical fertilization using osmocote. Highly significant increases in growth were evident for inoculated plants compared with uninoculated ones. *G. mosseae* doubled the root growth of the inoculated plantlets, compared to that of the fertilized plants. This change in the root: shoot ratio permitted greater utilization of soil resources and strengthened the plant's capacity to resist transplantation shock and water stress. Molecular techniques revealed the abundance of the two fungi in the roots of wild olives just as in the inoculated olives. Predominance of *G. intraradices* was observed amongst the natural ones.

5.11 Pomegranate (*Punica granatum* L.)

Ex-vitro Application

Singh *et al*. (2012) used four AMF strains namely, *G. mosseae, Acaulospora laevis, G. manihotis* and a mixed AMF strain as bio-hardening agents to improve survival and growth of *in-vitro* raised pomegranate plantlets. Plantlets inoculated with *G. mosseae* gave highest survival (90.40% and 88.00% at 60 and 90 DAI, respectively) and root colonization per cent (47.40 and 87.60 at 60 and 90 DAI, respectively). The predominant effect of *G. mosseae* was also evident on increased plant height (24.96 and 30.50 cm at 60 and 90 DAI, respectively) and root length (23.42 and 27.68 cm at 60 and 90 DAI, respectively) of the inoculated plantlets. *G. mosseae* and *G. manihotis* were found more effective in improving most of the growth, physiological and biochemical attributes of inoculated tissue culture raised plantlets. However, total phenol (24.94 and 28.62 _g/g at 60 and 90 DAI, respectively) and total chlorophyll (3.70 and 3.96 mg/g at 60 and 90 DAI, respectively) were found highest in mixed AMF inoculated plantlets.

5.12 Pineapple (*Ananas comosus* L. Merr.)

Ex-vitro Application

Guillemin *et al*. (1992) tested several AMF (*Glomus clarum* (LPA16), *Scutellospora pellucida* (LPA20), *Glomus* sp. (LPA21), *Glomus* sp (LPA22) and *Glomus* sp (LPA25)) for plant growth effects and infection development in Queen Tahiti, Smooth Cayenne (clone CY0) and Spanish varieties of micropropagated pineapple growing in an acid soil under growth chamber tropical conditions. Endomycorrhizal plants of all 3 pineapple varieties grew better than non-mycorrhizal plants. However, increase in plant growth was not related to infection development. Screening of different isolates of arbuscular endomycorrhizal fungi showed some specificity of fungi for promoting growth of the different pineapple varieties. Queen Tahiti and Smooth Cayenne pineapple plants associated with *Glomus* sp (LPA21) grew better than those infected with the other fungi, whilst best growth was obtained for the Spanish variety by inoculating plants with *Glomus* sp (LPA25). The root/shoot ratio was modified by endomycorrhizal inoculation, infected pineapple plants showing a greater increase in shoot production in comparison to root production. (Guillemin, *et al*., 1994).

Guillemin *et al*. (1994) inoculated micropropagated plants of two varieties, Queen Tahiti and Smooth Cayenne (clone CY0), during transplantation from axenic conditions with AMF to evaluate the importance of endomycorrhiza development for biological protection against *Phytophthora cinnamomi* which infects pineapple and causes severe crop loss. Growth and mineral nutrition of endomycorrhizal plants were not affected by different inoculum levels of *P. cinnamomi*, whilst they were reduced for non-mycorrhizal plants. Root/shoot ratio of endomycorrhizal plants was lower than that of non-mycorrhizal plants, and the pathogen did not modify this effect except at highest inoculum levels of

P.cinnamomi. Endomycorrhizal colonization was not altered by the pathogen; however symbiotic functioning was reduced by the highest concentration of inoculum of *P. cinnamomi.* The same inoculants were compared to efficient fungal isolates in micropropagated pineapple plants of 3 varieties grown in a controlled environment chamber with simulated tropical conditions. Plants grew better in acid than in alkaline soil, and *Glomus* sp. (isolate LPA21) was more efficient in acid soil than both commercial inoculants. An increased inoculant dose from 1% to 3% sometimes caused an increase in root infection with increases or decreases in plant growth depending on pineapple variety or type of inoculum used. One inoculant tended to improve growth in alkaline soil, while another was more efficient in acid soil.

5.13 Raspberry (*Rubus* spp.)

Ex-vitro Application

Taylor and Harrier (2000) investigated growth, development and nutrient status of micropropagated raspberry (Rubus idaeus cv. Glen Prosen) in response to inoculation with nine species of AMF from three different genera. The nine species of AMF included, *G.clarum, G. etunicatum, G. intraradices, G. rosea, Gi. gigantea, Gi. margarita, S. calospora, S. heterogama* and *S. persica.* Plant responses to AMF varied from growth enhancement to growth depression. Depressive growth effects were specific to *Gigaspora* species. Furthermore, particular species of AMF had unique effects on the mineral status of the raspberry plants. It is important to select isolates for inoculation of micropropagated raspberry plants by initial testing.

5.14 Strawberry (*Fragaria×ananassa* Duchesne)

Ex-vitro Application

Vestberg (1992) inoculated micropropagated plantlets of strawberry (*Fragaria* x *ananassa* Duch Senga Sengana) during the weaning stage with a *G. mosseae* strain from Rothamsted Experimental Station (UK) or with 8 Finnish *Glomus* isolates, including 1 *G. mosseae* isolate, 4 *G. intraradix* isolates and 3 isolates of an unidentified *Glomus* species. After 8 weeks (July 1990), inoculated plants were planted in the field: the most efficient fungi, i.e *G. mosseae* Rothamsted, *Glomus* sp. V3 and *Glomus* sp. V4,.increased shoot growth several-fold during the weaning stage. Growth responses persisted throughout the 1st year in the field, and partly throughout the 2nd year. Root colonization by different fungi was 0-60% during the weaning stage. In the 2nd year, after overwintering, root colonization of inoculated plants ranged from 8-35% as compared to 4% for the control plants. Preliminary studies were conducted on the effect of AMF against strawberry crown rot caused by *Phytophthora cactorum* (Leb. and Cohn) Schroet (Vestberg . 1994).

A micropropagated strawberry cultivar susceptible to the disease, 'Jonsok', was either inoculated with the Finnish AMF strains *G. mosseae* (Nicol. and Gerd.) Gerdemann and Trappe V57, Berch and Trappe V98 and *G. fistulosum* Skou and Jakobsen V128, or it was left un-inoculated. AMF inoculation at the beginning of the weaning stage, five weeks before the establishment of the pot or field experiment, did not decrease crown rot severity in either of the experiments. In the pot experiment, on the contrary, AMF lowered the plant health index when *P. cactorum* was added to the substrate in the form of infected plant residues. Results are discussed in relation to soil-borne vs. foliar disease, phosphorus concentration of the growth substrate and influence of weather conditions.

The plants produced by *in vitro* methods are free of any microflora contrary to natural systems where plants are colonized by symbiotic fungi. Borkowska (2000) evaluated the role of AMF in development of micropropagated strawberries and their photosynthetic activity (measured by chlorophyll fluorescence) under drought conditions. Mycorrhization strongly affected growth and tolerance to water deficiency of the plants cultivated in greenhouse.

6.0 Conclusion

Inoculation of AMF to the roots of micropropagated plantlets plays a beneficial role on their pre and post-transplanting performance. Mycorrhization of fruit crop seedlings is done in *in-vitro or ex-vitro*. As *in-vitro* mycorrhization, the inoculation of AMF to the roots of micropropagated plantlets growing on agar medium is a cumbersome and input intensive process, *ex-vitro* inoculation is most popular. Fruit crops micropropagated can be easily inoculated with AMF , hardened and planted in pots or in field for better establishment. Availability of pure culture for *in-vitro* inoculation and their regular maintenance on host plants is still a big challenge. However, AMF inocula are now available in a number of commercial outlets and there is lot of awareness about the utilization of AMF in hardening tissue cultured plantlets of fruit crops to harness their benefits.

7.0 Future strategies

Bioreactor production of AMF in dual culture should be concentrated upon. Large scale inoculum production and industrialization of technology is still a challenge.

The use of dual technology is promising as micropropagated plantlets are suitable platforms for understanding the mystery of host endophyte interaction, excessive production of secondary metabolites, heavy metal tolerance, bioprotection, bioremediation and growth-promoting activity.

Plate 14. 1. A) Response of micro-propagated tissue cultured banana to inoculation with AMF (C-Control; GF- Inoculated with *G. Fasiculatum*, GM- Inoculated with *G. Mosseae*). B) Micropropagated tissue cultured banana inoculated with *G. Mosseae*. Khade and Rodriguez .(2009).

References

1. Abdul-Khaliq,Gupta,M.L. and Alam, A., 2001. Biotechnological approaches for mass production of arbuscular mycorrhizal fungi: current scenario and future strategies. In: Mukerji, K.G., Manoharachary, C., Chamola, B.P. (Eds.), *Techniques in Mycorrhizal Studies*. Kluwer Academic Publishers, The Netherlands. 299–312.

2. Azcon-Aguilar, C., Barcelo, A., Vidal, M. T. and de la Vina, G.,1992. Further studies on the influence of mycorrhizae on growth and development of micropropagated avocado plants. *Agronomie*. 12,837-840.

3. Azcon-Aguillar, C. and Barea, J.A., 1996. Arbuscular mycorrhizas and biological control of soil-borne plant pathogens-an overview of the mechanisms involved. *Mycorrhiza*. 6,457–464.

4. Azcon.-Aguilar. C. and Barea.J.M.,1997. Departamento de Microbiologiu de1 Suelo y Sistemns Simbihicos, Estacio'n Experimental de1 Zaiilin, CSIC, ProjI Albaredu I, Grama, 18008, Spain; *Scientia Horticulturae* .68 , I-24.

5. Bagyaraj, D. J., Vesicular-arbuscular mycorrhiza: application in agriculture. In: Norris, J. R., Read, D. J., Varma, A. K., eds, 1992. *Methods in microbiology*, London: Academic Press, Harcourt Brace Jovanovich.24,450.

6. Barea, J.M., Azcon, R. and Azcon-Aguilar, C., 1993. Mycorrhiza and crops. In: Tommerup, I. (Ed.), Advances in Plant Pathology, Mycorrhiza: A Synthesis. Academic Press, London.167–189.

7. Biermann, B. and R.G. Linderman., 1983. Increased geranium growth using pretransplant inoculation with mycorrhizal fungus. *J. Am. Soc. Hort. Sci.* 108 , 972 - 976.

8. Borkowska, B., 2002. Growth and photosynthetic activity of micrropagated strawberry plants inoculated with endomycorrhizal fungi (AMF) and growing under drought stress. *Acta Physiol. Plant.* 24 (4), 365-370.

9. Brown, M.E., 1974. Seed and root bacterization. *Ann. Rev. Phypathol.*12, 181– 197.

10. Branzanti, B.,Gianinazzi-Pearson, V. and Gianinazzi, S., 1992. Influence of phosphate fertilization on the growth and nutrient status of micropropagated apple infected with endomycorrhizal fungi during the weaning stage. *Agronomie* . 12,841–5.

11. Caron. M., 1989. Potential use of mycorrhizae in control of soil borne diseases. *Can. J. Plant Pathol.* 11 , 177 - 179.

12. Carruthers, S., 1992. Aeroponics system review. *Practical Hydroponics*, July/August issue. 18-21.

13. Chellappan, P., Christy, S.A.A. and Mahadevan, A., 2001. Multiplication of arbuscular mycorrhizal fungi on roots. In: Mukerji, K.G., Manoharachary, C., Chamola, B.P. (Eds.), *Techniques in Mycorrhizal Studies*. Kluwer Academic Publishers, The Netherlands. 285–297.

14. Cordier, C., Trouvelot, A., Gianinazzi, S. and Gianinazzi-Pearson, V.,1996. Arbuscular mycorrhiza technology applied to micropropagated *Prunus avium* and to protection against *Phytophthora cinnamomi*. *Agronomie* .16 (10), 679–688.

15. Creus, C.M., Sueldo, R.J.and Barassi, C.A., 1998. Water relations in *Azospirillum* inoculated wheat seedlings under osmotic stress. *Can. J. Bot.* 76, 238–244.

16. Daniels, B. A., McCool, P. M. and Menge, J. A.,1981. Comparative inoculum potential of spores of six vesicular arbuscular mycorrhizal fungi. *New Phytol.* 89,385-391.

17. Davies, J.F.T., Potter, J.R. and R.G. Linderman., 1992. Mycorrhiza and repeated drought exposure affect, drought resistance and extraradical hyphae development of pepper plant independent of plant size and nutrient content. *J. Plant Physiol.* 139 , 289 - 294.

18. Davey, M.R., Webster, G., Manders, G., Ringrose, F.L., Power, J.B., and Cocking, E.C., 1993. Effective nodulation of micropropagated shoots of the nonlegume *Parasponia andersonii* by *Bradyrhizobium* L. *J. Exp. Bot.* 44, 863–867.

19. Estrada-Luna, A.A., Davies, F.T. and Egilla, J.N., 2000. Mycorrhizal fungi enhancement of growth and gas exchange of micropropagated guava plantlets (*Psidium guajava* L.) during *ex-vitro* acclimatization and plant establishment. *Mycorrhiza.* 10(1),1-8.

20. Estrada-Luna, A.A. and Davies, F.T., 2003. Arbuscular mycorrhizal fungi influence water relations, gas exchange, abscissic acid and growth of micropropagated chile ancho pepper (*Capsicum annum*) plantlets during acclimatization and post-acclimatization. *J. Plant Physiol.* 160, 1073–1083.

21. Elmeskaoui, A., Damont, J.P., Poulin, M.J., Piche, Y. and Desjardins, Y., 1995. A tripartite culture system for endomycorrhizal inoculation of micropropagated strawberry plantlets *in-vitro*. *Mycorrhiza.*5(5), 313-319.

22. Fortuna, P., Citernesi, A.S., Morini, S., Vitagliano, C. and Giovannetti, M., 1996. Influence of arbuscular mycorrhizae and phosphate fertilization on shoot apical growth of micropropagated apple and plum rootstocks. *Tree Physiology.* 16(9),757-763.

23. Gerdemann, J.W.and Nicolson, T.H., 1963. Spores of mycorrhizal Endogone species extracted from soil by wet sieving and decanting. *Trans. Br. Mycol. Soc.* 46, 235– 244.

24. Gianinazzi, S., Trouvelot, A. and V. Gianinazzi-Pearson., 1990. Role and use of mycorrhizas in horticultural crop production. *Proceedings of the Internaitonal Society for Horticultural Science,* August 1990, Firenze.

25. Guillemin,J.P., Gianinazzi, S.and Trouvelot, A., 1992. Screening of arbuscular endomycorrhizal fungi for establishment of micropropagated pineapple plants. *Agronomie.* 12, 831–836.

26. Guillemin, J.P., Gianinazzi-Pearson, V. and Marchal, J., 1994. Contribution of arbuscular mycorrhizas to biological protection of micropropagated pineapple *Ananas comosus* (L.) Merr) against *Phytophthora cinnamomi* Rands. *Agric. Sci. Finl.* 3, 241–251.

27. Guillon, C., St-Arnaud, M., Hamel, C. and Jabaji-Hare, S.H., 2002. Differential and systemic alteration of defence-related gene transcript levels in mycorrhizal bean plants infected with *Rhizoctonia solani. Can. J. Bot.* 80, 305–315.

28. Hayman, D. S.,1983. The physiology of VA endomycorrhizal symbiosis. *Can. J. Bot.* 61,944-963.

29. Harrier, L.A. and Watson, C.A., 2004. The potential role of arbuscular mycorrhizal (AM) fungi in the bioprotection of plants against soil-borne pathogens in organic and/or sustainable farming systems. *Pest Manag. Sci.* 60, 149–157.

30. HareKrishna,S.K.,Singh,R.R.,Sharma,R.N.,Khawale.,Minakshi Grover. andV.B.Patel.,2005.Biochemical changes in micropropagated grape (*Vitis vinifera* L.) plantlets due to arbuscular- mycorrhizal fungi (AMF) inoculation during *ex-vitro* acclimatization Division of Fruits and Horticultural Technology, Indian Agricultural Research Institute, New Delhi 110012, India . *Scientia Horticulturae* .106 ,554–567.

31. Hare Krishna ., R.K. Sairam ., S.K. Singh ., V.B. Patel. and R.R. Sharma .,2008. Mango explant browning: Effect of ontogenic age, mycorrhization and pre-treatments :*Scientia Horticulturae* .118,132-138.

32. Hendrickson, O.Q., Burges, D., Perinet, P., Tremblay, F. and Chatatpaul, L., 1993. Effects of Frankia on field performance of Alnus clones and seedlings. *Plant Soil* .150, 295–302.

33. Jarstfer, A.G. and Sylvia, D.M., 1995. Aeroponic culture of VAM fungi. In: Varma A, Hock B (eds.) Mycorrhiza – structure, function,molecular biology and biotechnology. *Springer*-Verlag Heidelberg. 427-441.

34. Jaizme-Vega, M.C., Rodriguez-Romero, A.S., Hermoso, C.M. and Declerck, S., 2003. Growth of micropropagated bananas colonized by root-organ culture produced arbuscular mycorrhizal fungi entrapped in Ca-alginate beads. *Plant and Soil*. 254(2),329-335.

35. Jones, H.G.,1998. Stomatal control of photosynthesis and transpiration. *Journal of Experimental Botany* .49, 387–398.

36. Juge, C., Samson, J., Bastien, C., Vierheilig, H., Coughlan, A. and Piche, Y., 2002. Breaking dormancy in spores of the arbuscular mycorrhizal fungus *Glomus intraradices*: a critical cold-storage period. *Mycorrhiza* .12, 37–42.

37. Kapoor,R.,D.Sharma and A.K.Bhatnagar., 2008. Arbuscular mycorrhizae in micropropagation systems and their potential applications. *Scientia Horticulturae*. 116,227–239.

38. Khade,S.W. and B.F. Rodrigues., 2009. Applications of Arbuscular Mycorrhizal Fungi in Agroecosystems *Tropical and Subtropical Agroecosystems*. 10 ,337 – 354 .

39. Krishna,H, Singh, S.K., Sharma, R.R., Khawale, R.N., Grover, M. and Patel, V.B.,2005. Biochemical changes in micropropagated grape (*Vitis vinifera* L.) plantlets due to arbuscular-mycorrhizal fungi (AMF) inoculation during *ex-vitro* acclimatization. *Scientia Horticulturae*. 106(4),554-567.

40. Krishna, H., Singh, S.K . and Patel, V.B., 2006a. Screening of arbuscular-mycorrhizal fungi for enhanced growth and survival of micropropagated grape (*Vitis vinifera*) plantlets. *Indian Journal Of Agricultural Sciences*. 76(5),297-301.

41. Krishna, H., Singh, S.K., Minakshi Patel, V.B., Khawale, R.N., Deshmukh, P.S. and Jindal, P.C., 2006b. Arbuscular-mycorrhizal fungi alleviate transplantation shock in micropropagated grapevine (*Vitis vinifera* L.). *Journal Of Horticultural Science and Biotechnology*. 81(2),259-263.

42. Lazarovits, G. and Nowak, J., 1997. Rhizobacteria for improvement of plant growth and establishment. *Hort. Sci.* 32, 188–192.

43. Lewu, F., Grierson, D. and Afolayan, A., 2006. Extracts from *Pelargonium sidoides* inhibit the growth of bacteria and fungi. *Pharm. Biol.* 44(4), 279-282.

44. Liu, L. H., J. A.Yao., M. Z.Wang., Y. M. Chen. and Z.W. Chong., 2006. Review and prospect of citrus Huanglongbing research. *Fujian Journal of Agricultural Science.* 21,317–320.

45. Lovato, P., Guillemin, J.P. and Gianinazzi, S.,1992. Application of commercial arbuscular endomycorrhizal fungal inoculants to the establishment of micropropagated grapevine rootstock and pineapple plants. *Agronomie.* 12,673-880.

46. Lovato, P.E., Hammatt, N., Gianinazzi-pearson, V. and Gianinazzi, S., 1994. Mycorrhization of Micropropagated mature wild cherry (*Prunus avium* L.) and common ash (fraxinus excelsior L.). *Agricultural Science in Finland.* 3,297-301.

47. Mathur, N. and Vyas, A., 1995. Influence of VA mycorrhizae on net photosynthesis and transpiration on *Ziziphus mauritiana. J. Plant Physiol.* 147, 328–330.

48. Marschner, H., 1995. Mineral nutrition of higher plants. London, UK, *Academic Press.*

49. Mathews,Ramakrishna V., Hegde and M.N. Sreenivasa., 2003. Influence of Arbuscular Mycorrhizae on the vigour and growth of micropropagated banana plantlets during acclimatization. Department of Horticulture University of Agricultural Sciences, Dharwad: Karnataka. *J. Agril. Sci.* 16 (3), 438-442.

50. Marie Chantal Koffi., Ivan Enrique de la Providencia1., Annemie Elsen., and Stéphane Declerck.,2009. Development of an *in-vitro* culture system adapted to banana mycorrhization : *African Journal of Biotechnology.* 8 (12), 2750-2756.

51. Meddad-Hamza, A., Beddiar, A., Gollotte, A., Lemoine, M.C., Kuszala, C. and Gianinazzi, S.,2010. Arbuscular mycorrhizal fungi improve the growth of olive trees and their resistance to transplantation stress. *African Journal of Biotech.* 9(8),1159-1167.

52. Morandi, D., Gianinazzi, S. and Gianinazzi-Pearson, V.,1979. Interet de l'endomycorrhization dans la reprise et la croissance des framboisiers issus de multiplication vegetative *in-vitro. Ann. Amelior. Plant.* 29,23-30.

53. Morte, M. A., Diaz, G. and Honrubia, M.,1996. Effect of arbuscular mycorrhizal inoculation on micropropagated Tetraclinis articulata growth and survival. *Agronomie* .16,633-637.

54. Mohammad, A., Khan, A.G. and Kuek, C., 2000. Improved Aeroponic culture of inocula of arbuscular mycorrhizal fungi. *Mycorrhiza.*9(6),337-339.

55. Moraes, R.M., De Andrade, Z., Bedir, E., Dayan, F.E., Lata, H., Khan, I. and Pereira, A.M.S., 2004. Arbuscular mycorrhiza improves acclimatization and increases lignan content of micropropagated mayapple (*Podophyllum peltatum* L.). *Plant Science.* 166(1),23-29.

56. Nishi Mathur and Anil Vysa.,2007. Flow diagram showing alternative strategies for mycorrhization of micro-plants *in-vitro* and *in-vivo.American journal of plant physicology.*2 (2),122-138.

57. Nripendra, V. Singh., Sanjay, K., Singh b , Anand, K., Singh B , Deodas, T., Meshrama , Sachin., S.Suroshea., Dwijesh C. and Mishrac., 2012. Arbuscular mycorrhizal fungi (AMF) induced hardening of micropropagated pomegranate (*Punica granatum* L.) plantlets: *Scientia Horticulturae.* 136 ,122–127.

58. Ozyigit, I. I., Kahraman, M. V. and Ercan, O., 2007. Relation between explant age, total phenols and regeneration response in tissue cultured cotton (*Gossypium hirsutum* L.). *Afr.J.Biotechnol.* 6(1), 3-8.

59. Pawlowska, T. E., Douds, D. D. and Charvat, I.,1999. *In vitro* propagation and life cycle of the arbuscular mycorrhizal fungus *Glomus etunicatum.* *Mycol. Res.* 103,1549-1556.

60. Pearson, J.N and I. Jakobsen., 1993. Symbiotic exchange of carbon and phosphorous between cucumber and three arbuscular mycorrhizal fungi. *New Phytol.* 124 , 481 - 488.

61. Pozo, M.J., Aacon-Aguilar, C., Dumas-Gaudot, E. and Barea, J.M., 1999. Beta-1,3-glucanase activities in tomato roots inoculated with arbuscular mycorrhizal fungi and/or *Phytophthora parasitica* and their possible involvement in bioprotection. *Plant Sci* .141, 149-157.

62. Pons, F., Gianinazzi-Pearson, V., Gianinazzi, S. and Navavatel, J. C.,1983. Studies of VA mycorrhizae *in-vitro*: mycorrhizal synthesis of axenically propagated wild cherry (*Prunus avium* L.) plants. *Plant Soil* .71,217-221.

63. Puthur, J.T., Prasad, K.V.S.K., Sharmila, P. and Pardhasaradhi, P., 1998. Vesicular Arbuscular mycorrhizal fungi improves establishment of micropropagated *Leucaena leucocephala* plantlets. *Plant Cell Tissue Organ Cult.* 53, 41–47.

64. Quatrini, P., Gentile, M., Carimi, F., De Pasquale, F., Puglia, A.M., 2003. Effect of native arbuscular mycorrhizal fungi and *Glomus mosseae* on acclimatization and development of micropropagated *Citrus limon* (L.) Burm. *Journal of Horticultural Science and Biotechnology.* 78(1),39-45.

65. Ravolanirina, F., S. Gianinazzi, A. Trouvelot and M. Carre., 1989. Production of endomycorrhizal explants of micropropagated grapevine rootstocks. *Agr. Ecosystems Environ.* 29,323-327.

66. Rai,M.K.,2000. Current Advances In Mycorrhization In Micropropagation *In-vitro Cell. Dev. Biol.ÐPlant* .37,158-167.

67. Rai, M., Acharya, D., Singh, A. and Varma, A., 2001. Positive growth responses of the medicinal plants Spilanthes calva and *Withania somnifera* to inoculation by *Piriformospora indica* in a field trial. *Mycorrhiza* . 11, 123–128.

68. Ruiz-Lozano, J.M., Roussel, H., Gianinazzi, S. and Gianinazzi-Perason, V., 1999. Defense genes are differentially induced by a mycorrhizal fungus and *Rhizobium* sp. in a wild-type and symbiosis-defective pea genotypes. *Mol. Plant-Microbe Interact.* 12,976-984.

69. Ruiz-Lozano, J.M., Collados, C., Barea, J.M. and Azcón, R., 2001. Clonig of cDNAs encoding SODs from lettuce plants which show differential regulation by arbuscular mycorhizal symbiosis and by drought stress. *J Exp Bot* . 52,2241–2242.

70. Rupam kapoor., Deepika Sharma and A.K. Bhatnagar.,2008. AMF in micropropagation system and their potential application.Environmental Biology Laboratory, Department of Botany, University of Delhi, Delhi 110007, India: *Scientia Horticulturae.* 116 , 227–239.

71. Salzer, P., Bonanomi, A., Beyer, K., Vogeli-Lange, R., Aescherbacher, R.A., Lange, J., Wiemken, A., Kim, D., Cook, D.R. and Boller, T., 2000. Differential expression of eight chitinase genes in *Medicago truncatula* roots during mycorrhiza formation, nodulation and pathogen infection. *Mol. Plant Microbe Inter.* 13, 763–777.

72. Sbrana, C., Giovannetti, M and Vitagliano, C., 1994. The effect of mycorrhizal infection on survival and growth renewal of micropropagated fruit rootstocks. *Mycorrhiza*, 5, 153-156.

73. Schubert, A., Bodrino, C. and Gribaudo, I., 1992. Vesicular-Arbuscular Mycorrhizal Inoculation Of Kiwifruit (*Actinidia-Deliciosa*) Micropropagated Plants. *Agronomie.* 12(10),847-850.

74. Schubert, A., Mazzitelli, M., Ariusso, O. and Eynard, I.,1990. Effects of vesicular arbuscular mycorrhizal fungi on micropropagated grapevines: influence of endophyte strain, P fertilization and growth medium. *Vitis.* 29,5-13.

75. Singh, G. and Tilak, K.U.B.R., 2001. Techniques of AM fungus inoculum production. In: Mukerji, K.G., Manoharachary, C., Chamola, B.P. (Eds.), *Techniques in Mycorrhizal Studies.* Kluwer Academic Publishers, The Netherlands. 273–283.

76. Sivaprasad,P. and K.K.Sulochana.,2004. Integration Of Arbuscular Mycorrhizal Technology with Micro Propagation. Kerala Agricultural University.13-15.

77. Slezack, S., E. Dumas-Gaudot., S. Rosendahl., R. Kjoller., M. Paynot., J. Negrel, and S. Gianinazzi., 1999. Endoproteolytic activities in pea roots inoculated with the arbuscular mycorrhizal fungus *Glomus mosseae* and/or *Aphanomyces euteiches* in relation to bioprotection. *New Phytol.* 142,517-529.

78. Sowmya, R and Sukhada Mohandas.,2002.Utilization of VAM fungi for improving the establishment of micropropagated plants, Dept. of Botany,Bengaluru University.156-162.

79. Subhan, S., Sharmila, P and Pardha Saradhi, P., 1998. *Glomus fasciculatum* alleviates transplantation shock of micropropagated *Sesbania sesban*. *Plant Cell Rep.* 17,268-272.

80. Sukhada Mohandas., Gowda, M.J.C. and Manamohan, M., 2004. Popularization of arbuscular mycorrhizal (AM) *in-vitro* inoculum production and application on-farm. *Acta Hort (ISHS)* .638,279–283.

81. Taylor,J and Harrier, L. A.,2000. A comparison of nine species of arbuscular mycorrhizal fungi on the development and nutrition of micropropagated *Rubus ideus* L. cv. Glen Prosen (red raspberry). *Plant Soil*. 255, 53-61.

82. Thaker, M.N. and Jasrai, Y.T., 2002, Increased Growth of Micropropagated Banana (*Musa paradisiaca*) with VAM Symbiont. *Plant Tiss. Cult.* 12(2) , 147-154.

83. Trouvelot, A., Kough, J.L. and Gianinazzi-Pearson, V.,1986. Masure du taux de mycorhizationnd un systeme radiculaire, Recherche de methods d estimation ayant une signification fonctionnelle. In: Physiological and General Aspects of Mycorrhizae, *I NRA*.217-221.

84. Vestberg, M.,1992. VAM-inoculation of Finnish strawberry. In: Micropropagation, root regeneration, and mycorrhizas. Joint meeting between COST 87 and COST 8.10, Dijion, France.46.

85. Vestberg, M. and Estaun, V., 1994. Micropropagated plants, an opportunity to positively manage mycorrhizal activities. In: Gianinazzi S, Schuepp H (eds.) Impact of arbuscular mycorrhizas on sustainable agriculture and natural ecosystems. Birkhauser, Basel.

86. Vestberg, M. and Uosukainen, M.,1996. Effect of AMF inoculation on rooting and subsequent growth of cuttings and microcuttings of greenhouse rose Mercedes. In: *Novel biotechnological approaches to plant production*: from sterile root to mycorrhizosphere. Joint COST action 8.21 and 8.22, Pisa, Italy.46.

87. Wang, H., Parent, S., Gosselin, A. and Y. Desjardins., 1993. Study of vesicular-arbuscular mycorrhizal peat-based substrates on symbioses establishment, acclimatization and Growth of three micropropagated species. *J . Am. Soc. Hort. Sci.* 118 , 896 - 901.

88. Wang, G. P., N. Hong. and H. Ahmed., 2005. Occurrence and research progress of fruit tree viroid diseases in China. *Journal of Fruit Science* .22,51–54.

89. Webster, G., Poulton, P.R., Cocking, E.C. and Davey, M.R., 1995. The nodulation of micropropagated plants of *Parasponia andersonii* by tropical legume rhizobia. *J. Exp. Biol.* 46, 1131–1137.

90. White,P.R., 1934.Potentially unlimited growth of excised tomato root tips in a liquid medium. *Plant Physiol* 9,585-600.

91. Wu, Q.S., Zou, Y.N. and He, X.H., 2011. Differences of hyphal and soil phosphatase activities in drought-stressed mycorrhizal trifoliate orange (*Poncirus trifoliata*) seedlings. *Scientia Horticulturae.* 129, 294-298.

92. Yano-Melo, A.M., Saggin, O.J., Lima, J.M., Melo, N.F. and Maia, L.C.,1999. Effect of arbuscular mycorrhizal fungi on the acclimatization of micropropagated banana plantlets. *Mycorrhiza.* 9, 2,119-123.

Chapter 15

Arbuscular Mycorrizal Fungi in Fruit Crop Seedling Production and Orchard Rejuvenation

Ramanathan. R[1], Bhuwanesvari . R[2], *Dhandapani. R[1] and Saritha. B[3]

[1]*Arignar Anna Government Arts College, Namakkal – 637 002, Tamilnadu, India.*

[2]*The H.H. Rajashs College(Autonomus), Pudukkottai, Tamilnadu, India.*

[3]*Indian Institute of Horticultural Research (ICAR), Bengaluru – 560 089, India*

*Adresses for the correspondance : paniroever2007@rediffmail.com

ABSTRACT

The availability of healthy and vigorous seedlings for field planting will ensure a healthy fruit orchard. The seedlings should be bioprimed with inocula of AMF before transplanting to field. The exact species of AMF to be used, the composition of the inoculum, its placement in the nursery and, dosage are important parameters to be considered before inoculation. This article discusses the methods of inoculation of important fruit crop seedlings with the AMF in the nursery and in the field at transplanting. The orchard rejuvenation in guava, replant problems in apple, peach and restoration of soil structure using AMF are also discussed.

Keywords: *AMF, biopriming with AMF, nursery inoculation, orchard rejuvenation, replant problem*

1.0 Introduction

In perennial horticultural crops, healthy seedlings make healthy plantations and an enterprising industry. The pre-requisites to establish a successful and viable plantation of fruit crops is availability of vigorous and healthy seedlings for field planting. When bio-enriched super seedlings are planted in main field they will perform better compared to plants not enriched. Nutrition of seedlings plays an important role in improving the vigour of the seedlings. Keeping these aspects in view, this chapter aims to provide the know how on composition of AMF inoculum, the suitable species to be used, its dosage and the nursery application methods to produce healthy seedlings of fruit crops for field planting.

The declining yield pattern in old guava orchard over the years is the major concern as farmers are shifting to other crops and cropping systems. The cause of decline are overcrowded shoots infested with insect and disease in branches and trunk, more wood mass and thin shoots in canopy and soil infestation with pathogens which adversely influence bearing quality fruits (Singh and Singh, 2003). Similarly in apple specific apple replant disease (SARD) is a widespread disease that impairs the growth and establishment of plants in replanted apple orchards (Bharat, 2013; Sabo *et al.*, 1998). The causes of this disease are both abiotic and biotic factors, such as low nutrients, phytotoxins, actinomycetes, fungal complexes and nematodes. (Utkhede and Smith, 1994). If apple crop has to be replanted in the same space it is difficult as the soil becomes sick with pathogens. So is the case with peach and few other fruit crops. AMF inoculation can change soil microbe species ratios, and impact the rhizospheric microbe community (Linderman, *et al.*, 1996). In this chapter we discuss the production of seedlings in fruit crops by biopriming with AMF in nurseries and thereafter out planting in glass house or field. A case study of guava orchards rejuvenation and apple, peach replanting problem is also presented.

2.0 AMF inoculum for fruit crop seedlings

2.1 Composition of Mycorrhizal inoculum and its carrier

The material that carries propagules of AMF is called inoculum. The mycorrhizal inocula include propagules such as spores, mycorrhizal root fragments, and pieces of mycelium. Spores are generally considered the most resistant to adverse environmental conditions, but are slower than other propagules to colonize new roots. Mycelial fragments are usually the fastest to colonize new roots. A significant limitation in practical use of mycorrhizal fungi is the size of the propagules. The spores are in the range of 1/10 millimeter in diameter, the largest of all fungal spores. Hyphal fragments that are large enough to constitute good inoculum may be of that size or larger. A result is that they quickly settle out of suspension and do not readily pass through apertures of small diameter. Thus endomycorrhizal inoculum does not suit itself to distribution through a liquid handling system. Material applied to the surface of soil or even a very open container mix is likely to remain on the surface, and may not reach the roots. This is in contrast to ectomycorrhizal inoculum, which works well when applied to the surface of a container mix. In addition to fungal structures, the inoculum usually includes a carrier. The carrier may be sand, soil, peat, clay, or other solid substrate. Suspensions of fungus plus carrier in a viscous liquid, such as certain polymer formulations, work well as root dips. Soil based inoculum is difficult to handle in bulk but is the best if prepared on-farm and stored for field applications. Soil rite and vermiculite based inoculum are easy to handle as they are lighter but expensive, useful for pot or nursery applications.

2.2 Choice of AMF in nursery

Endomycorrhizal fungi vary in their responses to soil properties, especially pH. While there is no specificity between fungus and host plant, there are preferences that can be expressed in field or greenhouse experiments (Brundrett 1991; Johnson *et al.* 1992). That is, some fungal species work better with particular host plants. The best way to assure a good fit between plant, soil, and fungus may be to isolate a mixture of native species from undisturbed vegetation on the same soil (Daft 1983; Perry *et al.* 1987). Trap crops can be used to isolate the strains. Unfortunately, native fungi are often more difficult to culture than the proven "generic" strains, and in any case require more time and expense to produce. Further, there is no assurance that the native fungi of the undisturbed soil will still be appropriate for the altered conditions after disturbance (Stahl *et al.* 1988). Most often, the nursery manager must select from a very short list of commonly available commercial strains. At the very minimum, the selected fungi must be suitable for the soil pH at both the nursery and the final planting site. For example, *G. intraradices* has provided good growth responses in a wide range of host plants, at soil pH from about 6 to 8.5 or higher and G. *etunicaturm* has been most effective in moderately acid soils.

2.3 Inoculum placement in nursery

A guiding principle in mycorrhizal work is that the inoculum must be applied to the root zone (Hayman *et al.*,, 1975; Ferguson and Menge, 1986) so that the roots of new seedlings must be able to use the inoculum efficiently, unlike the soil application. For plant seedlings raised from seeds in the nursery bags, the most cost-effective option may be to mix the inoculum into the nursery medium. This is more likely to use the inoculum at the best and requires less labour. Nursery farmers who will not be able to produce the container mixes may be able to persuade their medium suppliers to incorporate the inoculum.

Inoculation may be of greatest benefit to the plant when done at the earliest possible stage. However, germination and rooting stages can be difficult to inoculate because the facilities commonly have low light intensity, heavy use of chemicals, and very wet medium, making mycorrhizal colonization difficult. For plants that are moved one or more times during the production cycle, the first transplant may be a more practical time to inoculate. Most plants at some nurseries are inoculated by placing 2ml of granular inoculum, containing over 100 propagules, beneath the transplant. An additional possibility is a root dip. The nursery might dip bare root plants before delivering them to the customer, or might dip container plants that were not made mycorrhizal during production.

The slurry must contain ingredients that make the suspension viscous to keep the propagules from settling out rapidly, and must act as an adhesive. The root dip would also protect the root systems from desiccation. If the nursery could not supply mycorrhizal plants, the customer may wish to inoculate the plants at the field site. This may be done with a root dip or by dropping a pre-packaged "tea bag" of inoculum in each planting hole. Endomycorrhizal inoculum, packaged in tea bags with or without compatible fertilizer formulations, is now available. Finally, inoculum in a solid carrier like soil based inocula produced on-farm or in big containers can be incorporated into the soil by applying fixed quantity to the planting zone and mixing with the native soil.

2.4 Dosage of AMF in nursery

The amount of inoculum required for a particular application is an important question, because it directly influences both the cost of the operation and the chances for success. The recommendations come from inoculum suppliers and researchers, and the basis for each recommendation is not always clear. Obviously, motivations for recommending high doses are to be sure it works efficiently. Our own recommendations are based on spatial dispersion of propagules in the medium, and on empirical tests of dosage rates. Based on the intensive work carried out on fruit crops in field, certain recommendations have been made in mycorrhization of the seedlings before field transplanting which are given in **Table 15.1**

3.0 Fruit crop seedling production in the nursery

Fruit crops are raised directly through seeds, seedlings, cuttings, air layering, throgh grafting of scions of popular varieties on rootstocks, using, suckers and micro propagated plants. Mycorrhization may be carried out before planting seedlings in the soil.

3.1 Mycorrhization of seedlings

Seeds of papaya, sapota, and guava are mycorrhized either in the nursery beds, polybag nurseries or in pots. Propagation through seed is mostly done to raise open pollinated seedling rootstocks required for grafting/budding operations. Raising in polybag is recommended as this gives better establishment of plants in field on account of undisturbed root system **(Plate 15.1 ,15. 2 and 15.3).**

a. Polybag nursery: 10 X 20 cm polybags containing 500g of the sand: soil: FYM in 1:1; 1 mixture are prepared and 50g of soil based AMF inoculum is added around 5-6 cm below the surface as a thin band. Seeds of crops are sown in bag and irrigated normally. After the seeds germinate they are thinned to one per bag. Seedlings get colonized by AMF and will be ready for transplantation in 1 1/2 to 2 months in case of papaya and sapota. Guava seedlings will be ready in 4-5 months for transplanting or for grafting.

b. Nursery beds: Raised nursery beds are prepared out of sand: soil: FYM mixture (commonly used in India). 1 kg of AMF soil based inoculum containing around 50 spores g^{-1} soil for a bed of 1M x 1M is added uniformly in rows and seeds of papaya/guava/sapota are sown and covered with soil mixture and irrigated normally. When seeds germinate and grow into seedlings their roots get colonized with the AMF within.

Table.15.1 Standard recommended dosage of AMF inoculms in some selected fruit crops seedlings

Crop	Method of propagation	**AMF Inoculation (50 propagules g ⁻¹soil)	Source
Papaya,	Seedling	50g of AMF inoculum per 500 g soil mixture in polybag	IIHR, Extension literature ATIC series 10 (2004). Prepared by Sukhada Mohandas
Sapota, Guava Mango, and citrus (Rootstocks)	Grafted plants on rootstocks	50g of inoculum added to r 500 g soil holding rootstock seedlings in polybag. Scions are grafted on the rootstock seedlings when they are 4-6 months old	IIHR, Extension literature ATIC series 10 (2004). Prepared by Sukhada Mohandas Panneerselvam *et al.*, (2012, 2013)
Pomegranate	Air layering	20 g of inoculum added to 100g of air layering mixture	Sharma *et al.* 2009
Litchi	Air layering	20 g of inoculum containing of *G. fasciculatum* added to 100g of air layering mixture	Sharma *et al.*, 2009
Orange	Seedlings	400g of inoculum added to substrate mixture	Yao *et al.*, 2009
Peach	Seedlings	10 g each of soil based *Gigaspora margarita* inoculum during potting	Rutto and Mizutani, 2006
Apple	Saplings	100 gm⁻¹ per plant added in 5 cm deep furrow	Sharma *et al.*, 2012
Grapes	Plantlets	20 g inoculum containing spores are added to the potting mixture	Krishna *et al.*, 2005
Banana	micropropagated plantlets/Suckers	Micropropagated plants- 20 g of inoculum for 100g soil in polybag. For suckers 500 g of soil based inoculum is added per pit	IIHR, Extension literature ATIC series 10 (2004). Prepared by Sukhada Mohandas

**In organic farming AMF can be used along with *Azospirillum* and Phosphate solubilising bacteria without adding synthetic fertilizers.

3.2 Mycorrhization of rootstocks for grafting scions (Mango)

Mango stones are germinated in sand beds and germinated seedlings are shifted to polybags containing sand: soil: FYM mixture and AMF inoculum as mentioned above. One plant is maintained in each bag. Mango roots take around

5-6 months to show 50% colonization. Scions are grafted on these root sock seedlings in the nursery or after transplantation of the rootstocks to the field.

3.3 Mycorrhization of micropropagated plants

Micropropagated plants are inoculated with AMF when they are hardened **ex-vitro**. Sterile mixtures of coco peat or soil rite/perlite are prepared and about 20-50 g of AMF inoculum are added to the polybags used for hardening. Well rooted micropropagated. plants are carefully planted in the polybags. When plants get colonized with AMF within 1-2 months they are ready for transplantation.

3.4 Mycorrhization during Air-layering

Pomegranate and litchi plants are propagated through air layering.

In this method a 1-2 year old, healthy, vigorous, mature shoot of 45-60 cm in length and pencil thickness is selected. A circular strip of bark about 3 cm wide just below a bud is completely removed from the selected shoot. A rooting mixture consisting of 20 g of AMF is taken in polythene sheet and with moist *Sphagnum* moss is packed around this portion and tied with polyethylene sheet to prevent the loss of moisture. The rooted shoot is slowly detached by giving 2-3 successive cuts over a period of week before finally detaching from the parent plant. The polythene sheet is removed before planting them in pots. They are planted in pots and kept in nursery under shade.

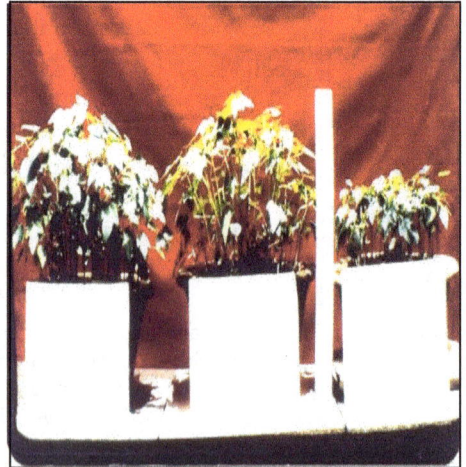

| *G.mosseae* inoculated | *G.fasciculatum* inoculated | Uninoculated control |

Plate.15.1 . Mycorrhization of Papaya (*Carica Papaya* L.) in the poly bag nursery and in pots. A. Application of AMF inoculum in papaya nursery. B showing the difference in plant growth obtained due to application of AMF inoculum.

Plate 15.2. Growth promotional effect of AMF inoculated with MAB on sapota root stock seedlings under nursery condition. C- Control, T- Treatment (Source: Sarita, Ph.D work under publication).

Plate 15.3.a *Streptomyces cinnamonensis*

Plate 15.3.b *Streptomyces canus*

Plate 15.3.c

Plate 15.3.d

Plate 15.3.e *Streptomyces cinnamonensis*

Plate 15.3.f *Streptomyces canus*

Plate 15.3 (a-e). Effect of actinomycetes inoculation on guava seed germination (a,b) and root growth of inoculated seedlings under nursery conditions (c,d). Growth promotion activity of aclinomycetes on 4 month old guava seedlings (e,f). (Source: Poovarasav Ph.D work under publication)

3.5 Plants propagated through corms (banana, gladiolus/tuberose etc.)

Corms or bulbs may be planted in the field or in pots. Around 250g or 100 g of AMF inoculum is added directly in the pits and pots (30 cm dia) respectively. When corms/bulbs produce roots they get colonized by AMF .

3.6 Checking Roots for colonization by AMF

Fruit crop seedlings in the nursery which are 30-45 day old depending on the crop are tested for colonisation by the fungus following Phillips & Hayman (1970). The roots are uprooted carefully during the time of harvesting and washed under running tap water. Fresh root sample are cut into 1 cm segments. The root bits bits are cleared by placing them in plastic vials containing 10 ml of 10% KOH for 48 hrs at room temperature. After decanting the alkali, the residual alkali id neutralized by HCl for 15 minutes. The acid solution is decanted, washed in fresh water and root segments are stained with 0.1% Tryphan Blue stain in Lactoglycerol for 24 hrs. The stained roots are observed under microscope by Gridline intersect method Giovannetti and Mosse (1980). Percent AMF colonization of the fungi is calculated Nicolson's simple formula (1995).

$$Present\ colonization = \frac{Number\ of\ root\ segments}{Total\ number\ of\ root\ segments} \times 100$$

3.7 Field Transplantation

As field application requires large quantity of AMF inoculum it may be produced **on-farm as detailed under chapter 13.**

Pits of size 45 x 45 cm x45cm are dug and neemcake and FYM are added as per standard practice. Around 250 g of AMF inoculums is added per pit. Fruit crop seedlings colonized with mycorrhizal fungi up to 40-50% in the nursery are transplanted in field. Plants are irrigated as usual. The dosage of P fertilizer may be reduced by 25% -50% (depending upon the soil P) as mycorrhizal fungi are efficient scavengers of P under low soil P. AMF inoculum is given once every year after pruning along with organic manure. After 15 days chemical fertilizer application may be made .

4.0 Guava (*Psidium guajava* L.) Orchard rejuvenation using AMF – A case study by Chandra *et al.* (2012)

Guava is grown over 60 countries throughout the tropics and subtropics including some Mediterranean areas. It is highly adoptable and grown in wide range of soil and climate. It is relished for its flavour, delicious taste and high nutritive value. The trees give more than one crop during the year. It is a rich source of vitamin C, calcium, fair in phosphorus and iron. The management of the guava orchard by Indian farmers is very poor as they do not fertilize them and hence it has resulted in low productivity (Chandra *et al.*, 2012) . The declining yield pattern from old guava orchard over the years is the major cause of farmers shifting their interest towards other crops and cropping system. The majority of orchards became old and senile characterized by intermingling, overcrowded, infestation of insect and disease in branches and trunk, more wood mass and thin shoots in canopy adversely affect fruiting quality (Singh and Singh, 2003). The fruits of guava are borne on new wood 9-11 months old and any treatments that encourage new growth influence the fruiting directly. In addition, researches are established that the fruiting potential of guava is largely governed by canopy architecture, density and photosynthetic efficiency (Kallo *et al.*, 2005). Rejuvenation of old senile orchards is advocated to allow new shoots on tree, elimination of infected branches, increase light penetration on floor for field crops along with higher fruit yield. This technology play significant role in conversion of old orchard into new ones but it requires at most care, skilled felling, scheduled irrigation and adequate nutrition.

A study was conducted to assess the influence of AMF in rejuvenated guava trees in farmers field and also analyze the gap between improved technology and actual adopted technology by Chandra *et al.* (2012) at KVK, Kumarganj, Faizabad U.P., India. Guava trees 48 in number were pruned regularly and applied with rejuvenation IPNM package consisting of 25 kg vermicompost + 1 kg neem cake + 1300 g urea 1800 g (46% N), single super phosphate (16% P_2O_5), 500 g muriate of potash (46% K_2O) plant[-1] and 50g micronutrient mixture in two split doses in the month of October and June. After six months of IPNM package AMF (*Glomus intraradics* and *Acaulospora scrobiculata*) 400g inoculum (Density 103 infective propagules with roots and spores) was applied per plant. Suitable controls were maintained. The compost, inorganic fertilizers and AMF inoculum were broadcasted 30 cm from the trunk and out to the edge of canopy of the tree.

In the guava orchard restoration studies clearly indicated that guava trees could be rejuvenated by adopting IPNM with AMF application. Tree rejuvenation with regular pruning decreased the tree height up to 50% compared to non rejuvenated trees. This initiated emergence of new shoots and stimulated shoots to convert in flowering shoots. It was clearly seen from the parameters measured in the present study that the application of IPNM with AMF inoculation was found to be most effective among other treatments. The shoot emergence increased 3.7 fold in rejuvenated tree followed with IPNM and AMF compared to tree without rejuvenation. The treatment of IPNM and AMF without combination but regular pruning observed beneficial in shoot emergence, flowering and fruiting **(Table 15.1)**. The yield increment was found maximum 140.55% with combined application of IPNM and AMF followed by IPNM (89.17%) and AMF (56.50%) over control **(Table 15. 1)**. The rejuvenation caused branching complexity resulted more fruiting shoots, profuse flowering in these shoots as was also reported by (Compbell and Wasielewski (2000). The growth of terminal and lateral shoots was stimulated by rejuvenation technique. Therefore initial yield was adversely affected on such trees but later year the yield increased 3.76 fold due to higher number of fruits and higher weight than control. The nutrient content in guava leaves found significantly higher N (60.95%), P (111.1%), K (31.66%), Mg (130.3%), Cu (45.24%), Zn (39.39%), Fe (28.82%) and Mn (16.12%) in trees applied with IPNM and AMF compared to the control **(Table 15. 2)**. Inoculation of AMF found beneficial in enhancing the nutrient particularly P (66.66%), N (37.14%) and Mg (48.48%) over control as compared to other elements. The authors have concluded that it is necessary to educate the farmers about the technology by conducting practical demonstrations in their field. It is possible to improve the productivity of the crop and socioeconomic condition of the farmers.

5.0 Apple replant problem

Specific apple replant disease (SARD) is a widespread disease that impairs the growth and establishment of plants in replanted apple orchards (Sabo *et al.*, 1998). Apple replant is a major problem to growers who are trying to establish new orchards on sites where apples have been grown previously. The disease is characterised by stunted growth, yellowing of foliage shortened internodes and small discoloured rotting root system. The cause of this disease has been suggested to include both abiotic factors and biotic factors, such as low nutrients, phytotoxins, actinomycetes, fungal complexes and nematodes. (Utkhede and Smith, 1994). These include harmful microorganisms (biotic) and nutritional deficiencies or excesses, soil pH, phytotoxins etc.(abiotic). Due to its complex etiology, replant problem is very difficult to manage. Only practice adopted worldwide for its management is soil fumigation. But soil fumigation with chemicals does not provide a sustainable growth of new seedlings as it also destroys beneficial soil microflora including AMF and biocontrol agents (BCAs) besides other environmental issues. However, fumigation of the soil has been shown to ameliorate the disease. Inoculation with AMF is known to have prevented SARD development in apple root stocks. AMF have proved their usefulness with regard to these problems, as it is known that these symbiotic organisms can improve nutrient and water uptake and pathogen resistance in their host.

Inoculation with AMF *Glomus etunicatum* was successful during the first 6 months of growth when apple seedlings were grown for the first three weeks in a sterile substrate (sand-soilpeat). When young grafted apple trees were inoculated directly in orchard soil with apple replant disease under non-sterile conditions, the effect of inoculation was negligible after 6 months. There were significant differences in the composition of the rhizosphere microflora after 6 months, especially between the soil with apple replant disease and the virgin soil (Catská and Taube-Baab, 1994). It has been shown that AMF can change soil microbe species and ratios, and impact on the rhizospheric microbe community (Linderman *et al.*, 1996). The rhizospheric microbes are influenced directly or indirectly by AMF, i.e. AMF is capable of changing the microbial population and establishing new equilibrium (Linderman, 1988; Kothari *et al.*, 1991). AMF have positive effects on replant diseases in grape or apple orchard (Camprubí *et al.*, 2008; Raj and Sharma, 2009). Bharath, (2013) conducted a pot culture experiment to see the effect of indigenous AMF viz., *Glomus* spp., *Acaulospora* spp., *Gigaspora* spp. and *Scutellospora* spp.and biocontrol agents (BCAs) *Trichoderma viridii* individually or in combinations on the growth of newly planted apple seedlings in apple replant diseased (ARD) soil with and without formaldehyde fumigation. The soil was collected from replant affected orchards at Kotkhai, district Shimla located in wet temperate region. The experiment showed that AMF plus *T.viridii* inoculation resulted in improvement in the growth of inoculated seedlings viz., seedling height, stem diameter, leaf area, seedling fresh and dry weights etc. than that of un-inoculated ones. The leaf nutrient status was also improved in inoculated seedlings. The effect of inoculation of AMF and BCAs was more prominent in growth improvement in the soil fumigated with formaldehyde.

Pre-inoculated peach seedlings transplanted in non-replant soils showed greater initial growth in the first year. Plant height, and lateral shoot length and number was highest in non-replant soils irrespective of mycorrhizal pre-inoculation. Similarly, biomass yield was significantly higher in seedlings in non-replant soils (Rutto and Mizutani, 2006). Inoculation with AMF can also change the bacteria and actinomycete numbers on the root surface and in the rhizosphere. The beneficial rhizospheric microbe which coexist with AMF is known to enhance AMF colonization (Panneerselvam *et al.*, 2012).

6.0 Effect of AMF on habitat restoration in nursery

AMF are best known for dramatic growth responses in seedling production and these growth responses are usually related to phosphorus nutrition as well as other micronutrients, and are most pronounced in soil of low fertility (Tinker, 1978). Other effects of AMF make less dramatic but may be more meaningful when nursery plants go out to a restoration or reforestation site. Drought tolerance is commonly higher in mycorrhizal than in non- mycorrhizal plants. Soil structure is a primary concern in restoration and reforestation. Some degraded soils are said to no longer support tree growth because they lack structure (Perry *et al.* 1987). Wild plants in healthy ecosystems suffer very little from root disease, because soil pathogens are in balance with other beneficial soil organisms (St. John 1993). The mycorrhizal symbiosis is a key player in this balance, since it selectively favours beneficial soil organisms (Linderman 1994). The beneficial effects of the mycorrhizal symbiosis are less evident in the favourable conditions of a nursery than in the harsh conditions of the final field site. The objective in habitat restoration is not simply to make plants grow, but to form a functional ecosystem. The characteristics of functional ecosystems include productivity, sustainability, retention of nutrients, resistance to invasive species, and biotic interactions (Ewel, 1987), which are highly dependent on mycorrhiza.

7.0 Conclusion and future strategies

For nursery seedlings of fruit crops, inoculation with AMF should be imperative for the following reasons *viz.*, better plant growth, increase the absorptive surface of roots, reduction in fertilizer application (particularly P, Zn and Cu), minimum transplant injury, support the plants under adverse conditions like abiotic stresses, increase resistance to some plant diseases and nematodes etc. Seedlings of fruit crops can be easily inoculated in the nursery and bio-enriched seedlings transplanted to field. The fruit crop seedlings responded differently to AMF when inoculated in rooting beds before transplant to soil containing a native AMF population. These results suggest that attention must be paid to the interaction between introduced and native AMF populations to achieve optimal benefits from the mycorrhizal symbiosis for specific seedlings or cultivars under field conditions. In horticultural seedlings production, the vegetative propagation is one the common techniques being followed in many fruit crops, hence future research works must be undertaken on the effects of these fungi on the performance of layering/ graft cuttings of seedlings under field conditions.

Restoration of senile orchards is an important area to be concentrated. An in depth study into the soil nutrient and pathogenic microbial assessment needs to be done before planning rejuvenation. Suitable AMF species needs to be found out for inoculation. Apple and peach replant problem to be addressed in the same way. In some of the places, the natural populations of AMF have been destroyed or damaged due to heavy fumigation, indescriminate use of chemicals/ fertilizers, dumping of industrial wastes and mining activities etc. The researchers should give special attention while restoration of these degraded areas by cultivation of different mycorrhized seedlings.

References

1. Bharat, N., 2013. Effect of indigenous AM fungi and BCAs on health of apple seedlings grown in replant disease soil. *Indian Phytopath*. 66 (4) , 381-386.

2. Brundrett, M. C., 1991. Mycorrhizas in natural ecosystems. In: *Advances in Ecological Research*. A. Macfaydn, M. Begon, and A. H. Fitter, (Eds.). Academic Press, London. 21, pp 171.

3. Catska, V. and Taube-Baab H., 1994. Biological control of replant problems. *Acta Horticulturae* .363, 115-120.

4. Camprubí, A., Estaún, V. Nogales, A., García-Figureueres, F. and Pitet, M., Calvet, C. 2008. Response of the grapevine rootstock Richter 110 to inoculation with native and selected arbuscular mycorrhizal fungi and growth performance in a replant vineyard. *Mycorrhiza*. 18(2),211-216.

5. Compbell, R.J. and Wasielewaski, J. 2000. Mango tree training for the hot tropics. *Acta Horticulture*. 509, 641- 651.

6. Chandra, K.K., Pandey,K. and Ajay K. Singh 2012. Influence of tree rejuvenation, IPNM and VA-Mycorrhizal fungi on shoot emergence, yield and fruit quality of *Psidium guajava* under farmers field condition. *International Journal of Biosciences (IJB)* .. 2(11),9-17.

7. Daft, M. J., 1983. The influence of mixed inocula on endomycorrhizal development. *Pl. soil* . 71, 1333- 1337.

8. Ewel, J. J., 1987. Restoration is the ultimate test of ecological theory. In: *Restoration Ecology*: a synthetic approach to ecological research.W. R. Jordan, M. E. Gilpin, and J. D. Aber (Eds.). Cambridge University Press, Cambridge. pp. 31-33

9. Ferguson, J. J. and Menge. J. A., 1986. Response of citrus seedlings to various field inoculation methods with *Glomus deserticola* in fumigated nursery soils. *J. Am. Soc. Hort. Sci.* 111, 288-292.

10. Giovannetti and Mosse 1980. An evaluation of techniques for measuring vesicular-arbuscular mycorrhizal infection in roots. *New Phytologist*. 84, 489-500.

11. Guissou, T., 2009. Contribution of arbuscular mycorrhizal fungi to growth and nutrient uptake by jujube and tamarind seedlings in a phosphate deficient soil. *Afr. J. Microbiol. Res.* 3(5), 297- 304.

12. Hayman, D. S., Johnson, A.M. and Ruddlesdin., 1975. The influence of phosphate and crop species on *Endogone* spores and vesicular-arbuscular mycorrhiza under field conditions. *Pl.soil.* 43, 489-495.

13. IIHR, Extension literature ATIC series 10 (2004). Prepared by Sukhada Mohandas.

14. Johnson, N. C., Tilman, D. and Wedin, D., 1992. Plant and soil controls on mycorrhizal fungal communities. *Ecol.* 73, 2034-2042.

15. Kallo, G., Reddy, B.M.C., Singh, G. and Lal, B., 2005. Rejuvenation of old and senile orchard. Pub. CISH, Lucknow, pp. 40.

16. Kothari, S.K., Marschner, H. and Romheld, V. 1991. Effect of a vesicular-arbuscular mycorrhizal fungus and rhizosphere microorganisms on manganese reduction in the rhizosphere and manganese concentrations in maize. *New Phytologist*, Vol. 117. No.4, pp. 649-655.117(4),649-655.

17. Krishna, H., Singh, S.K., Sharma, R.R., Khawale, R.N., Grover, M. and Patel, V.B., 2005. Biochemical changes in micropropagated grape (*Vitis vinifera* L.) plantlets due to arbuscular-mycorrhizal fungi (AMF) inoculation during *ex vitro* acclimatization. *Sci. Hort.* 106, 554–567.

18. Linderman, R. G. 1988. Mycorrhizal interaction with the rhizosphere microflora: the mcorrhizosphere effect. *Phytopathology.* Vol.78. No.3, pp. 366-371.78(3),366-371.

19. Linderman, R. G., 1994. Role of VAM fungi in biocontrol. In:. *Mycorrhizae and plant health.* F. L. Pfleger., R. G. Linderman (eds.). APS Press, St. Paul. P. 1-26.

20. Lindemann, B.,1996. Taste reception. *Physiol Rev* .76,719 –766.

21. Linderman, R. G., Marlow, J.L. and Davis, E. A. 1996. Contribution of microbial associates of VA mycorrhizae to mycorrhiza effects on plant growth and health. In: Abstracts of ICOM I. Berkeley campus, California University, USA.

22. Nicolson, T., 1995. Taxonomy Of Endomycorrhizal Fungi. In: E. Mukerji, B. Muthur; B. Chamola; P. Chitralekha: *Advances In Botany,.* Aph Publishing Corporation, New Delhi, 212-218.

23. Panneerselvam, P., Saritha, B., Sukhada Mohandas., Upreti, K.K., Poovarasan, S., Sulladmath, V.V. and Venugopalan, R.,2013. Effect of mycorrhiza associated bacteria on enhancing colonization and sporulation of *Glomus mosseae* and growth promotion in sapota (*Manilkara achras* (Mill.) Forsberg) seedlings. *Biological Agriculture and Horticulture.* 29(2),118–131.

24. Perry, D.A., Molina, R. and Amaranthus, M.P., 1987. Mycorrhizae, mycorrhizospheres, and reforestation: current knowledge and research needs. *Can. J. For. Res.*17, 929-940.

25. Periyasamy Panneerselvam., Sukhada Mohandas., Boya Saritha., Kaushal Kishore Upreti., Poovarasan., Ajay Monnappa. And Vijay Virupakshayy

Sulladmath., 2012. *Glomus mosseae* associated bacteria and their influence on stimulation of mycorrhizal colonization, sporulation, and growth promotion in guava (*Psidium guajava* L.) seedlings. *Biol. Agri. Hort.* (28)4, 267–279.

26. Phillips, J.M. and Hayman, D.S. 1970. Improved procedures for clearing roots and staining parasitic and vesicular- arbuscular mycorrhizal fungi for rapid assessment of infection. *Trans. Br. Mycol. Soc.* 55, 158-161.

27. Raj, H. and Sharma, S. D. 2009. Integration of soil solarization and chemical sterilization with beneficial microorganisms for the control of white root rot and growth of nursery apple. *Scientia Horticulturae.* 119(20),126-131..

28. Rutto, K. L. and Mizutani, F., 2006. Peach seedlings growth in replant and non-replant soils after inoculation with arbuscular mycorrhizal fungi. *Soil Biol. Biochem.* 38, 2536-2542.

29. Szabo winkler., and Marwitz, pet ., 1998. Evidence for pathogenicity of actinomycetes in in rootlets of apple seedlings from soils conducive to specific apple replant disease. *Acta Horticulturae.* 477,55-65.

30. Sharma.,Som Dev., Pramod Kumar., Harender Raj. and Satish Kumar Bhardwaj., 2009. Isolation of arbuscular mycorrhizal fungi and Azotobacter chroococcum from local litchi orchards and evaluation of their activity in the air-layers system. *Scientia Horticulturae* - 123(1),117-123.

31. Sharma,Som Dev., Sharma, N.C., Sharma, C.L. Pramod Kumar. and Ashu Chandel., 2012. *Glomus–Azotobacter* symbiosis in apple under reduced inorganic nutrient fertilization for sustainable and economic orcharding enterprise. *Sci. Hort.* 146, 175–181.

32. Singh, V.K. and Singh, G., 2003. Strategic approaches of precision technology for improvement of fruit production. In: Precision farming in horticulture. Singh, H.P., Singh, Gorakh., Samuel S.C., Pathak, R.K. (Eds.). NCPAH, DAC, MOA, PFDC, CISH, Lucknow, pp. 75- 91.

33. Stahl, P. D., Williams, S. E. and Christensen, M., 1988. Efficacy of native vesicular-arbuscular mycorrhizal fungi after severe soil disturbance. *New Phytol.* 110, 347-354.

34. St. John, T. V., 1993. The importance of mycorrhizal fungi and other beneficial organisms in biodiversity projects. In:. Proceedings, Western Forest Nursery Association. T. D. Landis(Ed.). USDA Forest Service General Technical Report RM-221. Pp. 99-105.

35. Tinker, P. B. H., 1978. Effects of vesicular-arbuscular mycorrhizas on plant nutrition and plant growth. *Physiol. Vegetal.* 16, 743-751.

36. Utkhede, R.S. and Smith E.M. 1994. Biotic and abiotic causes of replant problems of fruit trees. *Acta Hort.* 363, 25–31.

37. Yao, Q., Wang, L.R., Zhu, H.H. and Chen, J.Z., 2009. Effect of arbuscular mycorrhizal fungal inoculation on root system architecture of trifoliate orange (*Poncirus trifoliata* L. Raf.) seedlings. *Sci. Hort.* 121, 458-461.

Table 15.1: Effect of Mycorrhizal inoculation and IPNM on growth, flowering and yield of rejuvenated Guava cv. Allahabad safeda (Data in parenthesis shown standard deviation of 4 replicates) Source: Chandra et al., 2012.

Treatment	Tree height (Unit)	Emergence of new shoot (no)	Flowering shoots (%)	Mean fruit Yield (kg tree^{-1})	Precent increase in yield (q)	Quality parameter		
						No of fruit pl^{-1}	Fruit weight (g fruit^{-1})	TSS (brix0)
T1	3.75 (0.108)	11.80 (1.394)	59.00 (3.740)	73.44 (3.498)	140.55 (6.517)	410 (6.329)	176.50 (11.142)	13.00 (0.707)
T2	3.80 (0.216)	10.05 (0.521)	45.32 (3.905)	58.00 (3.544)	89.97 (9.487)	355 (10.801)	31. 166.00 (7.118)	13.00 (0.294)
T3	3.50 (0.216)	6.33 (0.270)	32.00 (3.265)	47.78 (4.709)	56.50 (9.475)	353 (5.165)	130.40 (5.746)	12.00 (0.248)
T4	7.4 (0.629)	3.15 (0.324)	15.66 (1.491)	30.53 (4.217)	–	300 (11.602)	93.33 (2.649)	66. 11.00 67.(0.355)
CD (p=0.05)	1.35	0.77	3.28	4.00	9.25	7.35	8.50	0.88

Table 15.2: Effect of AMF inoculation and IPNM on nutrient uptake in rejuvenated and non rejuvenated guava trees. (Data in parenthesis shown standard deviation of 4 replicates) Source: Chandra et al., 2012

Treatment	N %	P %	K %	Mg (ppm)	Cu (ppm)	Zn (ppm)	Fe (ppm)	Mn (ppm)
T1	1.69 (0.155)	0.19 (0.021)	1.58 (0.147)	0.76 (0.028)	16.50 (1.870)	230.00 (21.602)	143.00 (5.614)	144 (11.575)
T2	1.50 (0.217)	0.15 (0.016)	1.54 (0.329)	0.49 (0.069)	13.39 (0.795)	180.00 (22.730)	125.00 (3.560)	139 (4.898)
T3	1.44 (0.077)	0.15 (0.035)	1.25 (0.041)	0.49 (0.063)	13.41 (1.821)	181.50 (7.430)	121.00 (2.160)	130 (5.715)
T4	1.05 (0.131)	0.09 (0.014)	1.20 (0.243)	0.33 (0.021)	11.36 (1.160)	165.00 (26.920)	111.00 (2.160)	124 (6.133)
CD (p=0.05)	0.18	0.03	0.17	0.10	0.60	11.22	3.15	7.40

Chapter 16

Molecular Approaches in the study of Arbuscular Mycorrhizal Symbiosis

Poovarasan S[1], Sukhada Mohandas1 and Sita T[2]

[1]Indian Institute of Horticultural Research (ICAR),
Hessaraghatta,Bengaluru-560089,India.

[2]St, Martin's College of Engineering, Dollapally,
Hyderabad,India.

Address for the correspondence: Sukhada.mohandas@gmail.com

ABSTRACT

Arbuscular mycorrhizal fungi (AMF) form symbiotic association with more than 80% of plant system in the environment and still the mechanism of this symbiotic relationship is less understood. Limitations of manual methodologies have motivated the application of molecular tools to understand the various aspects of this symbiotic relationship. Basic molecular techniques like genomic DNA isolation and amplification of particular sequences using specific primers (ITS, rRNA and rDNA) facilitated the confirmation of morphological identification and the expression level of the genes responsible for agronomically important traits. Different tagging techniques have been used in on-field evaluation for monitoring the activities of the spores. Advancement in sequencing of full genome and other Pyrosequencing techniques have helped in identifying the difference between two identical species. The meta-transcriptomic approaches have helped to understand the principle of AMF symbiosis at the time of plant growth. Gene mining from AMF spores has resulted in isolating several important genes. Real-time PCR techniques have been effectively used to study expression level of the genes responsible for agronomically important traits. Molecular techniques may help to create database for taxonomic analysis and cDNA libraries followed by sequencing may help to obtain expressed sequence tags (ESTs) to analyze the gene expression.

Keywords: *Taxonomic diversity, DNA markers, ITS, NS1, EST, T-RFLP, Pyrosequencing technique*

1.0 Introduction

AMF-plant symbiosis which has unequivocally proved its significance in promoting plant growth, providing resistance against soil born diseases, helping in plant establishment, drought and salinity tolerance and even remedying metal tolerance is constrained by its inability to be cultured on a growth medium like other microbes. As AMF do not produce sexual states and exist in imperfect states characteristic features for studies on phylogeny and taxonomy are based on morphology of asexually produced propagules. The mycelium of the fungus during plant colonization penetrate deeply into the root and sometimes more than one species colonizes the endorhizosphere and rhizospere of the plant at the same time. Considering all these limitations, molecular techniques are inevitable in the study of AMF symbiosis (Simon *et al.*, 1992; Horton and Burns, 2001; Harrier, 2001; Redecker *et al.*, 2003). The advancement of the molecular methods help to estimate the distribution and abundance of the AMF at the intrageneric level (Helgason *et al.*, 1999). Molecular techniques are being used to clone the DNA fragments from the un-cultivable fungi and hybridize with homo/heterologous probes to identify the role of certain genes operating at different stages of life cycles of AMF (Singh, 2007).Therefore the focus of the study through molecular approach may lead to accurate conclusions and revolutionize the understanding of this symbiosis (Ram-Reddy et al., 2005). This technology would unravel the functioning of mycorrizae at molecular level which may help in crop improvement in a big way by understanding the plant microbial interactions. The present review elucidates the developments in the area of molecular biology in understanding the phylogeny, taxonomy and host-fungal interaction and the intricacies of the symbiosis.

2.0 Molecular approach in Taxonomic and diversity studies of AMF

2.1 Identification of spores using specific markers

The structural features of AMF spores are an important factor for identification and classification studies. Using classical morphology it is difficult to differentiate the of spores in a mixed population (Redecker *et al.*, 2003). Likewise the identification of spores colonized on the roots may be identified but the spores inside the roots could not be identified accurately due to the complexity of root structure (Helgason *et al.*, 1999). Basic PCR studies using conserved DNA regions as specific markers (ITS 1-4 and NS1-8) can effectively identify the spores colonized inside the root tissues (Edwards *et al.*, 1997; Redecker, 2000; Wubet *et al.*, 2006; Lee *et al.*, 2008; Shafiqua and Stephan, 2013) than by normal manual methods. Simon *et al.*, 1993 combinedly used PCR and single strand confirmation (SSC) technique for the identification of colonizing AMF in the root sample collected from the field. A competitive PCR technique was used to identify the relative abundance and spatial distribution of the mixed population of AMF colonized within the root system. This competitive PCR technique gives information with high sensitivity and accuracy (Di-bonito *et al.*, 1995). Gomez-Levya *et al.* (2008) designed a specific SCAR marker set (GIN680 F&R) through RAPD analysis of three genera (*Glomus, Giagaspora* and *Aculospora*) for the identification of specific fragment which are highly homologous to *G.intraradices*. Host specificity for AMF colonization was studied through RFLP technique by analyzing the ribosomal smaller subunit (SSU) of the specific species. The polymorphic pattern showed that the colonizing AMF community in *T. repens* was entirely different from the *A.capillaries* roots (Vandenkoornhuyse *et al.*, 2001). Four *AMF* species, *Archaeospora eptoticha, Scutellospora castanea, S. cerradensis, S. weresubiae* were identified first time in Korean soil through nested PCR by using specific primers AML1 and AML2 designed based on the 18S rDNA region of AMF. Croll *et al.* (2009) have tried the crossing between two genetically different AMF sp for the evaluation of new sp. The successful cytoplasmic fusion between the two different species was achieved and further the progeny DNA were subjected to the AFLP studies to confirm the genetic exchange.

Stockinger *et al.* (2009) reported the reconfirmation of the *G.intraradices* (DAOM197198) as *G.irregulare* based on the phylogenetic analysis of partial small sub unit (SSU), complete internal transcribed spacer (ITS) and partial large sub unit (LSU). Stockinger *et al.* (2010) reported the use of 1500bp sequence amplicon with kimura two-parameter distances to create AMF barcode database. The advantage of 1500 bp read covers all small subunit, ITS region, large subunit and nuclear ribosomal DNA for resolve the query at species level. Morphological observation did not reveal the exact status of the AMF community colonizing on two different citrus root stocks (trifoliate orange and red tangerine). The analysis of ribosomal small subunit DNA genes revealed the 10 discrete *Glomus* genus groups from 173 isolates. Among that G1 belongs to uncultured *Glomus* group which colonize 54.43% on trifoliate orange root stocks followed by G6 belongs to the *Glomus intraradices* group which colonize 35% on red tangerine root stocks. The results indicated that native AMF species in citrus rhizosphere had diverse colonization potential between two different rootstocks (Wang and Wang, 2014).

2.2 Molecular analysis of factors affecting AMF community

External forces are severely affecting AMF community in soil; soil tillage practices are indirectly affecting the AMF community and its structural features (Jansa *et al.*, 2001). Alguacil *et al.* (2007) studied the effect of agricultural practices on AMF community by the ribosomal DNA gene amplification followed by RFLP analysis. The sequence results of the study showed overall low AMF diversity in four agricultural practices, but the multi dimensional scaling and log-linear saturated model showed that AMF community was significantly affected by soil tillage method (Brito *et al.*, 2012). Real time PCR assay was used to check the AMF abundance based on the target of 18s rRNA or actin gene. The assays showed high sensitivity and accuracy for the quantification of *broad* range of templates from different parts of the spores and its environments (Gamper *et al.*, 2008). Kruger *et al.* (2009) reported a set of primer sequence which suits all arbuscular mycorrhizal fungi. To avoid the contamination problem a primer set which covers small subunit region- ITS region-large subunit region was used to get the species level resolution for taxonomical studies. Corradi *et al.*,(2007) studied the copy number polymorphism in AMF *Glomus intraradices* by using quantitative PCR. The amplification results showed the copy number polymorphism in rDNA genes, protein en-coding genes and pseudo genes. The effect of copy number polymorphism in AMF population is severely affecting the AMF community structure.

Boon *et al.* (2010) investigated the transcription of polymorphic nuclear genes in intra isolate AMF. The results showed the larger subunit rDNA and PLS-I like sequences had high sequence variations in genomic and transcriptomic level. Kruger *et al.* (2012) have developed a phylogenetic reference data set for the molecular systematic and environmental community analysis of AMF including deep sequencing studies. The dataset was created by over lapping the genomic sequence of LSU, ITS, SSU and rDNA sequences, A mix of 3kb fragment analysis gave sequence resolution up to species level. Specific factors for the AMF distribution in host plants were analyzed based on ITS rDNA sequence (Yang *et al.*, 2012). The sequence variance was observed in AMF collected from the plants from different places. Influence of bio-geographical, climatic changes on host specificity for AMF colonization and diversity was noticed.

High throughput 454 Pyrosequencing techniques was used to identify the fungal diversity in six different forest soils by analyzing ribosomal internal transcribed spacer (ITS[-1]) region. The application of Pyrosequencing on soil will help to study the spatiotemporal dynamics of fungal communities in the ecosystem (Buee *et al.*, 2009; Orgiazzi *et al.*, 2013). Two independent molecular techniques (T-RFLP and Pyrosequencing) were used to identify the differences of AMF diversity due to agronomic, climatic differences and physio-chemical properties of olive orchard soil. Pyrosequencing analysis of the study revealed the presence of five major groups of AMF spp (*Archaeospora, Diversispora* and *Paraglomus, Claroideoglome-raceae* and *Glomeraceae*) and the diversity difference in olive rhizosphere was highly influenced by the above said factors (Montes-Borrego *et al.*, 2014). A vineyard soil investigation revealed the presence of AMF species based on the AMF SSU rDNA sequences in both root and soil. The phylogenetic analysis showed a different distribution of sequence from the two sites in the main Glomeromycotan groups (Balestrini *et al.*, 2010). Tonge *et al.*, 2014 used a proton release sequencing platform which could read 400bp in length at significant depth for the identification of fungal community from the mixed sample. The method is based on a multi-region approach for conserved region of fungal genome because many of the fungal sp. may not amplify through the single specific primers for the conserved region.

Effect of inorganic and organic fertilizer with integrated pest management on AMF colonization was studied (Alguacil *et al.*, 2014) in Venezuelan agro-eco system. The phylogenetic analysis of AMF small subunit rRNA (SSU rRNA) genes revealed the richness of the AMF colonization. Diversity was observed in the treatment combined with inorganic, organic fertilizer alone and the treatment with chemical pesticide had lowest diversity and colonization level. The outcome of the study indicates that the application of chemical fertilizers have significant influence on AMF colonization. The diversity of AMF on the host plant may also influenced by the age of the host. High throughput sequencing (454) revealed that the percentage of AMF colonization was more in the long lived older perennial trees *Artocarpus altilis* (bread fruit tree) than the younger trees with different fungal communities (Hart *et al.*, 2014).

3.0 Molecular approach in AMF functional studies

For the effective application of AMF symbiosis in plant production, background of their functional interaction should be known. Due to limitation of traditional techniques in studies involving AMF functional symbiosis, the new molecular biology techniques have made the search for the genes which are actively controlling the symbiotic functions of AMF with its partner easy.

3.1 Studies on AMF root colonizing capacity

Amplification of AMF ribosomal large subunit variable domines (D1 and D2) was used to identify the variability in AMF capacity to colonize roots. The sequence of amplicon showed the interspecies variability, based on this variability the inter-discriminating primers were designed to identify the four AMF sp colonizing onion roots. The results showed majority of the root fragments were colonized by more than one of that four AMF. Interestingly the colonization of *S.castanea* and *Gi.rosae* was significantly increased due to the synergertic effect of the presence of two Glomus spp. *G.mosseae*, *G.intraradices* (Van-Tuinen *et al.*, 1998). Francoise Corbiere, (2002) studied the importance of sucrose synthase gene (MtSucS2) for the normal development of AMF symbiosis in *Medicus truncatula* plant. The transgenic approach on pea and maize plants with *MtSucS2* gene showed enhanced AMF colonization and arbuscules formation than the control plants which were not colonized by AMF. Genre *et al.* (2005) studied the mechanism of AMF infection in transgenic plant roots by using green fluorescent protein labeling technique. The results showed the expression of labeled ENOD11 gene in transgenic *Medicus truncatula* plant roots during the AMF infection and the real-time monitoring of the study showed internal bulging of root cells for AMF spore hypal penetration into the cell. Hohnjec *et al.* (2006) compared the untargeted transcriptome profiling with single cell profiling for the understanding of AMF symbiosis .Compared with untargeted approach the application of single cell profiling technique will help to clearly understand the response of targeted plant cells which are actively involved during the time of AMF interaction. Ferrol *et al.* (2000) identified the plasma membrane H^+-ATPase gene which act as a secondary transport system on fungal cell wall and maintain the growth and development of AMF networks from the AMF sp *Glomus mosseae*. Degenerative primer analysis revealed the presence of five sets of genes (GmH1-H5) in *G.mosseae* out of five, 1, 3 and 4 were identical, 2and 5 were highly divergent. Amann *et al.* (1990) reported the use of fluorescence *in situ* hybridization technique for the development of gene probes for the detection of microbial community in a mixed population. The labeled probes were developed based on large numbers of ribosome particularly from the fast growing young cells to enhance the sensitivity of the detection.

3.2 Analysis of Genes regulating the AMF Symbiosis

Development of arbuscular mycorrhizal symbiosis in any host plant is mediating by certain genes which are present in either host or in AMF. Differential mRNA display technique combined with RT-PCR technique was effectively used by Harrier *et al.*(1998) for the identification and quantification of fungal symbiotic genes expressed during functioning symbiosis. It compared the mRNA transcripts of colonized and non-colonized plant root system. Martin *et al.* (2008) carried out the transcriptomic analysis on *Laccaria bicolor* genome for the identification of genes encoding functional proteins. The results showed the secretion of small secreted protein (SSP) by the AMF spores playing a major role during the time of symbiosis. The excess secretion of SSP was found in AMF hyphae which helped to degrade the host plant cell wall during the time of infection and penetration. Bonfante and Genre, (2010) performed the analysis of genome and transcriptomes through high-throughput sequencing with advanced microscopy and identified the transporter molecules mediating the cell signaling between AMF and host plant root for symbiosis . Liu, (2012) studied the important role of Ca^{2+} channels, transporters and receptor genes in *G.intraradices* for the symbiotic relationship with host plant. The genes involved in regulating cytosolic Ca^{2+} in an AMF and consequently the homeostasis regulating changes in Ca^{2+} concentration drive downstream CCaMK-dependent signaling events that are essential for the establishment of mycorrhizal symbiosis. Ruzicka *et al.* (2013) studied the genes involved in functional response of AMF during symbiosis using meta-transcriptomic analysis. The results of the study showed the presence of hundreds of fungal and plant root specific genes responsible for nutrient transportation and cell wall remodeling during the symbiotic relationship. Salvioli *et al.* (2012) studied the effect of AMF (*G.mosseae*) colonization on tomato fruit development. The microarray combined with qRT-PCR techniques revealed the mycorrhizal colonization accelerated up regulation of eleven transcripts and enhanced nitrogen and carbohydrate metabolism during fruiting time and further, the mycorrhizal colonization increased the glutamine and aspargine content in fruit. The changes in the levels of plant proteins during the AMF interaction were studied by Cout *et al.*(2013), the proteomic analysis coupled with label free mass spectroscopy showed the changes in microsomal and core plasmalemma membrane bound proteins at the time of AMF interaction and network development.

3.3 Analysis of antifungal activity and stress management of AMF symbiosis

To determine the antifungal activity of AMF, a qRT-PCR analysis was demonstrated by Ismail, (2011). The mycotoxin production by *Fusarium* was gradually decreased in the presence of *G.irregulare* and, modification in gene (TR 13-16 and TR1101) regulation of *Fusarium* for the mycotoxin production was observed through real-time analysis. When AMF hyphae reached the *Fusarium mycelium* the production of toxin level was gradually decreased. Water stress management through AMF symbiosis was studied by Ruiz-lozano, (2003). The untargeted approach of c-DNA screening from AMF colonized plants and control plants revealed the activation of osmotic stress-regulated genes in AMF colonized plants under water stress conditions. Tonin *et al.* (2001) differentiated the heavy metal tolerant ability of AMF spores between the metal tolerant plant (*Viola calaminaria*, violet) and non-metal tolerant plant (*Trifolium subterraneum*, subterranean Clover) by using terminal restriction length polymorphism. Krupa and Seget, (2004) also isolated the heavy metal tolerance of AMF from the birch seedlings grown in metallurgic heap. The restriction analysis of ITS region of the isolates exhibited the region responsible for the heavy metal tolerance against Cd and Zn. Cicatelli *et al.* 2014 studied the changes in protein level for the heavy metal tolerance in plants under stress condition and colonized with/without AMF. The genome-vide proteomic analysis of the plants showed expression of genes responsible for the heavy metals tolerant (Cu, Zn) in AM symbiotic plants planted in metal contaminated soil.

3.4 Tracking of AMF in soil

For the tracking of active microbial population in the soil, Artursson and Jansson, (2003) demonstrated that bromodeoxyuridine (BrdU) incorporation and immuno capture method for identification of actively growing bacteria in soil inoculated with specific AMF. This method relies on incorporation of the thymidine analogue BrdU, into growing cells during DNA replication. This approach permits identification of specific populations that grow in response to specified stimuli. Incorporation of molecular fingerprinting technique (Artursson *et al.*, 2005) with this approach facilitates to identify bacterial species that were activated in the presence of specific AMF. The results of that study revealed distinct differences in active bacterial community compositions in response to *G. mosseae* inoculation. Berruti *et al.*, (2013) reported a new laser dissection technique for the identification of active AMF in the colonized roots. This technique is entirely different from the ITS region analysis, a micro dissection of arbuscules from the AMF colonized roots followed by sequencing revealed the information regarding the capacity of AMF community to produce arbuscules inside the roots.

4.0 Conclusion and future strategies

Application of molecular techniques in studies involving taxonomy, gene expression and AMF symbiosis makes it more reliable. Identification based on nucleotide differences can effectively differentiate the spores from the closely related species for taxonomic purposes. Further, structural changes in AMF spores showing host specificity interactions may also be monitored using real-time studies. The important stages of AMF symbiosis may be structured by identifying the expression molecules responsible for the AMF symbiosis. The level of AMF colonization and host response and interaction can also be quantified by using real-time imaging studies. In future the complete genomic analysis of AMF will facilitate to create a database with details of structural, taxonomical and functional aspects of AMF. Construction of cDNA libraries which may be sequenced to obtain ESTs and screening cDNA assays will be useful tools in gene expression studies which could be used for the sustainable improvement of crop plants. Studies on the molecular cross talk between symbiotic partners are an important area to concentrate in plant–microbe interactions. Franken and Requena (2001) suggested designing of a Glomalean- specific primer which could detect not only the presence of isolates but their activity, and also useful for making formulation for field inoculations. Hence, such types of approaches may have to be given top priority in future.

References

1. Alguacil, M.M., Lumini, E., Roldán, A., Salinas-García, J.R., Bonfante, P. and Bianciotto, V., 2007. The Impact of Tillage Practices on Arbuscular Mycorrhizal Fungal Diversity in Subtropical Crops. *Ecological Applications* .18,527–536.

2. Alguacil, M.D.M., Torrecillas, E., Lozano, Z., Torres, M.P. and Roldán A., 2014. *Prunus persica* Crop Management Differentially Promotes Arbuscular Mycorrhizal Fungi Diversity in a Tropical Agro-Ecosystem. *Plos One.* 9(2), 1-8.

3. Amann, R. I., Krumboltz, L. and Stahl, D.A., 1990. Fluorescent-oligonucleotide probing of whole cells for determinative, phylogenetic, and environmental studies in microbiology. *J. Bacteriol.* 172,762–770.

4. Artursson, V. and Jansson, J.K.2003. Use of bromodeoxyuridine immunocapture to identify active bacteria associated with arbuscular mycorrhizal hyphae. *Appl Environ Microbiol.* 69, 6208–6215.

5. Artursson, V., Finlay, R.D. and Jansson, J.K.,2005. Combined bromodeoxyuridine immunocapture and terminal restriction fragment length polymorphism analysis highlights differences in the active soil bacterial metagenome due to *Glomus mosseae* inoculation or plant species. *Environ Microbiol.* 7, 1952–1966.

6. Balestrini, R., Magurno, F., Walker, C., Lumini, E. and Bianciotto, V., 2010. Cohorts of arbuscular mycorrhizal fungi (AMF) in *Vitis vinifera*, a typical Mediterranean fruit crop. *Environ Microbiol Rep.* 2(4),594-604.

7. Berruti, A.,Berriello, R.,Lumini, E.,Scariot, V., Bianciotto, V. and Balestrini, R., 2013. iontoidentify the my corrhizal fungi that establish arbusculesins iderootcells, *Frontiers in plant science* 4(15), 1-10.

8. Bonfante, P. and Genre, A., 2010. Mechanisms underlying beneficial plant-fungus interaction in mycorrhizal symbiosis, *Nature Communication*. 1(48), 1-11.

9. Boon, E., Zimmerman, E., Lang, B.F. and Hijri, M., 2010. Intra-isolate genome variation in arbuscular mycorrhizal fungi persists in the transcriptome, *J.Evol. Biol.* 23, 1519-1527.

10. Brito, I, Goss, M.J., Carvalho, M.D., Chatagnier, D. and Tuinen, D.V., 2012. Impact of tillage system on arbuscular mycorrhiza fungal communities in the soil under Mediterranean conditions, *Soil and Tillage Research.* 121, 63-67.

11 Buee, M., Reich, M., Murat, C., Morin, E., Nilsson, R.H., Uroz, S. and Martin, F., 2009. 454 Pyrosequencing analyses of forest soils reveal an unexpectedly high fungal diversity, *New Phytologist*. 184, 449–456.

12 Cicatelli, A., Torrigiani, P.,Todeschini, V., Biondi, S., Castiglione, S. and Lingua, G., 2014. Arbuscular mycorrhizal fungi as a tool to ametiorate the phyto remediation potential of poplar: *biochemical and molecular aspects,* iForest, 7, 333-341.

13. Corradi, N., Croll, D., Colard, A.,Kuhn, G., Ehiner, M. and Sanders, I.R., 2007. Gene copy number polymorphisms in an arbuscular mycorrhizal fungal population, *Applied and Environmental Microbioloy*. 73(1), 366-369.

14. Couto, M.S.R., Lovato, P.E., Wipf, D. and Dumas-Gaudot, E., 2013. Proteomic studies of arbuscular mycorrhizal association, *Advances in biological chemistry*. 3, 48-58.

15. Croll, D., Giovannetti, M., Koch, A.M., Sbrana, C., Ehinger, M., Lammers, P.J. and Sanders, I.R., 2009. Nonself vegetative fusion and genetic exchange in the arbuscular mycorrhizal fungus *Glomus intraradices, New Phytologist,* 181, 924-937.

16. Di Bonito, R., Elliott, M. L. and Des Jardin, E. A., 1995. Detection of an arbuscular mycorrhizal fungus in roots of different plant species with the PCR. *Appl. Environ. Microbiol.* 61, 2809– 2810.

17. Edwards, S. G., Fitter, A. H. and Young, J. P. W., 1997. Quantification of an arbuscular mycorrhizal fungus, *Glomus mosseae* within plant roots by competitive polymerase chain reaction. *Mycol. Res.* 10, 1440–1444.

18. Ferrol, N., Barea, J.M. and Azcon-Aguilar, C., 2000. The plasma membrane H+-ATPase gene family in the arbuscular mycorrhizal fungus *Glomus mosseae, Curr Genet*. 37, 112-118.

19. Franken, P. and Requena, N. 2001. Analysis of gene expression in arbus mycorrhizas: a new approaches and challenges. *NewPhytol*. 517–523.

20. Francoise corbiere., H.L., 2002. The important of sucrose synthase for AM symbiosis in Maize, in pea and in medicago. *P.hD thesis.*

21. Gamper, H.A., Young, J.P.W., Jones, D.L. and Hodge, A., 2008. Real-time PCR and microscopy: Are the two methods measuring the same unit of arbuscular mycorrhizal fungal abundance?, *Fungal Genetics and Biology*. 45, 581–596.

22. Genre, A.,Chabaud, M.,Timmers, T.,Bonfante, P. and Barker, D.G., 2005. Arbuscular mycorrhizal fungi elicit a novel intracellular apparatus in *Medicago truncatula* root epidermal cells before infection, *The Plant Cell*. 17, 3489-3499.

23. Gomez-Levya, J.F., Lara-Renya, F., Hernandez-Cuevas, L.V. and Martinez-soriano, J.P., 2008. Specific polymerase chain reaction-based assay for the identification of arbuscular mycorrhizal fungus *Glomus intraradices*, *Journal of Biological Science*. 1-7.

24. Harrier, L. A., Wright, F. and Hooker, J. E., 1998. Isolation of the 3-phosphoglycerate kinase gene of the arbuscular mycorrhizal fungus *Glomus mosseae* (Nicol and Gerd.) Gerdemann and Trappe. *Curr. Genet*. 34, 386–392.

25. Harrier, L. A. 2001. The arbuscular mycorrhizal symbiosis: A molecular review of the fungal dimension. *J. Exp. Bot*. 52, 469– 478.

26. Hart, M.M., Gorzelak, M., Ragone, D. and Murch ,S.J., 2014. Arbuscular mycorrhizal fungal succession in a long-lived perennial. *Botany*. 92(4), 313-320.

27. Helgason, T., Fitter, A.H. and Young, J.P.W., 1999. Molecular diversity of arbuscular mycorrhizal fungi colonizing hyacinthoids non-scripta (bluebell) in semi natural woodland. *Molecular Ecology*. 8, 659-666.

28. Hohnjec, N., Henckel, K., Bakel, T., Gouzy, J., Dondrup, M., Goesmann, A. and Kuster, H., 2006. Transcriptional snapshots provide insights in to molecular basis of arbuscular mycorrhiza in the model legume *Medicago truncatula*, *Functional Plant Biology*. 33,737-748.

29. Horton, T. and Bruns, T.D., 2001. The molecular revolution in ectomycorrhizal ecology: peeking into the black-box. *Molecular Ecology*. 10, 1855–1871.

30. Ismail, Y., 2011. Molecular interaction of arbuscular mycorrhizal fungi with mycotoxin producing fungi and their role in plant defense response, Department de sciences, biologiques, Institute de recherché en biologie vegetale.

31. Jansa, J., Mozafar, A., Anken, T., Ruh, R., Sanders, I.R. and Frossard, E., 2001. Diversity and structure of AMF communities as affected by tillage in a temperate soil, *Mycorrhiza*. 12,225–234.

32. Krupa, P. and Seget, Z.P., 2004. Identification of Ectomycorrhizal fungi isolated from roots of Birch growing on Metallurgic Heap. *Polish Journal of Ecology*. 52(3), 353-357.

33. Kruger, M., Stockinger, H., Kruger, C. and Shubler, A., 2009. DNA-based species level detection of *Glomeromycota*: one PCR primer set for all arbuscular mycorrhizal fungi, *New Phytologist*. 183, 212-223.

34. Kruger, M., Kruger, C., Walker, C., Stockinger, H. and Shubler, A., 2012. Phylogenetic reference data for systematic and phylotaxonomy of arbuscular mycorrhizal fungi from phylum to species level, *New Phytologist*.193,970-984.

35. Lee, J., Lee, S. and Young, J.P., 2008. Improved PCR primers for the detection and identification of Arbuscular mycorrhizal fungi. *FEMS Microbiol Ecol.* 65, 339-49.

36. Lee, E., Eo, J.K., Ka, K.H. and Eom, A.H., 2013. Diversity of Arbuscular Mycorrhizal Fungi and Their Roles in Ecosystems. *Mycobiology.* 41(3), 121-125.

37. Liu, Y., 2012. Calcium-related fungal genes implicated in arbuscular mycorrhiza. Agricultural Sciences. Universite de Bourgogne; Huazhung agricultural university.

38. Martin, F., Aerts, A., Ahren, D., Brun, A. and Danchin, E.G.J ., 2008. The genome of *Laccaria bicolor* provides insight into mycorrhizal symbiosis, *Nature*.452, 88-92.

39. Montes-Borrego, M., Metsis, M. and Landa, B.B.,2014. Arbuscular Mycorhizal Fungi Associated with the Olive Crop across the Andalusian Landscape: Factors Driving Community Differentiation. *Plos One.* 9(5),1-12.

40. Orgiazzi, A.,Bianciotto, V., Bonfante, P.,Daghino, S., Ghignone, S., Lazzari, A., Lumini, E., Mello, A., Napoli, C., Perotto, S., Vizzini, A., Bagella, S., Murat, C. and Girlanda, M., 2013. 454 Pyrosequencing Analysis of Fungal Assemblages from Geographically Distant, Disparate Soils Reveals Spatial Patterning and a Core Mycobiome, *Diversity.* 5, 73-98.

41. Parrent, J.L., Peay, K., Arnold, A.E., Comas, L.H., Avis, P. and Tuininga, A., 2010. Moving from pattern to process in fungal symbioses: linking functional traits, community ecology and phylogenetics. *New Phytologist*.185, 882–886.

42. Ram-reddy, S., Pindi, P.K. and Reddy, S.M., 2005. Molecular methods for research on arbuscular mycorrhizal fungi in India: Problems and Prospects, *Current Science.* 89(10), 1699-1709.

43. Redecker, D., 2000. Specific PCR primers to identify arbuscular mycorrhizal fungi within colonized roots, *Mycorrhiza.* 10, 73–80.

44. Redecker, D., Hijri1, I. and Wiemken, A., 2003. Molecular Identification of Arbuscular Mycorrhizal fungi in Roots: perspectives and problems. *Folia Geobotanica* .38, 113–124.

45. Ruiz-Lozano, J.M., 2003. Arbuscular mycorrhizal symbiosis and alleviation of osmotic stress. New perspectives for molecular studies. *Mycorrhiza.* 13, 309-317.

46. Ruzicka, D., Chamala, S., Barrios-Masias, F.H., Martin, F. Smith, S., Jackson, L.E., Brad-Barbazuk, W. and Schachtman, D.P., 2013. Inside Arbuscular mycorrhizal roots-Molecular probes to understand the symbiosis, *The Plant Genome*. 6(2), 1-13.

47. Salvioli, A., Zouari, I., Chalot, M. and Bonfante, P., 2012. The arbuscular mycorrhizal status has an impact on the transcriptome profile and amino acid composition of tomato fruit, *BMC Plant Biology*. 12(44), 1-12.

48. Shafiqua, G. and Stephan, R., 2013. Molecular detection of arbuscular mycorrhizal fungi in calcareous soil and roots of *Zea mays* L., *J.Acad.Indus. Res*. 1(9), 547-549.

49. Simon, L., Lalonde, M. and Bruns, T.D., 1992. Specific Amplification of 18S Fungal Ribosomal Genes from Vesicular-Arbuscular Endomycorrhizal Fungi Colonizing Roots, *Applied and Environmental Microbiology*. 291-295.

50. Simon, L., Bousquet, J., Levesque, R. C. and Lalonde, M., 1993. Origin and diversification of endomycorrhizal fungi and coincidence with vascular land plants. *Nature*. 363, 67–69.

51. Singh, A., 2007. Molecular basis of plant-symbiotic fungi interaction: *An overview, Scientific World*. 5(5), 115-131.

52. Stockinger, H., Walker, C. and Schubler, A., 2009. *Glomus intraradices* DAOM197198', a model fungus in arbuscular mycorrhiza research, is not *Glomus intraradices, New Phytologist*. 183, 1176–1187.

53. Stockinger, H., Kruger, M. and Schubler, A., 2010. DNA barcoding of arbuscular mycorrhizal fungi, *New Phytologist*. 187,461-471.

54. Tonin, P., Vandenkoornhuyse, P., Straczek, E.J.J.J. and C. Leyval, C., 2001. Assessment of arbuscular mycorrhizal fungi diversity in the rhizosphere of *Viola calaminaria* and effect of these fungi on heavy metal uptake by clover, *Mycorrhiza*, 10, 161–168.

55. Tonge, D., Pashley, C. and Gant, T.W., 2014, Amplicon-based metagenomic analysis of mixed fungal samples using proton release amplicon sequencing, *Plos One*, 9(4), 1-8.

56. Van-Tuinen, D., Jacquot, E., Zhao, B.,Gollotte, A. and Gianinazzi-Pearson, V., 1998. Characterization of root colonization profiles by a microcosm community of arbuscular mycorrhizal fungi using 25s rDNA tagged nested PCR, *Molecular Ecology*. 7(7),879-887.

57. Vandenkoornhuyse, P., Husband, R., Daniell, T.J., Watson, I.J., Duck, J.M., Fitter, A.H. and Young, J.P.W., 2001. Arbuscular mycorrhizal community composition associated with two plant species in a grassland ecosystem, *Molecular Ecology*. 11, 1555–1564.

58. Wang, P. and Wang, Y., 2014. Community Analysis of Arbuscular Mycorrhizal Fungi in Roots of *Poncirus trifoliata* and *Citrus reticulata* Based on SSU rDNA. *The Scientific World Journal*. Article ID 562797. 8 pages.

59. Wubet, T., Weiss, M., Kottke, I., Teketay, D. and Oberwinkler F., 2006. Phylogenetic analysis of nuclear small subunit rDNA sequences suggests that the endangered African Pencil Cedar, *Juniperus procera*, is associated with distinct members of Glomeraceae. *Mycol Res* .110(9):1059-69.

60. Yang, H., Zang, Y., Yuan, Y., Tang, J. and Chen, X., 2012. Selectivity by host plants affects the distribution of arbuscular mycorrhizal fungi: evidence from ITS rDNA sequence metadata, *BMC Evolutionary Biology*. 12(50), 1-13.